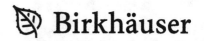 Birkhäuser

Operator Theory: Advances and Applications
Volume 246

Founded in 1979 by Israel Gohberg

Manfred Möller • Vyacheslav Pivovarchik

Spectral Theory of Operator Pencils, Hermite-Biehler Functions, and their Applications

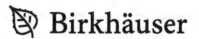 Birkhäuser

Manfred Möller
John Knopfmacher Center for Applicable
  Analysis and Number Theory
School of Mathematics
University of the Witwatersrand
Johannesburg, South Africa

Vyacheslav Pivovarchik
Department of Algebra and Geometry
South Ukrainian National Pedagogical University
Odessa, Ukraine

ISSN 0255-0156   ISSN 2296-4878 (electronic)
Operator Theory: Advances and Applications
ISBN 978-3-319-37567-0   ISBN 978-3-319-17070-1 (eBook)
DOI 10.1007/978-3-319-17070-1

Mathematics Subject Classification (2010): 47A56, 47E05, 34B07, 47B07, 34L20, 34A55, 34B45, 74K05, 74K10

Springer Cham Heidelberg New York Dordrecht London
© Springer International Publishing Switzerland 2015
Softcover reprint of the hardcover 1st edition 2015

Printed on acid-free paper

Springer International Publishing AG Switzerland is part of Springer Science+Business Media (www.birkhauser-science.com)

Contents

Part IV Background Material

Preface

A polynomial operator pencil, also called operator polynomial, is an expression of the form

$$L(\lambda) = \lambda^n A_n + \lambda^{n-1} A_{n-1} + \cdots + A_0, \tag{1}$$

where the A_k are operators acting in a Hilbert space and $\lambda \in \mathbb{C}$ is the spectral parameter. In the simplest case $L(\lambda) = \lambda I - A$, where I is the identity operator, we deal with the standard spectral problem. In case of $L(\lambda) = \lambda^2 A_2 + \lambda A_1 + A_0$ we have a quadratic operator pencil which we encounter in many applications.

Operator pencils acting in finite-dimensional Hilbert spaces are well known as matrix polynomials. There exists a vast amount of literature on this topic, see, e.g., [268], [92] and the references therein.

Historically, at its very beginning, spectral theory of operators in infinite-dimensional Hilbert spaces dealt with self-adjoint operators which appear in quantum mechanics and also in classical mechanics for conservative systems. The spectrum of a self-adjoint operator is real and the questions of interest are existence of lower bounds for the spectrum and for the essential spectrum, the number of negative eigenvalues, existence of spectral gaps, etc. Negative eigenvalues in quantum mechanics describe bound states while in classical mechanics they describe unstable modes of the system. However, even when dealing with conservative systems, sometimes it is more natural and convenient to consider a quadratic operator pencil. For example, this happens when gyroscopic forces are taken into account.

Mechanical systems with damping dissipate energy and are therefore nonconservative. In this case one deals with non-self-adjoint operators. Even in quantum mechanics, there exists an approach to deal with non-self-adjoint operators with real spectrum, see, e.g., [27].

In general, the spectrum of a non-self-adjoint operator can lie anywhere in the complex plane, and the problem of its location becomes more complicated. However, if the operator is of a concrete form or belongs to a particular class, then this information gives restrictions on the location of the spectrum, e.g., the spectrum of a dissipative operator lies in the upper half-plane, the spectrum of a bounded operator lies in a bounded domain, etc.

If the forces which make a mechanical system dissipative are caused by viscous friction, then the dynamical problem is described by the equation

$$M\frac{d^2u}{dt^2} + K\frac{du}{dt} + Au = 0,$$

where t is time, $M \geq 0$ is the operator describing mass distribution, $K \geq 0$ is the operator responsible for damping, A is the self-adjoint operator describing conservative forces, where A is bounded below, and u is the displacement vector. Substituting $u(x,t) = e^{i\lambda t}v(x)$, where λ is the spectral parameter, we arrive at a spectral problem $(\lambda^2 M - i\lambda K - A)v = 0$ for the quadratic operator pencil

$$L(\lambda) = \lambda^2 M - i\lambda K - A. \tag{2}$$

If the damping is at one point, then the operator K has rank one. Another case where the pencil is of the form (2) with an operator K of rank one occurs in operator realizations of boundary value problems for ordinary differential equations where one boundary condition depends linearly on the spectral parameter.

Surprisingly, one can reduce the spectral problem which occurs in a simple model describing nuclear interactions, proposed by T. Regge [238], to an operator pencil of the form (2) with rank one operator K. In this model it is supposed that the potential of interaction is finite or, in mathematical terms, has bounded support. Accordingly, the problem of S-wave scattering can be reduced to a boundary value problem on a finite interval with a spectral parameter dependent boundary condition. In this case the eigenvalues of the problem on the finite interval which are located in the lower half-plane describe bound states while those in the upper half-plane are responsible for the so-called resonances. The same situation occurs in the theory of quantum graphs.

Another source of quadratic operator pencils are dynamical problems with gyroscopic forces which are proportional to velocity and which lead to the operator pencil

$$L(\lambda) = \lambda^2 M - \lambda B - A, \tag{3}$$

where B is a self-adjoint operator describing gyroscopic forces.

In the quadratic operator pencils (2) and (3), the operators A, B, K, M are self-adjoint, and via the transformation $\lambda \mapsto i\lambda$, the quadratic pencils (2) and (3) are formally equivalent. In general, an operator pencil of the form (1), possibly with λ replaced by $i\lambda$, will be called a self-adjoint operator pencil if the operators $A_0, \ldots, A_n$ are self-adjoint.

Of course, by linearization one can reduce this problem to a linear operator pencil $L(\lambda) = \lambda T_1 - T_2$ acting in the direct sum of Hilbert spaces. But it is more convenient to investigate the quadratic (or polynomial) pencils directly, in particular, when the operators are self-adjoint.

There is an essential difference between finite-dimensional and infinite-dimensional cases. Unlike in the finite-dimensional case, in the infinite-dimensional case one may describe eigenvalue asymptotics. These asymptotics are important in order to establish basis properties of eigenvectors and associated vectors (root vectors) of a pencil. Furthermore, the asymptotics together with the general location of the spectrum help to determine if the corresponding dynamical system is stable. Some early results on asymptotics of eigenvalues of boundary value problems containing the spectral parameter in the boundary conditions have been obtained in [257], [258] without use of operator theory. However, those boundary value problems can be considered as spectral problems for polynomial operator pencils acting in an infinite-dimensional Hilbert spaces.

Renewed interest for further investigation of operator pencils was stimulated by M.V. Keldysh [138], see also [139]. Thereafter, many publications were devoted to completeness and basis properties of root vectors of quadratic and polynomial operator pencils, see [245], [246], [149], [237], [183], [184], [78]. These completeness and basis properties of the sets of eigenfunctions and associated functions are closely connected with asymptotics of eigenvalues. Here the monographs [273], [189] can be recommended for further reading.

The next important step towards a better understanding of quadratic operator pencils was achieved by M.G. Kreĭn and H. Langer [153], [154]. They considered monic quadratic operator pencils which have operator roots, i. e., operator pencils of the form $\lambda^2 I + \lambda B + C = (\lambda I - Z_1)(\lambda I - Z_2)$, where Z_1 and Z_2 have separated spectra. They proved that the root vectors of each Z_1 and Z_2 form a Riesz basis. Such separation of spectra of the operator roots can be done in case of so-called strongly damped and weakly damped pencils. After this many papers appeared on this topic. A detailed description of these and other results about polynomial operator pencils can be found in [182], which is the first monograph about operator pencils. Results on self-adjoint operator pencils can be found in [182, Chapter IV] and in [95, Section V.12] for quadratic self-adjoint operator pencils.

It was shown by S.G. Kreĭn [158] that the problem of small vibrations of a viscous liquid in a stable container having a free surface can be described by the equation

$$y = \lambda G y + \frac{1}{\lambda} H y,$$

where G and H are compact operators, $G > 0$, $H \geq 0$. Of course, the above equation can be reduced to an equation for a quadratic operator pencil with the exclusion of the point $\lambda = 0$. For an extention of this theory of small vibrations of a viscous liquid see [145], [146]. Some other applications of the theory of operator pencils can be found in [1] and [250].

Hermite–Biehler polynomials were first investigated by Ch. Hermite in [107] and by M. Biehler in [30]. This notion of Hermite–Biehler polynomial was generalized to entire functions by N.G. Cebotarev, L.S. Pontryagin, N.N. Meĭman, Ju.I. Neĭmark, M.G. Kreĭn and B.Ja. Levin, see [173]. A further generalization to shifted symmetric Hermite–Biehler or shifted symmetric generalized Hermite–Biehler functions was introduced in [227].

To explain the connection between quadratic operator pencils and shifted symmetric Hermite–Biehler functions we note that in the above examples, the operator pencil represents a boundary value problem for differential equations. Therefore, each such problem has a characteristic function, which is an entire function whose zeros form the spectrum of such a quadratic operator pencil. Its spectrum can therefore be described via the spectral theory of the quadratic operator pencil and via the zeros of the characteristic function. All characteristic functions which are considered in this monograph are shifted symmetric (generalized) Hermite–Biehler functions.

The spectral theory of quadratic operator pencils and the theory of shifted symmetric (generalized) Hermite–Biehler functions give roughly the same results on the general location of the spectrum. However, when we investigate spectral asymptotics, the more explicit form of the characteristic function is preferred over the spectral theory. Also, for the inverse problem, that is, for the problem of recovering parameters in the original problem from its spectral data, we will make use of the theory of entire functions.

A classical result in the theory of inverse problems states that two spectra of boundary value problems generated by the same Sturm–Liouville equation and different self-adjoint separated boundary conditions uniquely determine the Sturm–Liouville equation, see [32], [178], [180]. With the exception of Ambarzumian's case [12], one spectrum does not determine the equation uniquely. In [178], [180] one can also find necessary and sufficient conditions for two sequences of numbers to be the spectra of the above two problems. Furthermore, the method to recover the equation is presented. These classical inverse problems are related to a self-adjoint operator, i. e., to a monic linear self-adjoint operator pencil. In this monograph, we will solve inverse problems which are related to quadratic operator pencils.

We will now present an overview of results for some of the examples considered in this monograph.

The generalized Regge problem is defined as the boundary value problem

$$y'' + \lambda^2 y - q(x)y = 0,$$
$$y(\lambda, 0) = 0,$$
$$y'(\lambda, a) + (i\alpha\lambda + \beta)\, y(\lambda, a) = 0,$$

with $a > 0$, real-valued $q \in L_2(0, a)$, $\alpha > 0$ and $\beta \in \mathbb{R}$. The Regge problem is the special case $\alpha = 1$ and $\beta = 0$. The spectrum of the Regge problem may be empty as is easily seen for the case $\alpha = 1$, $\beta = 0$, $q = 0$. However, for the case $\alpha \neq 1$, on which we will focus in this part of the preface, the spectrum is well behaved. The generalized Regge problem can be represented by a self-adjoint quadratic operator pencil. Its characteristic function ϕ is a sine type function as well as a shifted Hermite–Biehler function which has the asymptotic representation

$$\phi(\lambda) = \cos \lambda a + i\alpha \sin \lambda a + \frac{M \sin \lambda a}{\lambda} - \frac{i\alpha N \cos \lambda a}{\lambda} + \frac{\psi(\lambda)}{\lambda}, \ \lambda \in \mathbb{C} \setminus \{0\},$$

where $\psi(\lambda)$ is "asymptotically small" with respect to $\cos \lambda a$ and $\sin \lambda a$. The coefficients M and N can be easily expressed in terms of the given parameters, namely, $N = \frac{1}{2} \int_0^a q(x)\, dx$ and $M = N + \beta$. The dominant term of ϕ is $\cos \lambda a$ for $0 < \alpha < 1$ and $i\alpha \sin \lambda a$ for $\alpha > 1$. For $\alpha > 1$, the asymptotic distribution of the sequence $(\lambda_k)_{k=-\infty}^{\infty}$ of the zeros of ϕ, and therefore the spectrum of the generalized Regge problem, is

$$\lambda_k = \frac{\pi k}{a} + \frac{i}{2a} \log \left(\frac{\alpha + 1}{\alpha - 1} \right) + \frac{P}{k} + \frac{\beta_k}{k}, \ k \in \mathbb{Z} \setminus \{0\},$$

where

$$P = \frac{1}{2\pi} \left(\int_0^a q(x)\, dx - \frac{2\beta}{\alpha^2 - 1} \right)$$

and $(\beta_k)_{k=-\infty}^{\infty} \in l_2$. A similar result holds in the case $0 < \alpha < 1$. There are at most finitely many eigenvalues in the closed lower half-plane, and they are all located on the nonpositive imaginary semiaxis. Furthermore, the pure imaginary eigenvalues have a certain pattern. All eigenvalues on the negative imaginary axis, if any, are simple, and denoting them by $-i\tau_1, \ldots, -i\tau_\kappa$, with $\tau_1 < \cdots < \tau_\kappa$, then the numbers $i\tau_k$ are no eigenvalues, the intervals $(i\tau_k, i\tau_{k+1})$ on the positive imaginary axis contain an odd number of eigenvalues, counted with multiplicity, whereas the interval $[0, i\tau_1)$ contains an even number of eigenvalues. Conversely, any sequence $(\lambda_k)_{k=-\infty}^{\infty}$ of the above form gives rise to a unique tuple of $a > 0$, $q \in L_2(0, a)$, $\alpha > 1$ and $\beta \in \mathbb{R}$ for which the sequence of eigenvalues of the corresponding generalized Regge problem coincides with $(\lambda_k)_{k=-\infty}^{\infty}$. We observe that it is sufficient to know only one spectrum to solve the inverse problem. The generalized Regge problem shares this behaviour with the problem of vibrations of a string with damping at one point, which was considered in [157] and [156].

Stieltjes strings, also called Sturm systems, describe massless threads bearing point masses. This notion was introduced in [85]. We consider the case that the string is fixed at both endpoints and has one point P of damping in the interior. The following picture describes the situation,

where $n_1 > 0$ and $n_2 > 0$, respectively, is the number of point masses to either side of the point of damping, the positive numbers $l_k^{(j)}$ denote the lengths of the individual threads, and the positive numbers $m_k^{(j)}$ denote the masses. If the positive coefficient of damping at the point P is denoted by ν, then vibrations of this Stieltjes string are governed by the system of equations

$$\frac{u_k^{(j)} - u_{k+1}^{(j)}}{l_k^{(j)}} + \frac{u_k^{(j)} - u_{k-1}^{(j)}}{l_{k-1}^{(j)}} - m_k^{(j)}\lambda^2 u_k^{(j)} = 0, \quad k = 1, \ldots, n_j, \, j = 1, 2,$$

$$u_0^{(j)} = 0, \qquad u_{n_1+1}^{(1)} = u_{n_2+1}^{(2)},$$

$$\frac{u_{n_1+1}^{(1)} - u_{n_1}^{(1)}}{l_{n_1}^{(1)}} + \frac{u_{n_2+1}^{(2)} - u_{n_2}^{(2)}}{l_{n_2}^{(2)}} + i\nu\lambda u_{n_1+1}^{(1)} = 0.$$

The same equations occur in the theory of synthesis of electrical circuits, see [102], [262]. The characteristic function Φ of the Stieltjes string is a generalized Hermite–Biehler polynomial of degree $2(n_1 + n_2) + 1$. Hence this problem has finitely many eigenvalues $(\lambda_k)_{k=-(n_1+n_2)}^{n_1+n_2}$. All eigenvalues lie in the closed upper half-plane and can be index in such a way that $\lambda_{-k} = -\overline{\lambda_k}$ for not pure imaginary λ_k. Furthermore, there are at most $2\min\{n_1, n_2\}$ real eigenvalues, all of which are simple and nonzero, and $\operatorname{Im}\Phi'(\lambda_k) = 0$ for each real eigenvalue λ_k. The inverse problem consists in finding parameters of a Stieltjes string from a given sequence $(\lambda_k)_{k=-n}^n$ so that it is the sequence of the eigenvalues of the Stieltjes string. The properties of this given sequence are that all of its terms lie in the closed upper half-plane, that it can be indexed in such a way that $\lambda_{-k} = -\overline{\lambda_k}$ for not pure imaginary λ_k, that the sequence has at most $2\lfloor \frac{n}{2} \rfloor$ real terms, that all real terms occur only once in the sequence and that $\operatorname{Im} \varphi'(\lambda_k) = 0$ for each real λ_k, where φ is any polynomial of degree $2n+1$ whose zeros, counted with multiplicity, are the numbers λ_k, $k = -n, \ldots, n$. This inverse problem has a solution, where the lengths of the two substrings from P to their endpoints can be arbitrarily prescribed. Furthermore, although $n_1 + n_2 = n$ is determined by the length $2n+1$ of the given sequence, the individual values of n_1 or n_2 can be arbitrarily chosen subject to the condition that the number of real terms in $(\lambda_k)_{k=-n}^n$ does not exceed $2 \min\{n_1, n_2\}$. Even with these values fixed, the solution of the inverse problem is not unique in general.

For the convenience of the reader we now list all our applications and the sections and subsections where we are dealing with each application.

- (Generalized) Regge problem: 2.1, 6.1, 7.2, 8.1
- Damped vibrations of a string: 2.2, 6.3, 7.3, 8.3
- Vibrations of star graphs with damping: 2.3, 6.4, 7.4, 8.4
- Sturm–Liouville problems on forked graphs: 2.4, 6.5
- Sturm–Liouville problems on lasso graphs: 2.5, 6.6
- Damped vibrations of Stieltjes strings: 2.6, 6.2, 8.2
- Damped vibrations of beams: 2.7, 6.7
- Vibrations of an elastic fluid-conveying pipe: 4.3.2

Now we briefly outline the contents of this monograph which is divided into four parts.

Part I, consisting of Chapters 1 to 4, deals with theory and applications of polynomial operator pencils.

In Chapter 1 we consider the spectrum of the quadratic operator pencil defined by (2), where the operators $M \geq 0$, $K \geq 0$ and $A \geq -\beta I$ ($\beta > 0$) satisfy additional conditions which guarantee that the spectrum of L consists of normal (isolated Fredholm) eigenvalues only. In Section 1.2 it is shown that the spectrum of L lies in the closed upper half-plane except for a finite number of eigenvalues, if any, which lie on the negative imaginary semiaxis. The properties and the distribution of the eigenvalues on the imaginary axis are thoroughly discussed in Sections 1.3 and 1.4. We also show in Section 1.3 that the total algebraic multiplicity of the eigenvalues in the open lower half-plane is independent of $K \geq 0$ and coincides with the total multiplicity of the negative eigenvalues of A. In Section 1.5 more specific results for the eigenvalues on the real and imaginary axes are obtained in the case that K is a rank one operator.

   In Chapter 2 we apply the results obtained in Chapter 1 to various physical problems. In Section 2.1 the spectrum of the Regge problem is described. In Section 2.2 we consider problems of small transversal vibrations of strings with damping. It appears that these problems can be considered as eigenvalue problems for quadratic operator pencils of the form (2). The same is true for the problems of vibrations of a star graph damped at the interior vertex, which we consider in Section 2.3. In Section 2.4 we consider a spectral problem for a quantum graph having the form of a fork, which is a star graph with one infinite edge. We assume that the potential on the half-infinite edge of such a graph is identically zero. Then the problem on the forked graph can be reduced to a certain Regge type problem on a finite interval and with spectral parameter dependent boundary conditions. In Section 2.5 we do the same for a lasso graph. In Section 2.6 we consider damped vibrations of a Stieltjes strings. In Section 2.7 we consider vibrations of beams with damping in a hinge at one of the ends.

   In Chapter 3 we present some results for operator pencils which do not satisfy the assumptions made in Chapter 1. In Section 3.1 we consider polynomial operator pencils of the form (1) where A_0 is self-adjoint, bounded below and can possess continuous spectrum on the semiaxis $[0, \infty)$, where $A_n = i^n I$, where $A_j = i^j K_j$ with $K_j \geq 0$, $k = 1, \ldots, n - 1$, and where the K_j are subordinate to A_0. The total algebraic multiplicity of the part of the spectrum of such a pencil which is located on the negative imaginary semiaxis coincides with the total multiplicity of the negative spectrum of A_0. In Section 3.2 we provide lower bounds for the number of eigenvalues of (2) on the negative imaginary semiaxis when $M = I$ and when the self-adjoint operator K is bounded below but not necessarily nonnegative.

   In Chapter 4 we investigate operator pencils of the form

$$L(\lambda) = \lambda^2 I - i\lambda K - \lambda B - A$$

with positive operator K and self-adjoint operators A and B. In Section 4.1 it is shown that under certain conditions the total algebraic multiplicity of the spectrum of this operator pencil in the open lower half-plane coincides with the total multiplicity of the negative spectrum of A. In Section 4.2 we compare this quadratic pencil with the linearized pencil in a Pontryagin space. In Section 4.3 we consider the pencil described by (3) which is associated with the problem of gyroscopic stabilization of a mechanical system. We describe necessary conditions for stabilization as well as certain sufficient conditions. Some applications are given.

   Part II, consisting of Chapters 5 and 6, deals with Hermite–Biehler functions and their applications to eigenvalue problems.

   In Chapter 5 we introduce Hermite–Biehler functions and shifted Hermite–Biehler functions and derive their main properties which will be used in Chapter 6. In Section 5.1 we obtain properties of Hermite–Biehler functions, whereas Section 5.2 is concerned with shifted Hermite–Biehler functions.

   In Chapter 6 we revisit the applications encountered in Chapter 2 and show that their characteristic functions are symmetric shifted Hermite–Biehler functions. Hence the spectra of these applications can be obtained as zeros of shifted Hermite–Biehler functions, which gives essentially the same results as in Chapter 2.

Part III, consisting of Chapters 7 and 8, deals with eigenvalue asymptotics of the applications considered in Parts I and II and with the corresponding inverse problems.

In Chapter 7 we derive eigenvalue asymptotics for some of the problems considered in Chapters 2 and 6, namely, for the generalized Regge problem in Section 7.2, for the damped string problem in Section 7.3, and for the star graph problem in Section 7.4.

In Chapter 8 we consider inverse problems for some of the applications encountered in Chapters 2, 6 and 7, where we have shown necessary properties of the sequences of eigenvalues of these problems. In this chapter we show for some applications that these properties are also sufficient, possibly under some mild additional conditions. That is, we solve the inverse problems. Section 8.1 deals with the generalized Regge problem. Here necessary and sufficient conditions for a sequence of complex numbers to be the spectrum of the generalized Regge problem are given, and it is shown that the parameters of the generalized Regge problem are uniquely determined by the corresponding spectrum. The inverse problem for a damped Stieltjes string, which has a finite spectrum, is considered in Section 8.2. The inverse problem for a damped smooth string is solved in Section 8.3. In Section 8.4 we consider the inverse problem on a star graph. Its spectrum is real, and therefore the spectrum of the star graph does not suffice to recover the Sturm–Liouville equations on the edges. As additional information we choose the spectra of the boundary value problems on the edges. If all spectra are mutually non-intersecting, the inverse problem has a unique solution.

In Part IV, consisting of Chapters 9 to 12, we have collected some background material which is used throughout Parts I, II, and III.

In Chapter 9 we consider analytic functions and analytic operator functions. We present results and proofs on the local dependence of their zeros, eigenvalues and eigenvectors on a parameter. Section 9.1 deals with analytic functions, whereas Section 9.2 is concerned with analytic operator functions.

In Chapter 10 we present results on differential operators which are used in Chapter 2 to verify that the operator pencils associated with our applications satisfy the assumptions made in Chapter 1. In Section 10.1 we briefly recall definitions and basic properties of Sobolev spaces on finite intervals. In Section 10.2 we write down the Lagrange identity and Green's formula, which will be used in Section 10.3 to prove criteria for self-adjointness of differential operators.

In Chapter 11 we collect various known results on meromorphic functions. In Section 11.1 we present basic properties of meromorphic Nevanlinna functions. In Section 11.2 we give a comprehensive account of sine type functions. In Section 11.3 perturbations of sine type functions are considered.

In Chapter 12 we present results on Sturm–Liouville operators in a form which is applied in various earlier chapters of this monograph. In Section 12.1 we prove Riemann's formula, which is used in Section 12.2 to give asymptotic representations of solutions of initial value problems of the Sturm–Liouville equation on a finite interval. Representations of particular sine type functions are given in Section 12.3. Sections 12.4 and 12.5 prepare for the main result of Section 12.6, the existence and uniqueness theorem for the inverse Sturm–Liouville problem with two spectra.

Part I

Operator Pencils

Chapter 1

Quadratic Operator Pencils

1.1 Operator pencils

In this chapter we will investigate the spectra of quadratic operator pencils L of the form

$$L(\lambda) = \lambda^2 M - i\lambda K - A$$

on a Hilbert space H with domain $D(L(\lambda)) = D(M) \cap D(K) \cap D(A)$, where $\lambda \in \mathbb{C}$ is the spectral parameter and the three operators M, K, A satisfy the following

Condition I. *The operators M, K, and A are self-adjoint operators on H with the following properties:*

(i) *$M \geq 0$, $K \geq 0$, and $A \geq -\beta I$ for some positive number β, i. e., A is bounded below;*

(ii) *the operator M is bounded on H, i. e., $M \in L(H)$;*

(iii) *for some $\beta_1 > \beta$, the operator $(A + \beta_1 I)^{-1} \in \mathcal{S}_\infty$, where $\mathcal{S}_\infty$ denotes the space of all compact operators on H;*

(iv) *the operator K is A-compact, i. e., $K(A + \beta_1 I)^{-1} \in \mathcal{S}_\infty$;*

(v) *$N(A) \cap N(K) \cap N(M) = \{0\}$.*

We will consider more general operator pencils in subsequent chapters. Therefore it is convenient to formulate some of the basic definitions and results for general polynomial operator pencils. To this end let H_1 be a Hilbert space and $V : H_1 \to H$ be a closed densely defined operator, e. g., a self-adjoint operator. Let n be a positive integer, let $T_1, \ldots, T_n$ be closable operators from H_1 to H with $D(T_j) \supset D(V)$ for $j = 1, \ldots, n$, and define a polynomial operator pencil T by

$$T(\lambda) = \sum_{j=0}^{n} \lambda^j T_j, \quad \lambda \in \mathbb{C}, \tag{1.1.1}$$

with $D(T(\lambda)) = D(V)$ for all $\lambda \in \mathbb{C}$. We will call T a V-bounded operator pencil. The formal derivative $\frac{dT}{d\lambda}$ is an operator pencil which will be denoted by T', and the general jth derivative by $T^{(j)}$.

Definition 1.1.1. The pencil T is said to be monic if $H_1 = H$ and $T_n = I$, where I is the identity operator.

Definition 1.1.2. The set of values $\lambda \in \mathbb{C}$ such that $T(\lambda)$ is invertible, i.e., $T(\lambda)$ is bijective and $T^{-1}(\lambda) := (T(\lambda))^{-1}$ is bounded from H to H_1, is said to be the resolvent set $\rho(T)$ of the pencil T. The spectrum of the pencil T is denoted by $\sigma(T)$, i.e., $\sigma(T) = \mathbb{C} \setminus \rho(T)$.

We recall that a closed densely defined operator (in Hilbert spaces) is called Fredholm if its nullspace has finite dimension and its range has finite codimension. The difference between the dimension of the nullspace and the codimension of the range is called the index of the Fredholm operator. The range of a Fredholm operator is closed, see [99, IV.1.13].

Definition 1.1.3.

1. A number $\lambda_0 \in \mathbb{C}$ is said to be an eigenvalue of the pencil T if there exists a vector $y_0 \in D(V)$, called an eigenvector of T, such that $y_0 \neq 0$ and $T(\lambda_0)y_0 = 0$. The vectors $y_1, \ldots, y_{m-1} \in D(V)$ are called associated to y_0 if

$$\sum_{s=0}^{k} \frac{1}{s!} T^{(s)}(\lambda_0) y_{k-s} = 0, \quad k = 1, \ldots, m-1. \tag{1.1.2}$$

   The number m is said to be the length of the chain of the eigenvector and associated vectors $(y_0, \ldots, y_{m-1})$.

2. The geometric multiplicity of an eigenvalue λ_0 is defined to be the number of the corresponding linearly independent eigenvectors, i.e., the dimension of the nullspace of $T(\lambda_0)$. The algebraic multiplicity of an eigenvalue is defined to be the greatest value of the sum of the lengths of chains corresponding to linearly independent eigenvectors. An eigenvalue is called semisimple if its algebraic multiplicity equals its geometric multiplicity. An eigenvalue is called simple if its algebraic multiplicity is 1.

3. An eigenvalue λ_0 is said to be isolated if there is a deleted neighbourhood of λ_0 which is contained in the resolvent set $\rho(T)$. An isolated eigenvalue of finite algebraic multiplicity is said to be normal. The set of all normal eigenvalues of T will be denoted by $\sigma_0(T)$.

4. A number $\lambda \in \mathbb{C}$ is called a regular point of T if it belongs to the resolvent set of T or if it is a normal eigenvalue of T.

5. The essential spectrum $\sigma_{\mathrm{ess}}(T)$ of T is the set of complex numbers λ such that $T(\lambda)$ is not a Fredholm operator.

Remark 1.1.4. It is well known and easy to see that (1.1.2) is satisfied if and only if λ_0 is a zero of order at least m of the vector polynomial

$$\lambda \mapsto T(\lambda) \sum_{j=0}^{m-1} (\lambda - \lambda_0)^j y_j \,.$$

Definition 1.1.5. The approximate spectrum of T is defined to be the set of all $\lambda \in \mathbb{C}$ such that there exists a sequence $(y_k)_{k=1}^\infty$ of vectors $y_k \in D(T)$ with $\|y_k\| = 1$ and $\lim_{k\to\infty} \|T(\lambda)y_k\| = 0$. We denote the approximate spectrum by $\sigma_{\mathrm{app}}(T)$.

Often it is more convenient to deal with bounded operator pencils. Hence we introduce the auxiliary pencil T_1 which coincides algebraically with the V-bounded operator pencil T, i.e., $T_1(\lambda)x = T(\lambda)x$ for all $\lambda \in \mathbb{C}$ and $x \in D(V)$, but $T_1(\lambda)$ is considered as an operator

$$T_1(\lambda) : D(V) \to H,$$

where $D(V)$ is equipped with the graph norm of V, which is defined by

$$\|x\|_V = \left(\|x\|^2 + \|Vx\|^2 \right)^{\frac{1}{2}}, \quad x \in D(V).$$

It is well known and easy to show that $D(V)$ becomes a Banach space when equipped with the graph norm of a closed operator V, that $D(V)$ is a Hilbert space if V is self-adjoint, and that the closability of T_j and $D(V) \subset D(T_j)$ implies that $T_j|_{D(V)}$ is bounded from $D(V)$ to H in view of the closed graph theorem. Hence $T_1(\lambda)$ will be bounded from $D(V)$ to H for all $\lambda \in \mathbb{C}$ if V is self-adjoint.

For a bounded operator S, $\|S\|$ will denote its norm, where the domain and range spaces will be clear from the context. For example, $T_1(\lambda)$ is considered as an operator from $D(V)$ with its graph norm into H, so that $\|T_1(\lambda)\|$ is the norm with respect to those spaces.

Lemma 1.1.6. *All spectral quantities of the two operator pencils T and T_1 coincide, i.e.,*

$$\sigma(T) = \sigma(T_1), \ \sigma_{\mathrm{ess}}(T) = \sigma_{\mathrm{ess}}(T_1),$$

if λ is a normal eigenvalue, then it is of the same algebraic and of the same geometric multiplicity for T and T_1. The vectors y_k, $k = 0, \ldots, m-1$, are a chain of an eigenvector and associated vectors at λ for T if and only if this is the case for T_1.

Proof. The statement of this lemma is obvious from the definition of resolvent set, spectrum, eigenvalues and associated vectors. See also [182, Lemma 20.1]. □

Definition 1.1.7. A domain $\Omega \subset \mathbb{C}$, bounded by a simple rectifiable closed curve $\partial\Omega$, is said to be normal for the operator pencil T if

(i) $$\partial\Omega \cap \sigma(T) = \emptyset,$$
(ii) $$\sigma_{\mathrm{ess}}(T) \cap \Omega = \emptyset.$$

If Ω is a normal domain for the pencil T, then $\sigma(T) \cap \Omega = \sigma_0(T) \cap \Omega$, and $\sigma_0(T) \cap \Omega$ is a finite or empty set. This can be deduced, e. g., from [189, Theorem 1.3.1 and Section 1.4] applied to the operator function T_1.

Definition 1.1.8. Suppose that the domain Ω is normal for the pencil T. The total algebraic multiplicity of the spectrum of the pencil T in Ω is the number $m(\Omega) = \sum\limits_{i=1}^{p} m_i$, where the positive integers m_i, $i = 1, \ldots, p$, denote the algebraic multiplicities of all eigenvalues of T lying in Ω.

The following lemma is a particular case of Rouché's theorem for finitely meromorphic operator functions.

Lemma 1.1.9. *Let Ω be a normal domain for the pencil T. Then*

$$\int_{\partial\Omega} T'(\lambda) T^{-1}(\lambda)\, d\lambda$$

is a finite rank operator, and

$$m(\Omega) = \frac{1}{2\pi i} \operatorname{tr} \int_{\partial\Omega} T'(\lambda) T^{-1}(\lambda)\, d\lambda =: \operatorname{ind}_\Omega T, \qquad (1.1.3)$$

where tr *denotes trace. If S is a V-bounded operator polynomial with*

$$\|S(\lambda) T^{-1}(\lambda)\| < 1, \quad \lambda \in \partial\Omega, \qquad (1.1.4)$$

then

$$\operatorname{ind}_\Omega(T + S) = \operatorname{ind}_\Omega T. \qquad (1.1.5)$$

Proof. Noting that $T'(\lambda) T^{-1}(\lambda) = T_1'(\lambda) T_1^{-1}(\lambda)$, the result follows from [97, Proposition 4.2.3 and Theorems 4.4.1 and 4.4.3], applied to the bounded operator pencil T_1 if we observe that in [97, (4.4.6)] the order in the product may be reversed. □

We now return to the operator pencil L satisfying Condition I.

Remark 1.1.10.

1. Parts (ii) and (iv) of Condition I imply that $D(A) \subset H = D(M)$ and $D(A) \subset D(K)$, so that $D(L(\lambda)) = D(A)$ for all $\lambda \in \mathbb{C}$.
2. If $(A + \beta_1 I)^{-1} \in S_\infty$ for some $\beta_1 > \beta$, then $(A + \lambda I)^{-1}$ and $K(A + \lambda I)^{-1}$ belong to S_∞ for all λ in the resolvent set of A.
3. Clearly, part (v) of Condition I is necessary for the pencil L to have a nonempty resolvent set, and in Lemma 1.2.1 below it will be shown that the resolvent set is nonempty if (v) is satisfied.

For the pencil L_1 with domain $D(A)$ equipped with the graph norm of A we write

$$L_1(\lambda) = \lambda^2 M_1 - i\lambda K_1 - A_1.$$

In general, the operators M_1, K_1 and A_1 will no longer be self-adjoint.

The following lemma summarizes some more or less obvious properties of the operator pencils L and L_1, in addition to those stated in Lemma 1.1.6.

Lemma 1.1.11.

1. *The operators M_1 and K_1 are compact operators and A_1 is a bounded Fredholm operator with index 0.*
2. *The operator pencils L and L_1 are Fredholm valued with index 0.*
3. *$L(\lambda)^* = L(-\overline{\lambda})$ for all $\lambda \in \mathbb{C}$. In particular, the spectrum of L is symmetric with respect to the imaginary axis.*

Proof. 1. By the closed graph theorem, $(A + \beta_1 I)^{-1}$ is an isomorphism from H onto $D(A)$ with its graph norm. Hence A_1 and $A(A + \beta_1 I)^{-1}$ are norm isomorphic and since

$$A(A + \beta_1 I)^{-1} = I - \beta_1(A + \beta_1 I)^{-1}$$

is a compact perturbation of the identity operator and thus a bounded Fredholm operator with index 0, see, e. g., [137, Theorem IV.5.26], it follows that A_1 has this property. Due to Condition I, a similar argument shows that M_1 and K_1 are compact.

2. The same perturbation arguments which were used in the proof of part 1 imply that L_1 is Fredholm valued with index 0. Since this property does not depend on the norm of $D(A)$, also L is Fredholm valued with index 0.

3. For all $\lambda \in \mathbb{C}$,

$$L(\lambda)^* = (\lambda^2 M - i\lambda K - A)^* = \overline{\lambda}^2 M + i\overline{\lambda}K - A = L(-\overline{\lambda}).$$

Observing that a closed operator is invertible if and only if its adjoint is invertible, it follows that $\lambda \in \sigma(L)$ implies $-\overline{\lambda} \in \sigma(L)$. Hence the spectrum of L is symmetric with respect to the imaginary axis. $\square$

1.2 Location of the spectrum of the pencil L

Lemma 1.2.1. *The spectrum of the pencil L consists of normal eigenvalues located in the closed upper half-plane and on the imaginary axis.*

Proof. Since the pencil L is an analytic Fredholm operator-valued function, its spectrum consists of eigenvalues of finite algebraic multiplicity and either $\sigma(L) = \mathbb{C}$ or all eigenvalues are normal, see, e. g., [94, Chapter XI, Corollary 8.4]. Hence it remains to show that any eigenvalue in the open lower half-plane must lie on the imaginary axis.

To this end let y_0 be an eigenvector corresponding to an eigenvalue λ_0. Then

$$(L(\lambda_0)y_0, y_0) = 0,$$

and consequently, taking real and imaginary parts,

$$((\operatorname{Re}\lambda_0)^2 - (\operatorname{Im}\lambda_0)^2)(My_0, y_0) + \operatorname{Im}\lambda_0(Ky_0, y_0) - (Ay_0, y_0) = 0 \qquad (1.2.1)$$

and
$$\operatorname{Re}\lambda_0[2\operatorname{Im}\lambda_0(My_0, y_0) - (Ky_0, y_0)] = 0. \tag{1.2.2}$$

If $\operatorname{Re}\lambda_0 \neq 0$, then (1.2.2) reduces to

$$2\operatorname{Im}\lambda_0(My_0, y_0) - (Ky_0, y_0) = 0, \tag{1.2.3}$$

and $\operatorname{Im}\lambda_0 \geq 0$ follows if $(My_0, y_0) \neq 0$ since $M \geq 0$ and $K \geq 0$. If $(My_0, y_0) = 0$, then (1.2.3) implies $(Ky_0, y_0) = 0$, i.e., $My_0 = 0$ as well as $Ky_0 = 0$. Then (1.2.1) would lead to $Ay_0 = 0$, which contradicts Condition I, part (v). Hence $\operatorname{Re}\lambda_0 \neq 0$ implies $\operatorname{Im}\lambda_0 \geq 0$.   □

Remark 1.2.2. The spectrum of a quadratic operator pencil satisfying Condition I may be empty. This happens for example in the trivial case when $M = 0$ and $K = 0$. Thus statements of the form "the spectrum consists of ..." have to be read bearing in mind that any or all listed components of the spectrum may be empty.

Lemma 1.2.3. *If $K > 0$, then the part of the spectrum of the pencil L located in the closed lower half-plane lies on the imaginary axis.*

Proof. In view of Lemma 1.2.1 we have to show that the pencil L has no nonzero real eigenvalues. Hence, assume that the pencil L has a nonzero real eigenvalue λ_0 with corresponding eigenvector y_0. Then (1.2.2) leads to $(Ky_0, y_0) = 0$, which contradicts $K > 0$.   □

Lemma 1.2.4.

1. *If $A \geq 0$, then the spectrum of the pencil L is located in the closed upper half-plane.*
2. *If $A > 0$ and $K > 0$, then the spectrum of the pencil L is located in the open upper half-plane.*
3. *If $A > 0$ and $\lambda^2 My - Ay \neq 0$ for all real λ and all nonzero $y \in N(K)\cap D(A)$, then the spectrum of the pencil L is located in the open upper half-plane.*

Proof. 1. In view of Lemma 1.2.1 we have to show that there are no eigenvalues on the negative imaginary semiaxis. Hence, let y_0 be an eigenvector corresponding to a pure imaginary eigenvalue λ_0. Then $\operatorname{Re}\lambda_0 = 0$, and in view of $M \geq 0$, $K \geq 0$, $A \geq 0$, equation (1.2.1) would imply that $(My_0, y_0) = 0$, $(Ky_0, y_0) = 0$, and $(Ay_0, y_0) = 0$ if $\operatorname{Im}\lambda_0 < 0$. But this contradicts Condition I, part (v).

2. From $A > 0$ it follows that $0 \in \rho(L)$, and hence the statement of part 2 follows in view of part 1 and Lemma 1.2.3.

3. The same proof as in part 2 above shows that 0 is not an eigenvalue of the pencil L. If the pencil L would have a nonzero real eigenvalue λ_0 with corresponding eigenvector y_0, then (1.2.2) would imply $Ky_0 = 0$, and therefore

$$\lambda_0^2 My_0 - Ay_0 = L(\lambda_0)y_0 = 0,$$

which contradicts the assumptions of this case.   □

Remark 1.2.5. We note that the assumption $A > 0$ is equivalent to $A \gg 0$, i.e., $A \geq \varepsilon I$ for some $\varepsilon > 0$. Indeed, Condition I, part (iii), implies that the spectrum of A consists of isolated eigenvalues, and therefore $A > 0$ leads to $A \gg 0$ by the spectral theorem.

Let us introduce the following parameter dependent operator pencil:

$$L(\lambda, \eta) = \lambda^2 M - i\lambda\eta K - A. \tag{1.2.4}$$

It is clear that $L(\lambda, 1) = L(\lambda)$. If we write L, we will always mean the operator function $\lambda \mapsto L(\lambda)$. The two parameter operator function will always be written as $L(\lambda, \eta)$, $L(\cdot, \eta)$ or $L(\lambda, \cdot)$.

Remark 1.2.6.

1. The number $m(\Omega, \eta)$ will denote the total algebraic multiplicity in Ω of the pencil $L(\cdot, \eta)$ defined in (1.2.4), see Definition 1.1.8.
2. For convenience we will also use the notations $m(\lambda)$ and $m(\lambda, \eta)$ to denote the algebraic multiplicities of the eigenvalues of the operator pencils L and $L(\cdot, \eta)$ at a point λ, where the multiplicity is zero if λ belongs to the resolvent set.

By Lemma 1.1.11 we know that the eigenvalues, eigenvectors and associated vectors of L and L_1 coincide, and that both L and L_1 are Fredholm operator valued. Since L_1 is a bounded operator function, Theorem 9.2.4 is applicable and we therefore have

Theorem 1.2.7. *Let $\eta_0 \in \mathbb{C}$ and let $\Omega \subset \mathbb{C}$ be a domain which contains exactly one eigenvalue λ_0 of the pencil $L(\cdot, \eta_0)$. Denote by m the algebraic multiplicity of λ_0. Then there exist numbers $\varepsilon > 0$ and $m_1 \in \mathbb{N}$, $m_1 \leq m$, such that the following assertions are true in a deleted neighbourhood $0 < |\eta - \eta_0| < \varepsilon$ of η_0:*

1. *The pencil $L(\cdot, \eta)$ possesses exactly m_1 distinct eigenvalues inside the domain Ω. Those eigenvalues can be arranged in groups $\lambda_{kj}(\eta)$, $k = 1, \ldots, l$, $j = 1, \ldots, p_k$, $\sum_{k=1}^{l} p_k = m_1$, in such a way that the following Puiseux series expansion*

$$\lambda_{kj}(\eta) = \lambda_0 + \sum_{n=1}^{\infty} a_{kn}(((\eta - \eta_0)^{\frac{1}{p_k}})_j)^n, \quad j = 1, \ldots, p_k, \tag{1.2.5}$$

holds, where, for $j = 1, \ldots, p_k$,

$$((\eta - \eta_0)^{\frac{1}{p_k}})_j = |\eta - \eta_0|^{\frac{1}{p_k}} \exp\left(\frac{2\pi i(j - 1) + i\arg(\eta - \eta_0)}{p_k}\right). \tag{1.2.6}$$

2. *A basis of the eigenspace corresponding to $\lambda_{kj}(\eta)$ can be written in the form*

$$y_{kj}^{(q)}(\eta) = b_{k0}^{(q)} + \sum_{n=1}^{\infty} b_{kn}^{(q)}(((\eta - \eta_0)^{\frac{1}{p_k}})_j)^n, \quad q = 1, \ldots, r_k, \tag{1.2.7}$$

where r_k is the geometric multiplicity of $\lambda_{kj}(\eta)$ and $b_{k0}^{(q)} \in N(L(\lambda_0, \eta_0)) \setminus \{0\}$. The vectors in the series (1.2.7) belong to $D(A)$ and the series (1.2.7) converges in the graph norm of $D(A)$.

Proof. For a suitable neighbourhood of λ_0 this follows from Theorem 9.2.4 if we observe that we may replace the exponent n in the series expansion (9.2.13) of $y_{kj}^{\iota 0}(\eta)$ by $n - \nu$, where $\nu = \min\{n \in \mathbb{N}_0 : b_{kn}^{\iota 0} \neq 0\}$, since multiplication by the nonzero constant $(((\eta - \eta_0)^{\frac{1}{p_k}})_j)^{-\nu}$ does not change the stated properties of the $y_{kj}^{\iota 0}(\eta)$ in (9.2.13). From $L(\lambda_{kj}(\eta), \eta) y_{kj}^{(q)}(\eta) = 0$ and a continuity argument it therefore follows that $L(\lambda_0, \eta_0) y_{k0}^{(q)} = 0$. $\qquad\square$

The vectors $b_{k0}^{(q)}$ in (1.2.7) are eigenvectors of $L(\cdot, \eta_0)$ with respect to the eigenvalue λ_0, but they may be linearly dependent.

We have seen in the last paragraph of the proof of Theorem 9.1.1 that for each k the least common multiple of p_k and the indices n for which $a_{kn} \neq 0$ in (1.2.5) is 1. Hence, if η is real and moves through η_0, then not all $\lambda_{kj}(\eta)$, $j = 1, \ldots, p_k$, can lie on a given line if $p_k > 1$. Therefore we have the following

Remark 1.2.8. If $\lambda_0 \neq 0$ is a real or pure imaginary eigenvalue of $L(\cdot, \eta_0)$ for some real η_0 and if the eigenvalues of $L(\cdot, \eta)$ near λ_0 for real η near η_0 are also real or pure imaginary, then $p_k = 1$ for all k in Theorem 1.2.7.

Remark 1.2.9. If both η and η_0 are real so that $\arg(\eta - \eta_0)$ is an integer multiple of π, then we can put

$$((\eta - \eta_0)^{\frac{1}{p_k}})_j = |\eta - \eta_0|^{\frac{1}{p_k}} \exp\left(\frac{2\pi i (j-1) + i\delta_k \arg(\eta - \eta_0)}{p_k}\right), \qquad (1.2.8)$$

where δ_k is an odd integer. This would change the indexing of the roots for $\eta - \eta_0 < 0$ if $\frac{\delta_k - 1}{p_k}$ is not an even integer but has the advantage that when taking $\delta_k = p_k$ if p_k is odd then $((\eta - \eta_0)^{\frac{1}{p_k}})_1$ would be real for all η and we would avoid a "vertex" when the roots move through η_0.

Although we will mostly deal with real η in $L(\lambda, \eta)$, it is advantageous to consider η as a complex eigenvalue parameter.

If we additionally assume that K is bounded and boundedly invertible, i. e., K bounded and $K \gg 0$, then we can write, setting $\lambda = i\tau$ and assuming $\lambda \neq 0$,

$$\begin{aligned}
L(\lambda, \eta) &= \lambda^2 M - i\lambda \eta K - A \\
&= -\tau^2 M + \tau \eta K - A \\
&= \tau K^{\frac{1}{2}} \left(\eta I - \tau K^{-\frac{1}{2}} M K^{-\frac{1}{2}} - \tau^{-1} K^{-\frac{1}{2}} A K^{-\frac{1}{2}}\right) K^{\frac{1}{2}}.
\end{aligned}$$

Hence

$$L(\lambda, \eta) = \tau K^{\frac{1}{2}} Q(\tau, \eta) K^{\frac{1}{2}}$$

where
$$Q(\tau, \eta) = \eta I - \tau K^{-\frac{1}{2}} M K^{-\frac{1}{2}} - \tau^{-1} K^{-\frac{1}{2}} A K^{-\frac{1}{2}}.$$

For $\tau \neq 0$ we note that if $N(Q(\tau, \eta)) \neq \{0\}$, then its dimension is the geometric multiplicity of the eigenvalue τ of the pencil $Q(\cdot, \eta)$ as well as the geometric multiplicity of the eigenvalue η of the pencil $Q(\tau, \cdot)$. The algebraic multiplicities will be different, in general, but for $\tau \in \mathbb{R}$, we have a standard spectral problem for a self-adjoint operator with the spectral parameter η, and hence all eigenvalues of $Q(\tau, \cdot)$ for real τ are real and semisimple. Therefore we have

Lemma 1.2.10. *Assume that K is bounded and that $K \gg 0$, let $\tau_0 \in \mathbb{R} \setminus \{0\}$ and let η_0 be an eigenvalue of the pencil $Q(\tau_0, \cdot)$ with (geometric) multiplicity l. Then there are $\varepsilon > 0$ and l real analytic functions*

$$\eta_k(\tau) = \eta_0 + \sum_{n=p_k}^{\infty} c_{kn} (\tau - \tau_0)^n, \quad k = 1, \ldots, l, \ |\tau - \tau_0| < \varepsilon, \tag{1.2.9}$$

where $p_k \in \mathbb{N}$, $c_{kp_k} \in \mathbb{R} \setminus \{0\}$, $c_{kn} \in \mathbb{R}$ for $n > p_k$, such that $(\eta_k(\tau))_{k=1}^{l}$ represents the eigenvalues near η_0 of the pencil $Q(\tau, \cdot)$, counted with multiplicity, for each $\tau \in \mathbb{C}$ with $|\tau - \tau_0| < \varepsilon$.

Proof. For real τ, the eigenvalues of the self-adjoint operator function $Q(\tau, \cdot)$ are real. Hence the lemma immediately follows from Theorem 1.2.7 and Remark 1.2.8. Alternatively, we could apply the theorem in [239, Section 136, p. 373]. We only have to observe that η_k cannot be constant, because otherwise $i\tau$ would be an eigenvalue of $L(\cdot, \eta_0)$ for all real τ near τ_0, which contradicts the discreteness of the spectrum of $L(\cdot, \eta_0)$, see Lemma 1.2.1. Therefore, at least one of the coefficients c_{kn} in (1.2.9) must be different from zero. $\qquad\square$

1.3　Spectrum of the pencil L in the lower half-plane

Lemma 1.3.1.

1. *All nonzero eigenvalues of the pencil L located in the closed lower half-plane are semisimple, i. e., they do not possess associated vectors.*
2. *If 0 is an eigenvalue of the pencil L, then its algebraic multiplicity is equal to $\dim N(A) + \dim(N(A) \cap N(K))$, and the maximal length of a chain of an eigenvector and associated vectors is 2.*
3. *If $K > 0$ on $N(A)$, then all eigenvalues of the pencil L located in the closed lower half-plane are semisimple.*
4. *If λ_0 is a real nonzero eigenvalue of L with eigenvector y_0, then $K y_0 = 0$.*

Proof. 1 and 4. Let λ_0 be an eigenvalue of the pencil L and assume there is a corresponding eigenvector y_0 with an associated vector y_1. By (1.1.2),

$$L(\lambda_0) y_1 + L'(\lambda_0) y_0 = 0, \tag{1.3.1}$$

and the inner product with y_0 gives

$$(L(\lambda_0)y_1, y_0) + (L'(\lambda_0)y_0, y_0) = 0. \tag{1.3.2}$$

First assume that $\lambda_0 \in \mathbb{C}^-$. Taking into account that λ_0 is pure imaginary by Lemma 1.2.1 we obtain with the aid of Lemma 1.1.11, part 3, that

$$(L(\lambda_0)y_1, y_0) = (y_1, L(-\overline{\lambda_0})y_0) = (y_1, L(\lambda_0)y_0) = 0.$$

This together with (1.3.2) leads to

$$i((2\operatorname{Im}\lambda_0 M - K)y_0, y_0) = (L'(\lambda_0)y_0, y_0) = 0. \tag{1.3.3}$$

Since $\operatorname{Im}\lambda_0 < 0$, $M \geq 0$ and $K \geq 0$, (1.3.3) gives $(My_0, y_0) = (Ky_0, y_0) = 0$, i.e., $My_0 = Ky_0 = 0$. Hence $Ay_0 = -L(\lambda_0)y_0 = 0$ and consequently $y_0 \in N(M) \cap N(K) \cap N(A)$. Due to Condition I, part (v), we have arrived at the contradiction $y_0 = 0$.

Now consider an eigenvalue $\lambda_0 \in \mathbb{R} \setminus \{0\}$. Then (1.2.3) implies $(Ky_0, y_0) = 0$, and, consequently, $Ky_0 = 0$. This proves part 4. Furthermore, $L(\lambda_0)^* y_0 = L(\lambda_0)y_0$ follows, and therefore

$$(L(\lambda_0)y_1, y_0) = (y_1, L(\lambda_0)^* y_0) = (y_1, L(\lambda_0)y_0) = 0.$$

This together with

$$L'(\lambda_0)y_0 = 2\lambda_0 My_0 - iKy_0 = 2\lambda_0 My_0,$$

$\lambda_0 \neq 0$ and (1.3.2) leads to

$$(My_0, y_0) = 0,$$

i.e., $My_0 = 0$. Hence, taking into account $Ky_0 = 0$ and $L(\lambda_0)y_0 = 0$ we obtain $Ay_0 = 0$, which contradicts Condition I, part (v).

2. Let y_0 be an eigenvector and y_1 an associated vector corresponding to the eigenvalue 0 of the pencil L. Then

$$Ay_0 = -L(0)y_0 = 0 \tag{1.3.4}$$

and (1.3.1) can be written as

$$Ay_1 + iKy_0 = 0. \tag{1.3.5}$$

Taking the inner product with y_0 gives

$$0 = (Ay_1, y_0) + i(Ky_0, y_0) = (y_1, Ay_0) + i(Ky_0, y_0) = i(Ky_0, y_0),$$

and therefore $Ky_0 = 0$ is necessary for an associated vector to exist. Conversely, if y_0 is an eigenvector corresponding to the eigenvalue 0 of the pencil L with $Ky_0 = 0$, we can choose $y_1 = 0$ in order to satisfy (1.3.5). Hence we have that the eigenvector y_0 has an associated vector if and only if $y_0 \in N(A) \cap N(K)$. The proof of part 2 will be complete if we show that the maximal length of a chain of an eigenvector and associated vectors equals 2. Hence assume by proof of contradiction that there exist $y_0, y_1, y_2 \in D(L)$, $y_0 \neq 0$, satisfying (1.3.4), (1.3.5) and

$$0 = \frac{1}{2}L''(0)y_0 + L'(0)y_1 + L(0)y_2 = My_0 - iKy_1 - Ay_2. \qquad (1.3.6)$$

We already know that $Ay_0 = 0$ and $Ky_0 = 0$. Hence, taking the inner product with y_0 in (1.3.6) leads to

$$0 = (My_0, y_0) - i(Ky_1, y_0) - (Ay_2, y_0) = (My_0, y_0).$$

Therefore we would obtain $My_0 = 0$, which contradicts Condition I, part (v), because of $y_0 \neq 0$.

3. In view of part 1 we only have to consider the case $\lambda_0 = 0$, and the statement immediately follows from part 2 since $N(A) \cap N(K = \{0\}$ implies that the algebraic multiplicity equals the geometric multiplicity $N(A)$. $\qquad \square$

Theorem 1.3.2.

1. *Assume that $M = I$. Then the total algebraic multiplicity of the spectrum of L in the open lower half-plane coincides with the total algebraic multiplicity (which is the same as the geometric multiplicity) of the negative spectrum of A.*

2. *Assume that $M = I$ and $K > 0$. Then the total algebraic multiplicity of the spectrum of L in the closed lower half-plane coincides with the total algebraic multiplicity of the nonpositive spectrum of A.*

Proof. 1. By Lemma 1.3.1, part 1, all eigenvalues of the pencil $L(\cdot, \eta)$ located in $\overline{\mathbb{C}^-} \setminus \{0\}$ are semisimple, and therefore their algebraic and geometric multiplicities coincide. Hence the total geometric multiplicity of the negative spectrum of A clearly coincides with the total algebraic multiplicity of the pencil $L(\cdot, 0)$ in the open lower half-plane since $L(\lambda, 0) = \lambda^2 I - A$, see (1.2.4).

We are going to show that $m(\mathbb{C}^-, \eta)$ for $L(\cdot, \eta)$ is independent of $\eta \in [0, 1]$. For this we observe that by Lemma 1.2.1 the spectrum of $L(\cdot, \eta)$ in the open lower half-plane is located on the imaginary axis. Thus, let us write $\lambda = -i\tau$, $\tau > 0$, for λ on the negative imaginary semiaxis. For any $\eta \in [0, 1]$, consider an eigenvalue $\lambda = -i\tau$, $\tau > 0$, of $L(\cdot, \eta)$ and a corresponding normed eigenvector y. Then

$$(-\tau^2 I - \tau\eta K - A)y = 0,$$

which leads to

$$\tau^2 = \tau^2(y, y) = -\tau\eta(Ky, y) - (Ay, y) \leq -(Ay, y) \leq \beta,$$

where β is the upper bound of $-A$ from Condition I, part (i). For each $\eta \in [0, 1]$ choose $\varepsilon(\eta) > 0$ such that the closure of

$$\Omega_\eta = \{\lambda \in \mathbb{C} : |\operatorname{Re} \lambda| < \varepsilon(\eta), \, -\beta^{\frac{1}{2}} - \varepsilon(\eta) < \operatorname{Im} \lambda < \varepsilon(\eta)\}$$

contains exactly those eigenvalues of $L(\cdot, \eta)$ which lie on the nonpositive imaginary semiaxis and such that the closure of

$$\Omega_\eta^0 = \{\lambda \in \mathbb{C} : |\operatorname{Re} \lambda| < \varepsilon(\eta), \, |\operatorname{Im} \lambda| < \varepsilon(\eta)\}$$

contains no nonzero eigenvalues of $L(\cdot, \eta)$. By Lemma 1.1.9, both $m(\Omega_{\eta_0}, \eta)$ and $m(\Omega_{\eta_0}^0, \eta)$ are independent of η near η_0 for each $\eta_0 \in [0, 1]$. But by Lemma 1.3.1, part 2, 0 is either in the resolvent set of $L(\cdot, \eta)$ for all $\eta \in [0, 1]$ or the algebraic multiplicity $m(0, \eta)$ of the eigenvalue 0 is independent of η for $\eta \in (0, 1]$ and may be larger for $\eta = 0$. Therefore,

$$m(\mathbb{C}^-, \eta_0) = m(\Omega_{\eta_0}, \eta_0) - m(\Omega_{\eta_0}^0, \eta_0) = m(\Omega_{\eta_0}, \eta_0) - m(0, \eta_0)$$

gives

$$m(\mathbb{C}^-, \eta) = m(\Omega_{\eta_0}, \eta) - m(0, \eta)$$

for η sufficiently close to any $\eta_0 \in (0, 1]$, and it follows that $m(\mathbb{C}^-, \eta)$ is independent of $\eta \in (0, 1]$. The proof of part 1 will be complete if we show that no eigenvalues can move from 0 onto the negative imaginary semiaxis when η increases from 0. Indeed, we will show that any eigenvalue on the negative imaginary semiaxis does not move away from zero as $\eta \in [0, 1]$ increases. To this end, we again use the notation $\lambda = -i\tau$, $\tau > 0$. Writing $\tilde{L}(\tau, \eta) = -L(-i\tau, \eta)$ we have

$$\tilde{L}(\tau, \eta) = \tau^2 I + \tau \eta K + A.$$

By Lemma 1.2.1, the eigenvalues τ of $\tilde{L}(\cdot, \eta)$ under consideration are positive for $\eta > 0$. By Theorem 1.2.7 and Remark 1.2.8, these eigenvalues and corresponding eigenvectors can be arranged locally in such a way that they are differentiable functions of η. Therefore, let $\tau(\eta)$ be such an eigenvalue depending on η with corresponding eigenvector $y(\eta)$. Then differentiation of

$$\tilde{L}(\tau(\eta), \eta)y(\eta) = 0$$

with respect to η gives

$$\frac{d\tau}{d\eta}\left(\frac{\partial}{\partial \tau}\tilde{L}(\tau, \eta)\right)y(\eta) + \left(\frac{\partial}{\partial \eta}\tilde{L}(\tau, \eta)\right)y(\eta) + \tilde{L}(\tau, \eta)\frac{d}{d\eta}y(\eta) = 0$$

at $\tau = \tau(\eta)$. The inner product with $y(\eta)$ leads to

$$\frac{d\tau}{d\eta}((2\tau I + \eta K)y(\eta), y(\eta)) + \tau(Ky(\eta), y(\eta)) + \left(\frac{d}{d\eta}y(\eta), \tilde{L}(\tau, \eta)y(\eta)\right) = 0,$$

where we have used that $\tilde{L}(\tau, \eta)$ is self-adjoint for real τ and η. Therefore

$$\frac{d\tau}{d\eta} = -\tau \frac{(Ky(\eta), y(\eta))}{((2\tau I + \eta K)y(\eta), y(\eta))} \le 0, \tag{1.3.7}$$

which proves that eigenvalues of the pencil $L(\cdot, \eta)$ on the negative imaginary semi-axis do not move away from 0 as η increases from 0.

2. By Lemma 1.3.1, part 2, $m(0) = \dim N(A)$. Hence the proof is complete in view of part 1 and Lemma 1.2.3. $\qquad\square$

Theorem 1.3.3. *Assume that $M + K \gg 0$. Then statement 1 of Theorem 1.3.2 is true, i. e., the total algebraic multiplicity of the spectrum of L in the open lower half-plane coincides with the total algebraic multiplicity of the negative spectrum of A.*

Proof. We consider

$$\tilde{L}(\lambda, \eta) = \lambda^2[(1 - \eta)I + \eta M] - i\lambda K - A.$$

For $\eta = 0$ we have the pencil considered in Theorem 1.3.2 and for $\eta = 1$ we have the pencil considered in this theorem. Hence it is sufficient to prove that the total algebraic multiplicity of the eigenvalues in the open lower half-plane is independent of $\eta \in [0, 1]$. Since $(1 - \eta)I + \eta M \ge 0$, $\tilde{L}(\lambda, \eta)$ satisfies Condition I for all $\eta \in [0, 1]$. The algebraic multiplicity of the eigenvalue 0 of the operator pencil $\tilde{L}(\cdot, \eta)$ is independent of η, see Lemma 1.3.1, part 2. Hence no eigenvalue can enter or leave the negative imaginary semiaxis through 0. Since eigenvalues depend continuously on η, it follows that there is a number $a > 0$ such that each eigenvalue $\lambda = -i\tau$ with $\tau > 0$ of $\tilde{L}(\cdot, \eta)$ for any $\eta \in [0, 1]$ satisfies $\tau > a$.

We are going to show that the eigenvalues $\lambda = -i\tau$, $\tau > 0$, have a bound which is independent of $\eta \in [0, 1]$. Hence, consider an eigenvalue $\lambda = -i\tau$ of $\tilde{L}(\cdot, \eta)$ and a corresponding normed eigenvector y. Then

$$(-\tau^2[(1 - \eta)I + \eta M] - \tau K - A)y = 0,$$

which leads to

$$\tau^2([(1 - \eta)I + \eta M]y, y) + \tau(Ky, y) + (Ay, y) = 0. \tag{1.3.8}$$

Since $M + K \gg 0$, there is $\varepsilon > 0$ such that $M + K \ge \varepsilon I$. If the coefficient of τ^2 equals 0, then $\eta = 1$ and $(My, y) = 0$, and therefore

$$(Ky, y) = ((M + K)y, y) \ge \varepsilon(y, y) = \varepsilon,$$

so that (1.3.8) gives

$$\tau = \frac{-(Ay, y)}{(Ky, y)} \le \frac{\beta}{\varepsilon}. \tag{1.3.9}$$

If the coefficient of τ^2 in (1.3.8) is different from 0 and hence positive, then $\tau > 0$ implies that $(Ay, y) \leq 0$, and

$$
\begin{aligned}
\tau &= \frac{-(Ky, y) + \sqrt{(Ky, y)^2 - 4([(1 - \eta)I + \eta M]y, y)(Ay, y)}}{2([(1 - \eta I)I + \eta M]y, y)} \\
&= \frac{-2(Ay, y)}{(Ky, y) + \sqrt{(Ky, y)^2 - 4([(1 - \eta)I + \eta M]y, y)(Ay, y)}}.
\end{aligned}
$$

Using that $(Ay, y) \leq 0$ and that $\sqrt{r + s} \geq \frac{1}{2}\sqrt{s}$ for $r, s \geq 0$, we obtain

$$
\tau \leq \frac{-2(Ay, y)}{(Ky, y) + \sqrt{-([(1 - \eta)I + \eta M]y, y)(Ay, y)}}.
$$

Observing that the function $t \mapsto \dfrac{t^2}{a + bt}$ with $a, b \geq 0$ and $a + b > 0$ is increasing in t on $[0, \infty)$ and that $0 \leq -(Ay, y) \leq \beta$, it follows that

$$
\tau \leq \frac{2\beta}{(Ky, y) + \sqrt{([(1 - \eta)I + \eta M]y, y)}\sqrt{\beta}},
$$

and it remains to show that the denominator has a positive lower bound. Putting $(My, y) = \gamma$ and observing $M + K \geq \varepsilon I$ gives

$$
(Ky, y) \geq \max\{\varepsilon - \gamma, 0\},
$$

and it follows that

$$
(Ky, y) + \sqrt{([(1 - \eta)I + \eta M]y, y)}\sqrt{\beta} \geq \varepsilon - \gamma_1 + \sqrt{1 - \eta + \eta\gamma_1}\sqrt{\beta},
$$

where $\gamma_1 = \min\{\gamma, \varepsilon\} \in [0, \varepsilon]$ and $\eta \in [0, 1]$. Here we have to note that γ depends on η and τ, but in the following we allow any $\gamma_1 \in [0, \varepsilon]$, so that γ_1 becomes independent of η. Since we may choose $\varepsilon < 1$, the function on the right-hand side is decreasing in η and therefore takes its minimum at $\eta = 1$, which gives

$$
(Ky, y) + \sqrt{([(1 - \eta)I + \eta M]y, y)}\sqrt{\beta} \geq \varepsilon - \gamma_1 + \sqrt{\gamma_1}\sqrt{\beta}.
$$

The right-hand side is a concave function of γ_1 and therefore takes its minimum on $[0, \varepsilon]$ at an endpoint, which finally implies

$$
(Ky, y) + \sqrt{([(1 - \eta)I + \eta M]y, y)}\sqrt{\beta} \geq \min\{\varepsilon, \sqrt{\varepsilon}\sqrt{\beta}\} > 0.
$$

We have seen above that there are positive numbers a and b such that for all $\eta \in [0, 1]$ each eigenvalue λ of $\tilde{L}(\cdot, \eta)$ in the open lower half-plane is of the form $\lambda = -i\tau$ with $a < \tau < b$. Let

$$
\Omega = \{\lambda \in \mathbb{C} : -b < \operatorname{Im} \lambda < a, \ -1 < \operatorname{Re} \lambda < 1\}.
$$

For $\lambda, \eta \in \mathbb{C}$ let $\tilde{L}_1(\lambda, \eta)$ be the operator $\tilde{L}(\lambda, \eta)$ acting from $D(A)$ to H with the graph norm on $D(A)$. The bounded operator $\tilde{L}_1(\lambda, \eta)$ depends continuously on λ and η, and therefore the set Λ of those (λ, η) for which $\tilde{L}(\lambda, \eta)$ is invertible is an open set and $\tilde{L}^{-1}(\lambda, \eta)$ depends continuously on $(\lambda, \eta) \in \Lambda$, see, e.g., [189, Proposition 1.1.4]. Since $\partial\Omega \times [0, 1]$ is a compact subset of Λ, there is a constant $C > 0$ such that

$$\|\tilde{L}_1^{-1}(\lambda, \eta)\| \le C \quad \text{for all } \lambda \in \partial\Omega \times [0, 1]. \tag{1.3.10}$$

Now let $\eta \in [0, 1]$. Since $\tilde{L}_1$ depends continuously on λ and η, for each $\lambda \in \partial\Omega$ one can choose a real number $\gamma_\lambda > 0$ such that

$$U_\lambda := \{\mu \in \mathbb{C} : |\mu - \lambda| < \gamma_\lambda\} \times \{\xi \in \mathbb{C} : |\xi - \eta| < \gamma_\lambda\} \subset \Lambda$$

and $\|\tilde{L}_1(\mu, \xi) - \tilde{L}_1(\lambda, \eta)\| < \frac{1}{2C}$ for all $(\mu, \xi) \in U_\lambda$. Since $\{U_\lambda : \lambda \in \partial\Omega\}$ is an open cover of the compact set $\partial\Omega \times \{\eta\}$, there is a finite subcover $\{U_{\lambda_1}, \ldots, U_{\lambda_n}\}$ of $\partial\Omega \times \{\eta\}$ with $\lambda_1, \ldots, \lambda_n \in \partial\Omega$. Let $\varepsilon_\eta := \min\{\gamma_{\lambda_1}, \ldots, \gamma_{\lambda_n}\}$. Let $(\lambda, \xi) \in \partial\Omega \times [\eta - \varepsilon_\eta, \eta + \varepsilon_\eta]$. There is an index k such that $(\lambda, \eta) \in U_{\lambda_k}$, and therefore also $(\lambda, \xi) \in U_{\lambda_k}$. It follows that

$$\|\tilde{L}_1(\lambda, \xi) - \tilde{L}_1(\lambda, \eta)\| \le \|\tilde{L}_1(\lambda, \xi) - \tilde{L}_1(\lambda_k, \eta)\| + \|\tilde{L}_1(\lambda, \eta) - \tilde{L}_1(\lambda_k, \eta)\| < \frac{1}{C}.$$

This together with (1.3.10) leads to

$$\|[\tilde{L}_1(\lambda, \xi) - \tilde{L}_1(\lambda, \eta)]\tilde{L}_1^{-1}(\lambda, \eta)\| \le 1.$$

Hence, by Lemma 1.1.9,

$$m(\mathbb{C}^-, \xi) = m(\Omega, \xi) = m(\Omega, \eta) = m(\mathbb{C}^-, \eta) \quad \text{for all } \xi \in (\eta - \varepsilon_\eta, \eta + \varepsilon_\eta) \cap [0, 1].$$

Therefore, both

$$S_0 = \{\eta \in [0, 1] : m(\mathbb{C}^-, \eta) = m(\mathbb{C}^-, 0)\}$$

and

$$S_1 = \{\eta \in [0, 1] : m(\mathbb{C}^-, \eta) \ne m(\mathbb{C}^-, 0)\}$$

are open subsets of $[0, 1]$. Since $0 \in S_0$ and $[0, 1]$ is connected, it follows that $S_0 = [0, 1]$, that is, the total algebraic multiplicity of the eigenvalues in the open lower half-plane is independent of $\eta \in [0, 1]$. $\qquad \square$

Notation 1.3.4. Let J be an index set and $(\lambda_j)_{j \in J}$ be a family of not necessarily distinct eigenvalues of an operator pencil. We say that the eigenvalues $(\lambda_j)_{j \in J}$ are distinct with multiplicity if for each eigenvalue λ of the pencil the number of indices $j \in J$ with $\lambda_j = \lambda$ does not exceed the algebraic multiplicity of the eigenvalue λ. We say that the finite or infinite sequence of eigenvalues $(\lambda_j)_{j \in J}$ of the pencil is an indexing of the eigenvalues in $\Omega \subset \mathbb{C}$ if $\lambda_j \in \Omega$ for all $j \in J$ and if for each eigenvalue $\lambda \in \Omega$ of the pencil the number of indices $j \in J$ with $\lambda_j = \lambda$ equals the algebraic multiplicity of the eigenvalue λ.

Lemma 1.3.5. *Let $M + K \gg 0$. If $\mathbb{C}^- \not\subset \rho(L)$, then there is an integer $\kappa > 0$ and for each $\eta \in [0,1]$ an indexing $(\lambda_{-k}(\mu))_{k=1}^{\kappa}$ of the eigenvalues of the pencil $L(\cdot, \eta)$ in $\mathbb{C}^-$ (which all lie on the negative imaginary semiaxis) such that each λ_{-k} is a continous function of η on $[0,1]$ which is analytic on $(0,1)$. We will also write λ_{-k} for $\eta = 1$, i. e., for the corresponding eigenvalues of L.*

Proof. From Lemma 1.2.1 we know that the eigenvalues of $L(\cdot, \eta)$ in the lower half-plane lie on the imaginary axis, and by Theorem 1.3.3 their total algebraic multiplicity κ is finite and independent of η. In view of Theorem 1.2.7 and Remark 1.2.8, for each $\eta_0 \in [0,1]$ there is an open interval I_{η_0} containing η_0 such that the eigenvalues can be arranged as κ continuous branches on I_{η_0} which are analytic when restricted to positive η. Since $[0,1]$ is compact, there is a finite subset F of $[0,1]$ such that $\bigcup_{\eta \in F} I_\eta \supset [0,1]$. Now choose $\mu_1 \in F$ such that $0 \in I_{\mu_1}$ and choose an indexing $(\lambda_{-k}(\mu))_{k=1}^{\kappa}$ of the eigenvalues on I_{μ_1} such that λ_{-k} is continuous on $I_{\mu_1} \cap [0,1]$ and analytic on $I_{\mu_1} \cap (0,1)$. If $I_{\mu_1} \not\supset [0,1]$, we can choose $\mu_2 \in F$ such that $I_{\mu_1} \cap I_{\mu_2} \cap (0,1) \neq \emptyset$ and $I_{\mu_2} \not\subset I_{\mu_1}$. By Theorem 1.2.7, there is $\eta_0 \in I_{\mu_1} \cap I_{\mu_2} \cap (0,1)$ such that the indexing of $(\lambda_{-k})_{k=1}^{\kappa}$ in I_{μ_1} and I_{μ_2}, respectively, can be chosen in such a way that it coincides near η_0. By the identity theorem for analytic functions, it would coincide on $I_{\mu_1} \cap I_{\mu_2} \cap (0,1)$, and it follows that we would have an indexing resulting in analytic functions on $(I_{\mu_1} \cup I_{\mu_2}) \cap (0,1)$. Continuing in this way, the statement of this lemma will be proved after a finite number of steps. $\square$

1.4   Spectrum of the pencil L in the upper half-plane

Lemma 1.4.1. *Assume that K is bounded and that $K \gg 0$. Let $\lambda_0 = i\tau_0$ with $\tau_0 > 0$. Then, with a slight abuse of notation, the Puiseux series $\lambda_{k,j}$ in (1.2.5) can be written in the form*

$$\lambda_{kj}(\eta) = \lambda_0 + \sum_{n=1}^{\infty} a_{kn}((\varepsilon_k(\eta - \eta_0)^{\frac{1}{p_k}})_j)^n, \quad j = 1, \ldots, p_k, \tag{1.4.1}$$

where $\varepsilon_k \in \{-1, 1\}$, $a_{k1} \neq 0$, and with a suitable indexing, all a_{kn} are pure imaginary and a_{k1} has positive imaginary part.

Proof. The eigenvalues λ near λ_0 of $L(\cdot, \eta)$ for η near η_0 are obtained by solving the equations (1.2.9) for τ with $\tau_0 = -i\lambda_0$ and thus for $\lambda = i\tau$. To this end, we fix some k and drop the index k. Then we have in view of Lemma 1.2.10 that

$$\eta - \eta_0 = \sum_{n=p}^{\infty} c_n(\tau - \tau_0)^n = c_p(\tau - \tau_0)^p h(\tau),$$

where $c_n \in \mathbb{R}$, $c_p \neq 0$ and h is analytic near τ_0 with $h(\tau_0) = 1$ and $h(\tau)$ is real for real τ. Near τ_0, h has a (unique) analytic pth root h_p with $h_p(\tau_0) = 1$, which is real for real τ. Hence the above identity becomes

$$\eta - \eta_0 = c_p((\tau - \tau_0)h_p(\tau))^p. \tag{1.4.2}$$

The implicit function theorem gives a unique analytic solution

$$\tau = \tau_0 + \sum_{n=1}^{\infty} \alpha_n z^n \tag{1.4.3}$$

of $z = (\tau - \tau_0)h_p(\tau)$, which is real for real z. Therefore all α_n are real. Since $\dfrac{dz}{d\tau}$ equals 1 at τ_0, $\dfrac{d\tau}{dz}$ equals 1 at $z = 0$, and hence $\alpha_1 = 1$. Observing that (1.4.2) has the p solutions

$$z = \tilde{c}_p(\varepsilon(\eta - \eta_0)^{\frac{1}{p}})_j, \quad j = 1, \dots, p, \tag{1.4.4}$$

where $\varepsilon = \operatorname{sgn}(c_p)$ and $\tilde{c}_p = |c_p|^{-\frac{1}{p}}$, a substitution of z from (1.4.4) into (1.4.3) completes the proof. $\qquad\square$

Lemma 1.4.2. *Let $M > 0$. If the eigenvalue λ_0 of the pencil L is not pure imaginary or if it is not semisimple, then $\operatorname{Im}\lambda_0 \in [m_1, m_2]$, where*

$$m_1 = \frac{1}{2} \inf_{0 \neq y \in D(A)} \frac{(Ky, y)}{(My, y)} \quad and \quad m_2 = \frac{1}{2} \sup_{0 \neq y \in D(A)} \frac{(Ky, y)}{(My, y)}.$$

Proof. If λ_0 is a not a pure imaginary eigenvalue of the pencil L, then the assertion of this lemma follows from (1.2.2). If λ_0 is a pure imaginary non-semisimple eigenvalue, then the assertion of this lemma follows from (1.3.3), which also holds for eigenvalues on the nonnegative imaginary semiaxis. $\qquad\square$

Remark 1.4.3. *If $M \gg 0$ and K is bounded, then $m_2 < \infty$, and if $K \gg 0$, then $m_1 > 0$.*

Lemma 1.4.4. *Assume that K is bounded and that $K \gg 0$. Let $\eta_0 > 0$ and let $\lambda_0 = i\tau_0$ with $\tau_0 > 0$ be an eigenvalue of $L(\cdot, \eta_0)$. Then there are $\varepsilon > 0$ and four nonnegative integers $\kappa_1, \kappa_2, \kappa_3, \kappa_4$, such that for $\eta \in (\eta_0 - \varepsilon, \eta_0 + \varepsilon)$ the eigenvalues of $L(\cdot, \eta)$ near λ_0 which lie on the imaginary axis, counted with multiplicity, can be divided into four classes $\Lambda_\uparrow(\lambda_0, \eta_0), \Lambda_\downarrow(\lambda_0, \eta_0), \Lambda_+(\lambda_0, \eta_0), \Lambda_-(\lambda_0, \eta_0)$, where*

- *$\Lambda_\uparrow(\lambda_0, \eta_0)$ consists of κ_1 increasing continuous functions from $(\eta_0 - \varepsilon, \eta_0 + \varepsilon)$ to $i\mathbb{R}$,*
- *$\Lambda_\downarrow(\lambda_0, \eta_0)$ consists of κ_2 decreasing continuous functions from $(\eta_0 - \varepsilon, \eta_0 + \varepsilon)$ to $i\mathbb{R}$,*
- *$\Lambda_+(\lambda_0, \eta_0)$ consists of κ_3 pairs of an increasing continuous function and a decreasing continuous function from $[\eta_0, \eta_0 + \varepsilon)$ to $i\mathbb{R}$,*
- *$\Lambda_-(\lambda_0, \eta_0)$ consists of κ_4 pairs of an increasing continuous function and a decreasing continuous function from $(\eta_0 - \varepsilon, \eta_0]$ to $i\mathbb{R}$.*

The difference of the number of increasing eigenvalues of $L(\cdot, \eta)$ on the positive imaginary semiaxis near λ_0 and the number of decreasing eigenvalues of $L(\cdot, \eta)$ on the positive imaginary semiaxis near λ_0 is $\kappa_1 - \kappa_2$ on $(\eta_0 - \eta, \eta_0 + \eta)$.

Proof. Consider a group of eigenvalue functions λ_{kj} with fixed k according to (1.4.1). Let $\eta \in \mathbb{R}$ be close to η_0, $\eta \neq \eta_0$. By Lemma 1.4.1, $\lambda_{k1}(\eta)$ is pure imaginary if $\varepsilon_k(\eta - \eta_0)$ is positive.

If p_k is odd, then $\lambda_{k1}(\eta)$ is the only $\lambda_{kj}(\eta)$ which is pure imaginary for $\varepsilon_k(\eta - \eta_0)$ positive, and $\lambda_{k1}(\eta)$ is increasing along the imaginary axis with $\varepsilon_k(\eta - \eta_0)$ since a_{k1} has positive imaginary part and $((\varepsilon_k(\eta - \eta_0))^{\frac{1}{p_k}})_1$ is positive. If $\varepsilon_k(\eta - \eta_0)$ is negative, then there is exactly one index s for which $\lambda_{ks}(\eta)$ is pure imaginary, and $((\varepsilon_k(\eta - \eta_0))^{\frac{1}{p_k}})_s$ is negative. Hence $\lambda_{ks}(\eta)$ is also increasing along the imaginary axis with $\varepsilon_k(\eta - \eta_0)$ if $\varepsilon_k(\eta - \eta_0)$ is negative. Therefore, for this group there is exactly one eigenvalue of $L(\cdot, \eta)$ on the imaginary axis for η in a deleted neighbourhood of η_0, and this eigenvalue depends continuously on η, belongs to $\Lambda_\uparrow(\lambda_0, \eta_0)$ if $\varepsilon_k = 1$ and to $\Lambda_\downarrow(\lambda_0, \eta_0)$ if $\varepsilon_k = -1$.

If p_k is even, $p_k = 2s$, and $\varepsilon_k(\eta - \eta_0)$ is positive, then $(\varepsilon_k(\eta - \eta_0)^{\frac{1}{p}})_{s+1} = -(\varepsilon_k(\eta - \eta_0)^{\frac{1}{p}})_1$ and hence both $\lambda_{k1}(\eta)$ and $\lambda_{k,s+1}(\eta)$ lie on the imaginary axis, where one is decreasing and the other one increasing with $\varepsilon_k(\eta - \eta_0)$, whereas all the other eigenvalues are not pure imaginary. If $\varepsilon_k(\eta - \eta_0)$ is negative, none of the $\lambda_{kj}(\eta)$ are pure imaginary. Hence these pairs contribute to $\Lambda_+(\lambda_0, \eta_0)$ if $\varepsilon_k = 1$ and to $\Lambda_-(\lambda_0, \eta_0)$ if $\varepsilon_k = -1$. $\qquad\square$

Theorem 1.4.5. *Let $M \gg 0$, $K \gg 0$ and K be bounded. Then there is a family of pure imaginary eigenvalues $(\lambda_k)_{k=1}^\kappa$ of the pencil L in the open upper half-plane which is distinct with multiplicity and which satisfies*

$$\operatorname{Im}(\lambda_k + \lambda_{-k}) \geq 0, \quad k = 1, \dots, \kappa,$$

where $(\lambda_{-k})_{k=1}^\kappa$ is an indexing of the eigenvalues of the pencil L in $\mathbb{C}^-$.

Proof. We consider the eigenvalues $(\lambda_{-k}(\eta))_{k=1}^\kappa$ of the pencil $L(\cdot, \eta)$, $\eta \in [0, 1]$, see Lemma 1.3.5. Since (1.3.7) also holds if the operator I there is replaced by M with $M \gg 0$, it follows that these eigenvalue functions λ_{-k} are nondecreasing along the imaginary axis when η increases. Since $L(\lambda, 0) = \lambda^2 M - A$, we can put $\lambda_k(0) = -\lambda_{-k}(0)$ for $k = 1, \dots, \kappa$ to obtain an indexing of the eigenvalues of $L(\cdot, 0)$ on the positive imaginary semiaxis. The proof will be complete if we show that for each $\eta \in (0, 1]$ there are pure imaginary eigenvalues $(\lambda_k(\eta))_{k=1}^\kappa$ in the upper half-plane which are distinct with multiplicity and increasing in $\eta \in [0, 1]$.

For small $\eta > 0$ we identify the $\lambda_k(\eta)$ as eigenvalues of $L(\cdot, \eta)$ according to Theorem 1.2.7 such that λ_k is continuous at 0 and analytic for $\eta > 0$. Again by Theorem 1.2.7, choose an eigenvector $y_k(\eta)$ corresponding to the eigenvalue $\lambda_k(\eta)$ of $L(\cdot, \eta)$ which is an analytic function for $\eta > 0$ and continuous at $\eta = 0$. Solving the eigenvalue equation

$$(L(\lambda_k(\eta), \eta)y_k(\eta), y_k(\eta)) = 0$$

for $\lambda_k(\eta)$ gives

$$\lambda_k(\eta)$$
$$= \frac{i\eta(Ky_k(\eta), y_k(\eta)) + i\sqrt{\eta^2(Ky_k(\eta), y_k(\eta))^2 - 4(Ay_k(\eta), y_k(\eta))(My_k(\eta), y_k(\eta))}}{2(My_k(\eta), y_k(\eta))}.$$
$$(1.4.5)$$

If we observe that y_k depends continuously on η in the graph norm of A, it follows that the inner products are continuous functions of η near $\eta = 0$. This, together with $M \gg 0$ and the fact that the radicant of the square root is positive for $\eta = 0$, because $\lambda_k(0)$ is pure imaginary with positive imaginary part, shows that $\lambda_k(\eta)$ is pure imaginary with positive imaginary part for sufficiently small $\eta > 0$.

Differentiating $L(\lambda_k(\eta), \eta)y_k(\eta) = 0$ with respect to η and taking the inner product with $y_k(\eta)$ leads to

$$\lambda_k'(\eta)((2\lambda_k(\eta)M - i\eta K)y_k(\eta), y_k(\eta)) - i\lambda_k(\eta)(Ky_k(\eta), y_k(\eta)) = 0, \quad (1.4.6)$$

and therefore

$$\lambda_k'(\eta) = \frac{i\lambda_k(\eta)(Ky_k(\eta), y_k(\eta))}{2\lambda_k(\eta)(My_k(\eta), y_k(\eta)) - i\eta(Ky_k(\eta), y_k(\eta))}. \quad (1.4.7)$$

$(My_k(\eta), y_k(\eta))$ is positive and depends continuously on η and $(Ky_k(\eta), y_k(\eta))$ depends continuously on η. Thus it follows that the denominator of (1.4.7) is pure imaginary with positive imaginary part for sufficiently small $\eta > 0$ and that the numerator is nonpositive real valued. We therefore have shown that the imaginary parts of λ_k and λ_{-k} are nondecreasing, which gives

$$\text{Im}(\lambda_k(\eta) + \lambda_{-k}(\eta)) \geq 0, \quad j = 1, \ldots, \kappa, \quad (1.4.8)$$

for $\eta \geq 0$ small enough.

Let $\eta_0 > 0$ be the supremum of all $\eta_1 \in (0, \infty)$ such that a family of increasing eigenvalue functions $(\lambda_k(\eta))_{k=1}^\kappa$ exists on $[0, \eta_1]$. The proof will be complete if we show that $\eta_0 > 1$. Therefore, assume that $\eta_0 \leq 1$. Since $M \gg 0$, K is bounded and A is bounded below, the pure imaginary eigenvalues of $L(\cdot, \eta)$ are uniformly bounded with respect to $\eta \in [0, 2]$ by (1.4.5). Choose $\varepsilon > 0$ such that Lemma 1.4.4 applies to each of the eigenvalues of $L(\cdot, \eta_0)$ on the positive imaginary semiaxis. By definition of η_0, there is a family of increasing eigenvalue functions $(\lambda_k(\eta))_{k=1}^\kappa$ on $[0, \eta_0 - \frac{\varepsilon}{2}]$. In view of Lemma 1.4.4, these functions can be extended to increasing eigenvalue functions on $[0, \eta_0]$. Now fix some $\lambda_0 \in \{\lambda_1(\eta_0), \ldots, \lambda_\kappa(\eta_0)\}$. If the number n of indices k with $\lambda_0 = \lambda_k(\eta_0)$ is at most as large as the number n_1 of those eigenvalue functions λ_{kj}, according to (1.2.5) with $\lambda_{kj}(\eta_0) = \lambda_0$, which are increasing through η_0 on the imaginary axis, then we choose n of those to represent these λ_k for $\eta \in (\eta_0, \eta_0 + \varepsilon)$. If however $n > n_1$, then by Lemma 1.4.4 there are $n - n_1$ increasing functions belonging to $\Lambda_-(\lambda_0, \eta_0)$, and therefore also $n - n_1$ decreasing eigenvalue functions for $\eta \in (\eta_0 - \varepsilon, \eta_0)$. We have seen at the beginning of the proof that, with the notation of Lemma 1.4.4, $\kappa_2 = \kappa_3 = \kappa_4 = 0$ if $\eta > 0$ is sufficiently small. Therefore a decreasing eigenvalue function must eventually come from some $\Lambda_+(\lambda_1, \eta_1)$ with $0 < \eta_1 < \eta_0$ and $\mathrm{Im}\,\lambda_0 < \mathrm{Im}\,\lambda_1$. Hence each of these decreasing eigenvalue branches would have been paired with an increasing branch, which might already have left the imaginary axis, but only after colliding with another decreasing eigenvalue branch. After a finite number of steps, we must meet a remaining increasing branch. Since we arrive at this branch by an upwards jump, it clearly follows that (1.4.8) is true for $\eta > \eta_0$ sufficiently close to η_0. Consequently, (1.4.8) holds for $\eta = 1$, that is, the theorem is proved. $\square$

1.5   The case when K has rank one

In this section we assume that in addition to Condition I the following holds:

Condition II. *The Hilbert space H is the orthogonal sum of a Hilbert space H_0 and $\mathbb{C}$, $H = H_0 \oplus \mathbb{C}$, and*

$$K = \begin{pmatrix} 0 & 0 \\ 0 & \varkappa \end{pmatrix}$$

with $\varkappa > 0$.

Lemma 1.5.1.

1. *Let $\tau \in \mathbb{R} \setminus \{0\}$ and assume that both $i\tau$ and $-i\tau$ are eigenvalues of the operator pencil $L(\cdot, \eta_0)$ for some $\eta_0 \in (0, 1]$. Then $i\tau$ and $-i\tau$ are eigenvalues of the operator pencil $L(\cdot, \eta)$ for all $\eta \in [0, 1]$.*

2. *Let $\lambda \in \mathbb{R} \setminus \{0\}$ be an eigenvalue of the operator pencil $L(\cdot, \eta_0)$ for some $\eta_0 \in (0, 1]$. Then λ and $-\lambda$ are eigenvalues of the operator pencil $L(\cdot, \eta)$ for all $\eta \in [0, 1]$.*

Proof. 1. Let $y_1 = \begin{pmatrix} y_{11} \\ y_{12} \end{pmatrix}$ be an eigenvector of $L(\cdot, \eta_0)$ corresponding to the eigenvalue $i\tau$ and let $y_2 = \begin{pmatrix} y_{21} \\ y_{22} \end{pmatrix}$ be and eigenvector of $L(\cdot, \eta_0)$ corresponding to the eigenvalue $-i\tau$. Then

$$(-\tau^2 M + \tau\eta_0 K - A)y_1 = 0, \qquad (1.5.1)$$

$$(-\tau^2 M - \tau\eta_0 K - A)y_2 = 0, \qquad (1.5.2)$$

and consequently

$$-\tau^2(y_2, My_1) + \tau\eta_0(y_2, Ky_1) - (y_2, Ay_1) = 0,$$
$$-\tau^2(My_2, y_1) - \tau\eta_0(Ky_2, y_1) - (Ay_2, y_1) = 0.$$

Taking into account that M, K and A are self-adjoint, the difference of the above equations gives

$$0 = (Ky_2, y_1) = \varkappa y_{22}\overline{y_{12}}.$$

Then one of the factors must be zero, say $y_{12} = 0$, which gives $Ky_1 = 0$. Hence (1.5.1) and (1.5.2) lead to $L(\pm i\tau, \eta)y_1 = 0$, which completes the proof of part 1.

2. Due to the symmetry of the spectrum, see Lemma 1.1.11, part 3, it follows that if $\lambda \in \mathbb{R} \setminus \{0\}$ is an eigenvalue of $L(\cdot, \eta_0)$, then also $-\lambda$ is an eigenvalue of $L(\cdot, \eta_0)$. Let y be an eigenvector of $L(\cdot, \eta_0)$ corresponding to the eigenvalue λ. Then

$$\lambda^2(My, y) - i\lambda\eta_0(Ky, y) - (Ay, y) = 0.$$

Since M, K and A are self-adjoint, all three inner products are real. Therefore $\lambda \neq 0$ and $\eta_0 > 0$ give $(Ky, y) = 0$ and thus $Ky = 0$ because $K \geq 0$. It follows that $L(\pm\lambda, \eta)y = L(\lambda, \eta_0)y = 0$ for all $\eta \in [0, 1]$. $\qquad \square$

Definition 1.5.2. Let $\eta_0 \in (0, 1]$ and let $m_I(\lambda) = \min_{\eta \in (0,1]} m(\lambda, \eta)$.

1. An eigenvalue λ of the pencil $L(\cdot, \eta_0)$ is said to be an eigenvalue of type I if λ is an eigenvalue of the pencil $L(\cdot, \eta)$ for each $\eta \in (0, 1]$, i.e., if $m_I(\lambda) > 0$.
2. For $\lambda \in \mathbb{C}$ let $m_0(\lambda) = \dim(N(L(\lambda)) \cap N(K))$. If $m_0(\lambda) > 0$, then each nonzero vector in $N(L(\lambda)) \cap N(K)$ is called an eigenvector of type I for L at λ.
3. An eigenvalue λ of the pencil $L(\cdot, \eta_0)$ is said to be an eigenvalue of type II if $m(\lambda, \eta_0) \neq m_I(\lambda)$.

Remark 1.5.3.

1. An eigenvalue can be both of type I and type II. If λ is both of type I and type II for some η, then we say that λ is an eigenvalue of the pencil $L(\cdot, \eta)$ of type I multiplicity $m_I(\lambda)$ and of type II multiplicity $m(\lambda, \eta) - m_I(\lambda)$.
2. If 0 is an eigenvalue of the pencil L, then it follows from Lemma 1.3.1, part 2, that 0 is an eigenvalue of $L(\cdot, \eta)$ for all $\eta \in (0, 1]$, and, if $\dim N(A) = n$, the algebraic multiplicity $m(\lambda, \eta)$ is $2n$ if $K_{N(A)} = 0$ and $2n - 1$ if $K_{N(A)} \neq 0$.
3. If $N(M) \cap N(A) = \{0\}$, then the pencil $L(\cdot, 0)$ satisfies Condition I. Since eigenvalues λ of type I are eigenvalues of the pencil $L(\cdot, \eta)$ for all $\eta \in (0, 1]$, it follows from (1.2.5) that $m_I(\lambda)$ branches of the eigenvalue λ are constant near $\eta = 0$, so that $m_I(\lambda) \leq m(\lambda, 0)$, whereas the remaining $m(\lambda, 0) - m_I(\lambda)$ branches are not constant.

Lemma 1.5.4. *Assume that $N(M) \cap N(A) = \{0\}$. Then the eigenvalues of type I of $L(\cdot, \eta)$, which are independent of $\eta \in (0, 1]$, are located on the imaginary and real axes, and are symmetric with respect to 0. If additionally $M + K \gg 0$, at most finitely many of the eigenvalues of type I are on the imaginary axis.*

Proof. In view of Remark 1.5.3, part 3, the first statement is proved if we show that all eigenvalues of $L(\cdot, 0)$ are located on the real and imaginary axes. If λ is an eigenvalue of $L(\cdot, 0)$ with corresponding eigenvector y, then

$$\lambda^2 My - Ay = 0.$$

Observe that $My \neq 0$ since $My = 0$ implies $Ay = 0$ which would contradict $N(M) \cap N(A) = \{0\}$. Since $M \geq 0$, it follows that $(My, y) \neq 0$, and therefore

$$\lambda^2 = -\frac{(Ay, y)}{(My, y)} \in \mathbb{R}.$$

This shows that eigenvalues of type I are real or pure imaginary. Since $L(\lambda, 0) = L(-\lambda, 0)$, the spectrum of $L(\cdot, 0)$ is symmetric with respect to the origin.

If $M + K \gg 0$, then also $M + \eta K \gg 0$, and the total multiplicity of the eigenvalues of $L(\cdot, \eta)$ which lie on the negative imaginary semiaxis equals the total number of negative eigenvalues of A by Lemma 1.2.1 and Theorem 1.3.3. Hence this multiplicity is finite, and since the spectrum of type I is symmetric with respect to the origin, there are at most finitely many eigenvalues of type I on the positive imaginary axis, and therefore at most finitely many of the eigenvalues of type I are on the imaginary axis.   □

Lemma 1.5.5.

1. *For all $\lambda, \eta \in \mathbb{C}$, $N(L(\lambda, \eta)) \cap N(K) = N(L(\lambda)) \cap N(K)$, that is, $N(L(\lambda, \eta)) \cap N(K)$ is independent of η. In particular, $m_0(\lambda) \leq m_I(\lambda)$.*
2. *Let $\lambda \neq 0$, $\eta \in (0, 1]$ and assume that $N(L(\lambda)) \cap N(K) \neq \{0\}$. Then no eigenvector $y_0 \in N(L(\lambda)) \cap N(K)$ of $L(\cdot, \eta)$ at λ has an associated vector.*
3. *If $N(M) \cap N(A) = \{0\}$, then $m_I(\lambda) = m_0(\lambda)$ for all $\lambda \in \mathbb{C} \setminus \{0\}$.*

Proof. 1. If $y \in N(K) \cap D(A)$, then

$$L(\lambda, \eta)y = \lambda^2 My - Ay = L(\lambda)y,$$

so that $L(\lambda, \eta)y = 0$ if and only if $L(\lambda)y = 0$. It follows that $m_0(\lambda) \leq m(\lambda, \eta)$, so that $m_0(\lambda) \leq m_I(\lambda)$ by definition of $m_I(\lambda)$.

2. In view of part 1 and $m_0(\lambda) = \dim(N(L(\lambda)) \cap N(K)) > 0$ we conclude that λ is an eigenvalue of type I. Assume that there is an eigenvector $y_0 \in N(L(\cdot, \eta)) \cap N(K)$ of $L(\cdot, \eta)$ at λ which has an associated vector y_1. Then

$$(L'(\lambda, \eta)y_0, y_0) = (2\lambda My_0 - i\eta Ky_0, y_0) = 2\lambda(My_0, y_0),$$

and, by Lemma 1.1.11, part 3, $Ky_0 = 0$ and the fact that eigenvalues of type I are real or pure imaginary and therefore satisfy $\lambda^2 = (-\bar{\lambda})^2$, it follows that

$$L(\lambda, \eta)^* y_0 = L(-\bar{\lambda}, \eta)y_0 = L(\lambda, \eta)y_0 = 0.$$

The equation for the associated vector,

$$L(\lambda, \eta)y_1 + L'(\lambda, \eta)y_0 = 0,$$

leads to

$$0 = (y_1, L(\lambda, \eta)^* y_0) + (L'(\lambda, \eta)y_0, y_0) = 2\lambda(My_0, y_0),$$

and therefore $My_0 = 0$. This together with $L(\lambda, \eta)y_0 = 0$ and $Ky_0 = 0$ gives $Ay_0 = 0$ and hence $y_0 = 0$, which contradicts the fact that y_0 is an eigenvector. Thus y_0 cannot have an associated vector.

3. From part 1 we know that $m_I(\lambda) \geq m_0(\lambda)$.

We are going to prove $m_I(\lambda) = m_0(\lambda)$. This is trivial if $m_I(\lambda) = 0$, so that we may assume that λ is an eigenvalue of type I. From Lemma 1.5.4 we know that λ is real or pure imaginary, so that $\lambda^2 \in \mathbb{R}$. By Remark 1.5.3, part 3, $m_I(\lambda) \leq m(\lambda, 0)$, so that λ is an eigenvalue of $L(\cdot, 0)$. Since $\lambda \neq 0$, it is a semisimple eigenvalues of $L(\cdot, 0)$. Indeed, assume there is an eigenvector y_0 with associated vector y_1 of $L(\cdot, 0)$ at λ. Then

$$\lambda^2 My_0 - Ay_0 = 0 \quad \text{and} \quad 2\lambda My_0 + \lambda^2 My_1 - Ay_1 = 0.$$

Taking the inner product with y_0 of the second identity and observing that $\lambda^2 \in \mathbb{R}$, it follows that $2\lambda(My_0, y_0) = 0$. This gives $My_0 = 0$, and the first identity would imply $Ay_0 = 0$, which contradicts $y_0 \neq 0$ and $N(M) \cap N(A) = \{0\}$. Since λ is semisimple,

$$m(\lambda, 0) = \dim N(L(\lambda, 0)),$$

and since K is a rank 1 operator, it follows from part 1 that

$$m(\lambda, 0) - 1 \leq \dim(N(L(\lambda, 0)) \cap N(K)) = \dim(N(L(\lambda)) \cap N(K)) = m_0(\lambda).$$

Together with Remark 1.5.3, part 3, we obtain the inequalities

$$m(\lambda, 0) - 1 \leq m_0(\lambda) \leq m_I(\lambda) \leq m(\lambda, 0). \tag{1.5.3}$$

If $m_0(\lambda) = m(\lambda, 0)$, then the proof is complete.

Therefore, consider the case $m_0(\lambda) = m(\lambda, 0) - 1$. Assume that also $m(\lambda, \eta) = m(\lambda, 0)$ for small $\eta > 0$. We will show that λ is a semisimple eigenvalue of $L(\cdot, \eta)$. Indeed, assume by proof of contradiction that the algebraic multiplicity of the eigenvalue is larger than its geometric multiplicity. Then

$$\dim N(L(\lambda, \eta)) \leq m(\lambda, \eta) - 1 = m_0(\lambda)$$

so that $N(L(\lambda, \eta)) = N(L(\lambda, \eta)) \cap N(K)$ by part 1. But then $L(\cdot, \eta)$ has no associated vector at λ by part 2, which leads to the contradiction that the geometric multiplicity equals the algebraic multiplicity.

In the notation of Theorem 1.2.7 we have $m_1 = 1$ and thus $p_1 = 1$. Hence it follows from (9.2.15) and its generalization to the infinite-dimensional case as well as (9.2.16) that the function of two variables

$$A(\mu, \eta) = (\mu - \lambda)^{m(\lambda, 0)} L_1^{-1}(\mu, \eta)$$

is analytic at $(\lambda, 0)$. Since the pole-order of the resolvent $L_1^{-1}(\cdot, \eta)$ is the length of the largest chain of an eigenvector and associated vectors, see, e. g., [189, Theorem 1.6.5] and since λ is a semisimple eigenvalue of $L(\cdot, \eta)$, the above identity leads to the Laurent expansion

$$L^{-1}(\mu, \eta) = \frac{1}{\mu - \lambda} A_{-1}(\eta) + \widetilde{A}(\mu, \eta),$$

where A_{-1} is analytic at 0 and $\widetilde{A}$ is analytic at $(\lambda, 0)$. Again from [189, Theorem 1.6.5] we can infer that

$$\operatorname{rank} A_{-1}(\eta) = \dim N(L(\lambda, \eta)) = m(\lambda, 0)$$

and

$$R(A_{-1}(\eta)) = N(L(\lambda, \eta)).$$

Since $m_0(\lambda) = m(\lambda, 0) - 1$ and λ is a semisimple eigenvalue of $L(\cdot, 0)$, there is an eigenvector y_0 of $L(\cdot, 0)$ at λ with $Ky_0 \neq 0$, that is, $(Ky_0, y_0) > 0$, see part 1. Let x be such that $y_0 = A_{-1}(0)x$ and put $y(\eta) = A_{-1}(\eta)x$. Then $y(\eta)$ is an eigenvector of $L(\cdot, \eta)$ at λ. Differentiating $L(\lambda, \eta)y(\eta) = 0$ with respect to η gives

$$L(\lambda, \eta)y'(\eta) - i\lambda K y(\eta) = 0.$$

Putting $\eta = 0$ and taking $L(\lambda, 0)^* = L(\lambda, 0)$ into account we arrive at the contradiction $-i\lambda(Ky_0, y_0) = 0$. Hence we have shown that $m(\lambda, \eta) < m(\lambda, 0)$ for some small $\eta > 0$, so that $m_I(\lambda) < m(\lambda, 0)$. In view of (1.5.3), $m_I(\lambda) = m_0(\lambda)$. This completes the proof of part 2.   □

Theorem 1.5.6. *Assume that $N(M) \cap N(A) = \{0\}$.*

1. *$\lambda \neq 0$ is an eigenvalue of type I of the pencil L if and only if λ is an eigenvalue of the pencil $L(\cdot, 0)$ having an eigenvector of the form $(y_0, 0)^\mathsf{T}$, and $m_I(\lambda)$ is the dimension of the space of eigenvectors of this form.*

2. *If $\lambda \neq 0$ is an eigenvalue of type I of the pencil L but not an eigenvalue of type II, then λ is semisimple.*

3. *If $\lambda \neq 0$ is an eigenvalue of the pencil L of type II, then $N(L(\lambda))$ has a basis consisting of $m_I(\lambda)$ eigenvectors of type I and one eigenvector y_0 with $Ky_0 \neq 0$ with maximal chain length $m(\lambda) - m_I(\lambda)$, that is, there is a chain $(y_j)_{j=0}^{m(\lambda)-m_I(\lambda)-1}$ of the eigenvector y_0 and, if $m(\lambda) - m_I(\lambda) > 1$, associated vectors of L at λ.*

4. *If $\lambda \neq 0$ is an eigenvalue of type II of the pencil L, then $-\lambda$ is not an eigenvalue of type II of the pencil L.*

Proof. Part 1 follows from Lemma 1.5.5.

2. By assumption and in view of Lemma 1.5.5,

$$m(\lambda) = m(\lambda, 1) = m_I(\lambda) = m_0(\lambda) \le \dim N(L(\lambda)),$$

so that the algebraic and geometric multiplicities of the eigenvalue λ of L coincide.

3. By Lemma 1.5.5 we know that

$$m_I(\lambda) = m_0(\lambda) = \dim(N(L(\lambda)) \cap N(K))$$

and that no eigenvector of L at λ in $N(L(\lambda)) \cap N(K)$ has an associated vector. Since λ is an eigenvalue of type II, its multiplicity, $m(\lambda) - m_I(\lambda)$, is positive. Hence there are eigenvectors y_0 of L at λ with $Ky_0 \ne 0$. For two such eigenvectors, a suitable nontrivial linear combination would be in $N(K)$. Therefore, $\dim N(L(\lambda)) = m_0(\lambda) + 1 =: k$, and there is a basis $z_1, \ldots, z_k$ of $N(L(\lambda))$ where each z_j is an eigenvector of a chain of length m_j with

$$\sum_{j=1}^{k} m_j = m(\lambda),$$

see [189, Proposition 1.6.4]. Assume there are two such eigenvectors with associated vectors. Then a nontrivial linear combination would be an eigenvector in $N(K)$ with an associated vector, which is impossible by Lemma 1.5.5, part 2. Hence there is at most one j with $m_j > 1$, and for this j, $m_j = m(\lambda) - m_I(\lambda)$. If $m(\lambda) - m_I(\lambda) > 1$, then $z_j \notin N(K)$, and if $m(\lambda) - m_I(\lambda) = 1$, then any eigenvector not in $N(K)$ has multiplicity $m(\lambda) - m_I(\lambda)$.

4. Assume that both λ and $-\lambda$ are eigenvalues of type II. In view of part 3 there would be eigenvectors y_1 of L at λ and y_2 of L at $-\lambda$ such that $Ky_1 \ne 0$ and $Ky_2 \ne 0$. But in the proof of Lemma 1.5.1, part 1, we have seen that this is impossible. $\qquad\square$

Theorem 1.5.7. *Assume that $N(M) \cap N(A) = \{0\}$ and that $M + K \gg 0$. Then the eigenvalues of type II of the operator pencil L possess the following properties.*

1. *Only a finite number, denoted by κ_2, of the eigenvalues of type II lie in the closed lower half-plane.*

2. *All eigenvalues of type II in the closed lower half-plane lie on the negative imaginary semiaxis and their type II multiplicities are 1. If $\kappa_2 > 0$, they will be uniquely indexed as $\lambda_{-j} = -i|\lambda_{-j}|$, $j = 1, \ldots, \kappa_2$, satisfying $|\lambda_{-j}| < |\lambda_{-(j+1)}|$, $j = 1, \ldots, \kappa_2 - 1$.*

3. *If $\kappa_2 > 0$, then the numbers $i|\lambda_{-j}|$, $j = 1, \ldots, \kappa_2$, are not eigenvalues of type II.*

4. *If $\kappa_2 \ge 2$, then the number of eigenvalues of type II, counted with type II multiplicity, in each of the intervals $(i|\lambda_{-j}|, i|\lambda_{-(j+1)}|)$, $j = 1, \ldots, \kappa_2 - 1$, is odd.*

5. *When $\kappa_2 > 0$, the interval $[0, i|\lambda_{-1}|)$ contains no or an even number of eigenvalues of type II, counted with type II multiplicity, if $N(A) \subset N(K)$, and an odd number of eigenvalues of type II, counted with type II multiplicity, if $N(A) \not\subset N(K)$.*

Proof. 1. and 2. We know from Theorem 1.3.3 that the total algebraic multiplicity of the spectrum of L in the lower half-plane is finite, and by Lemma 1.3.1, part 1, and Theorem 1.5.6, part 3, the eigenvalues λ of type II on the negative imaginary semiaxis satisfy $m(\lambda) - m_I(\lambda) = 1$ and therefore their type II multiplicity must be 1. Then the statement follows if we show that there are no real eigenvalues of type II. Indeed, if λ is a nonzero real eigenvalue of the pencil L with eigenvector y, then taking the imaginary part of the inner product of the eigenvalue equation

$$\lambda^2 My - i\lambda Ky - Ay = 0$$

with y we arrive at $(Ky, y) = 0$ and hence $Ky = 0$. Then Theorem 1.5.6, part 3, shows that λ is not an eigenvalue of type II. On the other hand, it follows from Lemma 1.3.1, part 2, that $m(0, \eta) = \dim N(A) + \dim(N(A) \cap N(K))$ for all $\eta \in (0, 1]$. Therefore $m_I(0) = m(0, \eta)$ for all $\eta \in (0, 1]$, so that 0 is not an eigenvalue of type II.

3. This statement follows from Theorem 1.5.6, part 4, since the λ_{-j} are eigenvalues of type II.

4. By Lemma 1.3.5 we can find continuous eigenvalue functions $\hat{\lambda}_{-k}$, $k = 1, \ldots, \kappa$, on $[0, 1]$ so that for each $\eta \in [0, 1]$, $(\hat{\lambda}_{-k}(\eta))_{k=1}^{\kappa}$ represents all eigenvalues of $L(\cdot, \eta)$ on the negative imaginary semiaxis, counted with multiplicity. Discarding eigenvalues of type I, we are left with κ_2 functions, and after suitable reorganization of these functions, if necessary, we may assume without loss of generality that the eigenvalues of type II at $L(\cdot, \eta)$ are $\hat{\lambda}_{-j}(\eta)$, $j = 1, \ldots, \kappa_2$. Part 2 of this theorem is clearly also true for $L(\cdot, \eta)$ with $\eta \in (0, 1]$. Furthermore, since K has rank 1, it follows that $m(\lambda, 0) \le m_0(\lambda) + 1$, and therefore Lemma 1.5.5 shows that altogether, the $\hat{\lambda}_{-j}(\eta)$, $j = 1, \ldots, \kappa_2$, are mutually distinct. Hence we may index them in such a way that $|\hat{\lambda}_{-j}(\eta)| < |\hat{\lambda}_{-(j+1)}(\eta)|$ for all $j = 1, \ldots, \kappa_2$ and all $\eta \in [0, 1]$. Clearly, $\lambda_{-j} = \hat{\lambda}_{-j}(1)$ for $j = 1, \ldots, \kappa_2$. These eigenvalues depend analytically on η in view of Theorem 1.2.7. Clearly, $m(-\lambda, 0) = m(\lambda, 0)$, $m_0(-\lambda) = m_0(\lambda)$, and therefore $m_I(-\lambda) = m_I(\lambda)$. Therefore there is $\varepsilon > 0$ such that for each $\eta \in (0, \varepsilon)$ there are exactly κ_2 eigenvalues of type II on $(\frac{1}{2}\hat{\lambda}_1(0), \hat{\lambda}_{\kappa_2}(0) + i)$, where $\hat{\lambda}_j(0) = -\hat{\lambda}_{-j}(0)$ for $j = 1, \ldots, \kappa_2$. In view of Theorem 1.2.7 we can write these eigenvalues as $\hat{\lambda}_j(\eta)$, $j = 1, \ldots, \kappa_2$, in such a way that these eigenvalue functions are analytic in η. Here we have to observe that the eigenvalues have to stay on the imaginary axis since eigenvalues can leave the imaginary axis only in symmetric pairs, see Lemma 1.1.11, part 3. Clearly, for small enough $\varepsilon > 0$, $0 < \eta < \varepsilon$ and $j = 1, \ldots, \kappa_2 - 1$ we have $|\hat{\lambda}_j(\eta)| < |\hat{\lambda}_{-(j+1)}(\eta)|$.

For each $j = \pm 1, \ldots, \pm \kappa_2$ we can now choose an analytic eigenvalue function y_k on $[0, \varepsilon)$, i.e., $L(\hat{\lambda}_j(\eta), \eta)y_j(\eta) = 0$ according to Theorem 1.2.7. By Theorem

1.5.6, part 3, we may assume that $Ky_j(\eta) \neq 0$ for at least one η and therefore for all $\eta \in (0, \varepsilon)$ and sufficiently small ε. Then (1.4.7) is also valid for negative indices j under the assumptions made in this section, which shows that $\operatorname{Im} \hat{\lambda}'_{-j}(\eta) > 0$ for all $\eta \in (0, \varepsilon)$. For positive indices j, equation (1.4.6) shows that $\hat{\lambda}'_j(\eta) \neq 0$ for $\eta \in (0, \varepsilon)$. For all indices $j = 1, \ldots, \kappa_2$, we therefore have

$$\frac{\hat{\lambda}_j(\eta)}{\hat{\lambda}'_j(\eta)} = \frac{2\hat{\lambda}_j(\eta)}{i} \frac{(My_j(\eta), y_j(\eta))}{(Ky_j(\eta), y_j(\eta))} - \eta. \tag{1.5.4}$$

If $(My_j(0), y_j(0)) = 0$, then $(Ky_j(0), y_j(0)) > 0$. Hence the right-hand side of (1.5.4) tends to 0 as $\eta \to 0$. However, the left-hand side does not tend to zero since $\hat{\lambda}_j$ is analytic at 0. This contradiction shows that $(My_j(0), y_j(0)) \neq 0$. Then the right-hand side of (1.5.4) is positive for sufficiently small $\eta > 0$, so that $\hat{\lambda}_j$ increases along the imaginary axis for small $\eta > 0$. We can therefore conclude that

$$\hat{\lambda}_j(\eta) \in (i|\hat{\lambda}_{-j}(\eta)|, i|\hat{\lambda}_{-(j+1)}(\eta)|), \quad j = 1, \ldots, \kappa_2 - 1, \quad |\hat{\lambda}_{\kappa_2}(\eta)| > |\hat{\lambda}_{-\kappa_2}(\eta)|, \tag{1.5.5}$$

for sufficiently small $\eta > 0$.

Hence, statement 4 is true for small $\eta > 0$. Due to the symmetry of the spectrum, see Lemma 1.1.11, part 3, eigenvalues can only leave or join the imaginary axis in pairs as η increases. Finally, we observe that these eigenvalues of type II cannot leave the interval $\lambda_j(\eta) \in (i|\lambda_{-j}(\eta)|, i|\lambda_{-j}(\eta)|)$ through its endpoints due to Theorem 1.5.6, part 4.

5. This is similar to part 4; we have to prove the statement for small $\eta > 0$. By Lemma 1.3.1, part 2, we know that $m(0, 0) = 2 \dim N(A)$, $m(0, \eta) = 2 \dim N(A)$ for $\eta > 0$ if $N(A) \subset N(K)$, and $m(0, \eta) = 2 \dim N(A) - 1$ for $\eta > 0$ if $N(A) \not\subset N(K)$. Hence, if $N(A) \subset N(K)$, then no eigenvalue branch of the eigenvalue 0 at $\eta = 0$ moves away from zero, whereas if $N(A) \not\subset N(K)$, then a simple eigenvalue moves away from zero. Due to the symmetry of the spectrum, this eigenvalue must stay on the imaginary axis. Since (1.3.7) is also true with the operator I there replaced by M, this zero cannot move onto the negative imaginary semiaxis. Therefore it moves onto the positive imaginary semiaxis. Since $m(0, 0)$ is zero or even, the result follows. □

Remark 1.5.8. Let us show how one can distinguish eigenvalues of L of type I from those of type II, without using η-dependence. We assume that $N(M) \cap N(A) = \{0\}$.

All eigenvalues $\lambda \neq 0$ such that $-\lambda$ is also an eigenvalue are eigenvalues of type I according to Lemma 1.5.4.

If λ is an eigenvalue of multiplicity p and $-\lambda$ is an eigenvalue of multiplicity $q \geq p$, then it follows from Theorem 1.5.6, part 4, that both λ and $-\lambda$ are eigenvalues of type I with $m_I(\lambda) = m_I(-\lambda) = p$, that λ is not an eigenvalue of type II, and that $-\lambda$ is an eigenvalue of type II if and only if $q > p$, in which case $q - p$ is the type II multiplicity of $-\lambda$.

The previous statements can be strengthened for nonzero real eigenvalues λ. Due to the symmetry of the operator pencil, see Lemma 1.1.11, part 3, it follows that the multiplies of the eigenvalues λ and $-\lambda$ coincide. Hence each nonzero real eigenvalue is an eigenvalue of type I but not an eigenvalue of type II.

If $M + K \gg 0$, then 0 is not an eigenvalue of type II by Theorem 1.5.7, part 2.

All eigenvalues which are not real and not pure imaginary lie in the upper half-plane and are not of type I and are therefore of type II.

The classification of the type I and type II eigenvalues of L stated in this section heavily depend on the fact that K is a rank 1 operator. If K were of finite rank larger than 1, then the classification would be much more involved.

Theorem 1.5.9. *Assume that $M \gg 0$.*

1. *If $\kappa_2 > 0$, then the interval $(i|\lambda_{-\kappa_2}|, i\infty)$ contains an odd number of eigenvalues of type II, counted with type II multiplicity.*

2. *If $\kappa_2 = 0$, then there is an even number of positive imaginary eigenvalues of type II, counted with type II multiplicity.*

Proof. This is shown as in the proof of Theorem 1.5.7, where we have to observe that due to the assumption $M \gg 0$ no eigenvalue can leave the imaginary axis at $i\infty$. Indeed, if $\lambda = i\tau$, $\tau > 0$ is an eigenvalue of L with corresponding eigenfunction y, then

$$((\lambda^2 M - i\lambda K - A)y, y) = 0$$

leads to

$$\tau^2(My, y) - \tau(Ky, y) + (Ay, y) = 0.$$

Choosing $\alpha > 0$ such that $M \geq \alpha I$ and $\gamma = \|K\|$, it follows that

$$\left(\tau - \frac{1}{2}\frac{(Ky, y)}{(My, y)}\right)^2 = \frac{1}{4}\frac{(Ky, y)^2}{(My, y)^2} - \frac{(Ay, y)}{(My, y)} \leq \frac{\gamma^2}{4\alpha^2} + \frac{\beta}{\alpha}.$$

Hence

$$\tau \leq \frac{\gamma}{2\alpha} + \sqrt{\frac{\gamma^2}{4\alpha^2} + \frac{\beta}{\alpha}},$$

so that the eigenvalues on the positive imaginary semiaxis are bounded. $\square$

1.6 Notes

Theorem 1.2.7 is a special case of Theorem 9.2.4. Part 1 of Theorem 9.2.4 is an application of Theorem 9.1.1, which in turn is based on the Weierstrass preparation theorem. In [136, Theorem 2], T. Kato proved the first result on eigenvalue and eigenspace dependence on a parameter for operator functions. He considered the case $(\lambda, \eta) \mapsto A(\eta) - \lambda I$, where A depends analytically on η. In [72], V.M. Eni extended T. Kato's result to polynomial operator pencils, and in [73], V.M. Eni generalized this result further to operator functions which are analytic in both parameters, thereby also showing the statements of Lemma 1.1.9 for the case of analytic operator functions. The general case of the logarithmic residue theorem is due to Gohberg and Sigal, see [98]. A comprehensive account of the results on dependence on a parameter of the eigenvalues and eigenvectors of operator pencils in finite-dimensional spaces can be found in [25] and [185]. The general Banach space case is briefly outlined in [25].

Theorem 1.3.2 is a special case of a theorem for monic polynomial operator pencils which was proved in [211, Theorem 2.1], see also Chapter 3. There exist many related results, see [247], [5], [4], [3]. This theorem remains true in the case of nonsymmetric operators K but such that $\operatorname{Re} K \gg 0$ (under some additional restrictions), see Theorem 4.1.8 and also [207], [208].

Theorem 1.4.5 was proved in [187] and Theorem 1.5.7 was proved in [220]. A review on instability index theory for quadratic pencils can be found in [43].

We note that all boundary value problems generated by the Sturm–Liouville equation with the spectral parameter λ^2 and one of the boundary conditions containing λ have an operator representation as quadratic operator pencils where K is a rank one operator as considered in Section 1.5. Under suitable assumptions on the boundary conditions, $K \geq 0$ and A is self-adjoint.

Chapter 2

Applications of Quadratic Operator Pencils

2.1 The Regge problem

The Regge problem occurs in quantum scattering theory [238] when the potential of interaction has finite support. The S-wave radial Schrödinger equation in physics, which is obtained after separation of variables in the three-dimensional Schrödinger equation with radial symmetric potential, is just the Sturm–Liouville equation on the semiaxis, see [166, §21]:

$$-y'' + q_1(x)y = \lambda^2 y \qquad (2.1.1)$$

where $\lambda^2 = \frac{2mE}{\hbar^2}$. Here m is the mass and E is the energy of the particle, $\hbar = \frac{h}{2\pi}$, h is the Plank constant, q_1 is the potential of interaction, and y is the radial component of the wave function. The boundary condition at $x = 0$ is

$$y(0) = 0. \qquad (2.1.2)$$

Since in nuclear physics the form of interaction is unknown, different models were proposed. One of the first of them is the Regge assumption that the potential has finite support, i. e.,

$$q_1(x) = \begin{cases} q(x) & \text{if } x \in [0, a], \\ 0 & \text{if } x \in (a, \infty), \end{cases}$$

where a is a positive number. We will assume, as is usual in quantum mechanics, that $q \in L_2(0, a)$ is real. It should be mentioned that different authors consider different classes of potentials, for example, $L_1(0, a)$ or $C(0, a)$. There exists the solution $s(\lambda, x)$ to (2.1.1) which satisfies the initial conditions $s(\lambda, 0) = 0$, $s'(\lambda, 0) = 1$. The Jost solution $e(\lambda, x)$ of (2.1.1) is the unique solution which behaves asymptotically as

$$e(\lambda, x) \underset{x \to \infty}{=} e^{-i\lambda x} + o(1).$$

Clearly,
$$e(\lambda, x) = e^{-i\lambda x} \quad \text{for } x \geq a, \tag{2.1.3}$$

and for $\operatorname{Im} \lambda < 0$ a solution to (2.1.1) belongs to $L_2(0, \infty)$ if and only if it is a multiple of the Jost solution. Note that the solution y of (2.1.1) is a multiple of e if and only if there is a complex number c such that

$$y(a) = ce^{-i\lambda a},$$
$$y'(a) = -ci\lambda e^{-i\lambda a}.$$

Hence for $\operatorname{Im} \lambda < 0$ the spectral problem (2.1.1), (2.1.2) on $L_2(0, \infty)$ is equivalent to the eigenvalue problem (2.1.1), (2.1.2),

$$y'(a) + i\lambda y(a) = 0 \tag{2.1.4}$$

on $L_2(0, a)$. The problem (2.1.1), (2.1.2), (2.1.4) on $L_2(0, a)$ is called the Regge problem. This problem was considered first in [238]. It should be mentioned that the eigenvalues of the Regge problem in the upper half-plane are identified by physicists as resonances for the corresponding scattering problem (2.1.1), (2.1.2).

Let us consider the operator theoretic approach to this problem. Introduce the operators A, K and M acting in the Hilbert space $H = L_2(0, a) \oplus \mathbb{C}$ according to

$$A \begin{pmatrix} y \\ c \end{pmatrix} = \begin{pmatrix} -y'' + qy \\ y'(a) \end{pmatrix},$$

$$D(A) = \left\{ \begin{pmatrix} y \\ c \end{pmatrix} : y \in W_2^2(0, a), \ y(0) = 0, \ c = y(a) \right\},$$

$$M = \begin{pmatrix} I & 0 \\ 0 & 0 \end{pmatrix}, \quad K = \begin{pmatrix} 0 & 0 \\ 0 & 1 \end{pmatrix}.$$

Then the eigenvalue problem (2.1.1), (2.1.2), (2.1.4) has the operator representation

$$L(\lambda) = \lambda^2 M - i\lambda K - A$$

in the sense that $y \in L_2(0, a)$ satisfies (2.1.1), (2.1.2), (2.1.4) if and only if $Y = (y, y(a))^\mathsf{T} \in D(L)$ and $L(\lambda)Y = 0$.

Proposition 2.1.1. *The operators A, K and M are self-adjoint, M and K are bounded, K has rank 1, and A is bounded below and has a compact resolvent. If $q \geq 0$, then $A \gg 0$.*

Proof. The statements about M and K are obvious, so that we turn our attention to A. In Section 10.3 we have provided two slightly different approaches to verify self-adjointness. Testing both of them, we will therefore give two proofs for the self-adjointness of A. For both cases, we have to observe that $y^{[1]} = -y'$ according to Definition 10.2.1.

First proof. Here we will use Theorem 10.3.4. Let $(y,c)^\mathsf{T} \in W_2^2(0,a) \oplus \mathbb{C}$ and $(z,d,e)^\mathsf{T} \in W_2^2(0,a) \oplus \mathbb{C} \oplus \mathbb{C}$ and put

$$\Delta = [y,z](a) - [y,z](0) + d^* V\hat{Y} - e^* U_2 \hat{Y}$$
$$= -\left(y(a)\overline{z^{[1]}(a)} - y^{[1]}(a)\overline{z(a)}\right) + \left(y(0)\overline{z^{[1]}(0)} - y^{[1]}(0)\overline{z(0)}\right)$$
$$+ \overline{d}y'(a) - \overline{e}y(a)$$
$$= y(a)\overline{z'(a)} - y'(a)\overline{z(a)} + y'(0)\overline{z(0)} + y'(a)\overline{d} - y(a)\overline{e}.$$

If $(z,d)^\mathsf{T} \in D(A)$ and $e = V\hat{Z} = z'(a)$, then

$$\Delta = y(a)\overline{z'(a)} - y'(a)\overline{z(a)} + y'(a)\overline{z(a)} - y(a)\overline{z'(a)} = 0.$$

Conversely, if $\Delta = 0$ for all $y \in W_2^2(0,a)$ with $y(0) = 0$, choose polynomials y with $y(0) = 0$ and exactly one of $y'(0)$, $y'(a)$, or $y(a)$ different from zero. Then we obtain, in turn, that

$$z(0) = 0, \ d = z(a) \quad \text{and} \quad e = z'(a),$$

which shows that $(z,d)^\mathsf{T} \in D(A)$ and $e = V\hat{Z}$. By Theorem 10.3.4 this means that A is self-adjoint.

Second proof. Here we will use Theorem 10.3.5, so that we have to find U_3 and U defined in (10.3.12) and (10.3.13). It is easy to see that

$$U_1 = \begin{pmatrix} 1 & 0 & 0 & 0 \end{pmatrix}, \quad U_2 = \begin{pmatrix} 0 & 0 & 1 & 0 \end{pmatrix}, \quad V = \begin{pmatrix} 0 & 0 & 0 & -1 \end{pmatrix},$$

so that

$$U_3 = \begin{pmatrix} 0 & -1 & 0 & 0 \\ 1 & 0 & 0 & 0 \\ 0 & 0 & 0 & 1 \\ 0 & 0 & -1 & 0 \\ 0 & 0 & 0 & -1 \\ 0 & 0 & -1 & 0 \end{pmatrix}, \quad U = \begin{pmatrix} 1 & 0 & 0 & 0 & 0 & 0 \\ 0 & 0 & 1 & 0 & -1 & 0 \\ 0 & 0 & 0 & -1 & 0 & -1 \end{pmatrix}.$$

Then it follows that

$$N(U_1) = \operatorname{span}\{e_2, e_3, e_4\}, \quad U_3(N(U_1)) = \operatorname{span}\{e_1, e_4 + e_6, e_3 - e_5\},$$
$$R(U^*) = \operatorname{span}\{e_1, e_3 - e_5, e_4 + e_6\},$$

so that $U_3(N(U_1)) = R(U^*)$. Hence A is self-adjoint by Theorem 10.3.5.

Now it follows from Theorem 10.3.8 that A has a compact resolvent and that A is bounded below.

Finally, for $Y = (y, y(a))^\mathsf{T} \in D(A)$ we conclude in view of (10.2.5) that

$$(AY, Y) = \int_0^a |y'(x)|^2 dx + \int_0^a q(x)|y(x)|^2 dx.$$

Hence $A \geq 0$ if $q \geq 0$. Furthermore if $AY = 0$, then $y' = 0$, and $y(0) = 0$ gives $y = 0$ and thus $Y = 0$. Therefore $A > 0$ has been shown. Since A has a compact resolvent and thus a discrete spectrum consisting only of eigenvalues, it follows that $A \gg 0$. $\square$

It is obvious that $M \geq 0$, $K \geq 0$, $M+K = I$ and $N(K) \cap N(M) = \{0\}$. Thus, the operator pencil L satisfies Condition I, and Condition II is obviously satisfied. Furthermore, also $N(M) \cap N(A) = \{0\}$. Indeed, if $Y \in N(M) \cap N(A)$, then $Y \in D(A)$, so that $Y = (y, y(a))^{\mathsf{T}}$ for some $y \in W_2^2(0, a)$. But $MY = 0$ means $y = 0$, so that also $y(a) = 0$, and $Y = 0$ follows. Hence the operator pencil L satisfies all assumptions of Theorem 1.5.7. Moreover, the geometric multiplicity of each eigenvalue of the pencil L is 1 because there exists only one linearly independent solution of (2.1.1) which satisfies the boundary condition (2.1.2). If there would be an eigenvalue λ of L with an eigenvector $Y \in N(K)$, then $s(\lambda, a) = 0$, and (2.1.4) would imply $s'(\lambda, a) = 0$, which is impossible since $s(\lambda, \cdot)$ is not identically zero. Therefore, Definition 1.5.2 and Lemma 1.5.5 show that we have only eigenvalues of type II in $\mathbb{C} \setminus \{0\}$. If 0 is an eigenvalue of L, then 0 is clearly of type I and the above reasoning and Lemma 1.3.1, part 2, show that also 0 is a simple eigenvalue with $N(A) \cap N(K) = \{0\}$. Thus Theorem 1.5.7 implies

Theorem 2.1.2. *The eigenvalues of the pencil L associated with the problem (2.1.1), (2.1.2), (2.1.4) on $L_2(0, a)$ possess the following properties.*

1. *Only a finite number of the eigenvalues lie in the closed lower half-plane.*
2. *All nonzero eigenvalues in the closed lower half-plane lie on the negative imaginary semiaxis and are simple. If their number κ is positive, they will be uniquely indexed as $\lambda_{-j} = -i|\lambda_{-j}|$, $j = 1, \ldots, \kappa$, satisfying $|\lambda_{-j}| < |\lambda_{-(j+1)}|$, $j = 1, \ldots, \kappa - 1$.*
3. *If $\kappa > 0$, then the numbers $i|\lambda_{-j}|$, $j = 1, \ldots, \kappa$, are not eigenvalues.*
4. *If $\kappa \geq 2$, then in each of the intervals $(i|\lambda_{-j}|, i|\lambda_{-(j+1)}|)$, $j = 1, \ldots, \kappa - 1$, the number of eigenvalues, counted with multiplicity, is odd.*
5. *If $\kappa > 0$, then the interval $[0, i|\lambda_{-1}|)$ contains no or an even number of eigenvalues, counted with multiplicity. If $N(A) = \{0\}$, they are all nonzero, otherwise one of them is the simple eigenvalue 0.*

In Section 6.1 we will consider the generalized Regge problem in which the boundary condition (2.1.4) is replaced by the condition

$$y'(a) + (i\lambda\alpha + \beta)y(a) = 0, \tag{2.1.5}$$

with $\alpha > 0$ and $\beta \in \mathbb{R}$.

2.2 Damped vibrations of strings

2.2.1 Problem identification

Small transversal vibrations of a smooth inhomogeneous string subject to viscous damping are described by the boundary value problem

$$\frac{\partial^2 u}{\partial s^2} - \sigma(s)\frac{\partial u}{\partial t} - \rho(s)\frac{\partial^2 u}{\partial t^2} = 0,$$

$$u(0,t) = 0,$$

$$\left.\left(\frac{\partial u}{\partial s} + m\frac{\partial^2 u}{\partial t^2} + \nu\frac{\partial u}{\partial t}\right)\right|_{s=l} = 0.$$

Here $u = u(s,t)$ is the transversal displacement of a point of the string which is as far as s from the left end of the string at time t, l is the length of the string, $\rho \geq \varepsilon > 0$ is its density, and $\sigma \geq 0$ is the coefficient of damping along the string. We will assume $\rho, \sigma \in L_\infty(0,l)$. The left end of the string is fixed while the right end is free to move in the direction orthogonal to the equilibrium position of the string subject to damping with damping coefficient $\nu > 0$. The right end bears a point mass $m > 0$.

Substituting $u(s,t) = v(\lambda,s)e^{i\lambda t}$ we arrive at

$$\frac{\partial^2 v}{\partial s^2} - i\lambda\sigma(s)v + \lambda^2\rho(s)v = 0, \tag{2.2.1}$$

$$v(\lambda, 0) = 0, \tag{2.2.2}$$

$$\left.\left(\frac{\partial v}{\partial s} - \lambda^2 m v + i\lambda\nu v\right)\right|_{s=l} = 0. \tag{2.2.3}$$

Like in Section 2.1 the eigenvalue problem (2.2.1)–(2.2.3) is described by the operator pencil

$$L(\lambda) = \lambda^2 M - i\lambda K - A,$$

where the operators A, K and M act in the Hilbert space $H = L_2(0,l) \oplus \mathbb{C}$ according to

$$A\begin{pmatrix} v \\ c \end{pmatrix} = \begin{pmatrix} -v'' \\ v'(l) \end{pmatrix},$$

$$D(A) = \left\{ \begin{pmatrix} v \\ c \end{pmatrix} : v \in W_2^2(0,l),\ v(0) = 0,\ c = v(l) \right\},$$

$$M = \begin{pmatrix} \rho & 0 \\ 0 & m \end{pmatrix}, \quad K = \begin{pmatrix} \sigma & 0 \\ 0 & \nu \end{pmatrix}.$$

Proposition 2.2.1. *The operators A, K and M are self-adjoint, $M \gg 0$ and $K \geq 0$ are bounded, and $A \gg 0$ has a compact resolvent. If $\sigma > 0$, then $K > 0$.*

Proof. The statements about M and K are obvious, and A is the particular case of the operator A in Section 2.1 with $q = 0$. Hence $A \gg 0$ by Proposition 2.1.1. $\square$

Proposition 2.2.1 and Lemma 1.2.4 lead to

Theorem 2.2.2. *All eigenvalues of* (2.2.1)–(2.2.3) *lie in the closed upper half-plane and are different from zero. If* $\sigma \neq 0$, *they lie all in the open upper half-plane.*

Proof. We still have to prove the last statement. Assume there exists a real nonzero eigenvalue λ with corresponding eigenvector $Y = (v, c)$. We conclude as in the proof of Lemma 1.2.3 that $(KY, Y) = 0$. Since $\sigma \geq 0$, this implies that $\sigma|v^2| = 0$ almost everywhere on $[0, a]$. We recall that $\sigma \neq 0$ means that $\sigma(x) \neq 0$ for all x in a set of positive Lebesgue measure. Hence v is zero on a set of positive Lebesgue measure. Due to the uniqueness of the solution of the initial value problem for linear ordinary differential equations it follows that $v = 0$. Then also $c = v(l) = 0$, which leads to the contradiction $V = 0$. $\square$

2.2.2   The location of the spectrum

In order to represent (2.2.1)–(2.2.3) in terms of an operator pencil of the form $\hat{L}(\tau) = \tau^2 \hat{M} - i\tau \hat{K} - \hat{A}$ with rank one operator $\hat{K}$, we assume that $\sigma(s) \equiv 2\varrho\rho(s)$ for some constant ϱ. Since $\rho > 0$ and $\sigma \geq 0$, it follows that $\varrho \geq 0$. We make use of the parameter transformation $\lambda = \pm\tau + i\varrho$ with $\lambda = \tau + i\varrho$ if $\nu - 2m\varrho \geq 0$ and with $\lambda = -\tau + i\varrho$ if $\nu - 2m\varrho < 0$. This parameter transformation gives rise to the operator pencil

$$L(\pm\tau + i\varrho) =: \hat{L}(\tau) = \tau^2 \hat{M} - i\tau \hat{K} - \hat{A},$$

where

$$\hat{M} = M, \quad \hat{K} = \pm(K - 2\varrho M), \quad \hat{A} = A + \varrho^2 M - \varrho K.$$

This gives the representations

$$\hat{A}\begin{pmatrix} v \\ c \end{pmatrix} = \begin{pmatrix} -v'' - \varrho^2 \rho v \\ v'(l) + (m\varrho^2 - \nu\varrho)v(l) \end{pmatrix}, \quad \begin{pmatrix} v \\ c \end{pmatrix} \in D(\hat{A}) = D(A) \subset L_2(0, l) \oplus \mathbb{C},$$

$$\hat{M} = \begin{pmatrix} \rho & 0 \\ 0 & m \end{pmatrix}, \quad \hat{K} = \begin{pmatrix} 0 & 0 \\ 0 & |\nu - 2m\varrho| \end{pmatrix}.$$

By Proposition 2.2.1, $\hat{A}$ is a relatively compact self-adjoint perturbation of A, and therefore $\hat{A}$ is self-adjoint, bounded below and has a compact resolvent. Also, observe that $\hat{M} \gg 0$ and that $\hat{K} \geq 0$, and that both $\hat{M}$ and $\hat{K}$ are bounded.

Three distinguished cases will be considered:

1) $\nu = 2mk$,   2) $\nu > 2mk$ and   3) $\nu < 2mk$.

The first case is trivial because $\hat{K} = 0$ and hence $\hat{L}(\tau) = \tau^2 \hat{M} - \hat{A}$. The spectrum of the pencil $\hat{L}(\tau)$ is linked to the spectrum of the operator $\hat{M}^{-\frac{1}{2}} \hat{A} \hat{M}^{-\frac{1}{2}}$ via the substitution $\tau^2 = \zeta$. The infinitely many positive eigenvalues of $\hat{M}^{-\frac{1}{2}} \hat{A} \hat{M}^{-\frac{1}{2}}$ give pairs of real eigenvalues of $\hat{L}$ which are symmetric with respect to the origin, and the at most finitely many negative eigenvalues of $\hat{M}^{-\frac{1}{2}} \hat{A} \hat{M}^{-\frac{1}{2}}$ give pairs of pure imaginary eigenvalues of $\hat{L}$ which are symmetric with respect to the origin. Taking Theorem 2.2.2 into account, we therefore conclude that the spectrum of problem (2.2.1)–(2.2.3) lies on the line $\operatorname{Im}\lambda = \varrho$ and on the interval $(0, 2i\varrho)$ of the imaginary axis, being symmetric with respect to the imaginary axis and with respect to the line $\operatorname{Im}\lambda = \varrho$, respectively.

In the second and third cases, $\hat{K} \geq 0$ is a rank one operator. Hence the pencil $\hat{L}$ satisfies Condition II if $\nu \neq 2m\varrho$, and we can apply all the statements of Theorems 1.5.6 and 1.5.7 to the pencil $\hat{L}$. Note that the pencil L associated with (2.2.1)–(2.2.3) is related to the pencil $\hat{L}$ via $L(\lambda) = \tilde{L}(\pm(\lambda - i\varrho))$. Observing Theorem 2.2.2, we obtain the following result as in Section 2.1, taking into account that an eigenvalue of type I can occur only at $\tau = 0$, i.e. at $\lambda = i\varrho$.

Theorem 2.2.3. *Let $\sigma(s) \equiv 2\varrho\rho(s)$ and $\nu > 2m\varrho$. Then:*

1. *Only a finite number of the eigenvalues of problem (2.2.1)–(2.2.3) lie in the closed half-plane $\operatorname{Im}\lambda \leq \varrho$.*
2. *All eigenvalues in the half-plane $\operatorname{Im}\lambda \leq \varrho$ which are different from $i\varrho$ lie on $(0, i\varrho)$ and are simple. Their number will be denoted by κ. If $\kappa > 0$, they will be uniquely indexed as $\lambda_{-j} = i\varrho - i|\lambda_{-j} - i\varrho|$, $j = 1, \ldots, \kappa$, satisfying $|\lambda_{-j} - i\varrho| < |\lambda_{-(j+1)} - i\varrho|$, $j = 1, \ldots, \kappa - 1$.*
3. *If $\kappa > 0$, then the numbers $i\varrho + i|\lambda_{-j} - i\varrho|$, $j = 1, \ldots, \kappa$, are not eigenvalues.*
4. *If $\kappa \geq 2$, then in each of the intervals $(i\varrho + i|\lambda_{-j} - i\varrho|, i\varrho + i|\lambda_{-(j+1)} - i\varrho|)$, $j = 1, \ldots, \kappa - 1$, the number of eigenvalues, counted with multiplicity, is odd.*
5. *If $\kappa > 0$, then the interval $[i\varrho, i\varrho + i|\lambda_{-1} - i\varrho|)$ contains no or an even number of eigenvalues, counted with multiplicity. If $N(\hat{A}) = \{0\}$, they are all different from $i\varrho$, otherwise one of them is the simple eigenvalue $i\varrho$.*

Theorem 2.2.4. *Let $\sigma(s) \equiv 2\varrho\rho(s)$ and $\nu < 2m\varrho$. Then:*

1. *Only a finite number of the eigenvalues of problem (2.2.1)–(2.2.3) lie in the closed half-plane $\operatorname{Im}\lambda \geq \varrho$, and all other eigenvalues lie in the open strip $0 < \operatorname{Im}\lambda < \varrho$.*
2. *All eigenvalues in the half-plane $\operatorname{Im}\lambda \geq \varrho$ which are different from $i\varrho$ lie on $(i\varrho, i\infty)$ and are simple. Their number will be denoted by κ. If $\kappa > 0$, they will be uniquely indexed as $\lambda_{-j} = i\varrho + i|\lambda_{-j} - i\varrho|$, $j = 1, \ldots, \kappa$, satisfying $|\lambda_{-j}| < |\lambda_{-(j+1)}|$, $j = 1, \ldots, \kappa - 1$.*
3. *If $\kappa > 0$, then the numbers $i\varrho - i|\lambda_{-j} - i\varrho|$, $j = 1, \ldots, \kappa$, are not eigenvalues.*
4. *If $\kappa \geq 2$, then in each of the intervals $(i\varrho - i|\lambda_{-(j+1)} - i\varrho|, i\varrho - i|\lambda_{-j} - i\varrho|)$, $j = 1, \ldots, \kappa - 1$, the number of eigenvalues, counted with multiplicity, is odd.*
5. *If $\kappa > 0$, then the interval $(i\varrho - i|\lambda_{-1} - i\varrho|, i\varrho]$ contains no or an even number of eigenvalues, counted with multiplicity. If $N(\hat{A}) = \{0\}$, they are all different from $i\varrho$, otherwise one of them is the simple eigenvalue $i\varrho$.*

We observe that Theorem 2.2.4 shows a priori in case $\nu < 2m\varrho$ that the spectrum lies in a horizontal strip of the complex plane. Furthermore, we have explicit upper and lower bounds for the imaginary parts of the eigenvalues if we possibly disregard finitely many eigenvalues on the positive imaginary axis.

2.2.3   Liouville transform for smooth strings

In this subsection we give an alternate approach under the assumption that $\rho \in W_2^2(0, l)$. Then the Liouville transform [57, p. 292]

$$x(s) = \int_0^s \rho^{\frac{1}{2}}(r)\, dr, \quad 0 \le s \le l,$$

$$y(\lambda, x) = \rho^{\frac{1}{4}}(s(x))v(\lambda, s(x)), \quad 0 \le x \le a, \ \lambda \in \mathbb{C},$$

leads to the equivalent boundary value problem

$$y'' - i\lambda\sigma(s(x))\rho^{-1}(s(x))y - q(x)y + \lambda^2 y = 0, \tag{2.2.4}$$

$$y(\lambda, 0) = 0, \tag{2.2.5}$$

$$y'(\lambda, a) + (-\lambda^2 \tilde{m} + i\lambda\tilde{\nu} + \beta)y(\lambda, a) = 0, \tag{2.2.6}$$

where

$$q(x) = \rho^{-\frac{1}{4}}(s(x))\frac{d^2}{dx^2}\rho^{\frac{1}{4}}(s(x)),$$

$$a = \int_0^l \rho^{\frac{1}{2}}(r)\, dr,$$

$$\tilde{m} = \rho^{-\frac{1}{2}}(s(a))m > 0,$$

$$\tilde{\nu} = \rho^{-\frac{1}{2}}(s(a))\nu > 0,$$

$$\beta = -\rho^{-\frac{1}{4}}(s(a))\frac{d\rho^{\frac{1}{4}}(s(x))}{dx}\bigg|_{x=a}.$$

As in Subsection 2.2.2 we assume that $\sigma(s) \equiv 2\varrho\rho(s)$ for some nonnegative constant ϱ, and we will make use of the parameter transformation $\lambda = \pm\tau + i\varrho$ with $\lambda = \tau + i\varrho$ if $\tilde{\nu} - 2\tilde{m}\varrho \ge 0$ and with $\lambda = -\tau + i\varrho$ if $\tilde{\nu} - 2\tilde{m}\varrho < 0$. Like in Subsection 2.2.2 the eigenvalue problem (2.2.4)–(2.2.6) is described by the operator pencil

$$\tilde{L}(\tau) = \tau^2\tilde{M} - i\tau\tilde{K} - \tilde{A},$$

where the operators $\tilde{A}$, $\tilde{K}$ and $\tilde{M}$ act in the Hilbert space $H = L_2(0, a) \oplus \mathbb{C}$ according to

$$\tilde{A}\begin{pmatrix} y \\ c \end{pmatrix} = \begin{pmatrix} -y'' + (q - \varrho^2)y \\ y'(a) + (\beta - \tilde{\nu}\varrho + \tilde{m}\varrho^2)\,y(a) \end{pmatrix},$$

$$D(\tilde{A}) = \left\{ \begin{pmatrix} y \\ c \end{pmatrix} : y \in W_2^2(0, a), \ y(0) = 0, \ c = y(a) \right\},$$

$$\tilde{M} = \begin{pmatrix} I & 0 \\ 0 & \tilde{m} \end{pmatrix}, \qquad \tilde{K} = \begin{pmatrix} 0 & 0 \\ 0 & |\tilde{\nu} - 2\tilde{m}\varrho| \end{pmatrix}.$$

It is clear that

$$\tilde{A} = A + \begin{pmatrix} -\varrho^2 & 0 \\ 0 & \beta - \tilde{\nu}\varrho + \tilde{m}\varrho^2 \end{pmatrix},$$

where A is defined in Section 2.1. Thus $\tilde{A}$ is a relatively compact symmetric perturbation of A, and therefore $\tilde{A}$ is self-adjoint, bounded below and has a compact resolvent.

Now it follows that Theorems 2.2.3 and 2.2.4 are also true if (2.2.1)–(2.2.3) there is replaced by (2.2.4)–(2.2.6), either by a direct proof referring to the results from Section 2.1, as we did in Subsection 2.2.2, or by observing that the problems (2.2.1)–(2.2.3) and (2.2.4)–(2.2.6) are equivalent.

2.3   Vibrations of star graphs with damping

2.3.1   Problem identification

Let us consider p inhomogeneous strings, labelled by subscripts $1,\ldots,p$, $p \geq 2$, each having one end joined at the interior vertex and the other end fixed. The interior vertex is free to move in the direction orthogonal to the equilibrium position of the star graph subject to damping at this interior vertex with damping coefficient $\nu > 0$. A point mass m may be present at the interior vertex. Small transverse vibrations of such a star graph are described by the following system of equations:

$$\frac{\partial^2}{\partial s_j^2}u_j(s_j,t) - \sigma_j(s_j)\frac{\partial}{\partial t}u_j(s_j,t) - \rho_j(s_j)\frac{\partial^2}{\partial t^2}u_j(s_j,t) = 0, \quad j = 1,\ldots,p, \quad (2.3.1)$$

$$u_j(0,t) = 0, \quad j = 1,\ldots,p, \tag{2.3.2}$$

$$u_1(l_1,t) = u_2(l_2,t) = \cdots = u_p(l_p,t), \tag{2.3.3}$$

$$\sum_{j=1}^{p} \frac{\partial}{\partial s_j}u_j(l_j,t) + m\frac{\partial^2}{\partial t^2}u_1(l_1,t) + \nu\frac{\partial}{\partial t}u_1(l_1,t) = 0. \tag{2.3.4}$$

Here $l_j > 0$ is the length of the jth string, $\rho_j \geq \varepsilon > 0$ its density and $\sigma_j \geq 0$ its damping, $u_j(s_j,t)$ stands for the transverse displacement of the jth string at position s_j and at time t. We will assume that $\rho_j \in L_\infty(0,a_j)$.

We mention two particular cases.

The first case is that we have a damped string with a mass at an interior point. In this case, $p = 2$ and $m > 0$.

The second particular case is that we have a star graph with undamped strings and without mass at the interior point. In this case, $\sigma_j = 0$ for all $j = 1,\ldots,p$ and $m = 0$.

Substituting $u_j(s_j, t) = v_j(\lambda, s_j)e^{i\lambda t}$ into (2.3.1)–(2.3.4) we obtain

$$\frac{\partial^2}{\partial s_j^2} v_j(\lambda, s_j) - i\lambda \sigma_j(s_j) v_j(\lambda, s_j) + \lambda^2 \rho_j(s_j) v_j(\lambda, s_j) = 0, \quad j = 1, \ldots, p, \quad (2.3.5)$$

$$v_j(\lambda, 0) = 0, \quad j = 1, \ldots, p, \quad (2.3.6)$$

$$v_1(\lambda, l_1) = v_2(\lambda, l_2) = \cdots = v_p(\lambda, l_p), \quad (2.3.7)$$

$$\sum_{j=1}^{p} \frac{\partial}{\partial s_j} v_j(\lambda, l_j) - \lambda^2 m v_1(\lambda, l_1) + i\lambda \nu v_1(\lambda, l_1) = 0. \quad (2.3.8)$$

The eigenvalue problem (2.3.5)–(2.3.8) is described by the operator pencil

$$L(\lambda) = \lambda^2 M - i\lambda K - A,$$

where A, K and M act in the Hilbert space $\bigoplus_{j=1}^{p} L_2(0, l_j) \oplus \mathbb{C}$. A is defined by

$$A \begin{pmatrix} v_1 \\ \vdots \\ v_p \\ c \end{pmatrix} = \begin{pmatrix} -v_1'' \\ \vdots \\ -v_p'' \\ \sum_{j=1}^{p} v_j'(l_j) \end{pmatrix},$$

$$D(A) = \left\{ \begin{pmatrix} v_1 \\ \vdots \\ v_p \\ c \end{pmatrix} : v_j \in W_2^2(0, l_j), \ v_j(0) = 0, \ v_j(l_j) = c, \ j = 1, \ldots, p \right\},$$

whereas

$$M = \begin{pmatrix} \rho_1 & \cdots & 0 & 0 \\ \vdots & \ddots & \vdots & \vdots \\ 0 & \cdots & \rho_p & 0 \\ 0 & \cdots & 0 & m \end{pmatrix}, \quad K = \begin{pmatrix} \sigma_1 & \cdots & 0 & 0 \\ \vdots & \ddots & \vdots & \vdots \\ 0 & \cdots & \sigma_p & 0 \\ 0 & \cdots & 0 & \nu \end{pmatrix}.$$

Proposition 2.3.1. *The operators A, K and M are self-adjoint, $M \geq 0$ and $K \geq 0$ are bounded, $M + K \gg 0$, and $A \gg 0$ has a compact resolvent. If $m > 0$, then $M \gg 0$, and if $\sigma_1 \geq \varepsilon, \ldots, \sigma_p \geq \varepsilon$ for some $\varepsilon > 0$, then $K \gg 0$.*

Proof. The statements about M and K are obvious, so that we turn our attention to A. We will use Theorem 10.3.5 to verify self-adjointness. Observe that $v_j^{[1]} = -v_j'$, $j = 1, \ldots, p$, according to Definition 10.2.1. We have to find U_3 and U defined in (10.3.12) and (10.3.13). Before doing so we note that the conditions $v_j(l_j) = c$, $j = 1, \ldots, p$, can be written as $v_1(l_1) = c$ and $v_1(l_1) - v_j(l_j) = 0$, $j = 2, \ldots, p$, which contribute to U_2 and U_1, respectively. It follows now that U_1 is a $(2p-1) \times 4p$

matrix, whose rows are e_{4j-3}^T, $j = 1, \ldots, p$ and $e_3^\mathsf{T} - e_{4j-1}^\mathsf{T}$, $j = 2, \ldots, p$. Hence $N(U_1)$ is spanned by e_{4j-2}, $j = 1, \ldots, p$, e_{4j}, $j = 1, \ldots, p$, and $e_3 + \sum_{j=2}^{p} e_{4j-1}$.

Furthermore,

$$U_2 = e_3^\mathsf{T}, \quad V = -\sum_{j=1}^{p} e_{4j}^\mathsf{T}.$$

We recall that

$$U_3 = \begin{pmatrix} J \\ V \\ -U_2 \end{pmatrix} \quad \text{where} \quad J = \bigoplus_{j=1}^{p} J_1 \quad \text{and} \quad J_1 = \begin{pmatrix} 0 & -1 & 0 & 0 \\ 1 & 0 & 0 & 0 \\ 0 & 0 & 0 & 1 \\ 0 & 0 & -1 & 0 \end{pmatrix}$$

and

$$U = \begin{pmatrix} U_1 & 0 & 0 \\ U_2 & -I & 0 \\ V & 0 & -I \end{pmatrix}.$$

It is now straightforward to verify that $U_3(N(U_1))$ is the subspace of $\mathbb{C}^{4p+2}$ spanned by e_{4j-3}, $j = 1, \ldots, p$, $e_{4j-1} - e_{4p+1}$, $j = 1, \ldots, p$, and $e_4 + e_{4p+2} + \sum_{j=2}^{p} e_{4j}$ and that $R(U^*)$ is the subspace of $\mathbb{C}^{4p+2}$ spanned by e_{4j-3}, $j = 1, \ldots, p$, $e_3 - e_{4j-1}$, $j = 2, \ldots, p$, $e_3 - e_{4p+1}$ and $\sum_{j=1}^{p} e_{4j} + e_{4p+2}$. This shows that $U_3(N(U_1)) = R(U^*)$.

Hence A is self-adjoint by Theorem 10.3.5. Furthermore, A has a compact resolvent by Theorem 10.3.8.

To prove that $A > 0$, we take $Y = (v_1, \ldots, v_p, c)^\mathsf{T} \in D(A) \setminus \{0\}$. Then at least one v_j is not constant, and (10.2.5) leads to

$$(AY, Y) = \sum_{j=1}^{p} \int_0^{l_j} |v_j'(s_j)|^2 \, ds_j + \sum_{j=1}^{p} [v_j, v_j]_1(l_j) + \sum_{j=1}^{p} v_j'(l_j)\overline{v_j(l_j)}$$

$$= \sum_{j=1}^{p} \int_0^{l_j} |v_j'(s_j)|^2 \, ds_j > 0.$$

Since A has a compact resolvent and thus a discrete spectrum consisting only of normal eigenvalues, it follows that $A \gg 0$. $\qquad\square$

Theorem 2.3.2.

1. *The eigenvalues of problem (2.3.5)–(2.3.8) lie in the closed upper half-plane and are nonzero. The real eigenvalues, if any, are semisimple eigenvalues whose multiplicities do not exceed $p - 1$.*

2. If $\sigma_1 \neq 0, \ldots, \sigma_p \neq 0$, all eigenvalues lie in the open upper half-plane.

3. If $\sigma_1 = 0, \ldots, \sigma_p = 0$, the real eigenvalues, if any, are of type I but not of type II.

Proof. Throughout this proof, we will use the results of Proposition 2.3.1.

1. Lemma 1.2.4, part 1, shows that all eigenvalues lie in the closed upper half-plane. Since $A \gg 0$, it follows that 0 is no eigenvalue, and all other real eigenvalues, if any, are semisimple by Lemma 1.3.1, part 1. If $Y = (v_1, \ldots, v_p, c)^\mathsf{T} \in D(A)$ satisfies $L(\lambda)Y = 0$, then $v_1, \ldots, v_p$ are solutions of second-order linear differential equations satisfying the initial conditions $v_j(0) = 0$, and since $v_1(l_1) = c$, there are at most p linearly independent solutions. Note, however, that (2.3.7) and (2.3.8) imply that $v_p(l_p)$ and $v_p'(l_p)$ are uniquely determined by $v_1, \ldots, v_{p-1}$. Hence v_p is the unique solution of an initial value problem and therefore uniquely determined by $v_1, \ldots, v_{p-1}$. Therefore we have at most $p - 1$ linearly independent solutions.

Part 2 can be proved like the corresponding statement of Theorem 2.2.2.

In part 3, K has rank one, and the statement follows from Lemma 1.1.11, part 3, Theorem 1.5.6, part 4, and the fact that 0 is no eigenvalue.   □

Due to arguments similar to those in Section 2.2 we assume in what follows that $\sigma_1(s_1) \equiv 2\varrho\rho_1(s_1), \ldots, \sigma_p(s_p) \equiv 2\varrho\rho_p(s_p)$ for some nonnegative constant ϱ, and we will make use of the parameter transformation $\lambda = \tau + i\varrho$ if $\nu - 2m\varrho \geq 0$ and $\lambda = -\tau + i\varrho$ if $\nu - 2m\varrho < 0$.

2.3.2   The location of the spectrum

The parameter transformation gives rise to the operator pencil

$$L(\pm\tau + i\varrho) =: \hat{L}(\tau) = \tau^2 \hat{M} - i\tau \hat{K} - \hat{A},$$

where

$$\hat{M} = M, \quad \hat{K} = \pm(K - 2\varrho M), \quad \hat{A} = A + \varrho^2 M - \varrho K.$$

The operators $\hat{A}$, $\hat{K}$ and $\hat{M}$ act in the Hilbert space $H = \bigoplus_{j=1}^p L_2(0, l_j) \oplus \mathbb{C}$ and have the representations

$$\hat{A} \begin{pmatrix} v_1 \\ \vdots \\ v_p \\ c \end{pmatrix} = \begin{pmatrix} -v_1'' - \varrho^2 \rho_1 v_1 \\ \vdots \\ -v_p'' - \varrho^2 \rho_p v_p \\ \sum_{j=1}^p v_j'(l_j) + (m\varrho^2 - \nu\varrho) v_1(l_1) \end{pmatrix},$$

$$D(\hat{A}) = \left\{ \begin{pmatrix} v_1 \\ \vdots \\ v_p \\ c \end{pmatrix} : v_j \in W_2^2(0, l_j), \ v_j(0) = 0, \ v_j(l_j) = c, \ j = 1, \ldots, p \right\},$$

$$\hat{M} = \begin{pmatrix} \rho_1 & \cdots & 0 & 0 \\ \vdots & \ddots & \vdots & \vdots \\ 0 & \cdots & \rho_p & 0 \\ 0 & \cdots & 0 & m \end{pmatrix}, \quad \hat{K} = \begin{pmatrix} 0 & \cdots & 0 & 0 \\ \vdots & \ddots & \vdots & \vdots \\ 0 & \cdots & 0 & 0 \\ 0 & \cdots & 0 & |\nu - 2m\varrho| \end{pmatrix}.$$

Proposition 2.3.3. *The operators $\hat{A}$, $\hat{K}$ and $\hat{M}$ are self-adjoint, $\hat{M} \geq 0$ and $\hat{K} \geq 0$ are bounded, and $\hat{A}$ is bounded below and has a compact resolvent. Furthermore, $N(\hat{M}) \cap N(\hat{A}) = \{0\}$ while $\hat{M} \gg 0$ if $m > 0$.*

Proof. In view of Proposition 2.3.1 we only have to show $N(\hat{M}) \cap N(\hat{A}) = \{0\}$. To this end let $Y = (v_1, \ldots, v_p, c)^{\mathsf{T}} \in N(\hat{M}) \cap N(\hat{A})$. Then $\hat{M}Y = 0$ implies $v_j = 0$, $j = 1, \ldots, p$, and $Y \in D(\hat{A})$ then leads to $c = 0$. □

Three distinguished cases will be considered:

1) $\nu = 2m\varrho$, 2) $\nu > 2m\varrho$ and 3) $\nu < 2m\varrho$.

The first case is trivial because $\hat{K} = 0$ and hence $\hat{L}(\tau) = \tau^2 \hat{M} - \hat{A}$. Note that $\nu > 0$ implies $m > 0$ and $\varrho > 0$ in this case. The spectrum of the operator pencil $\hat{L}$ is linked to the spectrum of the operator $\hat{M}^{-\frac{1}{2}} \hat{A} \hat{M}^{-\frac{1}{2}}$ via the substitution $\tau^2 = \zeta$. The infinitely many positive eigenvalues of $\hat{M}^{-\frac{1}{2}} \hat{A} \hat{M}^{-\frac{1}{2}}$ give pairs of real eigenvalues of $\hat{L}$ which are symmetric with respect to the origin, and the at most finitely many negative eigenvalues of $\hat{M}^{-\frac{1}{2}} \hat{A} \hat{M}^{-\frac{1}{2}}$ give pairs of pure imaginary eigenvalues of $\hat{L}$ which are symmetric with respect to the origin. Taking Theorem 2.3.2 into account, we conclude that the spectrum of problem (2.3.5)–(2.3.8) lies on the line $\operatorname{Im} \lambda = \varrho$ and on the interval $(0, 2i\varrho)$ of the imaginary axis, being symmetric with respect to the imaginary axis and to the line $\operatorname{Im} \lambda = \varrho$, respectively.

In the second and third cases, $\hat{K}$ is a rank one operator. Recall that $L(\lambda) = \hat{L}(\pm(\lambda - i\varrho))$. We have defined eigenvalues of type I and type II in Section 1.5. Below, we will use this notation relative to the pencil $\hat{L}$, that is with respect to the eigenvalue parameter τ.

Theorem 2.3.4. *Let $\sigma_1(s_1) \equiv 2\varrho\rho_1(s_1), \ldots, \sigma_p(s_p) \equiv 2\varrho\rho_p(s_p)$ and $\nu \neq 2m\varrho$. Then:*

1. *The geometric multiplicity of each eigenvalue of (2.3.5)–(2.3.8) does not exceed $p - 1$.*
2. *The eigenvalues of type I are located on the imaginary axis and on the line $\operatorname{Im} \lambda = \varrho$, and they are symmetric with respect to $i\varrho$.*
3. *The sets of eigenvalues of types I and II do not intersect.*

Proof. 1. The proof for the bound of the geometric multiplicity in Theorem 2.3.2 is valid for all eigenvalues.

Part 2 follows from Lemma 1.5.4.

3. Let $\lambda \neq i\varrho$ be an eigenvalue of type II. In view of Theorem 1.5.6, part 3, there is a corresponding eigenvector $Y = (v_1, \ldots, v_p, c)^\mathsf{T}$ with $c \neq 0$. Then $v_1(l_1) = \cdots = v_p(l_p) = c \neq 0$, so that none of the v_j, $j = 1, \ldots, p$ is identically zero. Assume that λ is also an eigenvalue of type I. In view of Theorem 1.5.6, part 1, there is an eigenvector $Z = (w_1, \ldots, w_p, d)^\mathsf{T}$ corresponding to the eigenvalue λ with $d = 0$, and it follows that $w_1(l_1) = \cdots = w_p(l_p) = 0$. Since v_j and w_j, $j = 1, \ldots, p$, satisfy the second-order differential equations (2.3.5) and the initial conditions (2.3.6), each w_j must be a multiple of v_j. But $v_j(l_j) = d_j \neq 0$ and $w_j(l_j) = d = 0$ then leads to $w_j = 0$ for $j = 1, \ldots, p$, so that $Z = 0$, which is impossible since Z is an eigenvector. Finally, Theorem 1.5.7, part 2, applied to the pencil $\hat{L}$ shows that $i\varrho$ is not an eigenvalue of type II.   □

Remark 2.3.5. Eigenvalues $\lambda \neq i\varrho$ of type I can exist. Let $p = 2$. By Theorem 1.5.6, part 1, an eigenvector $Y = (v_1, v_2, c)^\mathsf{T}$ of $\hat{L}$ at τ is of type I if and only if $c = 0$. Hence, if $v_1(l_1) = v_2(l_2) = c = 0$, then $v_1'(l_1) + v_2'(l_2) = 0$ which is possible, e. g., in case $l_1 = l_2$, $\rho_1 \equiv \rho_2 \equiv \varrho$ and $\tau^2 = \tau_j^2 - \varrho^2$ with real $\tau \neq 0$ and $\tau_j = \frac{\pi j}{l_1}$, $j \in \mathbb{N}$, so that $v_1(s_1) = \sin(\tau s_1)$ and $v_2(s_2) = -\sin(\tau s_2)$. The physical interpretation of this phenomenon is that there can be modes of vibration with a node at the point where the mass m is located. These modes are independent of $\nu - \varrho m$. If $\varrho = 0$, these modes do not depend on ν, and the amplitudes of corresponding modes do not decay with time since the corresponding eigenvalues are real as $\hat{A} = A \gg 0$.

We describe the eigenvalues of type II by applying Theorem 1.5.7 to the pencil $\hat{L}$ and by observing Proposition 2.3.1 and Theorems 2.3.2 and 2.3.4, part 3.

Theorem 2.3.6. *Let $\sigma_1(s_1) \equiv 2\varrho\rho_1(s_1), \ldots, \sigma_p(s_p) \equiv 2\varrho\rho_p(s_p)$ and $\nu > 2m\varrho$. Then:*

1. *Only a finite number, denoted by κ_2, of the eigenvalues of type II of problem (2.3.5)–(2.3.8) lie in the closed half-plane $\operatorname{Im} \lambda \leq \varrho$.*
2. *All eigenvalues of type II in the closed half-plane $\operatorname{Im} \lambda \leq \varrho$ lie on $(0, i\varrho)$ and their type II multiplicities are 1. If $\kappa_2 > 0$, they will be uniquely indexed as $\lambda_{-j} = i\varrho - i|\lambda_{-j} - i\varrho|$, $j = 1, \ldots, \kappa_2$, satisfying $|\lambda_{-j} - i\varrho| < |\lambda_{-(j+1)} - i\varrho|$, $j = 1, \ldots, \kappa_2 - 1$.*
3. *If $\kappa_2 > 0$, then the numbers $i\varrho + i|\lambda_{-j} - i\varrho|$, $j = 1, \ldots, \kappa_2$, are not eigenvalues.*
4. *If $\kappa_2 \geq 2$, then the intervals $(i\varrho + i|\lambda_{-j} - i\varrho|, i\varrho + i|\lambda_{-(j+1)} - i\varrho|)$, $j = 1, \ldots, \kappa_2 - 1$, contain an odd number of eigenvalues of type II, counted with multiplicity.*
5. *Let $\kappa_2 > 0$. Then the interval $[i\varrho, i\varrho + i|\lambda_{-1} - i\varrho|)$ contains no or an even number of eigenvalues of type II, counted with multiplicity, if $N(\hat{A}) \subset N(\hat{K})$, and an odd number of eigenvalues of type II, counted with multiplicity, otherwise.*

Proof. All statements except for part 3 are exactly the same as in Theorem 1.5.7. To complete the proof of part 3 we have to show that the numbers $i\varrho + i|\lambda_{-j} - i\varrho|$ are not eigenvalues of type I. Indeed, if any of those numbers were an eigenvalue of type I, then by Theorem 2.3.4, part 2, also $\lambda_{-j} = i\varrho - i|\lambda_{-j} - i\varrho|$ would be an eigenvalue of type I. But this is impossible by Theorem 2.3.4, part 3, since $i\varrho - i|\lambda_{-j} - i\varrho|$ is an eigenvalue of type II. $\qquad\square$

Theorem 2.3.7. *Let* $\sigma_1(s_1) \equiv 2\varrho\rho_1(s_1), \dots, \sigma_p(s_p) \equiv 2\varrho\rho_p(s_p)$ *and* $\nu < 2m\varrho$. *Then:*

1. *Only a finite number, denoted by* κ_2, *of the eigenvalues of type II of problem (2.3.5)–(2.3.8) lie in the closed half-plane* $\operatorname{Im}\lambda \geq \varrho$, *and all other eigenvalues lie in the open strip* $0 < \operatorname{Im}\lambda < \varrho$.
2. *All eigenvalues of type II in the closed half-plane* $\operatorname{Im}\lambda \geq \varrho$ *lie on* $(i\varrho, i\infty)$ *and their type II multiplicities are 1. If* $\kappa_2 > 0$, *they will be uniquely indexed as* $\lambda_{-j} = i\varrho + i|\lambda_{-j} - i\varrho|$, $j = 1, \dots, \kappa_2$, *satisfying* $|\lambda_{-j}| < |\lambda_{-(j+1)}|$, $j = 1, \dots, \kappa_2 - 1$.
3. *If* $\kappa_2 > 0$, *then the numbers* $i\varrho - i|\lambda_{-j} - i\varrho|$, $j = 1, \dots, \kappa_2$, *are not eigenvalues.*
4. *If* $\kappa_2 \geq 2$, *then the intervals* $(i\varrho - i|\lambda_{-(j+1)} - i\varrho|, i\varrho - i|\lambda_{-j} - i\varrho|)$, $j = 1, \dots, \kappa_2 - 1$, *contain an odd number of eigenvalues of type II, counted with multiplicity.*
5. *Let* $\kappa_2 > 0$. *Then the interval* $(i\varrho - i|\lambda_{-1} - i\varrho|, i\varrho]$ *contains no or an even number of eigenvalues of type II, counted with multiplicity, if* $N(\hat{A}) \subset N(\hat{K})$, *and an odd number of eigenvalues of type II, counted with multiplicity, otherwise.*

2.3.3 Liouville transform for smooth star graphs

We assume that $\rho_j \in W_2^2(0, l_j)$ for $j = 1, \dots, p$ and apply the Liouville transform as in Section 2.2, i.e.,

$$x_j(s_j) = \int_0^{s_j} \rho_j(r)^{\frac{1}{2}} \, dr,$$

$$y_j(\lambda, x_j) = \rho_j[x_j]^{\frac{1}{4}} v_j(\lambda, s_j(x_j)),$$

where we use the notation $\rho_j[x_j] =: \rho_j(s_j(x_j))$. Then (2.3.5)–(2.3.8) with $\sigma_j(s_j) \equiv \varrho\rho_j(s_j)$ for $j = 1, \dots, p$ becomes the Sturm–Liouville problem

$$y_j''(\lambda, x_j) - 2i\lambda\varrho y_j(\lambda, x_j) + \lambda^2 y_j(\lambda, x_j) - q_j(x)y_j(\lambda, x_j) = 0, \quad j = 1, \dots, p, \tag{2.3.9}$$

$$y_j(\lambda, 0) = 0, \quad j = 1, \dots, p, \tag{2.3.10}$$

$$\rho_1[a_1]^{-\frac{1}{4}} y_1(\lambda, a_1) = \cdots = \rho_p[a_p]^{-\frac{1}{4}} y_p(\lambda, a_p), \tag{2.3.11}$$

$$\sum_{j=1}^{p} \frac{\rho_j[a_j]^{\frac{1}{4}}}{\rho_1[a_1]^{\frac{1}{4}}} y_j'(\lambda, a_j) + (-\tilde{m}\lambda^2 + i\tilde{\nu}\lambda + \beta)y_1(\lambda, a_1) = 0, \tag{2.3.12}$$

where primes denote x_j-differentiation and where

$$q_j(x_j) = \rho_j[x_j]^{-\frac{1}{4}} \frac{d^2}{dx_j^2}(\rho_j[x_j]^{\frac{1}{4}}),$$

$$a_j = \int_0^{l_j} \rho_j(r)^{\frac{1}{2}} \, dr,$$

$$\tilde{\nu} = \nu \rho_1 [a_1]^{-\frac{1}{2}},$$

$$\tilde{m} = m \rho_1 [a_1]^{-\frac{1}{2}},$$

$$\beta = -\sum_{j=1}^p \frac{\rho_j [a_j]^{\frac{1}{4}}}{\rho_1 [a_1]^{\frac{1}{2}}} \frac{d}{dx_j} \rho_j [x_j]^{\frac{1}{4}} \bigg|_{x_j = a_j}.$$

Making use of the parameter transformation $\lambda = \pm \tau + i \varrho$, the eigenvalue problem
(2.3.9)–(2.3.12) has an operator pencil representation

$$\tilde{L}(\tau) = \tau^2 \tilde{M} - i \tau \tilde{K} - \tilde{A}$$

acting in the Hilbert space $\bigoplus_{j=1}^p L_2(0, a_j) \oplus \mathbb{C}$, with

$$D(\tilde{A}) = \left\{ \begin{pmatrix} y_1 \\ \vdots \\ y_p \\ c \end{pmatrix} : y_j \in W_2^2(0, a_j), \ y_j(0) = 0, \ \rho_j [a_1]^{-\frac{1}{4}} y_j(a_j) = c, \ j = 1, \ldots, p \right\},$$

and with operators

$$\tilde{A} \begin{pmatrix} y_1 \\ \vdots \\ y_p \\ c \end{pmatrix} = \begin{pmatrix} -y_1'' + (q_1 - \varrho^2) y_1 \\ \vdots \\ -y_p'' + (q_p - \varrho^2) y_p \\ \sum_{j=1}^p \frac{\rho_j [a_j]^{\frac{1}{4}}}{\rho_1 [a_1]^{\frac{1}{4}}} y_j'(a_j) + (\beta - \tilde{\nu} \varrho + \tilde{m} \varrho^2) y_1(a_1) \end{pmatrix},$$

$$D(\tilde{A}) = \left\{ \begin{pmatrix} y_1 \\ \vdots \\ y_p \\ c \end{pmatrix} : y_j \in W_2^2(0, a_j), \ y_j(0) = 0, \ \rho_j [a_1]^{-\frac{1}{4}} y_j(a_j) = c, \ j = 1, \ldots, p \right\},$$

$$\tilde{M} = \begin{pmatrix} I & \cdots & 0 & 0 \\ \vdots & \ddots & \vdots & \vdots \\ 0 & \cdots & I & 0 \\ 0 & \cdots & 0 & m \end{pmatrix}, \quad \tilde{K} = \begin{pmatrix} 0 & \cdots & 0 & 0 \\ \vdots & \ddots & \vdots & \vdots \\ 0 & \cdots & 0 & 0 \\ 0 & \cdots & 0 & |\tilde{\nu} - 2\tilde{m}\varrho| \end{pmatrix}.$$

As in Proposition 2.3.1 or Proposition 2.3.3 it now follows that the operators
$\tilde{A}$, $\tilde{K}$ and $\tilde{M}$ are self-adjoint, $\tilde{M} \geq 0$ and $\tilde{K} \geq 0$ are bounded, $\tilde{K}$ is a rank
one operator or the zero operator, and $\tilde{A}$ is bounded below and has a compact
resolvent.

Now it follows that Theorems 2.3.6 and 2.3.7 are also true if (2.3.5)–(2.3.8) there is replaced by (2.3.9)–(2.3.12), either by a direct proof similar to the proofs given in Subsection 2.3.2, or by observing that the problems (2.3.5)–(2.3.8) and (2.3.9)–(2.3.12) are equivalent.

2.3.4   General potentials

For later use we will consider a slight generalization of (2.3.5)–(2.3.8) in that we allow an additional real potential $q_j \in L_2(0, l_j)$ in each of the differential equations in (2.3.5), i.e.,

$$\frac{\partial^2}{\partial s_j^2} v_j(\lambda, s_j) - i\lambda \sigma_j(s_j) v_j(\lambda, s_j) + \lambda^2 \rho_j(s_j) v_j(\lambda, s_j) - q_j(s) v_j(\lambda, s_j) = 0 \quad (2.3.13)$$

for $j = 1, \ldots, p$. Here the q_j are distinct from those in the previous subsection; in both cases we have used the conventional notation for potentials, and no confusion should arise. Clearly, this additional potential does not change the domain of the operator A. Also, these terms do not enter into any of the quasi-derivatives $v_j^{[0]}$, $v_j^{[1]}$, $j = 1, \ldots, p$. Hence the corresponding operator A is self-adjoint and bounded below.

A review of the proof of Theorem 2.3.2 shows that

Theorem 2.3.8.

1. *The geometric multiplicities of the eigenvalues of the problem (2.3.13), (2.3.6)–(2.3.8) do not exceed $p - 1$. The nonzero real eigenvalues, if any, are semisimple.*

2. *If $\sigma_1 = 0, \ldots, \sigma_p = 0$, then the real eigenvalues, if any, are of type I, and the real nonzero eigenvalues, if any, are not of type II.*

Theorem 2.3.4 holds verbatim also for the eigenvalue problem given by the differential equation (2.3.13) and the boundary conditions (2.3.6)–(2.3.8), whereas Theorems 2.3.6 and 2.3.7 have the following weaker counterparts, with the operator $\hat{A}$ given by

$$\hat{A} \begin{pmatrix} v_1 \\ \vdots \\ v_p \\ c \end{pmatrix} = \begin{pmatrix} -v_1'' + q_1 v_1 - \varrho^2 \rho_1 v_1 \\ \vdots \\ -v_p'' + q_p v_p - \varrho^2 \rho_p v_p \\ \sum_{j=1}^{p} v_j'(l_j) + \left(m\varrho^2 - \nu\varrho \right) v_1(l_1) \end{pmatrix},$$

whereas the operators $\hat{M}$ and $\hat{A}$ are as in Subsection 2.3.2.

Theorem 2.3.9. *Let $\sigma_1(s_1) \equiv 2\varrho \rho_1(s_1), \ldots, \sigma_p(s_p) \equiv 2\varrho \rho_p(s_p)$ and $\nu > 2m\varrho$. Then:*

1. *Only a finite number, denoted by κ_2, of the eigenvalues of type II of problem (2.3.5)–(2.3.8) lie in the closed half-plane $\operatorname{Im} \lambda \leq \varrho$.*

2. *All eigenvalues of type II in the closed half-plane* $\operatorname{Im}\lambda \le \varrho$ *lie on* $(-i\infty, i\varrho)$ *and their multiplicities are 1. If* $\kappa_2 > 0$, *they will be uniquely indexed as* $\lambda_{-j} = i\varrho - i|\lambda_{-j} - i\varrho|$, $j = 1,\ldots,\kappa_2$, *satisfying* $|\lambda_{-j} - i\varrho| < |\lambda_{-(j+1)} - i\varrho|$, $j = 1,\ldots,\kappa_2 - 1$.

3. *If* $\kappa_2 > 0$, *then the numbers* $i\varrho + i|\lambda_{-j} - i\varrho|$, $j = 1,\ldots,\kappa_2$, *are not eigenvalues.*

4. *If* $\kappa_2 \ge 2$, *then the intervals* $(i\varrho + i|\lambda_{-j} - i\varrho|, i\varrho + i|\lambda_{-(j+1)} - i\varrho|)$, $j = 1,\ldots,\kappa_2 - 1$, *contain an odd number of eigenvalues of type II, counted with multiplicity.*

5. *Let* $\kappa_2 > 0$. *Then the interval* $[i\varrho, i\varrho + i|\lambda_{-1} - i\varrho|)$ *contains no or an even number of eigenvalues of type II, counted with multiplicity, if* $N(\hat{A}) \subset N(\hat{K})$, *and an odd number of eigenvalues of type II, counted with multiplicity, otherwise.*

Theorem 2.3.10. *Let* $\sigma_1(s_1) \equiv 2\varrho\rho_1(s_1),\ldots,\sigma_p(s_p) \equiv 2\varrho\rho_2(s_p)$ *and* $\nu < 2m\varrho$. *Then:*

1. *Only a finite number, denoted by* κ_2, *of the eigenvalues of type II of problem* (2.3.5)–(2.3.8) *lie in the closed half-plane* $\operatorname{Im}\lambda \ge \varrho$.

2. *All eigenvalues of type II in the half-plane* $\operatorname{Im}\lambda \ge \varrho$ *lie on* $(i\varrho, i\infty)$ *and their multiplicities are 1. If* $\kappa_2 > 0$, *they will be uniquely indexed as* $\lambda_{-j} = i\varrho + i|\lambda_{-j} - i\varrho|$, $j = 1,\ldots,\kappa_2$, *satisfying* $|\lambda_{-j}| < |\lambda_{-(j+1)}|$, $j = 1,\ldots,\kappa_2 - 1$.

3. *If* $\kappa_2 > 0$, *then the numbers* $i\varrho - i|\lambda_{-j} - i\varrho|$, $j = 1,\ldots,\kappa_2$, *are not eigenvalues.*

4. *If* $\kappa_2 \ge 2$, *then the intervals* $(i\varrho - i|\lambda_{-(j+1)} - i\varrho|, i\varrho - i|\lambda_{-j} - i\varrho|)$, $j = 1,\ldots,\kappa_2 - 1$, *contain an odd number of eigenvalues of type II, counted with multiplicity.*

5. *Let* $\kappa_2 > 0$. *Then the interval* $(i\varrho - i|\lambda_{-1} - i\varrho|, i\varrho]$ *contains no or an even number of eigenvalues of type II, counted with multiplicity, if* $N(\hat{A}) \subset N(\hat{K})$, *and an odd number of eigenvalues of type II, counted with multiplicity, otherwise.*

2.4   Sturm–Liouville problems on forked graphs

The following spectral problem describes one-dimensional scattering of a quantum particle when the path of propagation is a graph which consists of two finite intervals and one half-infinite interval, where the three edges have one common vertex:

$$-y_j'' + q_j(x)y_j = \lambda^2 y_j, \quad x \in [0, a], \quad j = 1, 2, \tag{2.4.1}$$

$$-y_3'' = \lambda^2 y_3, \quad x \in [a, \infty), \tag{2.4.2}$$

$$y_1(\lambda, a) = y_2(\lambda, a) = y_3(\lambda, a), \tag{2.4.3}$$

$$y_1'(\lambda, a) + y_2'(\lambda, a) - y_3'(\lambda, a) = 0, \tag{2.4.4}$$

$$y_1(\lambda, 0) = 0, \tag{2.4.5}$$

$$y_2(\lambda, 0) = 0, \tag{2.4.6}$$

where $\lambda^2 = \frac{2mE}{\hbar^2}$, E is the energy of the quantum particle, $\hbar = \frac{h}{2\pi}$, h is the Plank constant, q_j is the potential of interaction, and y_j is the radial component of the wave function on the corresponding edge. Here we assume the potentials to be real valued and to satisfy $q_j \in L_2(0,a)$ for $j = 1,2$. Conditions (2.4.5) and (2.4.6) describe complete reflection of the wave at the pendant vertices.

Similar to the Regge problem we assume that the potential is supported only on the finite edges of the graph. The Jost solution $e(\lambda, x)$ of (2.4.2) is

$$e(\lambda, x) = e^{-i\lambda x}.$$

Substituting the multiple $y_3(\lambda, x) = y_1(\lambda, a)e^{i\lambda a}e^{-i\lambda x}$ of the Jost function into (2.4.3) and (2.4.4), the part of (2.4.3) relating to y_3 is satisfied, and (2.4.1)–(2.4.6) is reduced to

$$-y_j'' + q_j(x)y_j = \lambda^2 y_j, \quad x \in [0,a], \quad j = 1,2, \tag{2.4.7}$$

$$y_1(\lambda, a) = y_2(\lambda, a), \tag{2.4.8}$$

$$y_1'(\lambda, a) + y_2'(\lambda, a) + i\lambda y_1(\lambda, a) = 0, \tag{2.4.9}$$

$$y_1(\lambda, 0) = 0, \tag{2.4.10}$$

$$y_2(\lambda, 0) = 0. \tag{2.4.11}$$

This problem is a particular case of problem (2.3.13), (2.3.6)–(2.3.8) considered in Subsection 2.3.4 with $p = 2$, $\rho_1 = 1$, $\rho_2 = 1$, $\sigma_1 = 0$, $\sigma_2 = 0$, $m = 0$, and $\nu = 1$. Hence $\varrho = 0$, and in the notation of Section 2.3, $A = \hat{A}$, $M = \hat{M}$, and $K = \hat{K}$ are given by

$$A\begin{pmatrix} y_1 \\ y_2 \\ c \end{pmatrix} = \begin{pmatrix} -y_1'' + q_1 y_1 \\ -y_2'' + q_2 y_2 \\ y_1'(a) + y_2'(a) \end{pmatrix}, \quad M = \begin{pmatrix} I & 0 & 0 \\ 0 & I & 0 \\ 0 & 0 & 0 \end{pmatrix}, \quad K = \begin{pmatrix} 0 & 0 & 0 \\ 0 & 0 & 0 \\ 0 & 0 & 1 \end{pmatrix}.$$

Then $M + K = I \gg 0$, Lemma 1.5.4, and Theorems 2.3.8 and 2.3.9 lead to

Theorem 2.4.1.

1. *All eigenvalues of (2.4.7)–(2.4.11) have geometrical multiplicity* 1.
2. *All eigenvalues of type I are located on the real and imaginary axes, are symmetric with respect to the origin, are not eigenvalues of type II, and at most finitely many of them lie on the imaginary axis.*
3. *Only a finite number, denoted by κ_2, of eigenvalues of type II lie in the closed lower half-plane.*
4. *All eigenvalues of type II in the closed lower half-plane lie on the negative imaginary semiaxis and are simple. If $\kappa_2 > 0$, they will be uniquely indexed as $\lambda_{-j} = -i|\lambda_{-j}|$, $j = 1, \ldots, \kappa_2$, satisfying $|\lambda_{-j}| < |\lambda_{-(j+1)}|$, $j = 1, \ldots, \kappa_2 - 1$.*
5. *If $\kappa_2 > 0$, then the numbers $i|\lambda_{-j}|$, $j = 1, \ldots, \kappa_2$, are not eigenvalues.*
6. *If $\kappa_2 \geq 2$, then in each of the intervals $(i|\lambda_{-j}|, i|\lambda_{-(j+1)}|)$, $j = 1, \ldots, \kappa_2 - 1$, the number of eigenvalues of type II, counted with multiplicity, is odd.*

7. *Let $\kappa_2 > 0$. Then the interval $[0, i|\lambda_{-1}|)$ contains no or an even number of eigenvalues of type II, counted with multiplicity, if $N(A) \subset N(K)$, and an odd number of eigenvalues of type II, counted with multiplicity, otherwise.*

Let us briefly discuss the meaning of the eigenvalues of problem (2.4.7)–(2.4.11) for problem (2.4.1)–(2.4.6). The eigenvalues of problem (2.4.7)–(2.4.11) in the open lower half-plane are normal eigenvalues or, in physical terms, bound states of problem (2.4.1)–(2.4.6). The real nonzero eigenvalues of problem (2.4.7)–(2.4.11) are bound states embedded in the continuous spectrum of problem (2.4.1)–(2.4.6). The zero eigenvalue of problem (2.4.7)–(2.4.11) describes the so-called virtual label (or virtual state) for problem (2.4.1)–(2.4.6). The eigenvalues of problem (2.4.7)–(2.4.11) located in the open upper half-plane are poles of the resolvent of the corresponding operator or, in physical terms, resonances of problem (2.4.1)–(2.4.6).

2.5   Sturm–Liouville problems on lasso graphs

The following spectral problem describes one-dimensional scattering of a quantum particle whose path of propagation is a graph consisting of one finite and one half-infinite interval where the three finite ends of the edges meet in a common vertex:

$$-y_1'' + q(x)y_1 = \lambda^2 y_1, \quad x \in [0, a], \tag{2.5.1}$$

$$-y_2'' = \lambda^2 y_2, \quad x \in [0, \infty), \tag{2.5.2}$$

$$y_1(\lambda, 0) = y_1(\lambda, a) = y_2(\lambda, 0), \tag{2.5.3}$$

$$y_1'(\lambda, a) - y_1'(\lambda, 0) - y_2'(\lambda, 0) = 0, \tag{2.5.4}$$

where $\lambda^2 = \frac{2mE}{\hbar^2}$, E is the energy of a quantum particle, m is the mass, $\hbar = \frac{h}{2\pi}$, h is the Planck constant.

As in the Regge problem we assume that the potential is supported only on the finite edge of the graph. The Jost solution $e(\lambda, x)$ of (2.5.2) is

$$e(\lambda, x) = e^{-i\lambda x}.$$

Substituting the multiple $y_2(\lambda, x) = y_1(\lambda, 0)e^{-i\lambda x}$ of the Jost function, the part of (2.5.3) relating to y_2 is satisfied, and (2.5.1)–(2.5.4) is reduced to

$$-y_1'' + q(x)y_1 = \lambda^2 y_1, \quad x \in [0, a], \tag{2.5.5}$$

$$y_1(\lambda, 0) = y_1(\lambda, a), \tag{2.5.6}$$

$$y_1'(\lambda, 0) - y_1'(\lambda, a) - i\lambda y_1(\lambda, 0) = 0. \tag{2.5.7}$$

The eigenvalue problem (2.5.5)–(2.5.7) is described by the operator pencil

$$L(\lambda) = \lambda^2 M - i\lambda K - A,$$

where A, K and M act in the Hilbert space $H = L_2(0, a) \oplus \mathbb{C}$ according to

$$A \begin{pmatrix} y \\ c \end{pmatrix} = \begin{pmatrix} -y'' + qy \\ y'(a) - y'(0) \end{pmatrix},$$

$$D(A) = \left\{ \begin{pmatrix} y \\ c \end{pmatrix} : y \in W_2^2(0, a), \ y(0) = y(a) = c \right\},$$

$$M = \begin{pmatrix} I & 0 \\ 0 & 0 \end{pmatrix}, \qquad K = \begin{pmatrix} 0 & 0 \\ 0 & 1 \end{pmatrix}.$$

To prove that the operator A is Hermitian, we will apply Theorem 10.3.5, so that we have to find U_3 and U defined in (10.3.12) and (10.3.13). Observing that $y(0) = c$ contributes to U_2 and $y(0) = y(a)$ contributes to U_1, it is easy to see that

$$U_1 = \begin{pmatrix} 1 & 0 & -1 & 0 \end{pmatrix}, \quad U_2 = \begin{pmatrix} 1 & 0 & 0 & 0 \end{pmatrix}, \quad V = \begin{pmatrix} 0 & 1 & 0 & -1 \end{pmatrix},$$

so that

$$U_3 = \begin{pmatrix} 0 & -1 & 0 & 0 \\ 1 & 0 & 0 & 0 \\ 0 & 0 & 0 & 1 \\ 0 & 0 & -1 & 0 \\ 0 & 1 & 0 & -1 \\ -1 & 0 & 0 & 0 \end{pmatrix}, \quad U = \begin{pmatrix} 1 & 0 & -1 & 0 & 0 & 0 \\ 1 & 0 & 0 & 0 & -1 & 0 \\ 0 & 1 & 0 & -1 & 0 & -1 \end{pmatrix}.$$

Then it follows that

$$N(U_1) = \operatorname{span}\{e_1 + e_3, e_2, e_4\},$$
$$U_3(N(U_1)) = \operatorname{span}\{e_2 - e_4 - e_6, e_1 - e_5, e_3 - e_5\},$$
$$R(U^*) = \operatorname{span}\{e_1 - e_3, e_1 - e_5, e_2 - e_4 - e_6\},$$

so that $U_3(N(U_1)) = R(U^*)$. Hence A is self-adjoint by Theorem 10.3.5.

By Theorem 10.3.8, A has a compact resolvent, and A is bounded below. To see the latter property, we observe that $U_1 \hat{Y} = y(0) - y(1)$ only contains quasi-derivatives of order 0 and hence satisfies condition (ii) of Theorem 10.3.8. Condition (iii) of Theorem 10.3.8 holds since $U_2 \hat{Y} = y(0)$ only has quasi-derivatives of order 0, and finally, condition (iv) is trivially satisfied since the order of the quasi-derivative in $U_2 \hat{Y}$ is less than 1.

It is also clear that $M \geq 0$, $K \geq 0$ and $M + K = I \gg 0$. One can easily show as in Section 2.1 that $N(M) \cap N(A) = \{0\}$. Therefore, we can apply Lemma 1.5.4 and Theorems 1.5.6 and 1.5.7. Thus we obtain

Theorem 2.5.1.

1. *All nonzero real eigenvalues of problem (2.5.5)–(2.5.7) have geometric multiplicity 1.*

2. *All eigenvalues of type I are located on the real and imaginary axes, are symmetric with respect to the origin, and at most finitely many of them lie on the imaginary axis.*

3. *Only a finite number, denoted by κ_2, of eigenvalues of type II lie in the closed lower half-plane.*

4. *All eigenvalues of type II in the closed lower half-plane lie on the negative imaginary semiaxis and their type II multiplicities are 1. If $\kappa_2 > 0$, they will be uniquely indexed as $\lambda_{-j} = -i|\lambda_{-j}|$, $j = 1,\ldots,\kappa_2$, satisfying $|\lambda_{-j}| < |\lambda_{-(j+1)}|$, $j = 1,\ldots,\kappa_2 - 1$.*

5. *If $\kappa_2 > 0$, then the numbers $i|\lambda_{-j}|$, $j = 1,\ldots,\kappa_2$, are not eigenvalues of type II.*

6. *If $\kappa_2 \geq 2$, then in each of the intervals $(i|\lambda_{-j}|, i|\lambda_{-(j+1)}|)$, $j = 1,\ldots,\kappa_2 - 1$, the number of eigenvalues of type II, counted with type II multiplicity, is odd.*

7. *Let $\kappa_2 > 0$. Then the interval $[0, i|\lambda_{-1}|)$ contains no or an even number of eigenvalues of type II, counted with type II multiplicity, if $N(A) \subset N(K)$, and an odd number of eigenvalues of type II, counted with type II multiplicity, if $N(A) \not\subset N(K)$.*

Proof. We still have to prove statement 1. To this end let λ be a nonzero real eigenvalue of problem (2.5.5)–(2.5.7). Since (2.5.5) has two linearly independent solutions, it suffices to show that the solution $y(\lambda, \cdot)$ of (2.5.5) with $y(\lambda, 0) = 1$ and $y'(\lambda, 0) = 0$ is not an eigenvector of (2.5.5)–(2.5.7). Indeed, since q is a real-valued function and λ a real number, it follows that y is a real-valued function, so that $y'(\lambda, a) \in \mathbb{R}$. But on the other hand, (2.5.7) gives

$$y'(\lambda, a) = y'(\lambda, 0) - i\lambda y(\lambda, 0) = -i\lambda,$$

and we have arrived at a contradiction. $\square$

Let us briefly discuss the meaning of the eigenvalues of the eigenvalue problem (2.5.5)–(2.5.7) for problem (2.5.1)–(2.5.4). The eigenvalues of problem (2.5.5)–(2.5.7) in the open lower half-plane are eigenvalues or, in physical terms, bound states of problem (2.5.1)–(2.5.4). The real nonzero eigenvalues of problem (2.5.5)–(2.5.7) are bound states embedded in the continuous spectrum of problem (2.5.1)–(2.5.4). The zero eigenvalue of problem (2.5.5)–(2.5.7) describes the so-called virtual label (or virtual state) for problem (2.5.1)–(2.5.4). The eigenvalues of problem (2.5.5)–(2.5.7) in the open upper half-plane are poles of the resolvent of the corresponding operator or, in physical terms, resonances of problem (2.5.1)–(2.5.4).

2.6 Damped vibrations of Stieltjes strings

The notion of Stieltjes string was introduced in [85, Supplement II]. Like in [85] we suppose the string to be a thread, i. e. a string of zero density, bearing a finite number of point masses. Assume that the string consists of two parts, which are joined at one end and fixed at the other end.

Starting indexing from the fixed ends, n_j masses $m_k^{(j)} > 0$, $k = 1, \ldots, n_j$, are positioned on the jth part, $j = 1, 2$, which divide the jth part into n_j+1 substrings, denoted by $l_k^{(j)} > 0$, $k = 0, \ldots, n_j$, again starting indexing from the fixed end. In particular, $l_0^{(j)}$ is the distance on the jth part between the fixed endpoint and $m_1^{(j)}$, $l_k^{(j)}$ for $k = 1, \ldots, n_j - 1$ is the distance between $m_k^{(j)}$ and $m_{k+1}^{(j)}$, and $l_{n_j}^{(j)}$ is the distance on the jth thread between the joined endpoint P and $m_{n_j}^{(j)}$. The tension of the thread is assumed to be equal to 1, and at the point P damping is assumed with coefficient of damping $\nu > 0$. The transversal displacement of the point masses $m_k^{(j)}$ at the time t is denoted by $v_k^{(j)}(t)$, where we assume the thread to be stretched by a force equal to 1. For convenience, we denote by $v_0^{(j)} = 0$ the transversal displacement at the fixed endpoints and by $v_{n_j+1}^{(j)}$ the transversal displacement at the joined endpoints.

However, the mathematical treatment will become easier if we use a unified treatment by considering just one thread with finitely many point masses with damping on it.

This means we have n points P_k, $k = 1, \ldots, n$, on the thread, either bearing a mass $m_k > 0$, or being a joint, in which case we set $m_k = 0$. At each of the points P_k we also allow damping with coefficient of damping $\nu_k > 0$. Absence of damping is, of course, indicated by $\nu_k = 0$. It may be assumed that damping is present at each joint, in which case, using dimensionless quantities, $m_k + \nu_k > 0$ for all $k = 1, \ldots, n$. The transversal displacement of the points P_k from the position of rest at time t is denoted by $v_k(t)$ for $k = 1, \ldots, n$, and for convenience we introduce $v_0(t)$ and $v_{n+1}(t)$ for the displacements of the left and right end points of the thread, respectively.

It is clear that the Stieltjes string introduced at the beginning of this section is a special case of this latter setting, with $n = n_1 + n_2 + 1$, $\nu_k = \nu \neq 0$ if and only if $k = n_1 + 1$, and $m_k = 0$ if and only if $k = n_1 + 1$.

Since the threads have zero density, the general solution of the string equation for each substring is a linear function of s at any time t. Therefore, each P_k is connected by straight line threads to its two neighbours, possibly one of them one of the fixed ends of the thread. Linearizing the forces exerted on each of the points P_k by the adjacent strings in terms of the displacements, Newton's law of motion gives the following equations of motion for the particles:

$$\frac{v_k(t) - v_{k+1}(t)}{l_k} + \frac{v_k(t) - v_{k-1}(t)}{l_{k-1}} + \nu_k v_k'(t) + m_k v_k''(t) = 0, \quad k = 1, \ldots, n. \quad (2.6.1)$$

At the fixed ends we have

$$v_0(t) = 0, \quad v_{n+1}(t) = 0. \quad (2.6.2)$$

For the particular problem we started with, the corresponding equations are

$$\frac{v_k^{(j)}(t) - v_{k+1}^{(j)}(t)}{l_k^{(j)}} + \frac{v_k^{(j)}(t) - v_{k-1}^{(j)}(t)}{l_{k-1}^{(j)}} + m_k^{(j)} v_k^{(j)''}(t) = 0, \quad k = 1, \ldots, n_j, \; j = 1, 2. \quad (2.6.3)$$

At the fixed ends we have

$$v_0^{(j)}(t) = 0, \quad j = 1, 2. \quad (2.6.4)$$

The joined ends give rise to

$$v_{n_1+1}^{(1)}(t) = v_{n_2+1}^{(2)}(t), \quad (2.6.5)$$

and the equation of damped motion at the joint is

$$\frac{v_{n_1+1}^{(1)}(t) - v_{n_1}^{(1)}(t)}{l_{n_1}} + \frac{v_{n_2+1}^{(2)}(t) - v_{n_2}^{(2)}(t)}{l_{n_2}} + \nu v_{n_1+1}^{(1)'}(t) = 0. \quad (2.6.6)$$

Returning to the general case, we substitute $v_k(t) = u_k e^{i\lambda t}$ into (2.6.1)–(2.6.2) and obtain

$$\frac{u_k - u_{k+1}}{l_k} + \frac{u_k - u_{k-1}}{l_{k-1}} + i\lambda \nu_k u_k - \lambda^2 m_k u_k = 0, \quad k = 1, \ldots, n, \quad (2.6.7)$$

$$u_0 = 0, \quad u_{n+1} = 0. \quad (2.6.8)$$

Then problem (2.6.1)–(2.6.2) can be written in matrix form

$$(\lambda^2 M - i\lambda K - A)Y = 0,$$

where $A = (a_{k,m})_{k,m=1}^n$ is an $n \times n$ Jacobi matrix with diagonal elements

$$a_{k,k} = l_{k-1}^{-1} + l_k^{-1}, \quad k = 1, \ldots, n,$$

subdiagonal and superdiagonal elements

$$a_{k+1,k} = a_{k,k+1} = -l_k^{-1}, \quad k = 1, \ldots, n-1,$$

and

$$M = \mathrm{diag}(m_1, \ldots, m_n), \quad K = \mathrm{diag}(\nu_1, \ldots, \nu_n), \quad Y = (u_1, \ldots, u_n)^\mathsf{T}.$$

Proposition 2.6.1. *The matrices $M \geq 0$, $K \geq 0$, and $A > 0$ are Hermitian, and $M + K > 0$ if $m_k + \nu_k > 0$ for all $k = 1, \ldots, n$.*

Proof. The statements about M and K are obvious, so that we now turn our attention to A. The matrix A is clearly Hermitian as it is a real Jacobi matrix. For $Y = (v_1, \ldots, v_n)^\mathsf{T} \in \mathbb{C}^n$ we estimate

$$Y^* A Y = l_0^{-1}|v_1|^2 + \sum_{k=1}^{n-1} l_k^{-1}|v_k|^2 + \sum_{k=2}^{n} l_{k-1}^{-1}|v_k|^2 + l_n^{-1}|v_n|^2$$

$$- \sum_{k=1}^{n-1} l_k^{-1}(v_k \overline{v_{k+1}} + v_{k+1}\overline{v_k})$$

$$= l_0^{-1}|v_1|^2 + l_n^{-1}|v_n|^2 + \sum_{k=1}^{n-1} l_k^{-1}\left(|v_k|^2 + |v_{k+1}|^2 - 2\,\mathrm{Re}(v_k\overline{v_{k+1}})\right)$$

$$\geq l_0^{-1}|v_1|^2 + l_n^{-1}|v_n|^2.$$

Hence $A \geq 0$. If $AY = 0$, then the above estimate gives $v_1 = 0$. Since A is tridiagonal and since all entries in the superdiagonal are nonzero, it follows by induction that $v_2 = 0, \ldots, v_n = 0$, so that $N(A) = \{0\}$. Altogether, we have shown that $A > 0$. $\qquad\square$

We note that the Stieltjes string introduced at the beginning of this section has the operator pencil representation with a rank one operator K and satisfies $M + K > 0$. Hence the following theorem holds for Stieltjes strings.

Theorem 2.6.2.

1. *All eigenvalues of (2.6.7), (2.6.8) lie in the closed upper half-plane.*
2. *If $\nu_m > 0$ for exactly one $m \in \{1, \ldots, n\}$ and $m_k > 0$ for $k \neq m$, then all eigenvalues of type II lie in the open upper half-plane, whereas all eigenvalues of type I are nonzero real numbers.*

Proof. 1. This follows from Proposition 2.6.1 and Lemma 1.2.4, part 1.
2. Since eigenvalues of type I lie on the real and imaginary axes and are symmetric with respect to the origin by Lemma 1.5.4, eigenvalues of type I can only be nonzero and real by part 1 and Proposition 2.6.1. By Lemma 1.3.1, part 3, all real eigenvalues are semisimple, and by Lemma 1.3.1, part 4, and Theorem 1.5.6, part 3, all real eigenvalues, counted with multiplicity, are of type I, so that eigenvalues of type II must be nonreal. $\qquad\square$

2.7   Damped vibrations of beams

2.7.1   Beams with friction at one end

The equation

$$\frac{\partial^4 u}{\partial x^4} - \frac{\partial}{\partial x}g(x)\frac{\partial u}{\partial x} + \frac{\partial^2 u}{\partial t^2} = 0 \tag{2.7.1}$$

describes small transverse vibrations of an elastic beam. Here g is a stretching or compressing distributed force, $u(x, t)$ is the transverse displacement of the point located at x and at time t. We assume a hinge connection at the left end of the beam, which is described by the boundary conditions

$$u(0, t) = \left.\frac{\partial^2 u}{\partial x^2}\right|_{x=0} = 0.$$

The boundary conditions at the right end $a > 0$,

$$u(a, t) = \left.\frac{\partial^2 u}{\partial x^2}\right|_{x=a} + \alpha \left.\frac{\partial^2 u}{\partial t \partial x}\right|_{x=a} = 0,$$

describe a hinge connection with viscous friction at the hinge, where $\alpha > 0$ is the coefficient of damping. We will assume that the real-valued function g belongs to $W_2^1(0, a)$.

Substituting $u(x, t) = y(\lambda, x)e^{i\lambda t}$ we obtain

$$y^{(4)}(\lambda, x) - (gy')'(\lambda, x) = \lambda^2 y(\lambda, x), \tag{2.7.2}$$

together with the boundary conditions

$$y(\lambda, 0) = 0, \tag{2.7.3}$$
$$y''(\lambda, 0) = 0, \tag{2.7.4}$$
$$y(\lambda, a) = 0, \tag{2.7.5}$$
$$y''(\lambda, a) + i\lambda\alpha y'(\lambda, a) = 0. \tag{2.7.6}$$

Now we establish the operator approach to this problem. Define the operators A, K and M by

$$D(A) = \left\{Y = \begin{pmatrix} y \\ c \end{pmatrix} : y \in W_2^4(0, a),\ y(0) = y''(0) = y(a) = 0,\ y'(a) = c\right\},$$

$$D(K) = D(M) = L_2(0, a) \oplus \mathbb{C},$$

$$A\begin{pmatrix} y \\ c \end{pmatrix} = \begin{pmatrix} y^{(4)} - (gy')' \\ y''(a) \end{pmatrix}, \quad K = \begin{pmatrix} 0 & 0 \\ 0 & \alpha \end{pmatrix}, \quad M = \begin{pmatrix} I & 0 \\ 0 & 0 \end{pmatrix}.$$

Then the eigenvectors of the operator pencil L given by

$$L(\lambda) = \lambda^2 M - i\lambda K - A \tag{2.7.7}$$

correspond to nontrivial solutions of (2.7.2)–(2.7.6).

Proposition 2.7.1. *The operators A, K and M are self-adjoint, M and K are bounded, K has rank 1, $M \geq 0$, $K \geq 0$, $M + K \gg 0$, $N(M) \cap N(A) = \{0\}$, and A is bounded below and has a compact resolvent.*

Proof. The statements about M and K are obvious. If $(y, c)^\mathsf{T} \in N(M) \cap N(A)$, then $(y, c)^\mathsf{T} \in N(M)$ gives $y = 0$, and $(y, c)^\mathsf{T} \in D(A)$ leads to $c = y'(a) = 0$. Hence $N(M) \cap N(A) = \{0\}$.

We are going to use Theorem 10.3.5 to verify that A is self-adjoint. Observe that $y^{[j]} = y^{(j)}$ for $j = 0, 1, 2$ and $y^{[3]} = y^{(3)} - gy'$ according to Definition 10.2.1. We have to find U_1, U_3 and U defined in (10.3.3), (10.3.12) and (10.3.13). First, it is straightforward to see that

$$U_1 = \begin{pmatrix} 1 & 0 & 0 & 0 & 0 & 0 & 0 & 0 \\ 0 & 0 & 1 & 0 & 0 & 0 & 0 & 0 \\ 0 & 0 & 0 & 0 & 1 & 0 & 0 & 0 \end{pmatrix},$$

$$U_2 = \begin{pmatrix} 0 & 0 & 0 & 0 & 0 & 1 & 0 & 0 \end{pmatrix},$$

$$V = \begin{pmatrix} 0 & 0 & 0 & 0 & 0 & 0 & 1 & 0 \end{pmatrix}.$$

In particular, $\begin{pmatrix} U_1 \\ U_2 \end{pmatrix}$ has rank 4. Clearly,

$$N(U_1) = \operatorname{span}\{e_2, e_4, e_6, e_7, e_8\} \subset \mathbb{C}^8.$$

We recall that

$$U_3 = \begin{pmatrix} J_2 \\ V \\ -U_2 \end{pmatrix} \quad \text{where} \quad J_2 = \begin{pmatrix} -J_{2,1} & 0 \\ 0 & J_{2,1} \end{pmatrix} \quad \text{and} \quad J_{2,1} = \begin{pmatrix} 0 & 0 & 0 & 1 \\ 0 & 0 & -1 & 0 \\ 0 & 1 & 0 & 0 \\ -1 & 0 & 0 & 0 \end{pmatrix},$$

and

$$U = \begin{pmatrix} U_1 & 0 & 0 \\ U_2 & -1 & 0 \\ V & 0 & -1 \end{pmatrix}.$$

It is now straightforward to verify that

$$U_3(N(U_1)) = \operatorname{span}\{e_3, e_1, e_7 - e_{10}, e_9 - e_6, e_5\} \subset \mathbb{C}^{10},$$
$$R(U^*) = \operatorname{span}\{e_1, e_3, e_5, e_6 - e_9, e_7 - e_{10}\} \subset \mathbb{C}^{10}.$$

This shows that $U_3(N(U_1)) = R(U^*)$. Hence A is self-adjoint by Theorem 10.3.5.

Finally, Theorem 10.3.8 shows that A has a compact resolvent and that A is bounded below. $\square$

It is clear that in this problem both eigenvalues of type I and type II can exist. Theorems 1.5.6 and 1.5.7 imply

Theorem 2.7.2.

1. *The geometric multiplicity of each eigenvalue of (2.7.2)–(2.7.6) does not exceed 2.*

2. *All eigenvalues of type I are located on the imaginary and real axes and are symmetric with respect to the origin. All nonzero eigenvalues of type I have type I multiplicity 1. If the geometric multiplicity of a nonzero eigenvalue is 2, then this eigenvalue is a pure imaginary eigenvalue of both type I and type II.*

3. *Only a finite number, denoted by κ_2, of the eigenvalues of type II lie in the closed lower half-plane.*

4. *All eigenvalues of type II in the closed lower half-plane lie on the negative imaginary semiaxis and their type II multiplicity is 1. If $\kappa_2 > 0$, they will be uniquely indexed as $\lambda_{-j} = -i|\lambda_{-j}|$, $j = 1, \ldots, \kappa_2$, satisfying $|\lambda_{-j}| < |\lambda_{-(j+1)}|$, $j = 1, \ldots, \kappa_2 - 1$.*

5. *If $\kappa_2 > 0$, then the numbers $i|\lambda_{-j}|$, $j = 1, \ldots, \kappa_2$, are not eigenvalues of type II.*

6. *If $\kappa_2 \geq 2$, then the number of eigenvalues of type II, counted with type II multiplicity, in each of the intervals $(i|\lambda_{-j}|, i|\lambda_{-(j+1)}|)$, $j = 1, \ldots, \kappa_2 - 1$, is odd.*

7. *Let $\kappa_2 > 0$. Then the interval $(0, i|\lambda_{-1}|)$ contains no or an even number of eigenvalues of type II, counted with type II multiplicity, if $N(A) \subset N(K)$, and an odd number of eigenvalues of type II, counted with type II multiplicity, otherwise.*

Proof. Parts 3 to 7 of this theorem immediately follow from Theorem 1.5.7.

1. Because of the initial conditions (2.7.3) and (2.7.4), each eigenvalue λ can have at most two linearly independent eigenvectors.

2. The first statement follows from Lemma 1.5.4. Now let λ be a nonzero eigenvalue of type I. Then every eigenvector of type I is of the form $(y(\lambda, \cdot), 0)^{\mathsf{T}}$ by Theorem 1.5.6, and therefore $y'(\lambda, a) = 0$. Hence $y(\lambda, \cdot)$ is a solution of the initial value problem (2.7.2), $y(\lambda, a) = y'(\lambda, a) = y''(\lambda, a) = 0$, and it follows from Theorem 1.5.6, part 1, that the type I multiplicity $m_I(\lambda)$ equals 1. If now $\lambda \neq 0$ has geometric multiplicity 2, then λ must be an eigenvalue of type I since the geometric multiplicity of a type II eigenvalue is 1 by Theorem 1.5.6, part 3. We already know that the type I multiplicity is 1, so that λ must also be of type II and pure imaginary since a type II eigenvalue cannot be real by part 4.   □

2.7.2 Damped beams

Physically better justified boundary conditions for the problem generated by a fourth-order differential equation were considered in [100]:

$$y^{(4)}(\lambda, x) - (gy')'(\lambda, x) + 2i\lambda\varrho y = \lambda^2 y(\lambda, x), \qquad (2.7.8)$$

$$y(\lambda, 0) = 0, \qquad (2.7.9)$$
$$y''(\lambda, 0) = 0, \qquad (2.7.10)$$
$$y''(\lambda, a) = 0, \qquad (2.7.11)$$
$$-y'''(\lambda, a) + g(a)y'(\lambda, a) + i\lambda\alpha y(\lambda, a) = \lambda^2 m y(\lambda, a). \qquad (2.7.12)$$

Here $g \in W_2^1(0, a)$ is a stretching or compressing distributed force, $2\varrho > 0$ is the constant damping coefficient due to viscous friction along the beam and m is a point mass at the right end $a > 0$. This right end can move with damping in the direction orthogonal to the equilibrium position of the beam, where $\alpha > 0$ is the coefficient of damping.

For the operator approach to this problem we define the operators A, K and M by

$$D(A) = \left\{ Y = \begin{pmatrix} y \\ c \end{pmatrix} : y \in W_4^2(0, a),\ y(0) = y''(0) = y''(a) = 0,\ y(a) = c \right\},$$

$$D(K) = D(M) = L_2(0, a) \oplus \mathbb{C},$$

$$A \begin{pmatrix} y \\ c \end{pmatrix} = \begin{pmatrix} y^{(4)} - (gy')' \\ -y'''(a) + g(a)y'(a) \end{pmatrix}, \quad K = \begin{pmatrix} 2\varrho & 0 \\ 0 & \alpha \end{pmatrix}, \quad M = \begin{pmatrix} I & 0 \\ 0 & m \end{pmatrix}.$$

Then the eigenvectors of the operator pencil L given by

$$L(\lambda) = \lambda^2 M - i\lambda K - A \qquad (2.7.13)$$

correspond to nontrivial solutions of (2.7.8)–(2.7.12).

Proposition 2.7.3. *The operators A, K and M are self-adjoint, M and K are bounded, $M \gg 0$, $K \gg 0$, and A is bounded below and has a compact resolvent.*

Proof. The statements about M and K are obvious. We will use Theorem 10.3.5 to verify that A is self-adjoint. Observe that $y^{[j]} = y^{(j)}$ for $j = 0, 1, 2$ and that $y^{[3]} = y^{(3)} - gy'$ according to Definition 10.2.1. We have to find U_1, U_3 and U defined in (10.3.3), (10.3.12) and (10.3.13). First, it is straightforward to see that

$$U_1 = \begin{pmatrix} 1 & 0 & 0 & 0 & 0 & 0 & 0 & 0 \\ 0 & 0 & 1 & 0 & 0 & 0 & 0 & 0 \\ 0 & 0 & 0 & 0 & 0 & 0 & 1 & 0 \end{pmatrix},$$

$$U_2 = \begin{pmatrix} 0 & 0 & 0 & 0 & 1 & 0 & 0 & 0 \end{pmatrix},$$

$$V = \begin{pmatrix} 0 & 0 & 0 & 0 & 0 & 0 & 0 & -1 \end{pmatrix}.$$

In particular, $\begin{pmatrix} U_1 \\ U_2 \end{pmatrix}$ has rank 4. Clearly,

$$N(U_1) = \operatorname{span}\{e_2, e_4, e_5, e_6, e_8\} \subset \mathbb{C}^8.$$

We recall that

$$U_3 = \begin{pmatrix} J_2 \\ V \\ -U_2 \end{pmatrix} \quad \text{where} \quad J_2 = \begin{pmatrix} -J_{2,1} & 0 \\ 0 & J_{2,1} \end{pmatrix} \quad \text{and} \quad J_{2,1} = \begin{pmatrix} 0 & 0 & 0 & 1 \\ 0 & 0 & -1 & 0 \\ 0 & 1 & 0 & 0 \\ -1 & 0 & 0 & 0 \end{pmatrix},$$

and

$$U = \begin{pmatrix} U_1 & 0 & 0 \\ U_2 & -1 & 0 \\ V & 0 & -1 \end{pmatrix}.$$

It is now straightforward to verify that

$$U_3(N(U_1)) = \operatorname{span}\{e_3, e_1, e_8 + e_{10}, e_7, e_5 - e_9\} \subset \mathbb{C}^{10},$$
$$R(U^*) = \operatorname{span}\{e_1, e_3, e_7, e_5 - e_9, e_8 + e_{10}\} \subset \mathbb{C}^{10}.$$

This shows that $U_3(N(U_1)) = R(U^*)$. Hence A is self-adjoint by Theorem 10.3.5.

Finally, Theorem 10.3.8 shows that A has a compact resolvent and that A is bounded below. $\square$

We will make use of the parameter transformation $\lambda = \tau + i\varrho$ if $\alpha - 2m\varrho \geq 0$ and $\lambda = -\tau + i\varrho$ if $\alpha - 2m\varrho < 0$. This parameter transformation gives rise to the operator pencil

$$L(\pm\tau + i\varrho) =: \hat{L}(\tau) = \tau^2 \hat{M} - i\tau \hat{K} - \hat{A},$$

where

$$\hat{M} = M, \quad \hat{K} = \pm(K - 2\varrho M), \quad \hat{A} = A + \varrho^2 M - \varrho K.$$

Thus $\hat{M}$, $\hat{K}$ and $\hat{A}$ act in $L_2(0, a) \oplus \mathbb{C}$ and are given as follows:

$$\hat{A}\begin{pmatrix} y \\ c \end{pmatrix} = \begin{pmatrix} y^{(4)} - (gy')' - \varrho^2 y(x) \\ -y^{(3)}(a) + g(a)y'(a) + (m\varrho^2 - \alpha\varrho)y(a) \end{pmatrix}, \tag{2.7.14}$$

$$D(\hat{A}) = \left\{ \begin{pmatrix} y \\ c \end{pmatrix} : y \in W_2^4(0, a), c = y(a), y(0) = y''(0) = y''(a) \right\}, \tag{2.7.15}$$

$$\hat{M} = \begin{pmatrix} I & 0 \\ 0 & m \end{pmatrix}, \quad \hat{K} = \begin{pmatrix} 0 & 0 \\ 0 & |\alpha - 2m\varrho| \end{pmatrix}. \tag{2.7.16}$$

Three distinguished cases will be considered:

1) $\alpha = 2m\varrho$,   2) $\alpha > 2m\varrho$ and   3) $\alpha < 2m\varrho$.

The first case is trivial because $\hat{K} = 0$ and hence $\hat{L}(\tau) = \tau^2 \hat{M} - \hat{A}$. The spectrum of the pencil L is linked to the spectrum of the operator $\hat{M}^{-\frac{1}{2}}\hat{A}\hat{M}^{-\frac{1}{2}}$ via the substitution $\tau^2 = \zeta$. The infinitely many positive eigenvalues of $\hat{M}^{-\frac{1}{2}}\hat{A}\hat{M}^{-\frac{1}{2}}$ give pairs of real eigenvalues of $\hat{L}$ which are symmetric with respect to the origin, and the at most finitely many negative eigenvalues of $\hat{M}^{-\frac{1}{2}}\hat{A}\hat{M}^{-\frac{1}{2}}$ give pairs of pure imaginary eigenvalues of $\hat{L}$ which are symmetric with respect to the origin.

Hence the spectrum of problem (2.7.8)–(2.7.12) lies on the line $\mathrm{Im}\,\lambda = \varrho$ and on a finite interval of the imaginary axis, being symmetric with respect to the imaginary axis and to the line $\mathrm{Im}\,\lambda = \varrho$, respectively.

In the second and third cases $\hat{K} \geq 0$ has rank one and we can apply the results of Section 1.5 to the pencil $\hat{L}$. Taking into account that the geometric multiplicities of the eigenvalues of problem (2.7.8)–(2.7.12) do not exceed 2 and that $L(\lambda) = \hat{L}(\lambda \pm i\varrho)$, we arrive at the following results.

From Lemma 1.5.4 we obtain

Theorem 2.7.4. *Assume that $\alpha \neq 2m\varrho$. Then the eigenvalues of type I of L are located on the imaginary axis and on the line $\mathrm{Im}\,\lambda = \varrho$, and are symmetric with respect to the point $\lambda = i\varrho$. At most finitely many of the eigenvalues of type I are on the imaginary axis.*

Theorem 1.5.7 implies the following two theorems.

Theorem 2.7.5. *Assume that $\alpha > 2m\varrho$. Then:*

1. *Only a finite number, denoted by κ_2, of the eigenvalues of type II of problem (2.7.8)–(2.7.12) lie in the closed half-plane $\mathrm{Im}\,\lambda \leq \varrho$.*
2. *All eigenvalues of type II in the closed half-plane $\mathrm{Im}\,\lambda \leq \varrho$ lie on $(-i\infty, i\varrho)$ and their type II multiplicities are 1. If $\kappa_2 > 0$, they will be uniquely indexed as $\lambda_{-j} = i\varrho - i|\lambda_{-j} - i\varrho|$, $j = 1, \ldots, \kappa_2$, satisfying $|\lambda_{-j} - i\varrho| < |\lambda_{-(j+1)} - i\varrho|$, $j = 1, \ldots, \kappa_2 - 1$.*
3. *If $\kappa_2 > 0$, then the numbers $i\varrho + i|\lambda_{-j} - i\varrho|$, $j = 1, \ldots, \kappa_2$, are not eigenvalues of type II.*
4. *If $\kappa_2 \geq 2$, then the intervals $(i\varrho + i|\lambda_{-j} - i\varrho|, i\varrho + i|\lambda_{-j-1} - i\varrho|)$, $j = 1, \ldots, \kappa_2 - 1$, contain an odd number of eigenvalues of type II, counted with type II multiplicity.*
5. *Let $\kappa_2 > 0$. Then the interval $[i\varrho, i\varrho + i|\lambda_{-1} - i\varrho|)$ contains no or an even number of eigenvalues of type II, counted with type II multiplicity, if $N(\hat{A}) \subset N(\hat{K})$, and an odd number of eigenvalues of type II, counted with type II multiplicity, otherwise.*

Theorem 2.7.6. *Assume that $\alpha < 2mk$. Then:*

1. *Only a finite number, denoted by κ_2, of the eigenvalues of type II of problem (2.7.8)–(2.7.12) lie in the closed half-plane $\mathrm{Im}\,\lambda \geq \varrho$.*
2. *All eigenvalues of type II in the closed half-plane $\mathrm{Im}\,\lambda \geq \varrho$ lie on $(i\varrho, i\infty)$ and their type II multiplicities are 1. If $\kappa_2 > 0$, they will be uniquely enumerated as $\lambda_{-j} = i\varrho + i|\lambda_{-j} - i\varrho|$, $j = 1, \ldots, \kappa_2$, satisfying $|\lambda_{-j}| < |\lambda_{-(j+1)}|$, $j = 1, \ldots, \kappa_2 - 1$.*
3. *If $\kappa_2 > 0$, then the numbers $i\varrho - i|\lambda_{-j} - i\varrho|$, $j = 1, \ldots, \kappa_2$, are not eigenvalues of type II.*
4. *If $\kappa_2 \geq 2$, then the intervals $(i\varrho - i|\lambda_{-(j+1)} - i\varrho|, i\varrho - i|\lambda_{-j} - i\varrho|)$, $j = 1, \ldots, \kappa_2 - 1$, contain an odd number of eigenvalues of type II, counted with type II multiplicity.*

5. *Let $\kappa_2 > 0$. Then the interval $(i\varrho - i|\lambda_{-1} - i\varrho|, i\varrho]$ contains no or an even number of eigenvalues of type II, counted with type II multiplicity, if $N(\hat{A}) \subset N(\hat{K})$, and an odd number of eigenvalues of type II, counted with type II multiplicity, otherwise.*

2.8 Notes

The Regge problem was considered in detail by T. Regge [238] in connection with the description of nuclear interaction as presented in Section 2.1. The results in Section 2.1 are taken from [186]. For further aspects of the theory related to the Regge problem we refer the reader to [122], [115] [255], [256], [150], [142], [243], [148] and [251]. In particular, statement 4 of Theorem 2.1.2 was obtained in [251, Theorem 6]. In general, the radial component of the Schrödinger equation has a correction term to the potential which is singular at 0. However, for zero orbital momentum, also called S-wave by physicists, this additional potential is zero. Therefore, the Schrödinger equation is a Sturm–Liouville equation with the original potential.

Similar equations occur in mechanics of string vibrations. The simplest model of string musical instrument leads to one-dimensional damping (viscous friction). In [19] a model was proposed to explain the playing of 'harmonics' on stringed instruments. This has a long history, see [59].

In Section 2.2 we consider strings with density from $\rho \in L_\infty(0, l)$, where $0 < l < \infty$ and $\rho \geq \varepsilon > 0$. This class does not include beaded strings, also called Stieltjes strings. These two types of strings may be regarded as extremal cases of a wider class of strings which was considered by I.S. Kac and M.G. Kreĭn in [127], see also [125], [71]. They studied the equation

$$\frac{\partial^2 u}{\partial M(s)\partial s} - \frac{\partial^2 u}{\partial t^2} = 0, \tag{2.8.1}$$

which describes small transversal vibrations of a stretched inhomogeneous string. Here t stands for the time, s for the longitudinal coordinate, $u(s, t)$ is the transverse displacement, M is a nonnegative nondecreasing function on $[a, b]$ describing the mass distribution. Recall that M is differentiable a. e. by Lebesgue's theorem, see, e. g., [108, (17.12)], and its derivative $\rho = M'$ is called the density of the string. Without loss of generality it is assumed that $M(s+0) = M(s)$ for all $s \in [a, b)$. If $M(s_0 - 0) < M(s_0)$, then there is a point mass $M(s_0) - M(s_0 - 0)$ at $s_0 \in (a, b]$. If $M(a) > 0$, then there is a point mass $M(a)$ at $s = a$. The intervals on which M is constant, if any, correspond to massless intervals, that is, intervals where the string is a thread.

Substituting $u(s, t) = y(\lambda, s)e^{i\lambda t}$ into (2.8.1) we obtain the following equation for the amplitude function $y(\lambda, s)$:

$$\frac{dy'}{dM(s)} + \lambda^2 y = 0. \tag{2.8.2}$$

The generalized differential operator $\frac{d}{dM(s)}$ is the Radon–Nikodým derivative. In the case when $M(s)$ is an absolutely continuous function and $M'(s) > 0$ almost everywhere on $[a, b]$, then this operator acts on absolutely continuous functions which have absolutely continuous first-order derivatives and the action of the generalized differential operator is given by

$$\frac{dy'}{dM(s)} \stackrel{a.e.}{=} \frac{y''}{M'(s)}. \tag{2.8.3}$$

Here and in the sequel a. e. means almost everywhere. In the general case the generalized differential operator is defined only on so-called prolonged functions $u[s]$ which are obtained from usual functions $u(s)$, $a < s < b$, by attaching two arbitrary numbers $u'_-(a)$ and $u'_+(b)$, which are called left derivative at $s = a$ and right derivative at $s = b$, respectively. Then $u[s] = (u(s), u'_-(a), u'_+(b))$. The domain of the generalized differential operator is the set D_M of all complex-valued functions $u[s]$ of the form

$$u(s) = \alpha + \beta s - \int_a^s (s - p)g(p)dM(p), \quad a < s < b, \tag{2.8.4}$$

$$u'_-(0) = \beta, \quad u'_+(b) = \beta - \int_a^b g(p)dM(p), \tag{2.8.5}$$

where α and β are complex numbers, and g is a complex-valued function which is summable on $[a, b]$ with respect to the measure dM. For each such pair $(u[s], g)$ we have the equation

$$-\frac{du'}{dM(s)} = g(s)$$

in view of (2.8.4).

In [168] it was noted that the Hamiltonian of a canonical system can be represented by "strings" whose mass distribution function is not necessarily nondecreasing. Therefore, in [168] generalized strings M are considered which may have exceptional points x_0 at which $M(x_0-) = \infty$, $M(x_0+) = -\infty$ and M is square integrable near x_0 or $M(x_0-)$ and $M(x_0+)$ are finite but $M(x_0+) - M(x_0-) < 0$.

Recent results on damped strings can be found in [90], who use a Dirac operator approach rather than separation of variables. The main results are trace formulas and completeness of eigenvectors and associated vectors. Smooth string vibrations with piecewise constant damping were considered in [141]. Related scattering problems for damped strings and further examples of physical configurations leading to such models can be found in [123, Section I.B, p. 1353]. Properties of spectra of finite-dimensional quadratic operator pencils in connection with the damped wave equation were studied in [81].

In quantum graph theory, see [28], [231], the Sturm–Liouville equation is assumed to be defined on a graph domain with boundary conditions and matching conditions at the vertices of a graph.

In most cases, continuity and Kirchhoff conditions are imposed at interior vertices especially in quantum mechanics. However, we have seen in Subsection 2.3.3, see (2.3.11), (2.3.12), that they can be more complicated. For more general conditions see [231]. In this monograph Sturm type theorems on number of nodes of eigenfunctions on graphs were established, see also [82], [21], [20], [205].

We have seen in Theorem 2.3.8 that the maximal multiplicity of an eigenvalue of a problem on a star graph is $p - 1$, where p is the number of edges. More generally, it was shown in [128] that an upper bound for the maximal multiplicity of eigenvalues can be deduced from the shape of the graph.

Theorem 2.4.1 was proved in [169]. Spectral problems on lasso graphs were considered in [88], [89], [77] with constant potential on the loop and identically zero potential on the tail.

The famous Stieltjes memoir [254] was devoted to infinite continued fractions

$$
\cfrac{c_0}{z + \cfrac{c_1}{1 + \cfrac{c_2}{z + \cfrac{c_3}{1 + \cfrac{\ddots}{+ \cfrac{c_{2n-1}}{1 + \cfrac{c_{2n}}{1 + \cfrac{1}{z + \ddots}}}}}}}} \tag{2.8.6}
$$

where the c_j are complex numbers and z is the complex variable. As far as we know, Stieltjes did not associate any physical object with this continued fraction. The first physical interpretation of finite continued fractions was given by W. Cauer [49] in connection with the theory of synthesis of electrical circuits. He gave a continued fraction representation for RC driving point impedances, see also [102, p. 119]. Here the coefficients are inductances and capacities.

An interpretation of Stieltjes' results in terms of problems in mechanics has been given by M.G. Kreĭn [151], [85]. These authors introduced the term *Stieltjes string* (another name is *Sturm system*) and considered also the question of convergence in case of infinite continued fractions corresponding to Stieltjes strings. It should be mentioned that there exists a nice review paper [58] containing the description of these problems and related experiments. The same equations as for transverse vibrations of a Stieltjes string appear when one considers longitudinal vibrations of point masses connected by springs [92], [181]. Vibrations of a star graph of Stieltjes strings were considered in [87], [34] and [224].

As shown in Section 2.6, spectral problems generated by Stieltjes string can be considered as problems for Jacobi matrices. In case of trees of Stieltjes strings the corresponding matrices are so-called indextree-patterned matrices tree-patterned matrices, see also the comments to Chapter 8.

Theorem 2.7.2 was proved in [195]. Some other spectral problems generated by fourth-order ordinary differential equation with dissipative terms can be found in [197], [198], [199].

While for star graphs of strings which are damped at the interior vertex the only point of intersection of the set of eigenvalues of type I and the set of eigenvalues of type II is $i\rho$, this intersection can have many points for beams which are damped at an end.

Chapter 3

Operator Pencils with Essential Spectrum

3.1 Monic polynomial operator pencils

In this section we will investigate the spectra of monic polynomial operator pencils

$$L(\lambda) = i^n \lambda^n I + \sum_{j=1}^{n-1} i^j \lambda^j K_j + A,$$

acting on a Hilbert space H with the domain $D(L(\lambda)) = \bigcap_{j=1}^{n-1} D(K_j) \cap D(A)$, where $\lambda \in \mathbb{C}$ is the spectral parameter and the operators K_j, A satisfy

Condition III. *The operators K_j, $j = 1, \ldots, n$ and A are self-adjoint operators on H with the following properties:*

(i) $K_j \geq 0$ *and* $A \geq -\beta I$ *for some positive number β;*

(ii) $K_j = \gamma_j A + K_j^{(0)}$, *where the γ_j are nonnegative numbers, the operators $K_j^{(0)}$ are subordinate to A, i. e., $D(A) \subset \bigcap_{j=1}^{n-1} D(K_j^{(0)})$ and the operators $K_j^{(0)}$, $j = 1, \ldots, n-1$, are bounded with respect to the operator $(A + \beta_1 I)^{\frac{1}{2}}$, where $\beta_1 > \beta$.*

The following lemma shows that the subordination in Condition III can be expressed in terms of an inequality.

Lemma 3.1.1. *The subordination property in Condition III, part (ii), is satisfied if and only if $D(A) \subset \bigcap_{j=1}^{n-1} D(K_j^{(0)})$ and there exist positive numbers a and b such that for all $y \in D(A)$*

$$\max_{j=1,\ldots,n-1} \|K_j^{(0)} y\|^2 \leq a\|y\|^2 + b(Ay, y). \tag{3.1.1}$$

Proof. Assume that the $K_j^{(0)}$ are bounded with respect to $(A + \beta_1 I)^{\frac{1}{2}}$. Then there is a number $c > 0$ such that

$$\max_{j=1,\dots,n-1} \|K_j^{(0)} y\|^2 \le c\|(A + \beta_1 I)^{\frac{1}{2}} y\|^2$$
$$= c((A + \beta_1 I)y, y)$$
$$= c\beta_1 \|y\|^2 + c(Ay, y)$$

for all $y \in D(A)$, which proves (3.1.1) with $a = c\beta_1$ and $b = c$.

Conversely, if (3.1.1) is satisfied, choose $\beta_1 > \max\left\{\frac{a}{b}, \beta\right\}$. Then $a < b\beta_1$ and thus

$$\max_{j=1,\dots,n-1} \|K_j^{(0)} y\|^2 \le a\|y\|^2 + b(Ay, y)$$
$$\le b\beta_1 \|y\|^2 + b(Ay, y)$$
$$= b((A + \beta_1 I)^{\frac{1}{2}} y, (A + \beta_1 I)^{\frac{1}{2}} y)$$

for all $y \in D(A)$. Thus $K_j^{(0)}|_{D(A)}$ is bounded from $D((A + \beta_1 I)^{\frac{1}{2}})$ with the graph norm of $(A + \beta_1 I)^{\frac{1}{2}}$ to H. Since $D(A) = D(A + \beta_1 I)$ is a core for $(A + \beta_1 I)^{\frac{1}{2}}$, see [137, Theorems VI.2.1 and VI.2.23], it follows that $K_j^{(0)}|_{D(A)}$ is bounded and densely defined on $D((A + \beta_1 I)^{\frac{1}{2}})$ with respect to the graph norm of $(A + \beta_1 I)^{\frac{1}{2}}$. Therefore, $K_j^{(0)}|_{D(A)}$ has a bounded closure $K_j^{(1)}$ from $D((A + \beta_1 I)^{\frac{1}{2}})$ to H with $D(K_j^{(1)}) = D((A + \beta_1 I)^{\frac{1}{2}})$. Since $K_j^{(0)}$ is self-adjoint and therefore closed, the graph of $K_j^{(1)}$ is a subset of the graph of $K_j^{(0)}$. Hence $D((A + \beta_1 I)^{\frac{1}{2}}) \subset D(K_j^{(0)})$ and $K_j^{(0)}(A + \beta_1 I)^{-\frac{1}{2}} = K_j^{(1)}(A + \beta_1 I)^{-\frac{1}{2}}$ is bounded, i.e., $K_j^{(0)}$ is subordinate to A. $\square$

Lemma 3.1.2.

1. *For each $b_1 > 0$ there exist $a_1 > 0$ such that*

$$\max_{j=1,\dots,n-1} \|K_j^{(0)} y\| \le a_1 \|y\| + b_1 \|Ay\| \tag{3.1.2}$$

   for each $y \in D(A)$.
2. *There exist $a_2 > 0$ and $b_2 > 0$ such that*

$$\max_{j \in 1,\dots,n-1} \|K_j y\| \le a_2 \|y\| + b_2 \|Ay\| \tag{3.1.3}$$

   for each $y \in D(A)$.
3. *There exist $a_3 > 0$ and $b_3 > 0$ such that*

$$\max_{j \in 1,\dots,n-1} |(K_j^{(0)} y, y)| \le a_3 \|y\|^2 + b_3(Ay, y) \tag{3.1.4}$$

   for each $y \in D(A)$.

4. *There exist $a_4 > 0$ and $b_4 > 0$ such that*

$$\max_{j \in 1,\ldots,n-1} (K_j y, y) \leq a_4 \|y\|^2 + b_4 (Ay, y) \tag{3.1.5}$$

for each $y \in D(A)$.

Proof. By Condition III and Lemma 3.1.1, the operators $K_j^{(0)}$ satisfy (3.1.1). Hence it suffices to show that (3.1.1) implies the corresponding inequality in each case.

1. Let $b_1 > 0$ and choose $a_1 > 0$ such that $b \leq 2a_1 b_1$ and $a \leq a_1^2$, i.e., $a_1 \geq \max\left\{\frac{b}{2b_1}, a^{\frac{1}{2}}\right\}$. Then

$$\begin{aligned}
\max_{j=1,\ldots,n-1} \|K_j^{(0)} y\|^2 &\leq a\|y\|^2 + b(Ay, y) \\
&\leq a_1^2 \|y\|^2 + 2a_1 b_1 \|Ay\| \, \|y\| \\
&\leq (a_1 \|y\| + b_1 \|Ay\|)^2 .
\end{aligned}$$

Taking square roots proves (3.1.2).

Part 2 follows from part 1 with $a_2 = a_1$ and $b_2 = b_1 + \max_{j=1,\ldots,n} \gamma_j$.

3. Since $(Ay, y) \geq -\beta(y, y) = -\beta \|y\|^2$ it follows that

$$\begin{aligned}
\|y\|^2 = 2\|y\|^2 - \|y\|^2 &= 2\|y\|^2 + \frac{1}{\beta}(-\beta\|y\|^2) \\
&\leq 2\|y\|^2 + \frac{1}{\beta}(Ay, y) \\
&= \frac{1}{b\beta}\left(2b\beta\|y\|^2 + b(Ay, y)\right).
\end{aligned}$$

Putting $a_3' = \max\{a, 2b\beta\}$ it follows that

$$\begin{aligned}
\max_{j=1,\ldots,n-1} \|K_j^{(0)} y\|^2 \|y\|^2 &\leq (a\|y\|^2 + b(Ay, y)) \|y\|^2 \\
&\leq (a_3'\|y\|^2 + b(Ay, y)) \frac{1}{b\beta}\left(a_3'\|y\|^2 + b(Ay, y)\right) \\
&= (a_3\|y\|^2 + b_3(Ay, y))^2 ,
\end{aligned}$$

where $a_3 = a_3'(b\beta)^{-1/2}$ and $b_3 = b^{1/2}\beta^{-1/2}$. This proves

$$\max_{j=1,\ldots,n-1} |(K_j^{(0)} y, y)| \leq \max_{j=1,\ldots,n-1} \|K_j^{(0)} y\| \|y\| \leq a_3 \|y\|^2 + b_3(Ay, y)$$

since the right-hand side is clearly nonnegative by the above estimates.

4. For all $k = 1,\ldots,n$ and $y \in D(A)$, the estimate (3.1.4) gives

$$\begin{aligned}
(K_k y, y) = (K_k^{(0)} y, y) + \gamma_k (Ay, y) &\leq a_3 \|y\|^2 + b_3(Ay, y) + \gamma_k[(Ay, y) + \beta\|y\|^2] \\
&\leq a_3 \|y\|^2 + b_3(Ay, y) + \max_{j=1,\ldots,n-1} \gamma_j[(Ay, y) + \beta\|y\|^2].
\end{aligned}$$

This proves (3.1.5) with $a_4 = a_3 + \beta \max_{j=1,\ldots,n-1} \gamma_j$ and $b_4 = b_3 + \max_{j=1,\ldots,n-1} \gamma_j$. $\square$

It is easy to see that the statement of Lemma 1.1.11, part 3, extends to this case:

Lemma 3.1.3. *For all $\lambda \in \mathbb{C}$, $L(\lambda)^* = L(-\bar{\lambda})$.*

Theorem 3.1.4. *If the domain Ω is normal for the pencil L, then there exists a number $\delta > 0$ depending on A and K_j with the following properties. If S_j, $j = 1, \ldots, n-1$ are closed operators on H with $D(A) \subset D(S_j)$ and if for all $y \in D(A)$*

$$\sum_{j=1}^{n-1} \|(S_j - K_j)y\| \le \delta(\|Ay\| + \|y\|), \tag{3.1.6}$$

then Ω is normal for the pencil $\mathcal{L}$ given by

$$\mathcal{L}(\lambda) = i^n \lambda^n I + \sum_{j=1}^{n-1} i^j \lambda^j S_j + A,$$

and the total algebraic multiplicities of the spectra of L and $\mathcal{L}$ in Ω coincide.

Proof. Let $\lambda_0 \in \overline{\Omega}$. Since $L(\lambda_0)$ is a Fredholm operator by assumption, it is a closed operator with domain $D(A)$, and hence the graph norms of $L(\lambda_0)$ and A are equivalent by the closed graph theorem. Therefore, there is $c > 0$ such that

$$\|Ay\| \le c(\|L(\lambda_0)y\| + \|y\|)$$

for all $y \in D(A)$. For each pair of positive numbers $a(\lambda_0)$ and $b(\lambda_0)$ there is $\delta = \delta(\lambda_0)$ such that for all $y \in D(A)$, λ in a sufficiently small neighbourhood of λ_0 and S_j satisfying (3.1.6) we estimate

$$\|(\mathcal{L}(\lambda) - L(\lambda_0))y\| \le \|(\mathcal{L}(\lambda) - L(\lambda))y\| + \|(L(\lambda) - L(\lambda_0))y\|$$
$$\le \sum_{j=1}^{n-1} |\lambda^j| \|(S_j - K_j)y\| + |\lambda^n - \lambda_0^n| \|y\| + \sum_{j=1}^{n-1} |\lambda^j - \lambda_0^j| \|K_j y\|$$
$$\le a(\lambda_0)\|y\| + b(\lambda_0)\|L(\lambda_0)y\|,$$

where we have also used (3.1.3). Observe that the reduced minimum modulus $\gamma(L(\lambda_0))$, defined in [137, p. 231], is positive since $L(\lambda_0)$ is a Fredholm operator, see [137, Theorem IV.5.2]. Choosing $\delta(\lambda_0)$ and the neighbourhood of λ_0 sufficiently small, we may assume that the positive numbers $a(\lambda_0)$ and $b(\lambda_0)$ are so small that the inequality $a(\lambda_0) < (1 - b(\lambda_0))\gamma(L(\lambda_0))$, i.e., [137, (IV.5.20)], is satisfied. Then the index stability theorem for Fredholm operators, see [137, Theorem IV.5.22] shows that $\mathcal{L}(\lambda)$ is a Fredholm operator for all λ in that neighbourhood of λ_0 and for all S_j satisfying (3.1.6) with $\delta = \delta(\lambda_0)$. Also, since $L(\lambda_0)$ is invertible for $\lambda_0 \in \partial\Omega$, we conclude again from [137, Theorem IV.5.22] that $\mathcal{L}(\lambda)$ is invertible for all λ in the corresponding neighbourhood of λ_0 and for all S_j satisfying (3.1.6) with $\delta = \delta(\lambda_0)$. A compactness argument shows that $\mathcal{L}(\lambda)$ is a Fredholm operator for all $\lambda \in \Omega$ and invertible for all $\lambda \in \partial\Omega$, with S_j satisfying (3.1.6) for a suitable positive δ (the minimum of all $\delta(\lambda_0)$ for the λ_0 used for the finite cover of $\overline{\Omega}$). Since $\mathcal{L}(\lambda)$ is invertible for $\lambda \in \partial\Omega$, it follows that Ω is a normal domain for $\mathcal{L}$ if the S_j satisfy (3.1.6).

Finally, introducing

$$\mathcal{L}(\lambda,t) = i^n \lambda^n I + \sum_{j=1}^{n-1} i^j \lambda^j (K_j + t(S_j - K_j)) + A$$

we have $\mathcal{L}(\lambda) = \mathcal{L}(\lambda,1)$ and $\mathcal{L}(\lambda,0) = L(\lambda)$. Rouché's theorem for finitely meromorphic operator functions, Lemma 1.1.9, also applies to $\mathcal{L}(\lambda,t)$ for all $t \in [0,1]$, and a compactness argument shows that $m(\Omega)$ is independent of t, which proves that the total algebraic multiplicities of the spectra of L and $\mathcal{L}$ in Ω coincide. $\quad\square$

For $\lambda, \eta \in \mathbb{C}$ we define

$$L(\lambda,\eta) = i^n \lambda^n I + \sum_{j=1}^{n-1} i^j \lambda^j \eta K_j + A,$$

$$\xi(\lambda,\eta) = \eta \sum_{j=1}^{n-1} \gamma_j i^j \lambda^j + 1,$$

$$\Xi(\eta) = \{\lambda \in \mathbb{C} : \xi(\lambda,\eta) = 0\}.$$

Since $\xi(\cdot,\eta)$ is a polynomial of degree less or equal $n-1$, $\Xi(\eta)$ consists of at most $n-1$ complex numbers.

Lemma 3.1.5. *Let $\lambda_0, \eta_0 \in \mathbb{C}$ such that $\lambda_0 \notin \Xi(\eta_0)$ and assume that the operator $L(\lambda_0, \eta_0)$ has the following properties:*

(i) *there exists a number $\varepsilon > 0$ such that*

$$\|L(\lambda_0, \eta_0)y\| \geq \varepsilon\|y\| \tag{3.1.7}$$

for all $y \in D(A)$;

(ii) $\operatorname{codim} R(L(\lambda_0, \eta_0)) = q$, $0 \leq q \leq \infty$.

Then for every $\varepsilon' \in (0, \varepsilon)$ there is some $\delta > 0$ such that the operators $L(\lambda, \eta)$ have the properties (i) and (ii) in the neighbourhood $\{(\lambda, \eta) \in \mathbb{C}^2 : |\lambda - \lambda_0| + |\eta - \eta_0| \leq \delta\}$ of (λ_0, η_0) with ε replaced by ε' in (3.1.7) but with the same q in (ii).

Proof. We can write

$$L(\lambda,\eta) = i^n \lambda^n I + \sum_{j=1}^{n-1} i^j \lambda^j \eta K_j^{(0)} + \xi(\lambda,\eta)A.$$

Because of $\xi(\lambda_0, \eta_0) \neq 0$, the operator

$$A_0 = i^n \lambda_0^n I + \xi(\lambda_0, \eta_0) A$$

is closed with $D(A_0) = D(A)$, and hence the graph norms of A and A_0 on $D(A)$ are equivalent by the closed graph theorem. By Lemma 3.1.2, part 1, there exists $a_1 > 0$ such that

$$\|(L(\lambda_0, \eta_0) - A_0)y\| \leq a_1 \|y\| + \frac{1}{2}\|A_0 y\|$$

for all $y \in D(A)$. Then [137, Theorem IV.1.1] implies that $L(\lambda_0, \eta_0)$ is closed. It follows from inequality (3.1.7) that the image $R(L(\lambda_0, \eta_0))$ is closed and that $L(\lambda_0, \eta_0)$ is injective, and hence $L(\lambda_0, \eta_0)$ is semi-Fredholm, see [137, p. 230]. By the closed graph theorem, the closedness of $L(\lambda_0, \eta_0)$ gives that the graph norms of A and $L(\lambda_0, \eta_0)$ are equivalent on $D(A)$. Hence, since all the operators $K_j^{(0)}$ are subordinate to A, we estimate for $y \in D(A)$ with the aid of (3.1.2)

$$
\begin{aligned}
&\|(L(\lambda,\eta) - L(\lambda_0, \eta_0))y\| \\
&= \left\| i^n(\lambda^n - \lambda_0^n)y + \sum_{j=1}^{n-1} i^j(\lambda^j \eta - \lambda_0^j \eta_0)K_j^{(0)}y + (\xi(\lambda, \eta) - \xi(\lambda_0, \eta_0))Ay \right\| \\
&\leq |\lambda^n - \lambda_0^n|\|y\| + \sum_{j=1}^{n-1} |\lambda^j \eta - \lambda_0^j \eta_0|\|K_j^{(0)}y\| + |\xi(\lambda, \eta) - \xi(\lambda_0, \eta_0)|\|Ay\| \\
&\leq a_0(\lambda, \eta)\|y\| + b_0(\lambda, \eta)\|L(\lambda_0, \eta_0)y\| \qquad\qquad (3.1.8)
\end{aligned}
$$

with non-negative functions a_0 and b_0 such that $a_0(\lambda, \eta) \to 0$ and $b_0(\lambda, \eta) \to 0$ as $(\lambda, \eta) \to (\lambda_0, \eta_0)$. Note that since $L(\lambda_0, \eta_0)$ is injective, the reduced minimum modulus $\gamma(L(\lambda_0, \eta_0))$ equals the inverse of the norm of the inverse operator, so that $\varepsilon \leq \gamma(L(\lambda_0, \eta_0))$. In other words, $\gamma(L(\lambda_0, \eta_0))$ is the supremum of all ε for which (3.1.7) holds. For any $\varepsilon' \in (0, \varepsilon)$ there exists a number $\delta > 0$ such that

$$a_0(\lambda, \eta) + \varepsilon' \leq (1 - b_0(\lambda, \eta))\varepsilon \qquad\qquad (3.1.9)$$

for $|\lambda - \lambda_0| + |\eta - \eta_0| \leq \delta$, and hence, for these λ and η, the operator $L(\lambda, \eta)$ is closed and semi-Fredholm by [137, Theorem IV.5.22], with

$$
\begin{aligned}
\dim N(L(\lambda, \eta)) &\leq \dim N(L(\lambda_0, \eta_0)), \\
\operatorname{codim} R(L(\lambda, \eta)) &\leq \operatorname{codim} R(L(\lambda_0, \eta_0)), \\
\operatorname{ind} L(\lambda, \eta) &= \operatorname{ind} L(\lambda_0, \eta_0).
\end{aligned}
$$

Since the dimension of the nullspace of $L(\lambda_0, \eta_0)$ is equal to 0, it follows that the dimension of the nullspace of $L(\lambda, \eta)$ is also equal to 0 for all λ and η satisfying the inequality $|\lambda - \lambda_0| + |\eta - \eta_0| \leq \delta$. Using that the indices of $L(\lambda, \eta)$ and $L(\lambda_0, \eta_0)$ are equal, we conclude that

$$\operatorname{codim} R(L(\lambda, \eta)) = \operatorname{codim} R(L(\lambda_0, \eta_0)) = q$$

for all λ and η satisfying the inequality $|\lambda - \lambda_0| + |\eta - \eta_0| \leq \delta$. Finally, for these (λ, η) and $y \in D(A)$, the inequalities (3.1.7), (3.1.8) and (3.1.9) lead to

$$
\begin{aligned}
\|L(\lambda, \eta)y\| &\geq \|L(\lambda_0, \eta_0)y\| - \|(L(\lambda, \eta) - L(\lambda_0, \eta_0))y\| \\
&\geq (1 - b_0(\lambda, \eta))\|L(\lambda_0, \eta_0)y\| - a_0(\lambda, \eta)\|y\| \\
&\geq (1 - b_0(\lambda, \eta))\varepsilon\|y\| - a_0(\lambda, \eta)\|y\| \\
&\geq \varepsilon'\|y\|.
\end{aligned}
$$

This completes the proof of the lemma.   □

Corollary 3.1.6. *For all $\eta \in \mathbb{C}$, $\partial\sigma(L(\cdot, \eta)) \setminus \Xi(\eta) \subset \sigma_{\mathrm{app}}(L(\cdot, \eta))$.*

Proof. Let $\lambda \in \partial\sigma(L(\cdot, \eta)) \setminus \Xi(\eta)$.

If $\dim N(L(\lambda, \eta)) > 0$, then choose $y \in N(L(\lambda, \eta))$ with $\|y\| = 1$ and put $y_k = y$. Then $L(\lambda, \eta)y_k = 0$, so that $\lambda \in \sigma_{\mathrm{app}}(L(\cdot, \eta))$ follows.

If $\dim N(L(\lambda, \eta)) = 0$ but there is no $\varepsilon > 0$ such that $\|L(\lambda, \eta)y\| \geq \varepsilon\|y\|$ for all $y \in D(A)$, then we can choose $y_k \in D(A)$ with $\|y_k\| = 1$ and $\|L(\lambda, \eta)y_k\| \leq \frac{1}{k}$. Hence $\lambda \in \sigma_{\mathrm{app}}(L(\cdot, \eta))$.

But if $\dim N(L(\lambda, \eta)) = 0$ and there is $\varepsilon > 0$ such that $\|L(\lambda, \eta)y\| \geq \varepsilon\|y\|$ for all $y \in D(A)$, then, by Lemma 3.1.5, either $\lambda' \in \rho(L(\cdot, \eta))$ for all λ' in a neighbourhood of λ or $\lambda' \in \sigma(L(\cdot, \eta))$ for all λ' in a neighbourhood of λ. Both properties contradict the assumption that $\lambda \in \partial\sigma(L(\cdot, \eta))$.   □

Let us introduce the notations

$$
\begin{aligned}
W_n &= \left\{\lambda \in \mathbb{C} \setminus \{0\} : 0 < \left|\arg\lambda + \frac{\pi}{2}\right| < \frac{\pi}{n}\right\}, \\
W_n^0 &= \left\{\lambda \in \mathbb{C} \setminus \{0\} : 0 \leq \left|\arg\lambda + \frac{\pi}{2}\right| < \frac{\pi}{n}\right\}.
\end{aligned}
$$

Lemma 3.1.7. $W_n \cap \sigma_{\mathrm{app}}(L(\cdot, \eta)) = \emptyset$ *for all $\eta \in [0, 1]$.*

Proof. Let $\lambda_0 \in W_n \cap \sigma_{\mathrm{app}}((L(\cdot, \eta_0))$, where $\eta \in [0, 1]$ and $-\frac{\pi}{2} < \arg\lambda_0 < -\frac{\pi}{2} + \frac{\pi}{n}$ and let $\{y_k\}_{k=1}^\infty$ be a corresponding approximate sequence. Then

$$\lim_{k \to \infty} (L(\lambda_0, \eta_0)y_k, y_k) = 0$$

and consequently

$$\operatorname{Im}((i\lambda_0)^n) + \eta_0 \sum_{j=1}^{n-1}(K_j y_k, y_k)\operatorname{Im}((i\lambda_0)^j) = \operatorname{Im}(L(\lambda_0, \eta_0)y_k, y_k) \underset{k \to \infty}{=} o(1). \quad (3.1.10)$$

Taking into account that $\eta_0 \geq 0$ and that for all $j = 1,\ldots,n$ the inequalities $\mathrm{Im}(i\lambda_0)^j > 0$ and $(K_j y_k, y_k) \geq 0$ are valid, we obtain that the left-hand side of (3.1.10) is not less than $\mathrm{Im}(i\lambda_0)^n > 0$, which means that (3.1.10) is false. The proof for the case $-\frac{\pi}{2} - \frac{\pi}{n} < \arg \lambda_0 < -\frac{\pi}{2}$ is analogous.   □

Lemma 3.1.8. $(-i\infty, -i\beta^{1/n}) \cap \sigma_{\mathrm{app}}(L(\cdot, \eta)) = \emptyset$ for all $\eta \in [0, 1]$.

Proof. Let $\lambda_0 \in (-i\infty, -i\beta^{1/n})$ and $\eta_0 \in [0, 1]$. Then for all $y \in D(A)$ the inequalities $(Ay, y) \geq -\beta\|y\|^2$ and $\eta_0 \sum_{j=1}^{n-1} (i\lambda_0)^j (K_j y, y) \geq 0$ hold. This leads to

$$\|L(\lambda_0, \eta)y\| \|y\| \geq (L(\lambda_0, \eta)y, y) \geq ((i\lambda_0)^n - \beta)\|y\|^2$$

for all $y \in D(A)$ and particularly to

$$\|L(\lambda_0, \eta)y\| \geq ((i\lambda_0)^n - \beta) > 0$$

for all $y \in D(A)$ with $\|y\| = 1$. But this shows that $L(\cdot, \eta)$ cannot have an approximate sequence at λ_0, i.e., $\lambda_0 \notin \sigma_{\mathrm{app}}(L(\cdot, \eta))$.   □

Lemma 3.1.9. $\sigma(L(\cdot, \eta)) \cap W_n^0 \subset [-i\beta^{1/n}, 0)$ for all $\eta \in [0, 1]$.

Proof. By Lemmas 3.1.7 and 3.1.8, $(W_n^0 \setminus [-i\beta^{1/n}, 0)) \cap \sigma_{\mathrm{app}}(L(\cdot, \eta))) = \emptyset$ for all $\eta \in [0, 1]$. Clearly, for $(\lambda, \eta) \in W_n^0 \times [0, 1]$ we have $\xi(\lambda, \eta) \geq 1$ or $\mathrm{Im}\,\xi(\lambda, \eta) > 0$ or $\mathrm{Im}\,\xi(\lambda, \eta) < 0$. Therefore $L(\lambda, \eta)$ satisfies the assumption of Lemma 3.1.5 for all $(\lambda, \eta) \in (W_n^0 \setminus [-i\beta^{1/n}, 0)) \times [0, 1]$. In view of Lemma 3.1.5, the sets

$$S_0 = \{(\lambda, \eta) \in (W_n^0 \setminus [-i\beta^{1/n}, 0)) \times [0, 1] : \mathrm{codim}\,R(L(\lambda_0, \eta_0)) = 0\}$$

and

$$S_1 = \{(\lambda, \eta) \in (W_n^0 \setminus [-i\beta^{1/n}, 0)) \times [0, 1] : \mathrm{codim}\,R(L(\lambda_0, \eta_0)) \neq 0\}$$

form a disjoint relatively open cover of $(W_n^0 \setminus [-i\beta^{1/n}, 0)) \times [0, 1]$. Since A is self-adjoint, the spectrum of the operator pencil $i^n \lambda^n I + A$ lies on the rays given by $\arg \lambda = \frac{\pi}{n}j - \frac{\pi}{2}$, $j = 0,\ldots, 2n - 1$, so that $W_n \times \{0\} \subset S_0$. Hence S_0 is nonempty, and since $(W_n^0 \setminus [-i\beta^{1/n}, 0)) \times [0, 1]$ is connected, it follows that $S_0 = (W_n^0 \setminus [-i\beta^{1/n}, 0)) \times [0, 1]$. But $L(\lambda, \eta)$ is invertible for each $(\lambda, \eta) \in S_0$, and the proof of the lemma is complete.   □

Taking Corollary 3.1.6 and Lemma 3.1.9 into account and observing that $\lambda \notin \Xi(\eta)$ for $\lambda \in W_n^0$ and $\eta \in [0, 1]$ we obtain

Corollary 3.1.10. For $\eta \in [0, 1]$, the spectrum of the pencil $L(\cdot, \eta)$ located on the interval $[-i\beta^{1/n}, 0)$ is approximate.

Lemma 3.1.11. Let $\lambda_0 \in [-i\beta^{1/n}, 0) \cap \sigma_0(L(\cdot, \eta_0))$ and let $\eta_0 \in [0, 1]$. Then the eigenvalue λ_0 does not possess associated vectors.

Proof. Assume there is a chain of an eigenvector y_0 and an associated vector y_1 corresponding to λ_0. The equation for the associated vector y_1 is

$$L(\lambda_0, \eta_0)y_1 + \frac{\partial}{\partial \lambda} L(\lambda_0, \eta_0)y_0 = 0.$$

This implies

$$(L(\lambda_0, \eta_0)y_1, y_0) + \left(\frac{\partial}{\partial \lambda} L(\lambda_0, \eta_0)y_0, y_0 \right) = 0.$$

Since $\xi(\lambda_0, \eta_0) \neq 0$ and $\eta \sum_{j=1}^{n-1} (i\lambda_0)^j K_j^{(0)}$ is subordinate to $(i\lambda_0)^n I + \xi(\lambda_0, \eta_0)A$, it follows from Lemma 3.1.2, part 1, and [137, Theorem V.4.3] that $L(\lambda_0, \eta_0)$ is self-adjoint. Hence we have

$$(L(\lambda_0, \eta_0)y_1, y_0) = (y_1, L(\lambda_0, \eta_0)y_0) = 0,$$

and

$$\left(\frac{\partial}{\partial \lambda} L(\lambda_0, \eta_0)y_0, y_0 \right) = 0$$

follows. Taking the imaginary part of the latter equation we obtain

$$n|\lambda_0|^{n-1}\|y_0\|^2 + \sum_{j=1}^{n-1} j|\lambda_0|^{j-1}(K_j y_0, y_0) = 0,$$

which is impossible due to $|\lambda_0| > 0$ and $(K_j y_0, y_0) \geq 0$. $\square$

Lemma 3.1.12. *Assume that $\sigma(A) \cap [-\beta, 0) \subset \sigma_0(A)$. Then*

$$[-i\beta^{1/n}, 0) \cap \sigma(L(\cdot, \eta)) \subset \sigma_0(L(\cdot, \eta))$$

for every $\eta \in [0, 1]$.

Proof. Let $\lambda \in [-i\beta^{1/n}, 0)$ and $\eta \in [0, 1]$. Then $L(\lambda, \eta)$ is symmetric with $L(\lambda, \eta) \geq (i\lambda)^n I + A$. From $\xi(\lambda, \eta) \neq 0$, Lemma 3.1.2, part 1, and [137, Theorem V.4.3] it follows that $L(\lambda, \eta)$ is self-adjoint. Choose $\varepsilon \in (0, (i\lambda)^n)$ and let P be the projection associated with the self-adjoint operator $L(\lambda, \eta)$ and its spectrum below ε. For all $y \in R(P)$ we have

$$(((i\lambda)^n I - \varepsilon I + A)y, y) \leq ((L(\lambda, \eta) - \varepsilon I)y, y) \leq 0$$

so that $R(P)$ is a nonpositive subspace of $((i\lambda)^n - \varepsilon)I + A$. From $(i\lambda)^n - \varepsilon > 0$ and the assumption on A it follows that every nonnegative subspace of the operator $((i\lambda)^n - \varepsilon)I + A$ must be finite dimensional. Hence $R(P)$ must be finite dimensional, and therefore the spectrum of $L(\lambda, \eta)$ below ε consists of finitely many eigenvalues of finite multiplicity. In particular, since $\varepsilon > 0$, $L(\lambda, \eta)$ is a Fredholm operator. In view of Lemma 3.1.9, $L(\lambda, \eta)$ is a Fredholm operator for all $\lambda \in W_n^0$. Since

$L(\cdot, \eta)$ is analytical on W_n^0 with respect to the graph norm of A on $D(A)$ and since $W_n \subset \rho(L(\cdot, \eta))$ by Lemma 3.1.9, it follows that $\sigma(L(\cdot, \eta)) \cap W_n^0$ is a discrete set, see, e. g., [189, Theorem 1.3.1]. Finally, by Lemma 3.1.11, the algebraic multiplicity of each eigenvalue equals its geometric multiplicity and is therefore finite. $\square$

Evidently, we obtain from Lemmas 3.1.9 and 3.1.12

Corollary 3.1.13. *If* $\sigma(A) \cap [-\beta, 0) \subset \sigma_0(A)$, *then* $\sigma(L(\cdot, \eta)) \cap W_n^0 \subset \sigma_0(L(\cdot, \eta))$ *for every* $\eta \in [0, 1]$.

Theorem 3.1.14. *Assume that* $\sigma(A) \cap [-\beta, 0) \subset \sigma_0(A)$. *Then the total algebraic multiplicity of the spectrum of* L *located on* $(-i\infty, 0)$ *coincides with the total algebraic (geometric) multiplicity of the negative spectrum of* A.

Proof. Let us denote by $N(\eta)$ the total algebraic multiplicity of the spectrum of $L(\cdot, \eta)$ located on $(-i\infty, 0)$. Since $(i\lambda)^n I + A \le L(\lambda, \eta)$ for $\lambda \in (-i\infty, 0)$, the minimax principle gives

$$N(\eta) \le N(0), \quad \eta \in [0, 1], \tag{3.1.11}$$

see the proof of Lemma 3.1.12. Consider the auxiliary operator pencil

$$L_0(\lambda, \zeta) = (i\lambda)^n I + \sum_{j=1}^{n-1} (i\lambda)^j [(1 - \zeta)(a_4 I + b_4 A) + \zeta K_j] + A,$$

where $\lambda \in \mathbb{C}$, $\zeta \in [0, 1]$, and $a_4 > 0$ and $b_4 > 0$ are the constants from Lemma 3.1.2, part 4, and where we may assume that $a_4 \ge \beta b_4$. Then $L_0(\cdot, \zeta)$ is an operator pencil satisfying Condition III with $K_j^{(0)}$ replaced by $\zeta K_j^{(0)} + (1 - \zeta)a_4 I$ and γ_j replaced by $\zeta \gamma_j + (1 - \zeta)b_4$. Applying Lemma 3.1.9 to the pencil $L_0(\cdot, \zeta)$ we obtain $\sigma(L(\cdot, \zeta)) \cap W_n^0 \subset [-i\beta^{1/n}, 0)$ for all $\zeta \in [0, 1]$. Clearly, Theorem 9.2.4 leads to a representation of the eigenvalues and eigenvectors of $L_0(\cdot, \zeta)$ on $(-i\infty, 0)$ as in Theorem 1.2.7, and Remark 1.2.8 also applies here. Then we have for $\zeta_0 \in [0, 1]$ and an eigenvalue $\lambda_0 \in [-i\beta^{1/n}, 0)$ of $L_0(\cdot, \zeta_0)$ the expansions

$$\lambda_\iota(\zeta) = \lambda_0 + \sum_{k=1}^{\infty} a_{\iota k}(\zeta - \zeta_0)^k, \quad \iota = 1, \dots, l,$$

$$y_\iota^{(q)}(\zeta) = b_{\iota 0}^{(q)} + \sum_{k=1}^{\infty} b_{\iota k}^{(q)}(\zeta - z\eta_0)^k, \quad q = 1, \dots, r_\iota,$$

of eigenvalues and eigenvectors near λ_0 of $L_0(\cdot, \zeta)$ for ζ near ζ_0. Differentiating $L_0(\lambda_\iota(\zeta), \zeta)y_\iota^{(q)}(\zeta) = 0$ with respect to ζ and taking the inner product of the resulting equation with $y_\iota(\zeta)$, we obtain for $\zeta = \zeta_0$ and with $\tau_0 = i\lambda_0 > 0$ that $ia_{\iota 1}$ is a quotient of real numbers, where the denominator

$$n\tau_0^{n-1}\|b_{\iota 0}^{(q)}\|^2 + \sum_{j=1}^{n-1} j\tau_0^{j-1}\left((1 - \zeta_0)(a_4\|b_{\iota 0}^{(q)}\|^2 + b_4(Ab_{\iota 0}^{(q)}, b_{\iota 0}^{(q)})) + \zeta_0(K_j b_{\iota 0}^{(q)}, b_{\iota 0}^{(q)})\right)$$

is positive because of $a_4 \geq \beta b_4$, whereas the numerator

$$\sum_{j=1}^{n-1} \tau_0^j \left(a_4 \|b_{\iota 0}^{(q)}\|^2 + b_4 (A b_{\iota 0}^{(q)}, b_{\iota 0}^{(q)}) - (K_j b_{\iota 0}^{(q)}, b_{\iota 0}^{(q)}) \right)$$

is nonnegative by Lemma 3.1.2, part 4. It follows that $i a_{\iota 1} \geq 0$ and consequently $N_0(\zeta) \geq N_0(0)$, where $N_0(\zeta)$ is the total algebraic multiplicity of the spectrum of $L_0(\cdot, \zeta)$ located on $(-i\infty, 0)$. Therefore

$$N_0(1) \geq N_0(0). \tag{3.1.12}$$

It is easy to check that the function f given by

$$f(t) = \frac{t^n + a_4 p(t)}{b_4 p(t) + 1}, \quad \text{where} \quad p(t) = \sum_{j=1}^{n-1} t^j,$$

maps $(0, \infty)$ onto $(0, \infty)$ and that $f'(t) > 0$ for all $t > 0$. Hence the parameter transformation $\tau = -f(i\lambda)$ maps the semiaxis $(-i\infty, 0)$ in the λ-plane bijectively onto the semiaxis $(-\infty, 0)$ in the τ-plane. Since

$$L_0(\lambda, 0) = ((i\lambda)^n + a_4 p(i\lambda)) I + (b_4 p(i\lambda) + 1) A,$$

it follows that there is a ono-to-one correspondence between the eigenvalues, with multiplicity, of $L_0(\cdot, 0)$ on $(-i\infty, 0)$ and those of A on $(-\infty, 0)$. Consequently, taking into account the absence of associated vectors, see Lemma 3.1.11, we obtain $N_0(0) = N_A$, where N_A is the total multiplicity of the negative spectrum of A. With a similar reasoning for $\tau = -(i\lambda)^n$ we obtain $N(0) = N_A$.

The evident identity $L(\lambda, 1) = L_0(\lambda, 1)$ implies $N(1) = N_0(1)$. Using (3.1.11) and (3.1.12) we obtain $N(0) \geq N(1) = N_0(1) \geq N_0(0) = N_A = N(0)$. This means that all numbers are equal, and in particular $N(1) = N_A$. $\qquad \square$

3.2 Quadratic pencils with indefinite linear term

In this section we consider the quadratic operator pencil

$$L(\lambda, \eta) = \lambda^2 I - i\lambda K - A,$$

where A and K are self-adjoint on H, $A \geq 0$ and $K \geq -\beta_2 I$ for some positive number β_2, and K is relatively compact with respect to A. Observe that $\sigma_{\text{ess}}(A) \subset \sigma(A) \subset [0, \infty)$.

Definition 3.2.1. Let p be a positive integer and assume that H_0 is a closed subspace of H. The operator K is said to have p negative squares on H_0 if $H_0 \subset D(K)$, if $\dim V \leq p$ for each subspace $V \subset H_0$ for which $K|_V < 0$, and there is at least one such subspace V with $\dim V = p$.

Theorem 3.2.2. *Let p be a positive integer and assume that K has p negative squares on $N(A)$. Then for each $\varepsilon > 0$ there exists $\delta > 0$ such that $L(\cdot, \eta)$ has at least p normal eigenvalues, counted with multiplicity, on $(-i\varepsilon, 0)$ for all $\eta \in (0, \delta)$.*

Proof. For $z \in \mathbb{C}$ and $\tau, \eta \in [0, \infty)$ define

$$\mathcal{L}(z, \tau, \eta) = zI - \tau\eta K - A.$$

Since K is A-compact, the essential spectrum of $\mathcal{L}(\cdot, \tau, \eta)$ does not depend on η and τ, see [137, Theorem IV.5.22]. In particular,

$$\sigma_{\mathrm{ess}}(\mathcal{L}(\cdot, \tau, \eta)) = \sigma_{\mathrm{ess}}(\mathcal{L}(\cdot, 0, 0)) = \sigma_{\mathrm{ess}}(A) \subset [0, \infty).$$

By assumption, there is a p-dimensional subspace V of $N(A)$ such that $K|_V < 0$.

For $k = 1, \ldots, p$ consider

$$z_k = \min_{M \subset D(A), \dim M = k} \max_{0 \neq y \in M} \frac{((\tau\eta K + A)y, y)}{\|y\|^2}$$

and

$$\alpha_k = \min_{M \subset V, \dim M = k} \max_{0 \neq y \in M} \frac{(Ky, y)}{\|y\|^2}.$$

For $\tau, \eta \in (0, \infty)$ it follows from $K|_V < 0$ and $A \geq 0$ that

$$z_k \leq \tau\eta\alpha_k < 0$$

and that

$$z_k \geq \tau\eta \min_{0 \neq y \in D(A)} \frac{(Ky, y)}{\|y\|^2} \geq -\tau\eta\beta_2.$$

Since $z_k < 0$ is below the essential spectrum of $\mathcal{L}(\cdot, \tau, \eta)$, the minimax principle says that, for $\eta > 0$ fixed, the number z_k is a (negative) eigenvalue of $\mathcal{L}(\cdot, \tau, \eta)$. This eigenvalue is a continuous function of $\tau \in (0, \infty)$ by Theorem 1.2.7. Its graph in the (τ, z)-plane is located in the angular region

$$\left\{ (\tau, z) \in \mathbb{R}^2 : \tau > 0,\ -\eta\beta_2 \leq \frac{z}{\tau} \leq \eta\alpha_k \right\}.$$

This curve intersects the parabola $z = -\tau^2$, and for all intersection points,

$$-\eta\alpha_k \leq \tau \leq \eta\beta_2. \tag{3.2.1}$$

Choose such an intersection point (τ_k, z_k). Then $\lambda_k = -i\tau_k$ will be an eigenvalue of the pencil $L(\cdot, \eta)$ located on the negative imaginary semiaxis. Notice that if z_k occurs with multiplicity larger than 1, then we may choose the same τ_k for these z_k, and every eigenvector of $\mathcal{L}(\cdot, \tau_k, \eta)$ at z_k is also an eigenvector of $L(\cdot, \eta)$ at λ_k. This shows that $L(\cdot, \eta)$ has at least p eigenvalues on the negative imaginary semiaxis, counted with multiplicity.

Now let $\varepsilon > 0$ and set $\delta = \varepsilon \beta_2^{-1}$. Then the right-hand side of the estimates (3.2.1) shows for $\eta \in (0, \delta)$ that

$$|\lambda_k| = \tau_k \le \eta \beta_2 < \delta \beta_2 = \varepsilon$$

for all $k = 1, \ldots, p$. $\qquad\qquad\qquad\qquad\qquad\qquad\qquad\qquad\qquad\qquad\qquad$ $\square$

3.3 Notes

Theorem 3.1.14 was obtained in [211]. Self-adjoint quadratic operator pencils occur in problems of plate and shell vibrations [149]. Further applications of polynomial operator pencils can be found in [193], [204].

The effect of inner damping is well known in mechanics and leads to an additional term in the equation of small motions. For transverse vibrations of a rod, (2.7.1) becomes

$$\gamma \frac{\partial^5 u}{\partial t \partial x^4} + \frac{\partial^4 u}{\partial x^4} - \frac{\partial}{\partial x} g(x) \frac{\partial u}{\partial x} + \frac{\partial^2 u}{\partial t^2} = 0 \tag{3.3.1}$$

where $\gamma > 0$ is the coefficient of inner damping, see [31, Section 3.6]. For pipes conveying fluid, the equation of small lateral motions was obtained in [206, (12)]. This equation is similar to (3.3.1), but has some additional lower-order terms.

The spectral problem corresponding to (3.3.1) is

$$(1 + i\gamma\lambda)y^{(4)}(\lambda, x) - (gy')'(\lambda, x) = \lambda^2 y(\lambda, x),$$
$$y(\lambda, 0) = 0,$$
$$y''(\lambda, 0) = 0,$$
$$y(\lambda, a) = 0,$$
$$y''(\lambda, a) + i\alpha\lambda y'(\lambda, a) = 0,$$

where $\alpha > 0$. We cannot apply the theory of Chapter 1 to this problem. But setting $n = 2$, Lemma 3.1.9 and Theorem 3.1.14 give that the spectrum of this problem in the open lower half-plane lies on the negative imaginary semiaxis and its total algebraic multiplicity does not depend on γ and on α.

A related problem describing vibrations of a rotating beam with inner damping leading to a non-self-adjoint operator pencil was considered in [6].

Lemma 3.1.9 is a generalization of [155, statement 2.4^0]. We recall that in [155], M.G. Kreĭn and H. Langer consider the pencil $\lambda^2 I + \lambda B + C$, where B is self-adjoint and bounded and $C > 0$ is compact. Replacing λ with $i\lambda$, this pencil becomes

$$(i\lambda)^2 I + i\lambda B + C,$$

which satisfies Condition III since the boundedness of the operators involved implies the subordination property. Hence Theorem 3.1.14 implies that the spectrum of the pencil $\lambda^2 I + \lambda B + C$ lies in the closed left half-plane if $B \geq 0$, as stated in [155, statement 2.4⁰].

Theorem 3.2.2 was proved in [147], where it was used to obtain a sufficient condition of instability of liquid convective motions in case of heating.

In the proof of Theorem 3.2.2 we have applied the minimax principle to a quadratic operator pencil. The minimax principle for more general operator functions was investigated in [74], [75], [167]. This method is the same as for finite-dimensional spaces or for compact operators, see, e. g., [137, p. 60] or [70, p. 908].

Chapter 4

Operator Pencils with a Gyroscopic Term

4.1 Quadratic operator pencils involving a gyroscopic term

In this section we will investigate the spectra of monic quadratic operator pencils which include a term corresponding to gyroscopic forces. We also admit presence of essential spectrum. Our operator pencil is

$$L(\lambda) = \lambda^2 I - i\lambda K - \lambda B - A,$$

acting in a Hilbert space H with domain $D(L(\lambda)) = D(K) \cap D(B) \cap D(A)$, where the operators K, B, A satisfy the following conditions.

Condition IV. *The operators K, B and A are self-adjoint operators on H with the following properties:*

(i) $K \geq \kappa I$, $\kappa > 0$, $A \geq -\beta I$ *for some positive number β, and $(-\gamma, 0) \subset \rho(A)$ for some $\gamma \in (0, \beta]$;*

(ii) *the operators B and K are subordinate to A, i. e., $D(A) \subset D(B) \cap D(K)$ and the operators B and K are bounded with respect to the operator $(A + \beta_1 I)^{\frac{1}{2}}$, where $\beta_1 > \beta$.*

We note that Condition IV, part (ii), implies $D(L(\lambda)) = D(A)$.

Proposition 4.1.1.

1. *Assume that part (i) of Condition IV holds. Then part (ii) of Condition IV is satisfied if and only if $D(A) \subset D(B) \cap D(K)$ and there exist positive numbers a and b such that the inequality*

$$\max\{\|By\|^2, \|Ky\|^2\} \leq a\|y\|^2 + b(Ay, y) \tag{4.1.1}$$

holds for each $y \in D(A)$.

2. *If Condition* IV *is satisfied, then*
   (i) *for each $b_1 > 0$ there exist $a_1 > 0$ such that*

$$\max\{\|By\|, \|Ky\|\} \leq a_1\|y\| + b_1\|Ay\| \qquad (4.1.2)$$

   for each $y \in D(A)$,
   (ii) *there exist $a_2 > 0$ and $b_2 > 0$ such that*

$$\max\{\|By\|^2, \|Ky\|^2\} \leq a_2\|y\|^2 + b_2\|Ay\|^2 \qquad (4.1.3)$$

   for each $y \in D(A)$,
   (iii) *there exist $a_3 > 0$ and $b_3 > 0$ such that*

$$\max\{|(By, y)|, (Ky, y)\} \leq a_3\|y\|^2 + b_3(Ay, y). \qquad (4.1.4)$$

Proof. The statements 1 and 2, parts (i) and (iii), are special cases of Lemmas 3.1.1 and 3.1.2 if we observe that the property $K_j \geq 0$ of Condition III is not used in the proof of Lemma 3.1.2. Hence we only have to prove 2, part (ii). From (4.1.1) we immediately obtain

$$\max\{\|By\|^2, \|Ky\|^2\} \leq a\|y\|^2 + b\|Ay\|\,\|y\|$$
$$\leq a\|y\|^2 + \frac{b}{2}\left(\|y\|^2 + \|Ay\|^2\right),$$

which proves (4.1.3) with $a_2 = a + \frac{b}{2}$ and $b_2 = \frac{b}{2}$. $\qquad \square$

In the same way as Theorem 3.1.4 one can prove

Theorem 4.1.2. *If the domain Ω is normal for the pencil L, then there exists a number $\delta > 0$ depending on A, B and K with the following properties. If B_1 and K_1 are closed operators in H which are subordinate to A and if for all $y \in D(A)$*

$$\|(B_1 - B)y\| + \|(K_1 - K)y\| \leq \delta(\|Ay\| + \|y\|), \qquad (4.1.5)$$

then Ω is normal for the pencil

$$\mathcal{L}(\lambda) = \lambda^2 I - i\lambda K_1 - \lambda B_1 - A,$$

and the total algebraic multiplicities of the spectra of L and $\mathcal{L}$ in Ω coincide.

Let us consider the operator pencil dependent on the parameter $\eta \in [0, 1]$:

$$L(\lambda, \eta) = \lambda^2 I - i\lambda K - \eta\lambda B - A.$$

Similar to Lemma 3.1.5 we have

Lemma 4.1.3. *Let* $\lambda_0, \eta_0 \in \mathbb{C}$ *and suppose that the operator* $L(\lambda_0, \eta_0)$ *has the following properties:*

(i) *there exists a number* $\varepsilon > 0$ *such that*

$$\|L(\lambda_0, \eta_0)y\| \geq \varepsilon \|y\| \qquad (4.1.6)$$

for all $y \in D(A)$;

(ii) $\operatorname{codim} R(L(\lambda_0, \eta_0)) = q, \ 0 \leq q \leq \infty.$

Then for every $\varepsilon' \in (0, \varepsilon)$ *there is some* $\delta > 0$ *such that the operators* $L(\lambda, \eta)$ *have these properties in the neighbourhood* $\{(\lambda, \eta) \in \mathbb{C}^2 : |\lambda - \lambda_0| + |\eta - \eta_0| \leq \delta\}$ *of* (λ_0, η_0) *with* ε *replaced by* ε' *in* (4.1.6) *but with the same* q *in* (ii).

Similar to Corollary 3.1.6 we have

Corollary 4.1.4. *For all* $\eta \in \mathbb{C}$, $\partial\sigma(L(\cdot, \eta)) \subset \sigma_{\mathrm{app}}(L(\cdot, \eta))$.

Lemma 4.1.5. *For* $\eta \in [0, 1]$, *the part of the spectrum of* $L(\cdot, \eta)$ *in the open lower half-plane is located in the set* $\{\lambda \in \mathbb{C} : 0 > \operatorname{Im}\lambda \geq -\beta^{1/2}, |\operatorname{Re}\lambda| \leq \beta^{3/4}\kappa^{-1/2}\}$.

Proof. Let $\lambda_0 \in \sigma_{\mathrm{app}}(L(\cdot, \eta_0))$, where $\eta_0 \in [0, 1]$ and $\operatorname{Im}\lambda_0 < 0$. Choose an approximate sequence $\{y_j\}$ of vectors $y_j \in D(A)$ such that $\|y_j\| = 1$ and

$$\lambda_0^2 - i\lambda_0(Ky_j, y_j) - \eta_0\lambda_0(By_j, y_j) - (Ay_j, y_j) = \xi_j, \qquad (4.1.7)$$

where $\lim_{j \to \infty} \xi_j = 0$. To shorten the notation we introduce $X = \operatorname{Re}\lambda_0$, $Y = \operatorname{Im}\lambda_0$, $k_j = (Ky_j, y_j)$, $b_j = (By_j, y_j)$, $a_j = (Ay_j, y_j)$. Taking the imaginary and the real parts of (4.1.7) we obtain

$$2XY - Xk_j - Y\eta_0 b_j = \gamma_j, \qquad (4.1.8)$$

$$X^2 - Y^2 + Yk_j - X\eta_0 b_j - a_j = \alpha_j \qquad (4.1.9)$$

with $\lim_{j \to \infty} \alpha_j = 0$ and $\lim_{j \to \infty} \gamma_j = 0$. Solving for X in (4.1.8) we obtain

$$X = \frac{Y\eta_0 b_j + \gamma_j}{2Y - k_j}$$

and thus

$$X - \eta_0 b_j = \frac{Y\eta_0 b_j + \gamma_j - 2Y\eta_0 b_j + k_j \eta_0 b_j}{2Y - k_j}$$

$$= \frac{(k_j - Y)\eta_0 b_j + \gamma_j}{2Y - k_j}.$$

Since (4.1.9) can be written as

$$X(X - \eta_0 b_j) - Y(Y - k_j) - a_j = \alpha_j,$$

substituting X and $X - \eta_0 b_j$ into this equation gives

$$\frac{(Y\eta_0 b_j + \gamma_j)[(k_j - Y)\eta_0 b_j + \gamma_j]}{(2Y - k_j)^2} - Y(Y - k_j) - a_j = \alpha_j.$$

This can be written as

$$\frac{-(Y - k_j)Y[(2Y - k_j)^2 + b_j^2 \eta_0^2]}{(2Y - k_j)^2} - a_j = \alpha_j - \frac{\eta_0 b_j k_j \gamma_j + \gamma_j^2}{(2Y - k_j)^2}. \tag{4.1.10}$$

We denote the left-hand side of (4.1.10) by c_j and the right-hand side by d_j. Hence

$$c_j \le -(Y - k_j)Y - a_j = -Y^2 + Yk_j - a_j.$$

Now assume additionally that $\operatorname{Im} \lambda_0 < -\beta^{1/2}$. The inequality $Y^2 > \beta > 0$ implies that there is $\delta > 0$ such that $Y^2 = (1 + \delta)\beta$. Because of $k_j > 0$ and $Y < 0$ we obtain the estimate

$$c_j \le -(1 + \delta)\beta + Yk_j - a_j \le -(1 + \delta)\beta - a_j.$$

On the other hand, in view of $Y < 0$, $k_j \ge \kappa > 0$ and Proposition 4.1.1, 2.(iii), there are positive constants a and b such that

$$d_j \ge \alpha_j - \frac{\eta_0 |b_j| k_j |\gamma_j|}{4|Y| k_j} - \frac{\gamma_j^2}{4Y^2}$$

$$= \alpha_j - \frac{\eta_0 |b_j| |\gamma_j|}{4|Y|} - \frac{\gamma_j^2}{4Y^2}$$

$$\ge \alpha_j - \frac{\eta_0 |\gamma_j|}{4|Y|}(a + ba_j) - \frac{\gamma_j^2}{4Y^2}.$$

Hence

$$0 = c_j - d_j \le -(1 + \delta)\beta - a_j - \alpha_j + \frac{b\eta_0 |\gamma_j|}{4|Y|} a_j + \frac{a\eta_0 |\gamma_j|}{4|Y|} + \frac{\gamma_j^2}{4Y^2}$$

$$\le -(1 + \delta)\beta - \left(1 - \frac{b\eta_0 |\gamma_j|}{4|Y|}\right) a_j - \alpha_j + \frac{a\eta_0 |\gamma_j|}{4|Y|} + \frac{\gamma_j^2}{4Y^2}.$$

Since $\gamma_j \to 0$ as $j \to \infty$, we have for all sufficiently large j that $b\eta_0 |\gamma_j| \le 4|Y|$, so that $-a_j \le \beta$ leads to

$$0 \le -(1 + \delta)\beta + \left(1 - \frac{b\eta_0 |\gamma_j|}{4|Y|}\right)\beta - \alpha_j + \frac{a\eta_0 |\gamma_j|}{4|Y|} + \frac{\gamma_j^2}{4Y^2}$$

$$\le -\delta\beta - \alpha_j + \frac{a\eta_0 |\gamma_j|}{4|Y|} + \frac{\gamma_j^2}{4Y^2}.$$

Letting $j \to \infty$ leads to the contradiction $0 \le -\delta\beta$. This means that there is no approximate spectrum of $L(\cdot, \eta_0)$ in the half-plane $\{\lambda \in \mathbb{C} : \operatorname{Im}\lambda < -\beta^{\frac{1}{2}}\}$.

Now let $0 > \operatorname{Im}\lambda_0 \ge -\beta^{1/2}$. Multiplying (4.1.8) by X and (4.1.9) by Y and taking the difference of the resulting equations we obtain

$$(X^2 + Y^2)(Y - k_j) + Y a_j = o(1). \tag{4.1.11}$$

Since $k_j \ge \kappa$, $-a_j \le \beta$ and $Y < 0$, it follows that

$$(X^2 + Y^2)(Y - k_j) + Y a_j \le -X^2 k_j + Y a_j \le -X^2 \kappa - Y\beta \le -X^2 \kappa + \beta^{3/2}. \tag{4.1.12}$$

Together with (4.1.11) this implies $|X| \le \beta^{3/4} \kappa^{-1/2}$.

Thus we have proved that for all $\eta \in [0, 1]$ the approximate spectrum of $L(\cdot, \eta)$ in the open lower half-plane is located in the domain

$$\Omega = \{\lambda \in \mathbb{C} : 0 > \operatorname{Im}\lambda \ge -\beta^{1/2}, \; |\operatorname{Re}\lambda| < \beta^{3/4} \kappa^{-1/2}\}.$$

Hence for each $(\lambda, \eta) \in (\mathbb{C}^- \setminus \Omega) \times [0, 1]$, $\lambda \notin \sigma_{\mathrm{app}}(L(\cdot, \eta))$, and therefore (4.1.6) holds for these (λ, η). Letting

$$S_0 = \{(\lambda, \eta) \in (\mathbb{C}^- \setminus \Omega) \times [0, 1] : \operatorname{codim} R(L(\lambda, \eta)) = 0\},$$
$$S_1 = \{(\lambda, \eta) \in (\mathbb{C}^- \setminus \Omega) \times [0, 1] : \operatorname{codim} R(L(\lambda, \eta)) \ne 0\},$$

it therefore follows from Lemma 4.1.3 that both S_0 and S_1 are open subsets of $(\mathbb{C}^- \setminus \Omega) \times [0, 1]$. Since $(\mathbb{C}^- \setminus \Omega) \times [0, 1]$ is connected, one of the sets S_0 or S_1 equals $(\mathbb{C}^- \setminus \Omega) \times [0, 1]$, whereas the other one is empty. Observe that $L(\cdot, 0)$ is a pencil of the form as considered in Chapter 3. Hence the spectrum of $L(\cdot, 0)$ in the open lower half-plane lies on the negative imaginary axis, see Lemma 3.1.9. Choosing $\lambda \in \mathbb{C}^- \setminus \Omega$ with $\operatorname{Re}\lambda \ne 0$ it follows that $(\lambda, 0) \in S_0$. Hence S_0 is nonempty, and from the above reasoning we obtain $(\mathbb{C}^- \setminus \Omega) \times [0, 1] = S_0$. Thus $R(L(\lambda, \eta)) = H$ for all $(\lambda, \eta) \in (\mathbb{C}^- \setminus \Omega) \times [0, 1]$. Since also (4.1.6) is satisfied for these (λ, η), it follows that $\mathbb{C}^- \setminus \Omega \subset \rho(L(\cdot, \eta))$ for all $\eta \in [0, 1]$. Thus we have shown that $\sigma(L(\cdot, \eta)) \cap \mathbb{C}^- \subset \Omega$ for all $\eta \in [0, 1]$. $\square$

With the notation in the proof of Lemma 4.1.5, an obvious modification of inequality (4.1.12) together with (4.1.11) leads to

$$0 \ge (X^2 + Y^2)(\kappa - Y) + Y\beta > (X^2 + Y^2)\kappa + Y\beta.$$

Hence we have

Remark 4.1.6. For $\eta \in [0, 1]$, the part of the spectrum of $L(\cdot, \eta)$ in the open lower half-plane is located in the disc $\{\lambda \in \mathbb{C} : 0 > [(\operatorname{Re}\lambda)^2 + (\operatorname{Im}\lambda)^2]\kappa + (\operatorname{Im}\lambda)\beta < 0\}$.

Lemma 4.1.7. *For each $\eta \in [0, 1]$ all points on the real axis, with the possible exception of 0, belong to $\rho(L(\cdot, \eta))$.*

Proof. The set of points $X + iY$ in the complex plane with $(X^2 + Y^2)\kappa + Y\beta < 0$ describes an open disc in the lower half-plane whose boundary touches the real axis at the origin. Hence it follows from Remark 4.1.6 and a reasoning as in the proof of Corollary 3.1.6 that $\sigma(L(\cdot, \eta)) \cap \mathbb{R} \subset \sigma_{\mathrm{app}}(L(\cdot, \eta))$. It is obvious that (4.1.11) also holds for $Y = 0$, so that $\lim_{j \to \infty} k_j X^2 = 0$, which is possible only if $X = 0$ due to $k_j \geq \kappa$. This means that $\sigma(L(\cdot, \eta)) \cap \mathbb{R} = \sigma_{\mathrm{app}}(L(\cdot, \eta)) \cap \mathbb{R} \subset \{0\}$. $\qquad\square$

Theorem 4.1.8. *Assume that the negative spectrum of A consists of at most finitely many eigenvalues of finite multiplicity. Then the total algebraic multiplicity of the spectrum of $L(\cdot, \eta)$ in the open lower half-plane does not depend on $\eta \in [0, 1]$ and equals the total multiplicity of the negative spectrum of A.*

Proof. According to Theorem 3.1.14 the part of the spectrum of $L(\cdot, 0)$ lying in the open lower half-plane consists of finitely many normal eigenvalues, and the total algebraic multiplicity of $L(\cdot, 0)$ in the open lower half-plane equals the total multiplicity of the negative eigenvalues of A. In view of Lemma 1.1.9, the fact that the spectrum in the open lower half-plane is located in the disc $\{\lambda \in \mathbb{C} : ((\mathrm{Re}\,\lambda)^2 + (\mathrm{Im}\,\lambda)^2)\kappa + Y\beta < 0\}$, see Remark 4.1.6, and a connectedness argument it is sufficient to show that this part of the spectrum is uniformly separated from the closed upper half-plane, that is, it remains to prove that there is no convergent sequence of pairs $(\lambda_s, \eta_s)_{s=1}^\infty$ with the following properties: $\eta_s \in [0, 1]$, $\mathrm{Im}\,\lambda_s < 0$, $\lim_{s \to \infty} \lambda_s = 0$, $\lambda_s \in \partial\sigma(L(\cdot, \eta_s))$.

By proof of contradiction, assume that such a sequence exists. In view of Corollary 3.1.6 it follows that $\lambda_s \in \sigma_{\mathrm{app}}(L(\cdot, \eta_s))$ for all $s \in \mathbb{N}$. Hence for each $s \in \mathbb{N}$ there exists a sequence $(y_{sn})_{n=1}^\infty$ of vectors $y_{sn} \in D(A)$ such that $\|y_{sn}\| = 1$ and $\lim_{n \to \infty} L(\lambda_s, \eta_s)y_{sn} = 0$. Let us choose for each s a vector $\psi_s = y_{sn_s}$ such that

$$\|L(\lambda_s, \eta_s)\psi_s\| < s^{-1}[\lambda_s]_1, \tag{4.1.13}$$

where $[\lambda_s]_1 = |\lambda_s|$ if $\mathrm{Re}\,\lambda_s = 0$ and $|\mathrm{Re}\,\lambda_s|$ otherwise. The Hilbert space H can be written as an orthogonal sum of two invariant subspaces D_1 and D_2 in such a way that if $f_1 \in D_1 \cap D(A)$, $f_2 \in D_2$ and $\|f_j\| = 1$, $j = 1, 2$, then $(Af_1, f_1) \geq 0$ and $-\beta \leq (Af_2, f_2) \leq -\gamma < 0$. We observe that D_2 is finite dimensional since it is assumed that the negative spectrum of A consists of at most finitely many eigenvalues of finite multiplicity.

The vectors ψ_s can be written in the form $\psi_s = \zeta_s \Phi_s + \theta_s \Psi_s$, where $\Phi_s \in D_2$, $\Psi_s \in D_1$, $\|\Psi_s\| = \|\Phi_s\| = 1$, $\zeta_s \geq 0$, $\theta_s \geq 0$, $\zeta_s^2 + \theta_s^2 = 1$. Write $X_s = \mathrm{Re}\,\lambda_s$ and $Y_s = \mathrm{Im}\,\lambda_s$. Below we will make use of the identities

$$\begin{aligned}
(L(\lambda_s, \eta_s)\psi_s, \Psi_s) &= \lambda_s^2\theta_s - i\lambda_s\zeta_s(K\Phi_s, \Psi_s) - i\lambda_s\theta_s(K\Psi_s, \Psi_s) \\
&\quad - \eta_s\lambda_s\zeta_s(B\Phi_s, \Psi_s) - \eta_s\lambda_s\theta_s(B\Psi_s, \Psi_s) - \theta_s(A\Psi_s, \Psi_s) \\
&= \theta_s\left[\lambda_s^2 - i\lambda_s(K\Psi_s, \Psi_s) - \eta_s\lambda_s(B\Psi_s, \Psi_s) - (A\Psi_s, \Psi_s)\right] \\
&\quad - \zeta_s\left[i\lambda_s(K\Phi_s, \Psi_s) + \eta_s\lambda_s(B\Phi_s, \Psi_s)\right]
\end{aligned} \tag{4.1.14}$$

and

$$\begin{aligned}
\mathrm{Re}(L(\lambda_s, \eta_s)\psi_s, \Phi_s) &= \mathrm{Re}[\lambda_s^2 \zeta_s - i\lambda_s(K\psi_s, \Phi_s) - \eta_s\lambda_s(B\psi_s, \Phi_s) - \zeta_s(A\Phi_s, \Phi_s)] \\
&= (X_s^2 - Y_s^2 - (A\Phi_s, \Phi_s))\zeta_s - \mathrm{Re}[i\lambda_s(K\psi_s, \Phi_s) + \eta_s\lambda_s(B\psi_s, \Phi_s)].
\end{aligned}$$

(4.1.15)

Inequality (4.1.13) implies

$$\mathrm{Re}(L(\lambda_s, \eta_s)\psi_s, \Phi_s) = o(1).$$

(4.1.16)

Since the operators B and K are bounded on the finite-dimensional space D_2, we conclude that

$$(L(\lambda_s, \eta_s)\Phi_2, \Psi_s) = -i\lambda_s(K\Phi_s, \Psi_s) - \eta_s\lambda_s(B\Phi_s, \Psi_s) = o(\lambda_s)$$

(4.1.17)

We consider two cases.

1. Assume there exists a positive number C such that $\|B\Psi_s\| < C$ and $\|K\Psi_s\| < C$ for all $s \in \mathbb{N}$. From (4.1.15), $X_s = o(1)$, $Y_s = o(1)$, $(A\Phi_s, \Phi_s) \le -\gamma$, $\zeta_s^2 + \theta_s^2 = 1$, $\theta_s \ge 0$, and (4.1.16) we deduce that

$$\lim_{s\to\infty} \zeta_s = 0, \quad \lim_{s\to\infty} \theta_s = 1.$$

(4.1.18)

In view of (4.1.13), (4.1.17) and (4.1.18), we obtain

$$\begin{aligned}
\lambda_s^2 - i\lambda_s(K\Psi_s, \Psi_s) &- \eta_s\lambda_s(B\Psi_s, \Psi_s) - (A\Psi_s, \Psi_s) \\
&= (L(\lambda_s, \eta_s)\Psi_s, \Psi_s) = \frac{1}{\theta_s}[(L(\lambda_s, \eta_s)\psi_s, \Psi_s) - \zeta_s(L(\lambda_s, \eta_s)\Phi_s, \Psi_s)] \\
&= o(\lambda_s).
\end{aligned}$$

We may assume, by choosing a suitable subsequence, if necessary, that either all X_s are zero or all X_s are different from zero. If all X_s are zero, then the real part of the previous identity gives

$$-Y_s^2 + Y_s(K\Psi_s, \Psi_s) - (A\Psi_s, \Psi_s) = o(Y_s).$$

Since all terms on the left-hand side are nonpositive, we arrive at

$$(K\Psi_s, \Psi_s) = o(1),$$

which contradicts $(K\Psi_s, \Psi_s) \ge \kappa$.

Now assume that $X_s \ne 0$ for all s. Taking the imaginary part in (4.1.14) and observing (4.1.13) and (4.1.18) it follows that

$$\begin{aligned}
\theta_s[-X_s(K\Psi_s, \Psi_s) &- \eta_s Y_s(B\Psi_s, \Psi_s)] + \zeta_s Y_s[\mathrm{Im}(K\Phi_s, \Psi_s) - \eta_s\,\mathrm{Re}(B\Phi_s, \Psi_s)] \\
&= \mathrm{Im}(L(\lambda_s, \eta_s)\psi_s, \Psi_s) - 2X_s Y_s + \zeta_s X_s[\mathrm{Re}(K\Phi_s, \Psi_s) + \eta_s\,\mathrm{Im}(B\Phi_s, \Psi_s)] \\
&= o(X_s)
\end{aligned}$$

and therefore, in view of $(K\Psi_s, \Psi_s) \geq \kappa > 0$,

$$X_s = -\frac{Y_s(\eta_s(B\Psi_s, \Psi_s) + o(1))}{(K\Psi_s, \Psi_s)} + o(X_s). \tag{4.1.19}$$

In particular, $X_s = O(Y_s)$ since $\{(B\Psi_s, \Psi_s) : s \in \mathbb{N}\}$ is bounded by assumption.
Taking the real part in (4.1.14) leads to

$$\theta_s[Y_s(K\Psi_s, \Psi_s) - \eta_s X_s(B\Psi_s, \Psi_s) - (A\Psi_s, \Psi_s)] = o(X_s) + o(Y_s). \tag{4.1.20}$$

Using (4.1.18) and substituting (4.1.19) into (4.1.20) we obtain

$$Y_s(K\Psi_s, \Psi_s) - (A\Psi_s, \Psi_s) + \eta_s^2 Y_s \frac{(B\Psi_s, \Psi_s)^2}{(K\Psi_s, \Psi_s)} = o(Y_s). \tag{4.1.21}$$

Since all summands on the left-hand side of (4.1.21) are nonpositive, it follows
that $(K\Psi_s, \Psi_s) = o(1)$, which contradicts $K \gg 0$.

2. If there exists no positive number C such that $\|B\Psi_s\| < C$ and $\|K\Psi_s\| < C$ for
all $s \in \mathbb{N}$, we may assume, by choosing a subsequence, if necessary, that

$$\|B\Psi_s\| + \|K\Psi_s\| \to \infty \text{ as } s \to \infty. \tag{4.1.22}$$

From (4.1.14) and (4.1.13) it follows that

$$\theta_s(A\Psi_s, \Psi_s) = \theta_s \left[\lambda_s^2 - i\lambda_s(K\Psi_s, \Psi_s) - \eta_s\lambda_s(B\Psi_s, \Psi_s)\right]$$
$$- \zeta_s \left[i\lambda_s(K\Phi_s, \Psi_s) + \eta_s\lambda_s(B\Phi_s, \Psi_s)\right] + o(1).$$

Hence there is a positive constant C_1 such that

$$\theta_s(A\Psi_s, \Psi_s) \leq C_1(1 + \|B\Psi_s\| + \|K\Psi_s\|).$$

In view of Proposition 4.1.1, part 1, there is a positive constant p such that

$$p(\|B\Psi_s\|^2 + \|K\Psi_s\|^2) \leq 1 + (A\Psi_s, \Psi_s),$$

where we have used that $(A\Psi_s, \Psi_s) \geq 0$. The last two inequalities lead to

$$\theta_s[p(\|B\Psi_s\|^2 + \|K\Psi_s\|^2) - 1] \leq C_1(1 + \|B\Psi_s\| + \|K\Psi_s\|),$$

and thus

$$\theta_s \leq \frac{C_1(1 + \|B\Psi_s\| + \|K\Psi_s\|\|)}{p(\|B\Psi_s\|^2 + \|K\Psi_s\|^2) - 1}.$$

In view of (4.1.22) and $\theta_s^2 + \zeta_s^2 = 1$, this gives

$$\theta_s = o(1), \quad \zeta_s = 1 + o(1). \tag{4.1.23}$$

From (4.1.16) and (4.1.15) we obtain

$$\zeta_s(A\Phi_s, \Phi_s) = (X_s^2 - Y_s^2)\zeta_s - \text{Re}[i\lambda_s(\psi_s, K\Phi_s) + \eta_s\lambda_s(\psi_s, B\Phi_s)] + o(1)$$
$$= o(1)$$

since K and B are bounded on the finite-dimensional space D_2. But this is a
contradiction since $\zeta_s = 1 + o(1)$ and $(A\Phi_s, \Phi_s) \leq -\gamma$. $\square$

4.2 Linearized pencils in Pontryagin spaces

Theorem 4.1.8 is related to indefinite inner product spaces. We consider the operator polynomial L from Section 4.1 satisfying Condition IV and, for simplicity, $0 \in \rho(A)$. The operator pencil L is linearized by introducing $z = \lambda y$. Then the equation $L(\lambda)y = 0$ can be rewritten as the system of equations

$$\lambda z - iKz - Bz - Ay = 0,$$
$$\lambda y - z = 0,$$

which has the operator matrix representation

$$(\lambda I - S)X = 0,$$

where $X = (z, y)^\mathsf{T}$, $z, y \in D(A)$, and

$$S = \begin{pmatrix} iK + B & A \\ I & 0 \end{pmatrix}. \tag{4.2.1}$$

We now use the spectral resolution of A to introduce an indefinite inner product on $D(|A|^{\frac{1}{2}})$. Let H_+ and H_- be the (invariant) spectral subspaces of H associated with the positive and negative spectrum of A, respectively. Then $H = H_+ \oplus H_-$, and $A_+ := A|_{H_+}$ and $A_- := A|_{H_-}$ are self-adjoint operators on H_+ and H_-, respectively, satisfying $A_+ \gg 0$ and $-A_- \gg 0$. Since $0 \in \rho(A)$ and since the negative spectrum of A consists of finitely many eigenvalues of finite multiplicity, H_- is finite dimensional and $|A|^{\frac{1}{2}}$ is bounded and invertible on H_-, whereas $|A|^{\frac{1}{2}}$ is invertible on H_+. Hence, for $x = x_+ + x_- \in D(|A|^{\frac{1}{2}})$ with $x_+ \in H_+$ and $x_- \in H_-$, and $y = y_+ + y_- \in D(|A|^{\frac{1}{2}})$,

$$\langle x, y \rangle := (|A|^{\frac{1}{2}}x_+, |A|^{\frac{1}{2}}y_+) - (|A|^{\frac{1}{2}}x_-, |A|^{\frac{1}{2}}y_-) \tag{4.2.2}$$

defines an inner product on $D(|A|^{\frac{1}{2}})$ which makes it a Pontryagin space. More precisely, $(D(|A|^{\frac{1}{2}}), \langle \cdot, \cdot \rangle)$ is a Π_κ-space, where κ, the dimension of H_-, equals the number of negative eigenvalues of A, counted with multiplicity. Note that $D(A_-) = H_-$ and that A_- is bounded. For $x \in D(A)$ we therefore have $x_+ \in D(A)$ and $x_- \in D(A)$, and thus, for $x \in D(A)$ and $y \in D(|A|^{\frac{1}{2}})$,

$$(|A|^{\frac{1}{2}}x_\pm, |A|^{\frac{1}{2}}y_\pm) = (|A|x_\pm, y_\pm) = \pm(Ax_\pm, y_\pm),$$

so that

$$\langle x, y \rangle = (Ax_+, y_+) + (Ax_-, y_-) = (Ax, y).$$

Therefore $H \oplus D(|A|^{\frac{1}{2}})$ equipped with the inner product

$$\langle X_1, X_2 \rangle = (z_1, z_2) + \langle y_1, y_2 \rangle, \quad X_1 = (z_1, y_1)^\mathsf{T}, \ X_2 = (z_2, y_2)^\mathsf{T} \in H \oplus D(|A|^{\frac{1}{2}}), \tag{4.2.3}$$

is a Π_κ-space. Note that $D(|A|^{\frac{1}{2}}) = D((A + \beta_1 I)^{\frac{1}{2}})$ so that we will use whichever appears to be more convenient.

Together with the operator S with domain $D(|A|^{\frac{1}{2}}) \times D(A)$ we consider the operator S_- with $D(S_-) = D(|A|^{\frac{1}{2}}) \times D(A)$ in the Pontryagin space $H \oplus D(|A|^{\frac{1}{2}})$ given by

$$S_- = \begin{pmatrix} -iK + B & A \\ I & 0 \end{pmatrix}. \tag{4.2.4}$$

Here we observe that $D((A + \beta_1 I)^{\frac{1}{2}})$ is a subset of $D(K)$ and $D(B)$, see the proof Lemma 3.1.1.

Proposition 4.2.1. *The operators S and S_- are closed in the Pontryagin space $H \oplus D(|A|^{\frac{1}{2}})$, and $S_- = S^*$.*

Proof. Let $X_1, X_2 \in D(|A|^{\frac{1}{2}}) \times D(A)$. Then

$$\begin{aligned} \langle SX_1, X_2 \rangle &= ((iK + B)z_1, z_2) + (Ay_1, z_2) + \langle z_1, y_2 \rangle \\ &= (z_1, (-iK + B)z_2) + \langle y_1, z_2 \rangle + (z_1, Ay_2) \\ &= \langle X_1, S_- X_2 \rangle \end{aligned}$$

shows that $S_- \subset S^*$. Now let $X_2 \in D(S^*)$. Then there is $X_3 \in H \oplus D(|A|^{\frac{1}{2}})$ such that $\langle SX_1, X_2 \rangle = \langle X_1, X_3 \rangle$ for all $X_1 \in D(S)$. Taking first $X_1 \in \{0\} \oplus D(A)$, i.e., $z_1 = 0$, it follows that

$$(Ay_1, z_2) = \langle SX_1, X_2 \rangle = \langle X_1, X_3 \rangle = \langle y_1, y_3 \rangle = (Ay_1, y_3).$$

Observing that $\{Ay_1 : y_1 \in D(A)\} = H$ since $0 \in \rho(A)$, we get $z_2 = y_3 \in D(|A|^{\frac{1}{2}})$. Now let $X_1 \in D(A) \times \{0\}$. Then it follows that

$$\begin{aligned} (Az_1, A^{-1}(z_3 + (iK - B)z_2)) &= (z_1, z_3 + (iK - B)z_2) \\ &= (z_1, z_3) - ((iK + B)z_1, z_2) \\ &= \langle X_1, X_3 \rangle - \langle SX_1, X_2 \rangle + \langle z_1, y_2 \rangle \\ &= (Az_1, y_2). \end{aligned}$$

Using again that A is invertible, we obtain $y_2 = A^{-1}(z_3 + (iK - B)z_2) \in D(A)$. We have thus shown that $D(S^*) \subset D(|A|^{\frac{1}{2}}) \times D(A) = D(S_-)$, and $S^* = S_-$ follows. Since we did not use $K \gg 0$ in this proof, we may replace K with $-K$, i.e., interchange S and S_-, in the above result. Then we obtain $S = S_-^* = S^{**}$, which means that S is closed. $\square$

Proposition 4.2.2. *The operator S is maximal dissipative, and the number $n_S(\mathbb{C}^-)$ of eigenvalues of the operator S, counted with multiplicity, in the open lower half-plane, does not exceed $\kappa = n_A(\mathbb{C}^-)$.*

Proof. For $X = (z, y)^\mathsf{T} \in D(S)$ we have

$$\operatorname{Im}\langle SX, X \rangle = \frac{1}{2i}(\langle SX, X \rangle - \langle X, SX \rangle) = \frac{1}{2i}(\langle SX, X \rangle - \langle S_- X, X \rangle) = (Kz, z) \geq 0.$$

It also follows immediately that

$$\operatorname{Im}\langle -S^* X, X \rangle = \operatorname{Im}\langle -S_- X, X \rangle = \operatorname{Im}\langle SX, X \rangle \geq 0$$

for all $X \in D(S^*) = D(S)$. Hence S is maximal dissipative by [18, Chapter 2, 2.7]. An application of [18, Chapter 2, Corollary 2.23] completes the proof, where we have to observe that the convention used in [18] is that in Pontryagin spaces the positive subspace is finite dimensional. Hence we have to replace the inner product and the operator S by their negatives, so that the upper half-plane $\mathbb{C}^+$ in [18, Chapter 2, Corollary 2.23] becomes the lower half-plane $\mathbb{C}^-$ in our settings. □

Remark 4.2.3. Proposition 4.2.2 states that the number $n_S(\mathbb{C}^-)$ does not exceed κ, whereas Theorem 4.1.8 states that this total multiplicity $n_S(\mathbb{C}^-) = m(\mathbb{C}^-)$ equals κ.

4.3 Gyroscopically stabilized operator pencils

4.3.1 General results

In Theorem 4.1.8 we saw that the condition $K \gg 0$ guarantees that the total algebraic multiplicity of the spectrum located in the lower half-plane is independent of the gyroscopic operator B.

Here we consider the case $K = 0$. This enables the so-called gyroscopic stabilization. Namely, the spectrum of the pencil

$$L(\lambda) = \lambda^2 I - \lambda B - A$$

may lie on the real axis and be simple while $\sigma(A) \cap (-\infty, 0) \neq \emptyset$.

We henceforth require in this section that the operators A and B in the pencil L satisfy

Condition V. *The operators A and B are self-adjoint operators on H with the following properties:*

(i) *$A \geq -\beta I$ for some positive number β;*

(ii) *$(A + \beta_1 I)^{-1} \in S_\infty$ for some $\beta_1 > \beta$;*

(iii) *the operator B is subordinate to A, i.e., $D(A) \subset D(B)$ and the operator B is bounded with respect to the operator $(A + \beta_1 I)^{1/2}$.*

Together with the pencil L we consider the family of pencils given by

$$L(\lambda, \eta) = \lambda^2 I - \lambda \eta B - A \tag{4.3.1}$$

with $D(L(\lambda, \eta)) = D(A)$, where $\eta \in [0,1]$ is the parameter of this family and λ is the spectral parameter.

Lemma 4.3.1. *For $\eta \in [0,1]$, the nonreal eigenvalues λ of $L(\cdot, \eta)$ satisfy $|\lambda| \le \beta^{\frac{1}{2}}$.*

Proof. If λ is a nonreal eigenvalue of $L(\cdot, \eta)$ with normalized eigenvector y, then $L(\lambda, \eta)y = 0$. Taking real and imaginary parts of $(L(\lambda, \eta)y, y) = 0$ leads to

$$(\mathrm{Re}\,\lambda)^2 - (\mathrm{Im}\,\lambda)^2 - (\mathrm{Re}\,\lambda)\eta(By, y) - (Ay, y) = 0,$$
$$2(\mathrm{Re}\,\lambda)(\mathrm{Im}\,\lambda) - (\mathrm{Im}\,\lambda)\eta(By, y) = 0.$$

Since we assume that $\mathrm{Im}\,\lambda \ne 0$, the second equation gives $\eta(By, y) = 2\,\mathrm{Re}\,\lambda$. Substituting this identity into the first equation leads to

$$|\lambda|^2 = (\mathrm{Re}\,\lambda)^2 + (\mathrm{Im}\,\lambda)^2 = -(Ay, y) \le \beta. \qquad \square$$

Remark 4.3.2. Since $(A + \beta_1 I)^{-1}$ is compact by Condition V, part (ii), the spectral theorem for self-adjoint operators gives that also the operator $(A + \beta_1)^{-\frac{1}{2}}$ is compact. In view of Condition V, part (iii), it follows that B is $(A + \beta_1 I)^{-1}$ compact. Therefore $L(\lambda, \eta)$ is a Fredholm operator for all $\lambda \in \mathbb{C}$ and $\eta \in [0,1]$, see [137, Theorem IV.5.26]. In view of Lemma 4.3.1 it follows that the spectrum of $L(\cdot, \eta)$ consists of normal eigenvalues.

Proposition 4.3.3. *If $\lambda = 0$ is a semisimple eigenvalue of the pencil $L(\cdot, \eta_0)$ for some $\eta_0 \in (0,1]$, then $\lambda = 0$ is a semisimple eigenvalue of the pencil $L(\cdot, \eta)$ for all $\eta \in (0,1]$, and the multiplicity of the eigenvalue $\lambda = 0$ of the pencil $L(\cdot, \eta)$ is independent of $\eta \in (0,1]$.*

Proof. It is clear that y_0 is an eigenvector of $L(\cdot, \eta)$ corresponding to the eigenvalue $\lambda = 0$ if and only if y_0 is a nonzero vector in $N(A)$. Assume there is $\eta \ne 0$ such that the eigenvector y_0 corresponding to the eigenvalue $\lambda = 0$ of $L(\cdot, \eta)$ has an associated vector y_1. Then $Ay_1 + \eta By_0 = 0$, and $\frac{\eta_0}{\eta} y_1$ would be a corresponding associated vector for the eigenvalue $\lambda = 0$ of $L(\cdot, \eta_0)$ with eigenvector y_0, which contradicts the assumption that $\lambda = 0$ is a semi-simple eigenvalue of $L(\cdot, \eta_0)$. $\qquad \square$

The following theorem and an outline of its proof can be found in [284].

Theorem 4.3.4. *Let $\eta \in [0,1]$. Then the total algebraic multiplicity of the spectrum of $L(\cdot, \eta)$ in the open lower (or, what is the same, upper) half-plane does not exceed the total geometric (or, what is the same, algebraic) multiplicity of the negative spectrum of the operator A.*

Proof. In view of $L(\lambda, \eta)^* = L(\overline{\lambda}, \eta)$ for all $\lambda \in \mathbb{C}$ and $\eta \in [0,1]$, it suffices to prove the statement about the spectrum in the lower half-plane. We have noted in Remark 4.3.2 that the spectrum of the operator pencil $L(\cdot, \eta)$ consists of normal eigenvalues. Also, the spectrum of A consists of normal eigenvalues.

We first consider the case that $0 \in \rho(A)$. In the notation of Section 4.2 we have

$$S = \begin{pmatrix} \eta B & A \\ I & 0 \end{pmatrix},$$

and $H \oplus D(|A|^{\frac{1}{2}})$ is a Π_κ-space with respect to the innner product $\langle \cdot, \cdot \rangle$ defined in (4.2.3), where κ is the number of negative eigenvalues of A, counted with multiplicity. Since for S_- defined in (4.2.4) we have $S_- = S$ in the present case, Proposition 4.2.1 shows that S is self-adjoint. Then the statement of our theorem follows from well-known results of L.S. Pontryagin [233], see also [121] or [18, Chapter 2, Corollary 3.15].

If $0 \in \sigma(A)$, then for small positive real numbers τ, $0 \in \rho(A + \tau I)$ and $A + \tau I$ has the same number of negative eigenvalues, counted with multiplicity, as A. Hence, the statement of this theorem holds for $L(\cdot, \mu)$ replaced with $\mathcal{L}(\cdot, \eta, \tau) = L(\cdot, \eta) - \tau I$, i.e., with A replaced by $A + \tau I$ for these small positive τ. Theorem 1.2.7 shows that the eigenvalues depend continuously on τ near $\tau = 0$, and this property is uniform in τ for all eigenvalues in the open lower half-plane since by Lemma 4.3.1 there are only finitely many eigenvalues in the open lower half-plane. Hence the number of eigenvalues of $L(\cdot, \eta)$ in the open lower half-plane, counted with multiplicity, does not exceed that of $\mathcal{L}(\cdot, \eta, \tau)$ for sufficiently small positive τ. But since $0 \in \rho(A + \tau I)$, we already know that this latter number does not exceed the total multiplicity of the negative spectrum of A. Thus we have shown that the total multiplicity of the spectrum of $L(\cdot, \eta)$ in the open lower half-plane does not exceed the total multiplicity of the negative spectrum of the operator A. $\qquad \square$

For the remainder of this section we write $B = iG$ and assume in addition to Condition V that the following holds.

Condition VI.

(i) *The Hilbert space H is a complexification of a real Hilbert space.*
(ii) *The operators A and G are real, i.e., Ax and Gx are real whenever $x \in D(A)$ or $x \in D(G)$, respectively, are real elements of the Hilbert space H.*

Remark 4.3.5. 1. The operator $G = -iB$ is a skew-symmetric (or antisymmetric) operator, i.e., $(Gy_1, y_2) = -(y_1, Gy_2)$ for all $y_1, y_2 \in D(A)$.

2. For all $\eta \in [0, 1]$, the spectrum of $L(\cdot, \eta)$ is symmetric with respect to the real and imaginary axes, with multiplicity. Indeed, by [189, Corollary 1.5.5] the statement on symmetry with respect to the real axis is obvious since $L(\lambda)^* = L(\bar{\lambda})$ due to the self-adjointness of A and B. Now let λ_0 be an eigenvalue of $L(\cdot, \eta)$ and $(x_0, \ldots, x_{m-1})$ be a corresponding chain of an eigenvector and associated vectors. Putting

$$x(\lambda) = \sum_{j=0}^{m-1} (\lambda - \lambda_0)^j x_j$$

it follows that

$$\overline{L(\lambda,\eta)x(\lambda)} = \left[\overline{(-\overline{\lambda})^2 I} - \overline{(-\overline{\lambda})iG} - A\right]\overline{x(\lambda)} = L(-\overline{\lambda},\eta)\overline{x(\lambda)}.$$

With $\mu = -\overline{\lambda}$ we can write

$$\overline{x(\lambda)} = \sum_{j=0}^{m-1}\overline{(-\overline{\mu}-\lambda_0)^j}\overline{x_j} = \sum_{j=0}^{m-1}(\mu+\overline{\lambda_0})^j(-1)^j\overline{x_j}.$$

Hence it follows from Remark 1.1.4 that $(\overline{x_0},\ldots,(-1)^{m-1}\overline{x_{m-1}})$ is a chain of an eigenvector and associated vectors of $L(\cdot,\eta)$ at $-\overline{\lambda_0}$. Therefore, the algebraic multiplicities of the eigenvalues λ_0 and $-\overline{\lambda_0}$ of $L(\cdot,\eta)$ are equal in view of Definition 1.1.3, part 2.

Theorem 4.3.6. *Assume that* $0 \in \rho(A)$. *Then the total algebraic (geometric) multiplicity of the spectrum of the pencil* $L(\cdot,0)$ *located in the open lower half-plane is odd [even] if and only if the total algebraic multiplicity of the spectrum of the pencil* $L(\cdot,\eta)$ *located in the open lower half-plane is odd [even] for all* $\eta \in (0,1]$.

Proof. Similar to Theorem 1.2.7 we conclude from Theorem 9.2.4, that the eigenvalues of the operator pencil $L(\cdot,\eta)$ are continuous and piecewise analytic functions of the parameter η, and the spectrum of the pencil $L(\cdot,\eta)$ is symmetric with respect to the real axis by Remark 4.3.5. The eigenvalues can leave or enter the upper half-plane only in pairs, taking multiplicities into account, if we observe that the assumption $0 \in \rho(A)$ implies that $0 \in \rho(L(\cdot,\eta))$ since $L(0,\eta) = -A$. Finally, since nonreal eigenvalues are bounded with bound independent of η by Lemma 4.3.1, nonreal eigenvalues cannot tend to infinity.   □

4.3.2 Vibrations of an elastic fluid conveying pipe

To solve the stability problem for small vibrations of a linear pipe conveying a stationary flow of an incompressible fluid, we must investigate the pipe frequencies or, equivalently, the location of the spectrum of the corresponding quadratic operator pencil. Small transversal vibrations of a horizontal elastic pipe conveying a stationary flow of an incompressible fluid are described in dimensionless variables by the equation, see [79],

$$\frac{\partial^4 u}{\partial x^4} + v^2\frac{\partial^2 u}{\partial x^2} + 2\eta v\frac{\partial^2 u}{\partial x \partial t} + \frac{\partial^2 u}{\partial t^2} = 0, \tag{4.3.2}$$

where $u(x,t)$ is the transversal displacement, t is the dimensionless time, $x = \frac{s}{l}$, s is the longitudinal coordinate, l is the pipe length, $v = \frac{m^{1/2}Ul}{(EI)^{1/2}}$, $\eta = (m/(m+m_r))^{1/2}$, m and m_r are linear densities of the fluid and the pipe, respectively, U is the fluid velocity, and EI is the bending stiffness of the pipe section. The boundary conditions

$$u(0, t) = \frac{\partial^2 u}{\partial x^2}\bigg|_{x=0} = u(1, t) = \frac{\partial^2 u}{\partial x^2}\bigg|_{x=1} = 0 \qquad (4.3.3)$$

represent hinge connection of the ends of the pipe.

We are going to give a definition of stability of such a problem. To this end we impose the initial conditions

$$u(x, 0) = u_0(x), \quad x \in [0, 1], \qquad (4.3.4)$$

$$\frac{\partial u(x, t)}{\partial t}\bigg|_{t=0} = u_1(x), \quad x \in [0, 1]. \qquad (4.3.5)$$

It follows from [272] that the initial-boundary value problem (4.3.2)–(4.3.5) possesses a unique solution for any $u_0 \in W_2^4(0, 1)$ with $u_0(0) = u_0''(0) = u_0(1) = u_0''(1) = 0$ and any $u_1 \in W_2^2(0, 1)$ with $u_0(0) = u_0(1) = 0$.

Definition 4.3.7. The initial-boundary value problem (4.3.2)–(4.3.5) is stable if there is a constant C such that for each $u_0 \in W_2^4(0, 1)$ with $u_0(0) = u_0''(0) = u_0(1) = u_0''(1) = 0$ and each $u_1 \in W_2^2(0, 1)$ with $u_1(0) = u_1(1) = 0$, the solution u of (4.3.2)–(4.3.5) satisfies

$$\|u\|_0(t) \le C\|u\|_0(0) \qquad (4.3.6)$$

for all $t \ge 0$, where

$$\|u\|_0^2(t) = \int_0^1 \left|\frac{\partial u}{\partial t}(x, t)\right|^2 dx + \int_0^1 \left|\frac{\partial^2 u}{\partial x^2}(x, t)\right|^2 dx.$$

Let us substitute $u(x, t) = y(x)e^{i\lambda t}$ into (4.3.2) and (4.3.3). Then we obtain

$$y^{(4)} + v^2 y'' + 2i\lambda \eta v y' - \lambda^2 y = 0, \qquad (4.3.7)$$

$$y(0) = y''(0) = y(1) = y''(0) = 0. \qquad (4.3.8)$$

We introduce the operators A_+, A_-, $A = A_+ + A_-$ and G acting in $L_2(0, 1)$ by setting

$$D(A_+) = D(A_-) = \{y \in W_2^4(0, 1) : y(0) = y(1) = y''(0) = y''(1) = 0\},$$

$$A_+ y = y^{(4)}, \quad A_- y = v^2 y'',$$

$$D(G) = \{y \in W_2^1(0, 1) : y(0) - y(1) = 0\}$$

$$Gy = 2vy'.$$

Proposition 4.3.8. The operators A_+ and A are self-adjoint, $A_+ \gg 0$, $A \ge -\frac{v^4}{4}I$, and $A_+^{-1}, (A + \beta_1 I)^{-1} \in S_\infty$ for $\beta_1 > \frac{v^4}{4}$. The operator iG is self-adjoint and subordinate to A.

Proof. We will use Theorem 10.3.5 to verify that A and A_+ are self-adjoint. Observe that $y^{[j]} = y^{(j)}$ for $j = 0, 1, 2$ according to Definition 10.2.1. We have to find U_1, U_3 and U defined in (10.3.3), (10.3.12) and (10.3.13), where in the present situation U_2 and V are zero dimensional, so that $U = U_1$. First, it is straightforward to see that

$$U_1 = \begin{pmatrix} 1 & 0 & 0 & 0 & 0 & 0 & 0 & 0 \\ 0 & 0 & 1 & 0 & 0 & 0 & 0 & 0 \\ 0 & 0 & 0 & 0 & 1 & 0 & 0 & 0 \\ 0 & 0 & 0 & 0 & 0 & 0 & 1 & 0 \end{pmatrix}.$$

In particular, U_1 has rank 4. Clearly,

$$N(U_1) = \operatorname{span}\{e_2, e_4, e_6, e_8\} \subset \mathbb{C}^8.$$

We recall that

$$U_3 = J_2 \quad \text{where} \quad J_2 = \begin{pmatrix} -J_{2,1} & 0 \\ 0 & J_{2,1} \end{pmatrix} \quad \text{and} \quad J_{2,1} = \begin{pmatrix} 0 & 0 & 0 & 1 \\ 0 & 0 & -1 & 0 \\ 0 & 1 & 0 & 0 \\ -1 & 0 & 0 & 0 \end{pmatrix}.$$

It is now straightforward to verify that

$$U_3(N(U_1)) = \operatorname{span}\{e_3, e_1, e_7, e_5\} \subset \mathbb{C}^8,$$
$$R(U^*) = R(U_1^*) = \operatorname{span}\{e_1, e_3, e_5, e_7\} \subset \mathbb{C}^8.$$

This shows that $U_3(N(U_1)) = R(U^*)$. Hence A_+ and A are self-adjoint by Theorem 10.3.5. Then we conclude with the aid of Theorem 10.3.8 that A_+ and A have compact resolvents and that A is bounded below.

Integration by parts shows that

$$v^{-4}\|A_-y\|^2 = (A_+y, y) \le \|A_+y\|\|y\|, \quad y \in D(A). \tag{4.3.9}$$

It follows immediately that $A_+ \ge 0$. Furthermore, it is easy to see that 0 is not an eigenvalue of A_+. Since A_+ has a compact resolvent, its spectrum is discrete, and altogether we have shown that $A_+ \gg 0$. From (4.3.9) we obtain for all $y \in D(A)$ that

$$(A_-y, y) \ge -\|A_-y\|\|y\| = -v^2(A_+y, y)^{\frac{1}{2}}\|y\| \ge -\left[(A_+y, y) + \frac{v^4}{4}\|y\|^2\right], \tag{4.3.10}$$

which leads to

$$(Ay, y) = (A_-y, y) + A_+(y, y) \ge -\frac{v^4}{4}(y, y).$$

For $y, z \in D(G)$ integration by parts leads to

$$(Gy, z) = 2v \int_0^1 y'(x)\overline{z(x)}\, dx$$

$$= 2v \left(y(1)\overline{z(1)} - y(0)\overline{z(0)} \right) - 2v \int_0^1 y(x)\overline{z'(x)}\, dx = -(y, Gz),$$

where we have used that $y(1) = y(0)$ and $z(1) = z(0)$. Hence iG is symmetric. Now let $z \in D(G^*)$ and $w = G^* z$. Then

$$(y', 2vz) = (Gy, z) = (y, G^*z) = (y, w)$$

for all $y \in D(G)$. In particular, this holds for all $y \in C_0^\infty(0, 1)$, and it follows that $w = -2vz'$ in the sense of distributions. Definition 10.1.1 therefore shows that $z \in W_2^1(0, 1)$. Taking now $y = 1 \in D(G)$, we calculate

$$0 = (2vy', z) = (Gy, z) = (y, w) = -2v(y, z') = -2v \int_0^1 \overline{z'(x)}\, dx$$

$$= -2v \left(\overline{z(1)} - \overline{z(0)} \right).$$

We conclude that $z(0) - z(1) = 0$, which means that $z \in D(G)$. We have thus shown that $D(G^*) \subset D(G)$, and the self-adjointness of iG follows in view of the symmetry of iG.

For $y \in D(A)$, integration by parts shows that

$$\|Gy\|^2 = -4(A_-y, y). \tag{4.3.11}$$

We observe that a slight modification of (4.3.10) gives

$$-(A_-y, y) \le \frac{1}{2}(A_+y, y) + \frac{v^4}{2}\|y\|^2$$

and thus

$$-\frac{1}{2}(A_-y, y) \le \frac{1}{2}(Ay, y) + \frac{v^4}{2}\|y\|^2.$$

This together with (4.3.11) leads to

$$\|Gy\|^2 \le 4(Ay, y) + 4v^4\|y\|^2$$

for all $y \in D(A)$. With the aid of Proposition 4.1.1, part 1, we therefore conclude that G is subordinate to A. $\square$

Proposition 4.3.9. *The numbers*

$$\tau = \pi^4 k^4 - \pi^2 v^2 k^2, \quad k \in \mathbb{N}, \tag{4.3.12}$$

are the eigenvalues of A, counted with multiplicity. The eigenvalues are simple with the exception of finitely many double eigenvalue in case $v^2 = (k_2^2 + k_1^2)\pi^2$ for distinct positive integers k_1 and k_2. For such a pair of distinct integers k_1 and k_2, the double eigenvalue is $\tau = -k_1^2 k_2^2 \pi^4$.

Proof. Since all coefficients of the differential equation (4.3.7) are constant, one can write down the eigenvalue equation explicitly and thus improve the lower bound of A and even find its spectrum explicitly. Indeed, if $\tau \in \mathbb{R}$, then $(A - \tau I)y = 0$ reads

$$y^{(4)} + v^2 y'' - \tau y = 0, \tag{4.3.13}$$

so that the characteristic function becomes

$$\rho^4 + v^2 \rho^2 - \tau = 0, \tag{4.3.14}$$

which can be rewritten as

$$\left(\rho^2 + \frac{v^2}{2}\right)^2 = \tau + \frac{v^4}{4}. \tag{4.3.15}$$

Hence

$$-\rho^2 = \frac{v^2}{2} \pm \sqrt{\tau + \frac{v^4}{4}},$$

which has four solutions for ρ, counted with multiplicity, of the form $\pm i\omega_1$, $\pm i\omega_2$ with complex numbers ω_1 and ω_2. If ω_1 and ω_2 are distinct and different from 0, then the functions $\sin \omega_1 x$, $\cos \omega_1 x$, $\sin \omega_2 x$, $\cos \omega_2 x$ form a basis for the solutions of the differential equation (4.3.13), and a straightforward calculation shows that

$$y_3(x) := \frac{\sin \omega_1 x}{\omega_1(\omega_2^2 - \omega_1^2)} + \frac{\sin \omega_2 x}{\omega_2(\omega_1^2 - \omega_2^2)},$$

$$y_2 := y_3', \quad y_1 := y_3'' + v^2 y_3, \quad y_0 := y_3''' + v^2 y_3'$$

defines a basis y_0, y_1, y_2, y_3 for the solutions of the differential equation (4.3.13) satisfying $y_k^{(j)}(0) = \delta_{kj}$ for $k, j = 0, 1, 2, 3$. Here δ_{kj} is the Kronecker symbol, which is 1 if $k = j$ and 0 otherwise. For the verification we note that one can use the differential equation to reduce differentiations of order greater or equal 4. Although this system is initially more complicated to derive, it has the advantage that it extends to the cases when $\omega_1 = 0$ or $\omega_2 = 0$ or $\omega_1 = \omega_2$ as these cases are removable singularities in the parameters ω_1 and ω_2 of the above solutions. The initial conditions $y(0) = 0$ and $y''(0) = 0$ show that a nontrivial solution of $(A - \tau I)y = 0$ must be a linear combination of y_1 and y_3.

Taking the boundary conditions $y(1) = 0$ and $y''(1) = 0$ into account, it follows that a nontrivial solution of $(A - \tau I)y = 0$ corresponds to a zero of the characteristic equation

$$c(\tau) = \det \begin{pmatrix} y_1(1) & y_3(1) \\ y_1''(1) & y_3''(1) \end{pmatrix} = 0, \tag{4.3.16}$$

where we have to note that the determinant also depends on τ; for brevity we have omitted the parameter τ in the functions ω_j and y_j. Also note that the multiplicity of τ as a zero of c equals the multiplicity of the eigenvalue τ of A, see,

e. g [189, Theorem 6.3.4], where it is shown that the structure of the eigenvector and associated vectors of a differential operator pencil and an associated characteristic matrix coincide; in particular, the (algebraic) multiplicities coincide and equal the multiplicity τ as a zero of the determinant of the characteristic matrix, see [189, Proposition 1.8.5]. In view of

$$c(\tau) = \det \begin{pmatrix} y_3''(1) + v^2 y_3(1) & y_3(1) \\ y_3^{(4)}(1) + v^2 y_3''(1) & y_3''(1) \end{pmatrix} = \det \begin{pmatrix} y_3''(1) & y_3(1) \\ y_3^{(4)}(1) & y_3''(1) \end{pmatrix},$$

a straightforward calculation shows that

$$c(\tau) = \frac{\sin \omega_1}{\omega_1} \frac{\sin \omega_2}{\omega_2}, \tag{4.3.17}$$

which also holds for $\omega_1 = 0$ and $\omega_2 = 0$. Indeed, since the corresponding limit would be 1, it follows that the cases $\omega_1 = 0$ and $\omega_2 = 0$ do not contribute towards eigenvalues of A.

It is convenient to introduce the variables $\varpi_j = \omega_j^2$, $j = 1, 2$. Then

$$\varpi_j = -\rho^2 = \frac{v^2}{2} + (-1)^{j-1} \sqrt{\tau + \frac{v^2}{4}} \tag{4.3.18}$$

and

$$\frac{\sin \omega_j}{\omega_j} = \sum_{m=0}^{\infty} \frac{(-1)^m}{(2m+1)!} \varpi_j^m =: s(\varpi_j), \quad j = 1, 2,$$

where s is an entire function with simple zeros at $k^2 \pi^2$, $k \in \mathbb{N}$, and no other zeros. Although the choice of the square root in the definition of ϖ_j is ambiguous, the uniqueness of the sets $\{\varpi_1(\tau), \varpi_2(\tau)\}$, $\tau \in \mathbb{C}$, shows that we can choose ϖ_j locally in such a way that it is analytic, except at the point $\tau_0 = -\frac{v^4}{4}$. If $c(\tau_0) = 0$, then $\varpi_1(\tau_0) = \varpi_2(\tau_0)$ shows that $s(\varpi_j(\tau_0)) = 0$ for $j = 1, 2$. By l'Hôpital's rule,

$$\lim_{\tau \to \tau_0} \frac{s(\varpi_j(\tau))}{\sqrt{\tau - \tau_0}} = \lim_{\tau \to \tau_0} \frac{s'(\varpi_j(\tau)) \varpi'(\tau)}{(-1)^{j+1} \varpi'(\tau)} = (-1)^{j-1} s'\left(\frac{v^2}{2}\right).$$

It follows that

$$c'(\tau_0) = \lim_{\tau \to \tau_0} \frac{s(\varpi_1(\tau)) s(\varpi_2(\tau))}{\tau - \tau_0} = \lim_{\tau \to \tau_0} \frac{s(\varpi_1(\tau))}{\sqrt{\tau - \tau_0}} \lim_{\tau \to \tau_0} \frac{s(\varpi_2(\tau))}{\sqrt{\tau - \tau_0}}$$

$$= -\left[s'\left(\frac{v^2}{2}\right)\right]^2 \neq 0,$$

and therefore, if τ_0 is a zero of c, then it must be simple. For $\tau \neq \tau_0$, $\varpi_j'(\tau) \neq 0$ is obvious, so that all zeros of $s \circ \varpi_j$ in $\mathbb{C} \setminus \{\tau_0\}$ are simple. In view of (4.3.14) and (4.3.18), $s(\varpi_j(\tau)) = 0$ if and only if

$$\tau = \varpi_j^2 - v^2 \varpi_j = \pi^4 k_j^4 - \pi^2 v^2 k_j^2, \quad k_j \in \mathbb{N}. \tag{4.3.19}$$

Since $\varpi_1(\tau) = \varpi_2(\tau)$ if and only if $\tau = \tau_0$, it follows that τ is a double zero of c if and only if $\tau \neq \tau_0$ and (4.3.19) holds for $j = 1$ and $j = 2$, where $\tau \neq \tau_0$ holds if and only if $k_1 \neq k_2$. In this case,

$$\pi^4 k_1^4 - \pi^2 v^2 k_1^2 = \tau = \pi^4 k_2^4 - \pi^2 v^2 k_2^2,$$

which means that $v^2 = \pi^2 (k_1^2 + k_2^2)$ and

$$\tau = \pi^4 k_1^4 - \pi^4 (k_1^2 + k_2^2) k_1^2 = -k_1^2 k_2^2 \pi^4. \qquad \square$$

Remark 4.3.10. 1. It is a rather curious fact that double eigenvalues can only be of the form $-k^2 \pi^4$ for integers $k \geq 2$, and letting $v^2 = (1 + k^2)\pi^2$ we see that each such number can indeed be a double eigenvalue for a suitably chosen $v > 0$. The sum of squares function, see [105, Section 16.9, Theorem 278] shows that for each positive integer n we can find v such that the number of double eigenvalues of A exceeds n.

2. Although in the original problem we assume $v \neq 0$, we may apply (4.3.12) to $v = 0$, in which case $A = A_+$. Hence $A_+ \geq \pi^4$. Then (4.3.9) leads to

$$v^{-4} \|A_- y\|^2 = (A_+ y, y) \geq \pi^4 \|y\|^2, \quad y \in D(A),$$

so that we have

$$\|A_- y\| \geq v^2 \pi^2 \|y\|, \quad y \in D(A). \tag{4.3.20}$$

We know by Remarks 4.3.2 and 4.3.5 that the spectra of the operator pencils (4.3.1) with $B = iG$, i.e.,

$$L(\lambda, \eta) = \lambda^2 I - i\lambda\eta G - A, \quad \eta \in (0, 1),$$

consist of normal eigenvalues which are symmetric with respect to the real and imaginary axes. Here we have to observe that by definition, the parameter η in (4.3.2) satisfies $\eta \in (0, 1)$.

The following criterion for stability was proved in [284].

Lemma 4.3.11 ([284]). *Problem* (4.3.2)–(4.3.5) *is stable if and only if the spectrum of the quadratic operator pencil* $L(\cdot, \eta)$ *is real and semisimple.*

It is this criterion which we will use below to investigate the stability of the pencil L. Consequently, we make the following definition.

Definition 4.3.12. Let $\eta \in (0, 1)$. A quadratic operator pencil $L(\cdot, \eta)$ of the form (4.3.1) is called stable if its spectrum is real and semisimple.

Remark 4.3.13. It should be mentioned that gyroscopic stabilization in terms of moving eigenvalues can be described as follows. Assume that the operator A is invertible and has negative eigenvalues. A symmetric pair of pure imaginary eigenvalues of $L(\cdot, 0)$ corresponds to each of them. When $\eta > 0$ grows the eigenvalues moving on the imaginary axis can collide. In this case they can leave the imaginary axis in symmetric pairs. Then they can join the real axis, colliding with another complex eigenvalue joining the real axis, and after a new collision they may again leave the real axis. Thus instead of two pairs of pure imaginary eigenvalues there appear two pairs of real eigenvalues. This can lead to gyroscopic stabilization. But the pure imaginary eigenvalues can disappear only in the case when the total algebraic multiplicity of the negative spectrum of A is even, otherwise one of the eigenvalues will not find a partner to leave the imaginary axis. Here we have used that eigenvalues of $L(\cdot, \eta)$ cannot cross the origin since $L(0, \eta) = -A$ is invertible for all $\eta \in (0, 1)$.

Proposition 4.3.14. *Assume that $A \gg 0$. Then all eigenvalues of $L(\cdot, \eta)$, $\eta \in (0, 1)$, are real and semisimple.*

Proof. The spectrum of $L(\cdot, \eta)$ is real in view of Theorem 4.3.4. To show that all eigenvalues are semisimple, assume that there is an eigenvalue λ of $L(\cdot, \eta)$ with an eigenvector y_0 and an associated vector y_1. Writing $L(\lambda, \eta)$ in the form (4.3.1), this means by Definition 1.1.3, part 1, of associated vectors that

$$(\lambda^2 I - \lambda \eta B - A)y_0 = 0, \tag{4.3.21}$$

$$(2\lambda I - \eta B)y_0 + (\lambda^2 I - \lambda \eta B - A)y_1 = 0. \tag{4.3.22}$$

Taking the inner product of (4.3.22) with y_0 and observing that λ is real and that A and B are self-adjoint, it follows in view of (4.3.21) that

$$2\lambda(y_0, y_0) - \eta(By_0, y_0) = 0.$$

Substituting $\eta(B_0 y_0, y_0)$ from this equation into the inner product of (4.3.21) with y_0, we arrive at

$$0 = \lambda^2(y_0, y_0) - \lambda \eta(By_0, y_0) - (Ay_0, y_0) = -\lambda^2(y_0, y_0) - (Ay_0, y_0),$$

which is impossible since the right-hand side is negative in view of $\lambda \in \mathbb{R}$ and $A \gg 0$. This contradiction proves that all eigenvalues of $L(\cdot, \eta)$ are semisimple. $\square$

Corollary 4.3.15. *Assume that $v < \pi$. Then problem (4.3.2)–(4.3.5) is stable.*

Proof. For $v < \pi$ we have $A \gg 0$ by (4.3.12). An application of Proposition 4.3.14 and Lemma 4.3.11 completes the proof. $\square$

Proposition 4.3.16. *Assume that $0 \in \sigma(A)$. Then $0 \in \sigma(L(\cdot, \eta))$ and 0 is not semisimple for all $\eta \in (0, 1)$.*

Proof. From Proposition 4.3.9 we know that all nonnegative eigenvalues of A have geometric multiplicity 1. Hence the algebraic multiplicity of the eigenvalue 0 of $L(\cdot, 0)$ is 2. Due to the symmetry of the spectrum of $L(\cdot, \eta)$, see Remark 4.3.5, eigenvalues can leave 0 only in pairs, which is impossible since 0 is an eigenvalue of geometric multiplicity 1 of $L(\cdot, \eta)$ for all η. It follows that the algebraic multiplicity of the eigenvalue 0 of $L(\cdot, \eta)$ is 2 for all η, and therefore 0 is not a semisimple eigenvalue of $L(\cdot, \eta)$ for any η. $\qquad\qquad\qquad\qquad\qquad\qquad\square$

We now consider the case that A has at least one negative eigenvalue. In view of the eigenvalue formula (4.3.12) this is true if and only if $v > \pi$. Denote by $\zeta_k(v^2)$ the kth eigenvalue of A, $k \in \mathbb{N}$, enumerated in increasing order, that is, $\zeta_k(v^2) \le \zeta_{k+1}(v^2)$. Then there is $k_1 \in \mathbb{N}$ such that $\zeta_{k_1}(v^2) < 0$ and $\zeta_{k_1+1}(v^2) \ge 0$. Clearly, $0 \in \sigma(A)$ if and only if $\zeta_{k_1+1}(v^2) = 0$. From (4.3.12) we immediately obtain that k_1 is the largest integer k such that $k < v\pi^{-1}$.

Let us introduce the operator function

$$\mathcal{L}(\tau, \eta) = L(\sqrt{\tau}, \eta) = \tau I - i\sqrt{\tau}\eta G - A.$$

We consider the term $i\sqrt{\tau}\eta G$ as a perturbation of the linear pencil $\mathcal{L}_0(\tau) = \tau I - A$.

Proposition 4.3.17. *Let $v > \pi$, $l \in (\zeta_1(v^2), 0) \cap \rho(A)$ and $d = \min\{\mathrm{dist}(l, \sigma(A)), |l|\}$, and define*

$$\eta_* = \frac{1}{4}\min\{d|l|^{-\frac{1}{2}}v^{-2}, d^{\frac{3}{4}}|l|^{-\frac{1}{2}}v^{-1}\}. \tag{4.3.23}$$

If $\eta \in (0, \eta_*)$, then the total algebraic multiplicity of the spectrum of the pencil $L(\cdot, \eta)$ in each of the two domains determined by the inequality $(\mathrm{Re}\,\lambda)^2 - (\mathrm{Im}\,\lambda)^2 \le l$ is equal to $k(l)$, the number of eigenvalues of A below l.

Proof. All points of the line $\tau = l + is$ ($s \in \mathbb{R}$) belong to the resolvent set of the operator A. We are going to show that for all $\eta \in (0, \eta_*)$, all points of the line $\tau = l + is$, $s \in \mathbb{R}$, also belong to the resolvent set of the operator function $\mathcal{L}(\cdot, \eta)$. Since the operator A is self-adjoint, we have the inequality

$$\|((l + is)I - A)y\| \ge (d^2 + s^2)^{\frac{1}{2}}\|y\| \tag{4.3.24}$$

for all $y \in D(A)$. Consider the following two cases for $s \in \mathbb{R}$ and $y \in D(A) \setminus \{0\}$.

1. Let the pair s, y satisfy the condition

$$\eta(l^2 + s^2)^{\frac{1}{4}}\|Gy\| < (d^2 + s^2)^{\frac{1}{2}}\|y\|. \tag{4.3.25}$$

Then it follows that

$$\|\mathcal{L}(l + is, \eta)y\| = \|((l + is)I - i\eta((l + is)^{\frac{1}{2}}G - A)y\|$$

$$\ge \|((l + is)I - A)y\| - \eta(l^2 + s^2)^{\frac{1}{4}}\|Gy\| > 0 \tag{4.3.26}$$

2. Now let the pair s, y satisfy the condition

$$\eta(l^2 + s^2)^{\frac{1}{4}}\|Gy\| \ge (d^2 + s^2)^{\frac{1}{2}}\|y\|. \tag{4.3.27}$$

We obtain from (4.3.9) and (4.3.11) for $\tau = l + is$ that

$$\|\mathcal{L}(\tau, \eta)y\| \geq \|A_+y\| - \|A_-y\| - |\tau|\|y\| - |\tau|^{\frac{1}{2}}\eta\|Gy\|$$
$$\geq v^{-4}\|A_-y\|^2\|y\|^{-1} - \|A_-y\| - (l^2 + s^2)^{\frac{1}{2}}\|y\|$$
$$- 2\eta(l^2 + s^2)^{\frac{1}{4}}\|A_-y\|^{\frac{1}{2}}\|y\|^{\frac{1}{2}}. \tag{4.3.28}$$

It follows from (4.3.11) and (4.3.27) that

$$\|A_-y\|^{\frac{1}{2}} \geq \frac{1}{2}\|Gy\|\|y\|^{-\frac{1}{2}}$$
$$\geq \frac{1}{2\eta}(l^2 + s^2)^{-\frac{1}{4}}(d^2 + s^2)^{\frac{1}{2}}\|y\|^{\frac{1}{2}}. \tag{4.3.29}$$

The estimate (4.3.29), $d \leq |l|$ and the fact that the function $t \mapsto (a + t)^p(b + t)^q$ takes its minimum on $[0, \infty)$ at $t = 0$ when a, b, p, q are real numbers satisfying $0 < a \leq b$, $p \geq 0$ and $p + q \geq 0$ lead to

$$\|A_-y\| \geq \frac{1}{4\eta^2}|l|^{-1}d^2\|y\|,$$

$$\|A_-y\|^2 \geq \frac{1}{16\eta^4}|l|^{-3}d^4(l^2 + s^2)^{\frac{1}{2}}\|y\|^2,$$

$$\|A_-y\|^{\frac{3}{2}} \geq \frac{1}{8\eta^3}|l|^{-2}d^3(l^2 + s^2)^{\frac{1}{4}}\|y\|^{\frac{3}{2}}.$$

For $\eta \in (0, \eta_*)$ the above estimates give

$$\frac{1}{3}v^{-4}\|A_-y\|^2\|y\|^{-1} > \|A_-y\|, \tag{4.3.30}$$

$$\frac{1}{3}v^{-4}\|A_-y\|^2\|y\|^{-1} > (l^2 + s^2)^{\frac{1}{2}}\|y\|, \tag{4.3.31}$$

$$\frac{1}{3}v^{-4}\|A_-y\|^2\|y\|^{-1} > 2\eta(l^2 + s^2)^{\frac{1}{4}}\|A_-y\|^{\frac{1}{2}}\|y\|^{\frac{1}{2}}. \tag{4.3.32}$$

Consequently, for all pairs s, y satisfying (4.3.27), the inequalities (4.3.28) and (4.3.30)–(4.3.32) show that $\|\mathcal{L}(\tau, \eta)y\| > 0$.

Combining cases 1 and 2 shows that $\mathcal{L}(\tau, \eta)$ is injective. Since the spectrum of $\mathcal{L}(\cdot, \eta)$ consists only of eigenvalues, the points of the line $\tau = l + is$, $s \in \mathbb{R}$, belong to the resolvent set of $\mathcal{L}(\tau, \eta)$ for all $\eta \in (0, \eta_*)$.

The transformation $\lambda^2 = \tau$ maps the half-plane $\mathrm{Re}\,\tau \leq l$ into two domains $\Omega_\pm$ in the open upper and lower half-planes, which are the sets of $\lambda \in \mathbb{C}$ determined by the inequality $(\mathrm{Re}\,\lambda)^2 - (\mathrm{Im}\,\lambda)^2 \leq l < 0$. Since $\mathcal{L}(\cdot, \eta)$ has no eigenvalues on the line $\mathrm{Re}\,\tau = l$ for $\eta \in [0, \eta_*)$, it follows that $L(\cdot, \eta)$ has no eigenvalues on the curves $(\mathrm{Re}\,\lambda)^2 - (\mathrm{Im}\,\lambda)^2 = l$ for $\eta \in [0, \eta_*)$. Bearing in mind the continuity in η of the eigenvalues of the operator function $L(\cdot, \eta)$ and that by Lemma 4.3.1 the nonreal eigenvalues have a bound which does not depend on η, it follows that the total algebraic multiplicity of the spectrum of the pencil $L(\cdot, \eta)$ in each of the two domains $\Omega_\pm$ does not depend on η for $\eta \in [0, \eta_*)$. Since this multiplicity is $k(l)$ for $\eta = 0$, the proof is complete.   □

Corollary 4.3.18. *Under the assumptions of Proposition 4.3.17 let $\eta \in (0, \eta_*)$. Then the total algebraic multiplicity κ of the spectrum of $L(\cdot, \eta)$ in the open lower half-plane satisfies the inequality $\kappa \geq k(l)$.*

Corollary 4.3.19. *Let $\eta \in (0, 1)$ and $v \geq \pi$. If $n\pi < v < (n+1)\pi$ for some even integer n, assume additionally that $\eta < \eta_*$, where η_* is defined in (4.3.23). Then problem (4.3.2)–(4.3.5) is unstable.*

Proof. For $v > \pi$ and $\eta \in (0, \eta_*)$ the statement follows from Definition 4.3.12 and Corollary 4.3.18 if we observe that $k(l) > 0$ for any suitably chosen l. If v does not belong to an interval of the form $(n\pi, (n+1)\pi)$ with even n, then $v \in (n\pi, (n+1)\pi)$ for some odd integer n or $v\pi^{-1}$ is an integer. The first case means that $0 \in \rho(A)$ and the total algebraic multiplicity of the negative spectrum of A is n and therefore odd. Hence it follows from Remark 4.3.13 that the problem is unstable in this case. Finally, if $v\pi^{-1}$ is an integer, then $0 \in \sigma(A)$, and Proposition 4.3.16 shows that the problem is unstable.   □

Theorem 4.3.20. *Let $v > \pi$, let $d = \frac{1}{2}|\zeta_{k_1}(v^2)|$ and define*

$$\eta_{*1} = \frac{1}{4}\min\{d^{\frac{1}{2}}v^{-2}, d^{\frac{1}{4}}v^{-1}\}.$$

Then for $\eta \in (0, \eta_{*1})$ the total algebraic multiplicity of the spectrum of the pencil $L(\cdot, \eta)$ in the open lower half-plane is equal to the total algebraic multiplicity k_1 of the negative spectrum of the operator A.

Proof. The number η_{*1} is the special case of the number η_* defined in (4.3.23) for $l = \frac{1}{2}|\zeta_{k_1}(v^2)|$. Let $\eta \in (0, \eta_{*1})$. Then the total algebraic multiplicity of the spectrum of the pencil $L(\cdot, \eta)$ in the open lower half-plane is at least as large as the total algebraic multiplicity of the negative spectrum of the operator A in view of Proposition 4.3.17. Since it cannot be larger by Theorem 4.3.4, the proof is complete.   □

Above we have found a reasonably simple constant η_*. We are now going to find the optimal constant based on the estimate (4.3.29). Here we consider now $l \in (\zeta_1(v^2), 0] \cap \rho(A)$ and $d = \mathrm{dist}(l, \sigma(A))$. Observe that we now allow $d > |l|$ and also $l = 0$ when $0 \in \rho(A)$. Similar to above, let $m(a, b, p, q)$ be the infimum of the function $f : (0, \infty) \to \mathbb{R}$ defined by $f(t) = (a+t)^p(b+t)^q$, where a, b, p, q are real numbers satisfying $a > 0$, $b \geq 0$, $p > 0$, $q < 0$, and $p + q > 0$. If $bp + aq \geq 0$, then $b > 0$ and f is also defined at 0. It is easy to see that the minimum of f is then taken at 0, and we have

$$m(a, b, p, q) = a^p b^q \quad \text{if } bp + aq \geq 0. \tag{4.3.33}$$

If $bp + aq < 0$, then the minimum is taken at $-\frac{bp+aq}{p+q}$, and we have

$$m(a,b,p,q) = \left(\frac{a-b}{p+q}\right)^{p+q} p^p(-q)^q \quad \text{if } bp + aq < 0. \tag{4.3.34}$$

Then the estimate (4.3.29) leads to

$$\|A_-y\| \geq \frac{1}{4\eta^2} m\left(d^2, l^2, 1, -\frac{1}{2}\right) \|y\|,$$

$$\|A_-y\|^2 \geq \frac{1}{16\eta^4} m\left(d^2, l^2, 2, -\frac{3}{2}\right) (l^2 + s^2)^{\frac{1}{2}} \|y\|^2,$$

$$\|A_-y\|^{\frac{3}{2}} \geq \frac{1}{8\eta^3} m\left(d^2, l^2, \frac{3}{2}, -1\right) (l^2 + s^2)^{\frac{1}{4}} \|y\|^{\frac{3}{2}}.$$

For α, β, γ in $(0,1)$ such that $\alpha + \beta + \gamma = 1$ we obtain

$$\alpha v^{-4} \|A_-y\|^2 \|y\|^{-1} > \|A_-y\|,$$

$$\beta v^{-4} \|A_-y\|^2 \|y\|^{-1} > (l^2 + s^2)^{\frac{1}{2}} \|y\|,$$

$$\gamma v^{-4} \|A_-y\|^2 \|y\|^{-1} > 2\eta (l^2 + s^2)^{\frac{1}{4}} \|A_-y\|^{\frac{1}{2}} \|y\|^{\frac{1}{2}}$$

if $\eta \in (0, \hat{\eta}(l, v, \alpha, \beta, \gamma))$, where

$$\hat{\eta}(l, v^2, \alpha, \beta, \gamma) = \min\{\alpha_1 a_1, \beta_1 b_1, \gamma_1 c_1\}$$

with $\alpha_1 = \alpha^{\frac{1}{2}}$, $\beta_1 = \beta^{\frac{1}{4}}$, $\gamma_1 = \gamma^{\frac{1}{4}}$ and

$$a_1 = \frac{1}{2v^2} m\left(d^2, l^2, \frac{1}{2}, -\frac{1}{4}\right), \tag{4.3.35}$$

$$b_1 = \frac{1}{2v} m\left(d^2, l^2, \frac{1}{2}, -\frac{3}{8}\right), \tag{4.3.36}$$

$$c_1 = \frac{1}{2v} m\left(d^2, l^2, \frac{3}{8}, -\frac{1}{4}\right). \tag{4.3.37}$$

It is clear that this minimum is maximal as a function of α, β, γ when the three numbers $\alpha_1 a_1, \beta_1 b_1, \gamma_1 c_1$ are equal, because for any other choice of the parameters α, β, γ, at least one of these numbers will be smaller. It is easy to see that $\alpha_1 a_1 = \beta_1 b_1 = \gamma_1 c_1$ if and only if

$$\alpha_1 = \frac{\delta_1 b_1 c_1}{a_1 b_1 + a_1 c_1 + b_1 c_1}, \quad \beta_1 = \frac{\delta_1 a_1 c_1}{a_1 b_1 + a_1 c_1 + b_1 c_1}, \quad \gamma_1 = \frac{\delta_1 a_1 b_1}{a_1 b_1 + a_1 c_1 + b_1 c_1},$$

where $\delta_1 = \alpha_1 + \beta_1 + \gamma_1$, and in this case,

$$\hat{\eta}(l, v) = \alpha_1 a_1 = \beta_1 b_1 = \gamma_1 c_1 = \frac{\delta_1 a_1 b_1 c_1}{a_1 b_1 + a_1 c_1 + b_1 c_1}. \tag{4.3.38}$$

We still have to find δ_1. We observe that from

$$\alpha_1^2 + \beta_1^4 + \gamma_1^4 = \alpha + \beta + \gamma = 1$$

we obtain the equation

$$1 = \frac{\delta_1^2 b_1^2 c_1^2}{(a_1 b_1 + a_1 c_1 + b_1 c_1)^2} + \frac{\delta_1^4 a_1^4 (c_1^4 + b_1^4)}{(a_1 b_1 + a_1 c_1 + b_1 c_1)^4},$$

which can be written in the form

$$g_1 \delta_1^4 + f_1 \delta_1^2 = e_1$$

with

$$e_1 = (a_1 b_1 + a_1 c_1 + b_1 c_1)^4, \quad f_1 = b_1^2 c_1^2 (a_1 b_1 + a_1 c_1 + b_1 c_1)^2, \quad g_1 = a_1^4 (c_1^4 + b_1^4).$$

It follows that

$$\delta_1 = \frac{1}{\sqrt{2g_1}} \sqrt{-f_1 + \sqrt{4 e_1 g_1 + f_1^2}}$$

$$= \frac{\sqrt{-b_1^2 c_1^2 + \sqrt{4 a_1^4 (b_1^4 + c_1^4) + b_1^4 c_1^4}} \, (a_1 b_1 + a_1 c_1 + b_1 c_1)}{\sqrt{2} a_1^2 \sqrt{b_1^4 + c_1^4}}$$

$$= \frac{\sqrt{2} \, (a_1 b_1 + a_1 c_1 + b_1 c_1)}{\sqrt{b_1^2 c_1^2 + \sqrt{4 a_1^4 (b_1^4 + c_1^4) + b_1^4 c_1^4}}}.$$

Substitution into (4.3.38) gives

$$\tilde{\eta}(l, v) = \frac{\sqrt{2} a_1 b_1 c_1}{\sqrt{b_1^2 c_1^2 + \sqrt{4 a_1^4 (b_1^4 + c_1^4) + b_1^4 c_1^4}}}. \tag{4.3.39}$$

Hence we may replace the number η_* in (4.3.23) by the optimal number $\tilde{\eta}(l, v)$ given by (4.3.39). Although the calculations to find $\tilde{\eta}(l, v)$ are quite involved, they are explicit for each given pair v and l. The number $\tilde{\eta}(l, v)$ can therefore be easily calculated with computer algebra programmes or numerically.

Finally, we define

$$\eta^*(v) = \sup_{l \in (\zeta_1(v^2), 0] \cap \rho(A)} \tilde{\eta}(l, v), \quad v > 2\pi. \tag{4.3.40}$$

Then we have the following improvement of Corollary 4.3.18.

Corollary 4.3.21. *Assume that v satisfies $n\pi < v < (n+1)\pi$ for some positive even integer n and that $0 < \eta < \eta^*(v)$. Then problem (4.3.2)–(4.3.5) is unstable.*

Similarly, we define

$$\eta_0^*(v) = \sup_{l \in (\zeta_{k_1}(v^2), 0]} \tilde{\eta}(l, v), \quad v > 2\pi, \; v\pi^{-1} \notin \mathbb{Z}. \tag{4.3.41}$$

Then we have the following improvement of Theorem 4.3.20.

Theorem 4.3.22. *Assume that v satisfies $n\pi < v < (n+1)\pi$ for some positive integer n and that $0 < \eta < \eta_0^*(v)$. Then for $\eta \in (0, \eta_0^*)$ the total algebraic multiplicity of the spectrum of the pencil $L(\cdot, \eta)$ in the open lower half-plane is equal to the total algebraic multiplicity k_1 of the negative spectrum of the operator A.*

Although the numbers $\tilde{\eta}(l, v)$ are explicitly given, it is not easy to find the corresponding suprema (4.3.40) and (4.3.41). For $j = 1, \ldots, k_1 - 1$, let $l_j(v^2)$ be the midpoint between $\zeta_j(v^2)$ and $\zeta_{j+1}(v^2)$, let $l_{k_1}(v^2)$ be the midpoint between $\zeta_{k_1}(v^2)$ and 0, and let $l_0(v^2) = 0$. Then we define $\tilde{\eta}_j(v) = \tilde{\eta}(l_j, v)$ for $l = 0, \ldots, k_j$. It should be reasonable to expect that the maximum of $\tilde{\eta}_0(v)$ and $\tilde{\eta}_{k_1}(v)$ is relatively close to η_0^*. Furthermore, the maximum of $\tilde{\eta}_j(v)$ for $j = 0, \ldots, k_1$ might be quite close to $\eta^*(v)$. Here we have to note that the values of the function m in (4.3.35)–(4.3.37) are given by (4.3.34) for $l = 0$ and by (4.3.33) for $l_j(v)$ with $l = 1, \ldots, k_j$ since $d \leq |l|$ if l is the midpoint of two nonpositive real numbers and d the distance of the midpoint from these two numbers.

A simpler expression will be obtained if we estimate the denominator in $\tilde{\eta}(l, v)$ as follows:

$$(4a_1^4 b_1^4 + 4a_1^4 c_1^4 + 2b_1^4 c_1^4)^{\frac{1}{4}} \leq \sqrt{b_1^2 c_1^2 + \sqrt{4a_1^4(b_1^4 + c_1^4) + b_1^4 c_1^4}}$$

$$\leq 2^{\frac{1}{4}}(4a_1^4 b_1^4 + 4a_1^4 c_1^4 + 2b_1^4 c_1^4)^{\frac{1}{4}}.$$

Then the numbers

$$\tilde{\eta}_1(l, v) = \left(\frac{1}{2a_1^4} + \frac{1}{b_1^4} + \frac{1}{c_1^4} \right)^{-\frac{1}{4}}$$

satisfy

$$2^{-\frac{1}{4}} \tilde{\eta}_1(l, v) \leq \tilde{\eta}(l, v) \leq \tilde{\eta}_1(l, v). \tag{4.3.42}$$

We are now going to present explicit η-bounds for the case that $v \in (4\pi, 5\pi)$. We have seen in (4.3.12) that A has 4 negative eigenvalues, which we will denote by $\tau_n(v^2) = \pi^4 n^4 - \pi^2 v^2 n^2$, $n = 1, 2, 3, 4$, and that the smallest positive eigenvalue is $\zeta_5(v^2) = 625\pi^4 - 25\pi^2 v^2$. It is easy to see that $\tau_1(v^2) > \tau_2(v^2) > \tau_3(v^2)$ and that $\tau_4(v^2) > \tau_1(v^2)$ if $4\pi < v < \sqrt{17}\pi$, $\tau_1(v^2) > \tau_4(v^2) > \tau_2(v^2)$ if $\sqrt{17}\pi < v < \sqrt{20}\pi$, and $\tau_2(v^2) > \tau_4(v^2) > \tau_3(v^2)$ if $\sqrt{20}\pi < v < 5\pi$. Numerical calculation starting at $v = 4.01\,\pi$ up to $v = 4.99\,\pi$ with step size $0.02\,\pi$ give the following graphs.

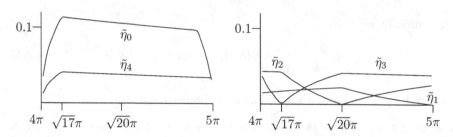

This shows the somewhat surprising fact that $\tilde{\eta}_0(v)$ is the best of these five values, except for v very close to 4π or 5π, respectively, in which case $\tilde{\eta}_2$ and $\tilde{\eta}_3$, respectively, appear to be best.

Further numerical calculations with 200 values for l show that the maximum over the corresponding values of $\tilde{\eta}(l, v)$ does not exceed the maximum of $\tilde{\eta}_j(v)$, $j = 0, \ldots, 4$, except for values of v which are very close the 5π. But even in these cases, the improvement is marginal.

Another necessary condition of instability was obtained in [217].

For a positive number ξ_1 we denote by $\kappa(i\xi_1, \eta)$ the total algebraic multiplicity of the spectrum of $L(\lambda, \eta)$ located on the interval $(i\xi_1, i\infty)$.

Lemma 4.3.23. *Let $\eta_1 > 0$, $\xi_1 > 0$ and assume that $i\xi_1 \in \rho(L(\cdot, \eta))$ for all $\eta \in [0, \eta_1]$. Then the parity of $\kappa(i\xi_1, \eta)$ is independent of $\eta \in [0, \eta_1]$, that is, if $\kappa(i\xi_1, 0)$ is odd (even), then $\kappa(i\xi_1, \eta)$ is odd (even) for all $\eta \in [0, \eta_1]$.*

Proof. Making use of the fact that the spectrum of $L(\cdot, \eta)$ is symmetric with respect to the imaginary axis by Remark 4.3.5, the proof is similar to the proof of Theorem 4.3.6, if we additionally observe that eigenvalues on the imaginary axis cannot move through $i\xi$ because of the assumption $i\xi_1 \in \rho(L(\cdot, \eta))$ for all $\eta \in [0, \eta_1]$.   □

Lemma 4.3.24. *Let $v > 2\pi$. If*

$$0 < \eta \le 2^{-\frac{3}{2}} v^{-2} (\zeta_2(v^2) - \zeta_1(v^2)) |\zeta_1(v^2) + \zeta_2(v^2)|^{-\frac{1}{2}}, \tag{4.3.43}$$

then the points $\pm\lambda_1 = \pm i 2^{-\frac{1}{2}} |\zeta_1(v^2) + \zeta_2(v^2)|^{\frac{1}{2}}$ belongs to $\rho(L(\cdot, \eta))$.

Proof. Let $\eta > 0$ and $\xi > 0$ and assume that $i\xi$ or $-i\xi$ is an eigenvalue of $L(\cdot, \eta)$. Let y be a corresponding eigenvector, which can be chosen to be real. Then

$$-\xi^2 y + \eta\xi G y - A_+ y - A_- y = 0. \tag{4.3.44}$$

The operator G is real and antisymmetric, and therefore

$$(Gy, y) = 0. \tag{4.3.45}$$

Using (4.3.44) and (4.3.45) we obtain

$$(A_+ y, y) = -(A_- y, y)| - \xi^2 \|y\|^2 \le \|A_- y\| \|y\| - \xi^2 \|y\|^2. \tag{4.3.46}$$

In view of (4.3.9) and (4.3.46) we arrive at

$$v^{-4}\|A_-y\|^2 \le \|A_-y\|\|y\| - \xi^2\|y\|^2,$$

which implies

$$\|A_-y\| \le \left(\frac{1}{2}v^4 + \frac{1}{2}\sqrt{v^8 - 4v^4\xi^2}\right)\|y\|. \tag{4.3.47}$$

Combining (4.3.11) with (4.3.47) gives

$$\|Gy\| \le v^2\sqrt{2 + 2\sqrt{1 - 4v^{-4}\xi^2}}\,\|y\|. \tag{4.3.48}$$

From (4.3.48) we obtain

$$0 = \|L(\pm i\xi, \eta)y\| \ge \| - \xi^2 y - Ay\| - \eta\xi\|Gy\|$$
$$\ge \left(d(i\xi) - \eta v^2\xi\sqrt{2 + 2\sqrt{1 - 4v^{-4}\xi^2}}\right)\|y\|, \tag{4.3.49}$$

where $d(i\xi)$ is the distance between the point $-\xi^2$ and the spectrum of A. Thus we have proved that

$$\eta > \frac{1}{2}v^{-2}\xi^{-1}d(i\xi). \tag{4.3.50}$$

Since $v > 2\pi$, both $\zeta_1(v^2)$ and $\zeta_2(v^2)$ are negative, and therefore $(\pm i\lambda_1)^2$ is the midpoint between $\zeta_1(v^2)$ and $\zeta_2(v^2)$. Evidently, $d(\pm\lambda_1) = \frac{1}{2}(\zeta_1(v^2) - \zeta_2(v^2))$. Therefore, the right-hand sides of (4.3.50) and (4.3.43) are equal for $i\xi = \lambda_1$. Thus (4.3.50) is false, which proves that $\pm\lambda_1$ cannot be an eigenvalue of $\mathcal{L}(\cdot, \eta)$. □

Theorem 4.3.25. *Under condition* (4.3.43) *the total algebraic multiplicity of the spectrum of the operator pencil* $L(\cdot, \eta)$ *in the interval* $(i2^{-\frac{1}{2}}|\zeta_1(v^2) + \zeta_2(v^2)|^{\frac{1}{2}}, i\infty)$ *is odd.*

Proof. For $\eta = 0$ the total algebraic multiplicity of the spectrum of $L(\lambda, \eta)$ located in the interval $(i2^{-\frac{1}{2}}|\zeta_1(v^2) + \zeta_2(v^2)|^{\frac{1}{2}}, i\infty)$ is 1. Hence the statement of this theorem immediately follows from Lemmas 4.3.23 and 4.3.24. □

Corollary 4.3.26. *Under condition* (4.3.43), *problem* (4.3.2)–(4.3.5) *is unstable.*

In (4.3.50) and (4.3.43) we have simplified the condition for instability derived from the inequality (4.3.49). Using the full strength of the inequality (4.3.49), we may divide the right-hand side of (4.3.43) by $\frac{1}{2}\sqrt{2 + 2\sqrt{1 - 4v^{-4}\xi^2}}$ to obtain a slightly larger η-interval of instability.

The contrapositive of Corollary 4.3.26 gives the following necessary condition for gyroscopic stabilization:

$$\eta > 2^{-\frac{3}{2}}v^{-2}(\zeta_2(v^2) - \zeta_1(v^2))|\zeta_0(v^2) + \zeta_1(v^2)|^{-\frac{1}{2}}.$$

Inequality (4.3.43) can be written in explicit form by finding $\zeta_1(v^2)$ and $\zeta_2(v^2)$ explicitly. We already know from Corollary 4.3.19 that the problem is unstable if $v \in [(2n-1)\pi, 2n\pi]$, $n \in \mathbb{N}$. Hence we may restrict our calculations to the case that $v \in (2\pi n, \pi(2n+1))$, $n \in \mathbb{N}$. For example, we will consider $n = 1$ and $n = 2$. If $v \in (2\pi, 3\pi)$, then A has exactly two negative eigenvalues, and therefore

$$\{\zeta_1(v^2), \zeta_2(v^2)\} = \{-\pi^2 v^2 + \pi^4, -4\pi^2 v^2 + 16\pi^4\}.$$

Substituting these values into (4.3.43) we obtain the following domains of instability:

$$\begin{cases} 2\pi \le v \le 3\pi, \\ 0 < \eta < 2^{-\frac{3}{2}} v^{-2} |5\pi^4 - 3\pi^2 v^2| (5\pi^2 v^2 - 17\pi^4|^{-\frac{1}{2}}. \end{cases}$$

For $v \in (4\pi, 5\pi)$, we have noted on page 109 that $\zeta_1(v^2) = 81\pi^4 - 9\pi^2 v^2$, whereas $\zeta_2(v^2) = 16\pi^4 - 4\pi^2 v^2$ if $v \le \sqrt{20}\pi$ and $\zeta_2(v^2) = 256\pi^4 - 16\pi^2 v^2$ if $v \ge \sqrt{20}\pi$. This gives the following domains of instability:

$$\begin{cases} 4\pi \le v \le \sqrt{20}\pi, \\ 0 < \eta < 2^{-\frac{3}{2}} v^{-2} (5\pi^2 v^2 - 65\pi^4)(13\pi^2 v^2 - 97\pi^4)^{-\frac{1}{2}} \end{cases}$$

and

$$\begin{cases} \sqrt{20}\pi \le v \le 5\pi, \\ 0 < \eta < 2^{-\frac{3}{2}} v^{-2} (175\pi^4 - 7\pi^2 v^2)(25\pi^2 v^2 - 337\pi^4)^{-\frac{1}{2}}. \end{cases}$$

Numerical calculations show that, even with the slight improvement mentioned after Corollary 4.3.26, the values of the upper bound is smaller than 0.0005 for $v \in (4\pi, 5\pi)$. This indicates that the value $\eta^*(v)$ is (much) better than the value on the right-hand side of (4.3.43).

4.4 Notes

In the papers [65], [261], [133], [283] the authors considered, in our notation, a finite-dimensional operator pencil $\lambda^2 M - \lambda B - A$ with Hermitian matrices A, B, and $M > 0$. It was shown that the zeros of the characteristic polynomial $\det(\lambda^2 M - \lambda B - A)$ can be all real and simple, i.e., the corresponding dynamical system described by the equation

$$M\frac{d^2 u}{dt^2} - iB\frac{du}{dt} + Au = 0,$$

can be stable even if some of the eigenvalues of A are negative. Here u is a vector whose components are small angular displacements of a moving shell, M is the inertia matrix, the real antisymmetric matrix iB describes gyroscopic forces and A describes potential forces. This effect, called gyroscopic stabilization, was pointed out by W.T. Kelvin and P.G. Tait [140]. They made the assertion, see [140, 345^VI]:

"When there is any dissipativity the equilibrium in the zero position is stable or unstable according as the same system with no motional forces, but with the same positional forces, is stable or unstable. The gyroscopic forces which we now proceed to consider may convert instability into stability" ... "*when there is no dissipativity:* – but when there is any dissipativity gyroscopic forces may convert rapid falling away from an unstable configuration into falling by (as it were) exceedingly gradual spirals, but they cannot convert instability into stability if there be any dissipativity."

That is, the character of A alone determines the stability of the system containing both dissipative and gyroscopic terms.

Ziegler made a similar conclusion following [286, Example 16.1] when he states, see [286, p. 117],

"*Theorem 16b. Dissipative forces, applied to other than purely nongyroscopic systems, may have a destabilizing effect.*"

W.T. Kelvin and P.G. Tait's statement was proved by N.G. Četaev, see [52, p. 89] with the aid of Ljapunov function techniques. It is now commonly referred to as the Kelvin–Tait–Četaev theorem. For the finite-dimensional case another proof of the Kelvin–Tait–Četaev theorem was given in [283], who also proved that the number of eigenvalues, with account of multiplicity, of the matrix pencil $\lambda^2 I - i\lambda K - \lambda B - A$ is equal to the total multiplicity of the negative spectrum of A. The matrix formulation of the Kelvin–Tait–Četaev–Zajac theorem has been extended to operators in Hilbert spaces by A.I. Miloslavskii, see [191], [192]. Theorem 4.1.8, which was proved in [208], is a generalization of the Kelvin–Tait–Četaev–Zajac–Miloslavskii theorem to quadratic operator pencils which are allowed to possess essential spectrum.

The Kelvin–Tait–Četaev theorem agrees with experience. Consider a child's top. The unspun top is unstable. With sufficient spin and with no friction the upright orientation of the top is stable. But the top falls down due to damping at the support.

If $K = 0$, existence of gyroscopic stabilization is clear in the finite-dimensional case if, for example, the matrix B is positive. Then the pencil $\lambda^2 I - \lambda\eta B - A$ has only real and semisimple eigenvalues for $\eta > 0$ large enough. More refined results on this topic can be found in [165], [116], [118], [117], [161], [160], [267].

The results in Subsection 4.3.2 are based on work in [209], [210], [217]. It was shown in [200] that the pipe is stable for $0 \le v < \pi$ and unstable for $v = \pi n$, $n \in \mathbb{N}$.

Problem (4.3.2)–(4.3.5) and related problems were studied in [206], [68], [67], [191] and [284] by numerical methods. It was shown in [206] that gyroscopic stabilization is possible, i. e., it was shown that although the pipe is unstable for some v and $\eta = 0$, it can be stable for the same v and some $\eta > 0$. In [284] gyroscopic stabilization was obtained for this problem, where it is stated that for $\eta > 3^{-1/2}$

and $v = 2\pi + \delta^2$, where δ is small enough, problem (4.3.2)–(4.3.3) is stable. But the proof is probably not correct.

The instability index

$$\kappa = \sum_{j=1}^{N} n_{-j} + \sum_{k=1}^{\infty}(n_k - p_k),$$

was investigated in [164], where n_{-j} are the algebraic multiplicities of the eigenvalues of the pencil located in the lower half-plane, and n_k are the algebraic and p_k are the geometric multiplicities of the eigenvalues of the pencil located on the real axis.

In [23] and [24] another situation was considered where a different version of gyroscopic stabilization occurred. In these papers the operators were supposed to be self-adjoint and bounded, and the pencil is given by $L(\lambda) = \lambda^2 I + \lambda B + A$. The quadratic operator pencil L is said to be (almost) gyroscopically stabilized if $A > 0$, B is invertible and indefinite and $|B| - kI - k^{-1}A$ is positive (semi-)definite for some $k > 0$, where $|B|$ denotes the positive square root of B^2. It was shown in [24] that the spectrum of an almost gyroscopically stabilized pencil is real. However, this result cannot be applied to the operator pencils considered in Subsection 4.3.2. Whereas the unboundedness of our operators may be more of a technical issue, the definiteness of A is crucial as the following simple example shows. For

$$A = \begin{pmatrix} 0 & 1 \\ 1 & 0 \end{pmatrix}, \quad B = \begin{pmatrix} 2\eta & 0 \\ 0 & -2\eta \end{pmatrix}$$

it is easy to see that the pencil L is almost gyroscopically stabilized if $\eta \geq 1$. But the eigenvalues are the four numbers λ satisfying

$$\lambda^2 = 2\eta^2 \pm \sqrt{4\eta^2 + 1},$$

which shows that the pencil has two real eigenvalues and two nonreal eigenvalues.

A collection of problems for finite-dimensional quadratic operator pencils, including those with gyroscopic terms, together with algorithms for finding eigenvalues can be found in [29].

We now return to the pencil with a damping term but without gyroscopic term, i.e., $\lambda^2 M - i\lambda K - A$, where we have introduced a bounded self-adjoint operator $M \gg 0$ in accordance with the publications we will discuss below. Clearly, this operator pencil is similar to a monic operator pencil. In [182, Section 31] the pencil $\tau^2 M + \tau K + A$ is called *hyperbolic* if

$$(Ky, y)^2 - 4(Ay, y)(My, y) > 0, \quad y \in D(A) \setminus \{0\}.$$

Clearly, the pencil is equivalent to our pencil with respect to the parameter transformation $\lambda = -i\tau$. With the physical meaning of this condition in mind, this condition is also called *overdamped*, see [69] and [268, Chapter 14] for the finite-dimensional case. It is clear that under this condition the pencil has only pure imaginary eigenvalues. Indeed, if λ is an eigenvalue and y is a corresponding eigenvector, then

$$\lambda = \frac{i(Ky, y) \pm \sqrt{-(Ky, y)^2 + 4(My, y)(Ay, y)}}{2(My, y)}$$

is pure imaginary. This result can be proved by variational principles, see [1, Ch. 5, Theorem 5.4]. Conversely, the eigenvalues are not pure imaginary in the case of a weakly damped pencil, i.e., when $(Ky, y)^2 < 4(My, y)(Ay, y)$, $y \in D(A) \setminus \{0\}$.

Rotating the eigenvalue parameter, hyperbolic problems are linked to gyroscopically stabilized problems. For more details we refer the reader to [23] and [24].

Part II

Hermite–Biehler Functions

Chapter 5

Generalized Hermite–Biehler Functions

5.1 S-functions and Hermite–Biehler functions

In this section we present some useful definitions and classic results on entire functions due to B.Ya. Levin, N.I. Ahiezer, M.G. Kreĭn, N.N. Meĭman as they are described in [173].

If Ω is an open subset of the complex plane $\mathbb{C}$, then $\Omega^* = \{\lambda \in \mathbb{C} : \overline{\lambda} \in \Omega\}$ denotes its conjugate complex set. For a function ω which is defined and analytic on a domain Ω, its conjugate complex $\overline{\omega}$ is defined on Ω^* by

$$\overline{\omega}(\lambda) = \overline{\omega(\overline{\lambda})}, \quad \lambda \in \Omega^*.$$

Then also $\overline{\omega}$ is analytic.

In the following, we will not take explicit care of the domains of analytic functions. For example, if we take a sum of two analytic functions, then its domains will be the intersection of the domains of the summands, possibly enlarged through analytic continuation. In particular, when both ω and $\overline{\omega}$ occur in an algebraic expression such as a sum and quotient, then it will be implicitly assumed that $\Omega \cap \Omega^*$ is dense in Ω; often Ω and Ω^* will coincide or differ in a discrete set of points. In particular, since Ω is connected, $\Omega \cap \mathbb{R}$ is an open nonempty subset of $\mathbb{R}$.

Definition 5.1.1 ([173, pp. 305, 317]). The real and imaginary parts $\mathrm{R}\,\omega$ and $\mathrm{I}\,\omega$ of the analytic function ω are defined by

$$\mathrm{R}\,\omega = \frac{1}{2}(\omega + \overline{\omega}), \quad \mathrm{I}\,\omega = \frac{1}{2i}(\omega - \overline{\omega}).$$

An analytic function ω is said to be real if it takes real values on the real axis.

Remark 5.1.2. 1. By the identity theorem it is clear that ω is real if and only if $\omega = \overline{\omega}$.

2. Let Ω be the domain of the analytic function ω and assume that $\Omega \cap \Omega^*$ is dense in Ω. Then the real and imaginary parts $R\omega$ and $I\omega$ are real, and

$$\omega = R\omega + iI\omega.$$

If ω is written in the form

$$\omega = P + iQ \tag{5.1.1}$$

with real analytic functions P and Q, then P and Q are uniquely determined by ω, and $P = R\omega$ and $Q = I\omega$. We will often write P and Q instead of $R\omega$ and $I\omega$.

Definition 5.1.3 ([173, p. 314]). The pair of real entire functions P and Q is said to be a *real pair* if P and Q have no common zeros and if all zeros of any of their nontrivial real linear combinations $\mu P + \vartheta Q$, $\mu, \vartheta \in \mathbb{R}$, $\mu^2 + \vartheta^2 > 0$, are real.

Definition 5.1.4 ([173, p. 307]). The entire function ω is said to be a function of Hermite–Biehler class (HB class) if it has no zeros in the closed lower half-plane $\overline{\mathbb{C}^-}$ and if

$$\left| \frac{\omega(\lambda)}{\overline{\omega}(\lambda)} \right| < 1 \quad \text{for } \operatorname{Im} \lambda > 0. \tag{5.1.2}$$

Remark 5.1.5. For nonconstant polynomials, (5.1.2) is redundant since it follows from the absence of zeros in the closed lower half-plane.

We recall the Weierstrass Factorization Theorem, see, e. g., [55, Theorem VII.5.14]: every nonzero entire function ω has a product representation of the form

$$\omega(\lambda) = \lambda^m e^{g(\lambda)} \prod_{k=1}^{(\infty)} \left(1 - \frac{\lambda}{a_k} \right) e^{P_k\left(\frac{\lambda}{a_k}\right)}, \tag{5.1.3}$$

where m is the multiplicity of the zero of ω at 0, with $m = 0$ if $\omega(0) \neq 0$, (∞) indicates that the product may end at a finite number or be void, g is an entire function, the a_k are the nonzero zeros of ω, counted with multiplicity, and where the P_k are polynomials of the form

$$\sum_{j=1}^{l_k} \frac{\lambda^j}{j} \tag{5.1.4}$$

of suitable degree l_k, with the zero polynomial if $l_k = 0$. Convergence of the infinite product is guaranteed if the degree of P_k is chosen to be k.

If convenient, also nonpositive integers may be used in the indexing of the zeros in (5.1.3), the P_k may be chosen to be zero for finitely many k, and $1 - \frac{\lambda}{a_k}$ may be replaced with $\lambda - a_k$ for finitely many k.

For convenience, we introduce the following notations.

Remark 5.1.6. Let $\mathfrak{a}$ and $\mathfrak{b}$ be countable sets of real numbers, at least one of them nonempty, without limit points. Assume that $\mathfrak{a}$ and $\mathfrak{b}$ are interlacing sets, i.e., $\mathfrak{a} \cap \mathfrak{b} = \emptyset$, if $\mathfrak{a}$ has at least two elements, then between any two elements of $\mathfrak{a}$ there is an element of $\mathfrak{b}$, and if $\mathfrak{b}$ has at least two elements, then between any two elements of $\mathfrak{b}$ there is an element of $\mathfrak{a}$. With $\mathfrak{a}$ and $\mathfrak{b}$ we associate index sets $I_{\mathfrak{a}}$ and $I_{\mathfrak{b}}$ which are sets of consecutive integers such that we can write $\mathfrak{a} = \{a_k : k \in I_{\mathfrak{a}}\}$ and $\mathfrak{b} = \{b_k : k \in I_{\mathfrak{b}}\}$, where both a_k and b_k are strictly increasing with k, and such that a_k is the largest element in $\mathfrak{a}$ smaller than b_k and b_k is the smallest element in $\mathfrak{b}$ larger than a_k, with the possible exception of a largest a_{k_+} in $\mathfrak{a}$ and a smallest b_{k_-} in $\mathfrak{b}$. We can choose the index sets in such a way that $a_k < 0$ and $b_k < 0$ whenever $k < 0$, and $a_k > 0$ and $b_k > 0$ whenever $k > 0$. Whenever possible, we will assume that $0 \in I_{\mathfrak{a}} \cap I_{\mathfrak{b}}$, but we will always assume that $0 \in I_{\mathfrak{a}} \cup I_{\mathfrak{b}}$.

We say that the zeros of two entire functions interlace if all zeros are real and simple and if the sets of their zeros interlace. An analogue definition is used for interlacing poles and zeros of a meromorphic function on $\mathbb{C}$.

Below we will consider products of functions containing the factors $\lambda - a_0$ and $\lambda - b_0$, respectively. If $0 \notin I_{\mathfrak{a}}$, then $\lambda - a_0$ has to be replaced with 1, whereas if $0 \notin I_{\mathfrak{b}}$, then $\lambda - b_0$ has to be replaced with -1. In order to remind the reader of this convention, we will use the notations $(\lambda - a_0)_+$ and $(\lambda - b_0)_-$, respectively.

Finally, we observe that if $I_{\mathfrak{a}} = I_{\mathfrak{b}} = \mathbb{Z}$, then the ordering conditions can be written as $a_k < b_k < a_{k+1}$ for all $k \in \mathbb{Z}$ and $b_{-1} < 0 < a_1$, and this extends to the general case whenever the corresponding numbers exist in the sets $I_{\mathfrak{a}}$ and $I_{\mathfrak{b}}$.

The following criterion is due to N.N. Meĭman, see [173, Theorem 3, p. 311].

Theorem 5.1.7. *Let $\omega = P + iQ$, where P and Q are real entire functions. For the function ω to be of HB class it is necessary and sufficient that the following conditions are satisfied:*

(i) *all zeros of the functions P and Q are real, at least one of the functions P or Q has at least one zero, and the zeros of the functions P and Q interlace;*

(ii) *the functions P and Q have the following expansions into infinite products:*

$$P(\lambda) = A e^{u(\lambda)} (\lambda - a_0)_+ \prod_{k \in I_{\mathfrak{a}} \setminus \{0\}} \left(1 - \frac{\lambda}{a_k}\right) e^{P_k\left(\frac{\lambda}{a_k}\right)}, \quad u(0) = 0, \qquad (5.1.5)$$

$$Q(\lambda) = B e^{v(\lambda)} (\lambda - b_0)_- \prod_{k \in I_{\mathfrak{b}} \setminus \{0\}} \left(1 - \frac{\lambda}{b_k}\right) e^{Q_k\left(\frac{\lambda}{b_k}\right)}, \quad v(0) = 0, \qquad (5.1.6)$$

where u and v are entire functions and $I_{\mathfrak{a}}$ and $I_{\mathfrak{b}}$ are as defined in Remark 5.1.6.

(iii) *the constants A and B are nonzero real numbers and have the same sign;*

(iv) *the entire functions u and v and the polynomials P_k and Q_k satisfy*

$$u(\lambda) - v(\lambda) + \sum_{k=-\infty, k\neq 0}^{\infty} \left[P_k\left(\frac{\lambda}{b_k}\right) - Q_k\left(\frac{\lambda}{a_k}\right) \right] = 0, \qquad (5.1.7)$$

where $P_k = 0$ or $Q_k = 0$ if $k \notin I_a$ or $k \notin I_b$, respectively.

Proof. First assume that ω is of HB class. Then ω has no zero in the closed lower half-plane, i.e., ω and $\overline{\omega}$ and thus P and Q have no common zeros. By Lemma 11.1.1, $\theta = QP^{-1}$ is a real meromorphic function which maps the open upper half-plane into itself. Since P and Q have no common zeros, the zeros of θ are the zeros of Q and the poles of θ are the zeros of P. Hence property (i) immediately follows from Theorem 11.1.6. Clearly, the product representations (5.1.3) of P and Q can be written in the form (5.1.5) and (5.1.6), respectively, where A and B are nonzero numbers. From

$$A = \lim_{\lambda \to 0} \frac{P(\lambda)}{(\lambda - a_0)_+}$$

we see that $A \in \mathbb{R}$; similarly, $B \in \mathbb{R}$. Furthermore, with the notations from Theorem 11.1.6 and this theorem we have that

$$\frac{B}{A} e^{w(\lambda)} = C,$$

where $C > 0$ and

$$w(\lambda) = v(\lambda) - u(\lambda) + \sum_{k=-\infty, k\neq 0}^{\infty} \left[Q_k\left(\frac{\lambda}{b_k}\right) - P_k\left(\frac{\lambda}{a_k}\right) \right].$$

Therefore w is constant, and $u(0) = v(0) = P_k(0) = Q_k(0) = 0$ shows that $w = 0$ and thus A and B have the same sign.

Conversely, assume that the conditions (i)–(iv) are satisfied. Then $\theta = QP^{-1}$ is a real meromorphic function θ of the form (11.1.2). By Theorem 11.1.6 it follows that θ maps the open upper half-plane into itself, and Lemma 11.1.1 shows that ω has no zeros in the open lower half-plane and that $F = \omega\overline{\omega}^{-1}$ maps the open upper half-plane into the open unit disc. Finally, since a real zero of ω would be a common real zero of the real entire functions P and Q, it follows that ω cannot have real zeros. $\qquad\square$

Definition 5.1.8 ([173, p. 313]). An entire function ω which has no zeros in the open lower half-plane and satisfies the condition

$$\left| \frac{\omega(\lambda)}{\overline{\omega}(\lambda)} \right| \leq 1 \quad \text{for } \operatorname{Im} \lambda > 0 \qquad (5.1.8)$$

is said to be a function of the generalized Hermite–Biehler class ($\overline{\text{HB}}$ class).

The next proposition shows that generalized Hermite–Biehler functions can be characterized in terms of Hermite–Biehler functions and real analytic functions with real zeros.

Theorem 5.1.9. *An entire function ω is of $\overline{HB}$ class if and only if $\omega = \omega_1\omega_2$, where ω_2 is a real entire function with only real zeros and ω_1 is either of HB class or a nonzero constant.*

Proof. It is easy to see that if $\omega = \omega_1\omega_2$, where ω_2 is a real function with only real zeros and ω_1 is either of HB class or a nonzero constant, then ω is of $\overline{HB}$ class.

Conversely, let ω be of $\overline{HB}$ class. If $|\omega(\lambda)\overline{\omega}^{-1}(\lambda)| = 1$ for some λ in the upper half-plane, then $\omega\overline{\omega}^{-1}$ would be a constant function in the open upper half-plane by the maximum modulus principle. Hence there is a complex number α_1 with $|\alpha_1| = 1$ such that $\alpha_1\omega = \overline{\omega}$ in the open upper half-plane and hence also in $\mathbb{C}$ by the identity theorem. Choosing $\alpha \in \mathbb{C}$ such that $\alpha^2\alpha_1 = 1$ and observing that $\alpha^{-1} = \overline{\alpha}$, it follows that

$$\omega_2 := \alpha^{-1}\omega = \alpha\alpha_1\omega = \alpha\overline{\omega} = \overline{\alpha^{-1}\omega} = \overline{\omega_2},$$

so that ω_2 is real and $\alpha\omega_2 = \omega$. Furthermore, since ω has no zeros in the open lower half-plane, so has the function ω_2. Since the zeros of a real function are symmetric with respect to the real axis, ω_2 therefore has only real zeros.

Now let ω be of $\overline{HB}$ class satisfying (5.1.2). Using a product representation we see that there is a real analytic function ω_2 which has the same real zeros as ω, counted with multiplicity. Then $\omega_1 := \omega\omega_2^{-1}$ has no zeros in the closed lower half-plane and satisfies (5.1.2), i.e., ω_1 is of HB class. □

Theorems 5.1.7 and 5.1.9 give a first characterization of functions in the generalized Hermite–Biehler class, see also [173, p. 314].

Proposition 5.1.10. *The closure of the HB class in the topology of uniform convergence on compact subsets of $\mathbb{C}$ consists of all functions in the $\overline{HB}$ class and the zero function.*

Proof. If $(\omega_n)_{n=1}^{\infty}$ is a sequence in HB which converges uniformly on compact subset of $\mathbb{C}$ to some ω, then either ω has no zeros in the open lower half-plane or ω is identically zero by Hurwitz' theorem, see [55, Corollary VII.2.6]. Clearly, (5.1.2) for ω_n implies (5.1.8) for ω if ω is not identically zero.

Conversely, if ω is a real analytic function of $\overline{HB}$ class and if it has at least one real zero, then we replace all real zeros a_k by the complex zeros $a_k + i\alpha_{kn}$ in the product representation of ω. For suitably chosen small positive numbers α_{kn} the product representation converges to a function ω_n of HB class, and the ω_n converge to ω uniformly on compact sets. If ω is a real analytic function without zeros, then $\omega_n(\lambda) = (1 + i\frac{\lambda}{n})\omega(\lambda)$ is a sequence of functions of HB class converging to ω. An application of Theorem 5.1.9 shows that every function of $\overline{HB}$ class is the uniform limit on compact subsets of functions of HB class. Finally, $\chi_n(\lambda) = \frac{1}{n}(\lambda - i)$ defines a sequence $(\chi_n)_1^{\infty}$ in HB which converges to 0 uniformly on compact sets. □

The following theorem is due to M.G. Kreĭn, see [173, Theorem 6, p. 318]. We note that a polynomial P_k of the form (5.1.4) is a real polynomial. Below the notation $\operatorname{R} P_k\left(\frac{\lambda}{\alpha_k}\right)$ for $\alpha_k \in \mathbb{C} \setminus \{0\}$ means the real part of the function $\lambda \mapsto P_k\left(\frac{\lambda}{\alpha_k}\right)$.

Theorem 5.1.11. *In order that the entire function ω be of $\overline{\mathrm{HB}}$ class, it is necessary and sufficient that it can be represented in the form*

$$\omega(\lambda) = \lambda^m e^{u(\lambda)+i(\nu\lambda+\delta)} \prod_{k=1}^{(\infty)} \left(1 - \frac{\lambda}{\alpha_k}\right) e^{RP_k\left(\frac{\lambda}{\alpha_k}\right)}, \tag{5.1.9}$$

where $m \in \mathbb{N}_0$, $\nu \geq 0$, $\delta \in \mathbb{R}$, u is a real entire function and

$$\sum_{k=1}^{(\infty)} \left|\operatorname{Im} \frac{1}{\alpha_k}\right| < \infty, \qquad \operatorname{Im} \alpha_k \geq 0 \text{ for all } k. \tag{5.1.10}$$

Proof. For the convenience of the reader we recall Levin's proof. First let ω be of $\overline{\mathrm{HB}}$ class and denote the sequence of its nonzero zeros by $(\alpha_k)_{k=1}^{(\infty)}$. The function ψ defined by

$$\psi(\lambda) = \frac{\omega(\lambda)}{\overline{\omega}(\lambda)}$$

satisfies the assumptions of Lemma 11.1.9. Hence (5.1.10) holds since the zeros of ω and ψ in the upper half-plane coincide and since the real zeros of ω do not contribute to the sum in (5.1.10). Therefore the series

$$\sum_{k=1}^{(\infty)} \lambda \left(1 - \frac{\lambda}{\overline{\alpha_k}}\right)^{-1} \left(\frac{1}{\overline{\alpha_k}} - \frac{1}{\alpha_k}\right) = -2i \sum_{k=1}^{(\infty)} \lambda \left(1 - \frac{\lambda}{\overline{\alpha_k}}\right)^{-1} \operatorname{Im} \frac{1}{\alpha_k}$$

converges absolutely and uniformly on each compact subset of $\mathbb{C}$ which does not contain any of the points $\overline{\alpha_k}$. Since

$$1 + \lambda \left(1 - \frac{\lambda}{\overline{\alpha_k}}\right)^{-1} \left(\frac{1}{\overline{\alpha_k}} - \frac{1}{\alpha_k}\right) = \left(1 - \frac{\lambda}{\alpha_k}\right)\left(1 - \frac{\lambda}{\overline{\alpha_k}}\right)^{-1},$$

it follows that

$$\chi(\lambda) = \prod_{k=1}^{(\infty)} \left(1 - \frac{\lambda}{\alpha_k}\right)\left(1 - \frac{\lambda}{\overline{\alpha_k}}\right)^{-1} \tag{5.1.11}$$

converges absolutely and uniformly on each such set to a function without zeros, see, e. g., [55, Corollary VII.5.6]. Let

$$\omega(\lambda) = \lambda^m e^{g(\lambda)} \prod_{k=1}^{(\infty)} \left(1 - \frac{\lambda}{\alpha_k}\right) e^{P_k\left(\frac{\lambda}{\alpha_k}\right)} \tag{5.1.12}$$

be the Weierstrass factorization of ω. Defining

$$\chi_n(\lambda) = \prod_{k=1}^{n} \left(1 - \frac{\lambda}{\alpha_k}\right) \left(1 - \frac{\lambda}{\overline{\alpha_k}}\right)^{-1},$$

$$\omega_n(\lambda) = \lambda^m e^{g(\lambda)} \prod_{k=1}^{n} \left(1 - \frac{\lambda}{\alpha_k}\right) e^{P_k\left(\frac{\lambda}{\alpha_k}\right)},$$

it follows that

$$\frac{\omega_n(\lambda)}{\overline{\omega_n}(\lambda)} [\chi_n(\lambda)]^{-1} = \exp 2i \left[\mathrm{I}\, g(\lambda) + \sum_{k=1}^{n} \mathrm{I}\, P_k\left(\frac{\lambda}{\alpha_k}\right)\right], \tag{5.1.13}$$

so that

$$\sum_{k=1}^{n} \mathrm{I}\, P_k\left(\frac{\lambda}{\alpha_k}\right)$$

converges uniformly on each compact subset of $\mathbb{C}$ as $n \to \infty$. Defining

$$v(\lambda) = \mathrm{I}\, g(\lambda) + \sum_{k=1}^{(\infty)} \mathrm{I}\, P_k\left(\frac{\lambda}{\alpha_k}\right) \quad \text{and} \quad u(\lambda) = \mathrm{R}\, g(\lambda), \tag{5.1.14}$$

we can write

$$\omega(\lambda) = \lambda^m e^{u(\lambda)+iv(\lambda)} \prod_{k=1}^{(\infty)} \left(1 - \frac{\lambda}{\alpha_k}\right) e^{\mathrm{R}\, P_k\left(\frac{\lambda}{\alpha_k}\right)}. \tag{5.1.15}$$

Since

$$\chi_n(\lambda) \to \prod_{k=1}^{n} \frac{\overline{\alpha_k}}{\alpha_k} \neq 0 \text{ as } \lambda \to \infty,$$

the maximum modulus principle gives that $\lambda \mapsto \omega(\lambda)[\overline{\omega}(\lambda)]^{-1}[\chi_n(\lambda)]^{-1}$ is a bounded function in the closed upper half-plane. Since it has modulus 1 on the real axis, it follows from the Phragmén–Lindelöf principle, see [173, Theorem 20, p. 48] or [55, Corollary VI.4.4], that

$$\left|\frac{\omega(\lambda)}{\overline{\omega}(\lambda)} [\chi_n(\lambda)]^{-1}\right| \leq 1 \quad \text{for} \quad \mathrm{Im}\, \lambda \geq 0. \tag{5.1.16}$$

Passing to the limits as $n \to \infty$ in (5.1.13) and (5.1.16) and taking the definition of v in (5.1.14) into account, it follows that $|e^{2iv(\lambda)}| \leq 1$ for $\mathrm{Im}\, \lambda \geq 0$, which means that v maps the closed upper half-plane into itself. If v is not constant, it must map the open upper half-plane into itself by the open mapping theorem, and since v has no poles, Theorem 11.1.6 shows that $v(\lambda) = C(\lambda - c_0)$ for some $c_0 \in \mathbb{R}$ and $C > 0$. Thus, in any case, $v(\lambda) = \nu\lambda + \delta$ with $\nu \geq 0$ and $\delta \in \mathbb{R}$, and (5.1.9) follows.

Conversely, if (5.1.9) holds subject to (5.1.10), then ω is an entire function without zeros in the open lower half-plane, and

$$\frac{\omega(\lambda)}{\overline{\omega}(\lambda)} = e^{2i(\nu\lambda+\delta)}\chi(\lambda)$$

implies that, for $\operatorname{Im}\lambda > 0$,

$$\left|\frac{\omega(\lambda)}{\overline{\omega}(\lambda)}\right| = \left|e^{2i\nu\lambda}\right|\,|\chi(\lambda)| \le |\chi(\lambda)| \le 1, \qquad (5.1.17)$$

which shows that ω is of $\overline{\mathrm{HB}}$ class. $\qquad\square$

Theorem 5.1.12. *In order that the entire function ω be of HB class, it is necessary and sufficient that it can be represented in the form*

$$\omega(\lambda) = e^{u(\lambda)+i(\nu\lambda+\delta)} \prod_{k=1}^{(\infty)} \left(1 - \frac{\lambda}{\alpha_k}\right) e^{\mathrm{R}\,P_k\left(\frac{\lambda}{\alpha_k}\right)}, \qquad (5.1.18)$$

where $\nu \ge 0$, δ is a real number, u is a real entire function,

$$\sum_{k=1}^{(\infty)} \left|\operatorname{Im}\frac{1}{\alpha_k}\right| < \infty, \qquad \operatorname{Im}\alpha_k > 0 \text{ for all } k, \qquad (5.1.19)$$

and $\nu > 0$ if ω has no zeros.

Proof. Assume that ω is of HB class. Then ω is of $\overline{\mathrm{HB}}$ class without real zeros, so that (5.1.18) and (5.1.19) follow from Theorem 5.1.11. If ω has no zeros, then χ defined by (5.1.11) satisfies $\chi = 1$ since it is defined by an empty product, and (5.1.17) shows that $\nu > 0$ is necessary for (5.1.2) to hold.

Conversely, if (5.1.18) and (5.1.19) hold, then ω has no zeros in the lower half-plane and is of $\overline{\mathrm{HB}}$ class by Theorem 5.1.11. Finally, since $|\chi(\lambda)| < 1$ for $\operatorname{Im}\lambda > 0$ if ω has zeros and since $|e^{2i\nu\lambda}| < 1$ for $\operatorname{Im}\lambda > 0$ if $\nu > 0$, (5.1.2) follows from (5.1.17), taking into account that $\nu > 0$ if ω has no zeros. $\qquad\square$

We will use also other criteria for an entire function to be of HB class.

Theorem 5.1.13 ([173, Theorem 4, p. 315]). *Let $\omega = P + iQ$, where P and Q are real entire functions. Then ω is of HB class if and only if P and Q form a real pair and at any point x_0 of the real axis the inequality*

$$Q'(x_0)P(x_0) - Q(x_0)P'(x_0) > 0 \qquad (5.1.20)$$

holds.

Proof. Assume that P and Q are a real pair satisfying (5.1.20). Then the real meromorphic function

$$\theta = \frac{Q}{P}$$

does not have zeros or poles in the open upper and open lower half-planes. In particular, θ must map the open upper half-plane into the open upper half-plane or into the open lower half-plane. Hence θ or $-\theta$ maps the open upper half-plane into itself, so that $\theta' > 0$ or $-\theta' > 0$, respectively, on the real axis by Lemma 11.1.3. Since

$$\theta' = \frac{Q'P - P'Q}{P^2}, \qquad (5.1.21)$$

and $Q'(x_0)P(x_0) - P'(x_0)Q'(x_0) > 0$ for at least one $x_0 \in \mathbb{R}$, it follows that $\theta' > 0$, and therefore θ maps the open upper half-plane into itself. From the equivalence of properties (ii) and (iii) in Lemma 11.1.1, part 1, it follows that ω is of HB class.

Conversely, if ω is of HB class, then we know from Lemma 11.1.1 that θ maps the open upper half-plane into itself, and thus the open lower half-plane into itself since θ is a real meromorphic function. Therefore, P and Q form a real pair. Again by Lemma 11.1.3 and by (5.1.21), $Q'(x)P(x) - P'(x)Q(x) > 0$ for all $x \in \mathbb{R}$ which are not zeros of P. But if $P(x) = 0$, then $Q'(x)P(x) - P'(x)Q(x) = P'(x)Q(x) \neq 0$ since the zeros of P and Q are simple and interlace by Lemma 11.1.3. By continuity, $Q'(x)P(x) - P'(x)Q(x) > 0$ follows for all $x \in \mathbb{R}$. $\qquad\square$

Definition 5.1.14. An entire function of $\overline{\text{HB}}$ class is said to be nontrivial if it is not a constant multiple of a real function.

It is clear that a function ω of $\overline{\text{HB}}$ class is nontrivial if and only if its real and imaginary parts are not multiples of each other.

Corollary 5.1.15. *Let $\omega = P + iQ$ be an entire function of $\overline{\text{HB}}$ class, where P and Q are real entire functions. Then*

$$Q'(x)P(x) - Q(x)P'(x) \geq 0 \qquad (5.1.22)$$

holds for all $x \in \mathbb{R}$, and

$$Q'(x_0)P(x_0) - Q(x_0)P'(x_0) > 0$$

for some $x_0 \in \mathbb{R}$ if ω is nontrivial.

Proof. By Proposition 5.1.10, ω is the uniform limit on compact subsets of functions of HB class. Hence also their real and imaginary parts have this property, and therefore converge, together with their derivatives, uniformly, see [55, Theorem VII.2.1]. Therefore (5.1.22) follows from (5.1.20).

Now assume that ω is nontrivial. Then $\theta = \frac{Q}{P}$ is not constant, so that $Q'P - QP'$ has only isolated zeros and therefore some nonzero values on the real axis. From (5.1.22) it follows that these nonzero values must be positive. $\qquad\square$

Definition 5.1.16. The function ω of HB class ($\overline{\text{HB}}$ class) is said to be a function of symmetric Hermite–Biehler (SHB) class (a function of symmetric generalized Hermite–Biehler class ($\overline{\text{SHB}}$ class)) if $\omega(-\bar{\lambda}) = \overline{\omega(\lambda)}$ for all $\lambda \in \mathbb{C}$.

Remark 5.1.17. In the sequel we will frequently deal with functions which are symmetric with respect to the imaginary axis, that is, entire functions ω which satisfy $\omega(-\bar{\lambda}) = \overline{\omega(\lambda)}$ for all $\lambda \in \mathbb{C}$. This property can be rewritten in the form $\overline{\omega}(-\lambda) = \omega(\lambda)$ for all $\lambda \in \mathbb{C}$. For an arbitrary entire function ω define ω_I by $\omega_I(\lambda) = \omega(i\lambda)$, $\lambda \in \mathbb{C}$. Observing that

$$\overline{\omega_I}(\lambda) = \overline{\omega_I(\bar{\lambda})} = \overline{\omega(i\bar{\lambda})} = \overline{\omega}(-i\lambda), \quad \lambda \in \mathbb{C},$$

holds, it is immediately clear that ω is symmetric with respect to the imaginary axis if and only if ω_I is a real entire function, that is, if and only if ω is real on the imaginary axis.

Proposition 5.1.18. *Let ω be an entire function which is symmetric with respect to the imaginary axis, i. e., $\omega(-\bar{\lambda}) = \overline{\omega(\lambda)}$ for all $\lambda \in \mathbb{C}$. Then $P(-\bar{\lambda}) = \overline{P(\lambda)}$ and $Q(-\bar{\lambda}) = -\overline{Q(\lambda)}$ for all $\lambda \in \mathbb{C}$. In particular, $Q(0) = 0$ and all real zeros of P and Q are symmetric with respect to the origin. Furthermore, there are unique real entire functions P_s and Q_s such that*

$$\omega(\lambda) = P_s(\lambda^2) + i\lambda Q_s(\lambda^2), \quad \lambda \in \mathbb{C}. \tag{5.1.23}$$

Proof. Since

$$P = \frac{1}{2}(\omega + \overline{\omega}), \quad Q = \frac{1}{2i}(\omega - \overline{\omega}),$$

it follows in view of Remark 5.1.17 that P and iQ are real on the imaginary axis, and thus $P(-\bar{\lambda}) = \overline{P(\lambda)}$ and $Q(-\bar{\lambda}) = -\overline{Q(\lambda)}$ for all $\lambda \in \mathbb{C}$. Since P and Q are real entire functions, the statement about the real zeros of P and Q follows from $P(-x) = P(x)$ and $Q(-x) = -Q(x)$ for $x \in \mathbb{R}$.

For $\lambda \in \mathbb{C}$ define

$$\tilde{P}(\lambda) = \frac{\omega(\lambda) + \omega(-\lambda)}{2} = P(\lambda), \quad \tilde{Q}(\lambda) = \frac{\omega(\lambda) - \omega(-\lambda)}{2i\lambda} = \frac{Q(\lambda)}{\lambda}. \tag{5.1.24}$$

Clearly, $\tilde{P}$ and $\tilde{Q}$ are even entire functions, so that there are entire functions P_s and Q_s such that

$$P_s(\lambda^2) = \tilde{P}(\lambda), \quad Q_s(\lambda^2) = \tilde{Q}(\lambda), \quad \lambda \in \mathbb{C}. \tag{5.1.25}$$

Hence (5.1.23) holds. Furthermore, $\tilde{P}$ and $\tilde{Q}$ are real on the real and imaginary axes, so that P_s and Q_s are real entire functions.

Clearly, $\mathrm{R}\,\omega(\lambda) = P_s(\lambda^2)$ and $\mathrm{I}\,\omega(\lambda) = \lambda Q_s(\lambda^2)$, $\lambda \in \mathbb{C}$. The uniqueness of $\mathrm{R}\,\omega$ and $\mathrm{I}\,\omega$ shows that the real entire functions P_s and Q_s are uniquely determined by ω. $\qquad\square$

Proposition 5.1.19. *If ω is of* SHB *class, then (5.1.5) and (5.1.6) can be rewritten in the form*

$$P(\lambda) = A_0 e^{u(\lambda^2)} \prod_{k=1}^{(\infty)} \left(1 - \frac{\lambda^2}{a_k^2}\right) e^{P_k\left(\frac{\lambda^2}{a_k^2}\right)}, \qquad (5.1.26)$$

$$Q(\lambda) = B_0 \lambda e^{v(\lambda^2)} \prod_{k=1}^{(\infty)} \left(1 - \frac{\lambda^2}{b_k^2}\right) e^{Q_k\left(\frac{\lambda^2}{b_k^2}\right)}, \qquad (5.1.27)$$

the exponents satisfy the condition

$$u(\lambda^2) - v(\lambda^2) + \sum_{k=1}^{(\infty)} \left(P_k\left(\frac{\lambda^2}{b_k^2}\right) - Q_k\left(\frac{\lambda^2}{a_k^2}\right)\right) = 0, \qquad (5.1.28)$$

and the numbers A_0 and B_0 have the same sign.

Proof. We use the notation of Theorem 5.1.7. Since $Q(0) = 0$, we have $b_0 = 0$, and hence $b_{-j} = -b_j$. In the Weierstrass product, we may replace each Q_k with a corresponding polynomial of higher order, so that we may assume that Q_k and Q_{-k} have the same even order $2l_{k'}$ and thus are identical. Then we can write (5.1.6) as

$$Q(\lambda) = B e^{v(\lambda)} \lambda \prod_{k=1}^{(\infty)} \left(1 - \frac{\lambda}{b_k}\right)\left(1 - \frac{\lambda}{-b_k}\right) e^{Q_k\left(\frac{\lambda}{b_k}\right) + Q_k\left(\frac{\lambda}{-b_k}\right)} = 0,$$

where

$$Q_k\left(\frac{\lambda}{b_k}\right) + Q_k\left(\frac{\lambda}{-b_k}\right) = \sum_{j=1}^{l_k'} \left(\frac{\lambda^j}{j b_k^j} + (-1)^j \frac{\lambda^j}{j b_k^j}\right)$$

$$= \sum_{j=1}^{l_k'} \frac{(\lambda^2)^j}{j(b_k^2)^j}.$$

With some abuse of notation we write the right-hand side as $Q_k\left(\frac{\lambda^2}{a_k^2}\right)$, so that

$$Q(\lambda) = B e^{v(\lambda)} \lambda \prod_{k=1}^{(\infty)} \left(1 - \frac{\lambda^2}{b_k^2}\right) e^{Q_k\left(\frac{\lambda^2}{b_k^2}\right)}.$$

Then

$$Q(-\bar\lambda) = -B e^{v(-\bar\lambda)} \bar\lambda \prod_{k=1}^{(\infty)} \left(1 - \frac{\bar\lambda^2}{b_k^2}\right) e^{Q_k\left(\frac{\bar\lambda^2}{b_k^2}\right)},$$

and Proposition 5.1.18 gives

$$e^{v(-\bar{\lambda})} = \overline{e^{v(\lambda)}}.$$

Since v is a real function, it follows that $v(-\bar{\lambda}) = \overline{v(\lambda)}$, which means that v is an even function.

A similar reasoning holds for P. We only have to observe that $a_{-j+1} = -a_j$ for $j = -1, -2, \ldots$, and that, in case P has zeros, the factor $\lambda - a_0$ has to be written in the form $-a_0 \left(1 - \frac{\lambda}{a_0}\right)$ with $-a_0 > 0$. $\square$

In the sequel we will use the notion of Nevanlinna function, also called R-function in [127] and function of negative imaginary type in [16, Appendix II], and the notion of S-function [127].

Definition 5.1.20. The function θ is said to be a Nevanlinna function, or an R-function, or an $\mathcal{N}$-function ($\hat{\mathcal{N}}$-function) if:

(i) θ is analytic in the half-planes $\operatorname{Im} \lambda > 0$ and $\operatorname{Im} \lambda < 0$;
(ii) $\theta(\bar{\lambda}) = \overline{\theta(\lambda)}$ if $\operatorname{Im} \lambda \neq 0$;
(iii) $\operatorname{Im} \lambda \operatorname{Im} \theta(\lambda) \geq 0$ for $\operatorname{Im} \lambda \neq 0$ ($\operatorname{Im} \lambda \operatorname{Im} \theta(\lambda) > 0$ for $\operatorname{Im} \lambda \neq 0$).

Remark 5.1.21. Let $\theta \in \mathcal{N}$. We recall that the open mapping theorem gives that θ is either constant, where this constant is real, or maps the open upper half-plane into itself. That is, $\mathcal{N} = \hat{\mathcal{N}} \cup \mathbb{R}$, where $\mathbb{R}$ represents here constant functions in $\mathbb{C}$.

Lemma 5.1.22. If $\theta \neq 0$ is a Nevanlinna function, then so are the functions $-\frac{1}{\theta}$ and $(\frac{1}{\theta} + c)^{-1}$ for each real constant c.

Proof. This easily follows from Remark 5.1.21,

$$\operatorname{Im}\left(-\frac{1}{\theta(\lambda)}\right) = \frac{\operatorname{Im} \theta(\lambda)}{|\theta(\lambda)|^2} \quad \text{and} \quad \operatorname{Im}\left(\frac{1}{\theta(\lambda)} + c\right)^{-1} = \frac{-\operatorname{Im} \frac{1}{\theta(\lambda)}}{\left|\frac{1}{\theta(\lambda)} + c\right|^2}. \qquad \square$$

The relation between $\hat{\mathcal{N}}$-functions and functions of HB class is given by the following simple lemma, see [173, p. 307].

Lemma 5.1.23. *For the function*

$$\omega = P + iQ,$$

where P and Q are real entire functions without common nonreal zeros, the statement $P^{-1}Q$ is an $\hat{\mathcal{N}}$-function is equivalent to the condition (5.1.2).

Proof. This follows from Lemma 11.1.1 if we observe that part 2 of Definition 5.1.20 is satisfied since $P^{-1}Q$ is real analytic. $\square$

Definition 5.1.24. The function θ is said to be an S-function or a function of Stieltjes class ($\hat{S}$-function) if:

  (i) θ is analytic in $\mathbb{C} \setminus [0, \infty)$;
 (ii) $\theta(\bar{\lambda}) = \overline{\theta(\lambda)}$ for $\mathrm{Im} \neq 0$;
(iii) $\mathrm{Im}\, \theta(\lambda) \geq 0$ for $\mathrm{Im}\, \lambda > 0$ ($\mathrm{Im}\, \theta(\lambda) > 0$ for $\mathrm{Im}\, \lambda > 0$);
 (iv) $\theta(\lambda) \geq 0$ for all $\lambda \in (-\infty, 0)$.

Definition 5.1.25. An S-function which is meromorphic on $\mathbb{C}$ is said to be of S_0 class if it has no pole at the origin.

Definition 5.1.26.

1. The class $\mathcal{N}^{\mathrm{ep}}$ of essentially positive Nevanlinna functions is the set of all functions $\theta \in \mathcal{N}$ which are analytic in $\mathbb{C} \setminus [0, \infty)$ with the possible exception of finitely many poles.
2. The class $\mathcal{N}_+^{\mathrm{ep}}$ is the set of all functions $\theta \in \mathcal{N}^{\mathrm{ep}}$ such that for some $\gamma \in \mathbb{R}$ we have $\theta(\lambda) > 0$ for all $\lambda \in (-\infty, \gamma)$.
3. The class $\mathcal{N}_-^{\mathrm{ep}}$ is the set of all functions $\theta \in \mathcal{N}^{\mathrm{ep}}$ such that for some $\gamma \in \mathbb{R}$ we have $\theta(\lambda) < 0$ for all $\lambda \in (-\infty, \gamma)$.
4. The class S^{-1} is the set of all functions $\theta \in \mathcal{N}$ such that $\theta(\lambda) \leq 0$ for all $\lambda \in (-\infty, 0)$.

Relations between $\mathcal{N}$-functions and S-functions were obtained by I.S. Kac and M.G. Kreĭn in [126] and [127]. However, we do not need these results in their full generality. In the next section we will deal with meromorphic functions in the whole complex plane, and statements and proofs will be given there for the subclasses consisting of functions which are meromorphic in $\mathbb{C}$.

5.2 Shifted Hermite–Biehler functions

The field of meromorphic functions in the whole complex plane will be denoted by $\mathcal{M}$.

First we recall some properties of meromorphic functions which belong to the classes of Nevanlinna functions defined in Section 5.1. General results without the restriction to $\mathcal{M}$ can be found, e. g., in [127], [13] and [14].

Remark 5.2.1. Let $\theta \in \mathcal{N} \cap \mathcal{M}$. If θ is not constant, it follows from Remark 5.1.21 and Theorem 11.1.6 that θ has a representation of the form (11.1.2). In particular, the zeros and poles of θ interlace.

Lemma 5.2.2. Let $\theta \in \mathcal{N} \cap \mathcal{M}$ be nonconstant. Then $\theta \in S$ if and only if θ has no negative zeros or poles, has at least one pole, and the smallest pole is less than all zeros, if any.

Proof. In view of Remark 5.2.1 and Lemma 11.1.3, θ has at least one pole or zero. Clearly if $\theta \in S$, then all zeros and poles are nonnegative, and in view of Remark 5.2.1 it remains to show that, under this assumption, $\theta(\lambda) > 0$ for all $\lambda < 0$ if and only if θ has at least one pole and the smallest pole is less than all zeros. Clearly, for $\lambda < 0$, the representation (11.1.2) shows that $\theta(\lambda) > 0$ if and only if

$$\frac{(\lambda - b_0)_-}{(\lambda - a_0)_+} > 0.$$

If θ has at least one zero and if the smallest zero is less than any pole, then we have that $0 \notin \mathfrak{a}$, in the notation of Theorem 11.1.6, so that

$$\frac{(\lambda - b_0)_-}{(\lambda - a_0)_+} = \lambda - b_0 < 0 \quad \text{for } \lambda < 0,$$

and it follows that $\theta \notin S$. If, however, θ has at least one pole and the smallest pole is less than any zero, then we have that

$$\text{either} \quad \frac{(\lambda - b_0)_-}{(\lambda - a_0)_+} = \frac{\lambda - b_0}{\lambda - a_0} > 0 \quad \text{or} \quad \frac{(\lambda - b_0)_-}{(\lambda - a_0)_+} = \frac{-1}{\lambda - a_0} > 0 \quad \text{for } \lambda < 0,$$

and it follows that $\theta \in S$. $\qquad\square$

Corollary 5.2.3. *Let $\theta \in \mathcal{N} \cap \mathcal{M}$ be nonconstant. Then $\theta \in \mathcal{N}_+^{\text{ep}}$ if and only if the set of poles of θ is nonempty and bounded below, and the smallest pole of θ is less than all of its zeros, if any.*

Proof. If $\theta \in \mathcal{N}_+^{\text{ep}}$, choose $r \geq 0$ such that θ is positive on $(-\infty, -r]$. Then the function $\lambda \mapsto \theta(\lambda - r)$ belongs to S_0. By Lemma 5.2.2, this function has at least one pole, and the smallest pole of this function is less than all zeros, if any. Clearly, the same properties hold for θ.

Conversely, if a is the smallest pole of θ and all zeros of θ, if any, are larger that a, then $\lambda \mapsto \theta(\lambda - a)$ belongs to S, and it follows that $\theta \in \mathcal{N}_+^{\text{ep}}$. $\qquad\square$

Lemma 5.2.4. *Let $\theta \in \mathcal{M}$. Then*

1. $\theta \in \mathcal{N}^{\text{ep}} \Leftrightarrow -\frac{1}{\theta} \in \mathcal{N}^{\text{ep}}$;
2. $\theta \in \mathcal{N}_+^{\text{ep}} \Leftrightarrow -\frac{1}{\theta} \in \mathcal{N}_-^{\text{ep}}$;
3. $\theta \in S \Leftrightarrow -\frac{1}{\theta} \in S^{-1}$;
4. $\theta \in S \Leftrightarrow (\theta \in \mathcal{N} \text{ and } \lambda \mapsto \lambda\theta(\lambda^2) \in \mathcal{N}) \Leftrightarrow \lambda \mapsto \lambda\theta(\lambda^2) \in \mathcal{N}$;
5. $\mathcal{N}^{\text{ep}} \cap \mathcal{M} = (\mathcal{N}_+^{\text{ep}} \cap \mathcal{M}) \cup (\mathcal{N}_-^{\text{ep}} \cap \mathcal{M})$;
6. $S \cap \mathcal{M} \subseteq \mathcal{N}_+^{\text{ep}} \cup \{0\}$, $(-S^{-1}) \cap \mathcal{M} \subseteq \mathcal{N}_-^{\text{ep}} \cup \{0\}$.

Proof. Observing Remark 5.2.1 and Lemma 5.1.22 and the obvious fact that $S \subseteq \mathcal{N}$, parts 1–3 follow immediately.

4. Let $\theta \in \mathcal{S}$. If θ is constant, then this constant is a nonnegative real number C, and $C\lambda$ is a Nevanlinna function. If θ is not constant, then, by Remark 5.2.1 and Lemma 5.2.2,

$$\theta_1(\lambda) := \lambda\theta(\lambda^2) = C\lambda\frac{(\lambda^2 - b_0)_-}{\lambda^2 - a_0} \prod_{k \in I'_{a,b}} \left(1 - \frac{\lambda^2}{b_k}\right)\left(1 - \frac{\lambda^2}{a_k}\right)^{-1}$$

$$= C\psi(\lambda) \prod_{k \in I'_{a,b}} \left(1 + \frac{\lambda}{b_k^{\frac{1}{2}}}\right)\left(1 + \frac{\lambda}{a_k^{\frac{1}{2}}}\right)^{-1}\left(1 - \frac{\lambda}{b_k^{\frac{1}{2}}}\right)\left(1 - \frac{\lambda}{a_k^{\frac{1}{2}}}\right)^{-1},$$

where

$$\psi(\lambda) = \lambda\frac{(\lambda + b_0^{\frac{1}{2}})_+(\lambda - b_0^{\frac{1}{2}})_-}{(\lambda + a_0^{\frac{1}{2}})(\lambda - a_0^{\frac{1}{2}})}.$$

It remains to be shown that ψ can be written such that it fits the form (11.1.2). We have to distinguish the cases $a_0 > 0$ and $a_0 = 0$. If $a_0 > 0$, then clearly poles and zeros of θ_1 are simple and interlace, and

$$\psi(\lambda) = \frac{1}{a_0^{\frac{1}{2}}}\frac{\lambda}{\lambda + a_0^{\frac{1}{2}}}\left(1 - \frac{\lambda}{a_0^{\frac{1}{2}}}\right)^{-1}$$

if θ has no zeros, and

$$\psi(\lambda) = \frac{b_0}{a_0^{\frac{1}{2}}}\frac{\lambda}{\lambda + a_0^{\frac{1}{2}}}\left(1 + \frac{\lambda}{b_0^{\frac{1}{2}}}\right)\left(1 - \frac{\lambda}{b_0^{\frac{1}{2}}}\right)\left(1 - \frac{\lambda}{a_0^{\frac{1}{2}}}\right)^{-1}$$

otherwise. If now $a_0 = 0$, then clearly 0 is a simple pole of θ_1, so that again zeros and poles of θ_1 interlace, and

$$\psi(\lambda) = \frac{(\lambda + b_0^{\frac{1}{2}})_+(\lambda - b_0^{\frac{1}{2}})_-}{\lambda},$$

which equals $-\lambda^{-1}$ if θ has no zeros and

$$b_0^{\frac{1}{2}}\frac{\lambda - b_0^{\frac{1}{2}}}{\lambda}\left(1 + \frac{\lambda}{b_0^{\frac{1}{2}}}\right)$$

otherwise.

To complete the proof of part 4, assume that $\theta_1 \in \mathcal{N}$, where $\theta_1(\lambda) = \lambda\theta(\lambda^2)$. Since θ_1 is an odd function, its zeros and poles are symmetric, so that we can write

$$\theta_1(\lambda) = C\lambda^m \prod_{k \in I'_{a,b}} \left(1 - \frac{\lambda^2}{b_k}\right)\left(1 - \frac{\lambda^2}{a_k}\right)^{-1}$$

with some abuse of notation as we might not have identified all terms with index
0, and where $C \neq 0$, but not necessarily positive. Here $m = 1$ if 0 is no pole of θ
and $m = -1$ if 0 is a pole of θ, all a_k and b_k are positive and interlace. Also, since
the zeros of θ_1 are simple, 0 cannot be a zero of θ. Hence we have

$$\theta(\lambda) = C\lambda^{\frac{m-1}{2}} \prod_{k \in I'_{a,b}} \left(1 - \frac{\lambda}{b_k}\right)\left(1 - \frac{\lambda}{a_k}\right)^{-1},$$

where in case $m = -1$, 0 is a pole of θ, and therefore smaller than the smallest
zero of θ, and in case $m = 1$, 0 is a zero of θ_1, so that the smallest positive pole
of θ_1 is smaller than the smallest positive zero of θ_1, which again means that the
smallest pole of θ is smaller than the smallest zero of θ. Since the case of constant
θ is trivial, by Lemma 5.2.2 it remains to show that $\theta \in \hat{\mathcal{N}}$. In view of Theorem
11.1.6, either $\theta \in \hat{\mathcal{N}}$ or $-\theta \in \hat{\mathcal{N}}$. But $-\theta \in \hat{\mathcal{N}}$ and the location of poles and zeros
of θ would imply by Lemma 5.2.2 that $-\theta \in \mathcal{S}$. Therefore $-\theta_1 \in \hat{\mathcal{N}}$ by what we
have already proved, which contradicts $\theta_1 \in \hat{\mathcal{N}}$.
5. Since poles and zeros of $\theta \in \mathcal{N}^{\text{ep}}$ interlace by Remark 5.2.1, there is $\gamma \leq 0$ such
that θ has no poles or zeros in $(-\infty, \gamma)$, so that either $\theta(\lambda) > 0$ for all $\lambda \in (-\infty, \gamma)$
or $\theta(\lambda) < 0$ for all $\lambda \in (-\infty, \gamma)$.
6. is obvious.   □

Proposition 5.2.5. *Let $n \in \mathbb{N}$. Then every rational $\mathcal{S}_0$-function θ with n poles has
a continued fraction expansion*

$$\theta(\lambda) = a_0 + \cfrac{1}{-b_1\lambda + \cfrac{1}{a_1 + \cdots + \cfrac{1}{-b_n\lambda + \cfrac{1}{a_n}}}}$$

with $b_j > 0$, $j = 1, \ldots, n$, $a_0 \geq 0$ and $a_j > 0$ for $j = 1, \ldots, n$.

Proof. We recall from the definition of $\mathcal{S}_0$ that $\theta(\lambda) > 0$ for $\lambda \in (-\infty, 0]$ and
that all poles and zeros of θ are real. By Theorem 11.1.6, the poles and zeros
of θ interlace, and by Lemma 5.2.2, the smallest pole of θ is smaller than any
zero of θ. Hence θ has n or $n - 1$ zeros. Therefore we write $\theta = \frac{q}{p}$, where p is
a polynomial of degree n and q is a polynomial of degree n or $n - 1$. We may
assume, for definiteness, that $p(\lambda) > 0$ for $\lambda \in (-\infty, 0]$, so that also $q(\lambda) > 0$ for
$\lambda \in (-\infty, 0]$. If λ_0 is the smallest pole of θ, that is, the smallest zero of p, then
$q(\lambda_0) > 0$. Performing long division, we can write

$$\theta(\lambda) = a_0 + \frac{q_1(\lambda)}{p(\lambda)}, \tag{5.2.1}$$

where q_1 is a polynomial of degree less than n and $a_0 \geq 0$ by taking the limit as $\lambda \to -\infty$ in the above identity. Multiplying (5.2.1) by $p(\lambda)$, it follows that $q_1(\lambda_0) = q(\lambda_0) > 0$.

If $n = 1$, we may assume that $q_1 = 1$, so that $p(\lambda) = -b_1\lambda + c$ where $b_1 > 0$ and $c > 0$ since $p(\lambda) > 0$ for $\lambda \in (-\infty, 0]$. Writing $a_1 = \frac{1}{c}$ proves the proposition for $n = 1$.

Now let $n \geq 2$. If λ_1 and λ_2 are two consecutive zeros of p, then $q_1(\lambda_j) = q(\lambda_j)$ for $j = 1, 2$, and since q has exactly one (simple) zero between λ_1 and λ_2 as the zeros of p and q interlace, $q_1(\lambda_1)$ and $q_1(\lambda_2)$ have opposite signs, so that q_1 has at least one zero between any two zeros of p. Since the degree of q_1 is at most $n - 1$, it must be exactly $n - 1$, and the zeros and q_1 and p are real and interlace, with both q_1 and q being positive on $(-\infty, 0]$. Hence long division gives

$$\frac{p(\lambda)}{q_1(\lambda)} = -b_1\lambda + \frac{p_1(\lambda)}{q_1(\lambda)}, \tag{5.2.2}$$

where $b_1 > 0$ and p_1 is a polynomial of degree less than n. Defining $\theta_1 = \frac{q_1}{p_1}$, we have shown that

$$\theta(\lambda) = a_0 + \cfrac{1}{-b_1\lambda + \cfrac{1}{\theta_1(\lambda)}}.$$

Then the statement of this proposition follows by induction on n provided we show that $\theta_1 \in S_0$ and θ_1 has $n - 1$ poles and zeros.

To prove this property, let μ_0 be the smallest zero of q_1. We already know that the zeros of p and q_1 interlace, that p has degree n and q_1 has degree $n - 1$. Therefore it follows that $\lambda_0 < \mu_0$. Since p is positive on $(-\infty, 0]$ we have $p(\mu_0) < 0$. Substituting λ_0 into (5.2.2) and observing that $\lambda_0 > 0$ and $q_1(\lambda_0) > 0$ shows that $p_1(\lambda_0) > 0$, and substituting μ_0 gives $p_1(\mu_0) = p(\mu_0) < 0$. Hence p_1 has a zero $\nu_0 \in (\lambda_0, \mu_0)$. We know that q_1 has $n - 1$ zeros, and the same argument we used to show that the zeros of p and q_1 interlace, now applied to (5.2.2), shows that between any two zeros of q_1 there is a zeros of p_1. Since the degree of p_1 does not exceed the degree of q_1, it follows that the zeros of p_1 and q_1 interlace, that p_1 and q_1 have both degree $n - 1$, and that the smallest zero ν_0 of p_1 is positive and less than the smallest zero μ_0 of q_1. Finally, both p_1 and q_1 are positive on the interval $(-\infty, \lambda_0)$, so that indeed $\theta_1 \in S_0$ by Theorem 11.1.6 and Remark 5.2.1. $\qquad\square$

Definition 5.2.6. Let P_s and Q_s be real entire functions with no common zeros such that $\frac{Q_s}{P_s}$ belong to $\mathcal{N}_+^{\mathrm{ep}}$. Then the function ω defined by $\omega(\lambda) = P_s(\lambda^2) + i\lambda Q_s(\lambda^2)$ is said to belong to the class of symmetric shifted Hermite–Biehler functions. The class of all symmetric shifted Hermite–Biehler functions is denoted by SSHB. If the number of negative zeros of P_s is κ, then we say that ω belongs to SSHB_κ. We note that a shifted Hermite–Biehler function ω is symmetric with respect to the imaginary axis, i. e., $\omega(-\bar{\lambda}) = \overline{\omega(\lambda)}$ for $\lambda \in \mathbb{C}$.

If ω denotes such a shifted Hermite–Biehler function, then $P_s(\lambda^2) = \mathrm{R}\,\omega(\lambda)$ and $Q_s(\lambda^2) = \lambda^{-1}\,\mathrm{I}\,\omega(\lambda)$, so that P_s and Q_s are uniquely determined by ω. Also recall that $P_s^{-1}Q_s \in \mathcal{N}_+^{\mathrm{ep}}$ implies that the set of zeros of P is bounded below.

Definition 5.2.7. Let P_* and Q_* be real entire functions with no common zeros such that $\frac{Q_*}{P_*}$ belong to $\mathcal{N}_+^{\mathrm{ep}}$ and let R_* be a real entire function whose zeros are real and bounded below. Then the function ω given by $\omega(\lambda) = R_*(\lambda^2)(P_*(\lambda^2) + i\lambda Q_*(\lambda^2))$ is said to belong to the generalized class of symmetric shifted Hermite–Biehler functions ($\overline{\mathrm{SSHB}}$ class). If the number of negative zeros of R_*P_* is κ, then we say that the function ω belongs to $\overline{\mathrm{SSHB}}_\kappa$.

If ω is of $\overline{\mathrm{SSHB}}$ class, then P_*, Q_* and R_* are uniquely determined by ω up to multiplication by real entire functions without zeros.

The following Proposition shows some relations between shifted (generalized) Hermite–Biehler functions and (generalized) Hermite–Biehler functions.

Proposition 5.2.8.

1. *An entire function ω is of SSHB_0 class if and only if there are an entire function ω_1 of HB class and $m \in \{0,1\}$ such that $\omega(\lambda) = \lambda^m \omega_1(\lambda)$, $\lambda \in \mathbb{C}$, and such that $i^m \omega_1$ is symmetric with respect to the imaginary axis.*

2. *An entire function ω is of $\overline{\mathrm{SSHB}}_0$ class if and only if ω is a nontrivial function of $\overline{\mathrm{SHB}}$ class.*

Proof. If ω belongs to any of the classses in the statement of this proposition, then ω is symmetric with respect to the imaginary axis, and ω can be written in the form (5.1.23). Let $\theta_s = \frac{Q_s}{P_s}$, and observing the notation (5.1.24) and the property (5.1.25), we define

$$\theta(\lambda) = \frac{\mathrm{I}\,\omega(\lambda)}{\mathrm{R}\,\omega(\lambda)} = \frac{Q(\lambda)}{P(\lambda)} = \frac{\lambda \tilde{Q}(\lambda)}{\tilde{P}(\lambda)} = \lambda \theta_s(\lambda^2).$$

1. If ω is of SSHB_0 class, then $P_s \neq 0$, $Q_s \neq 0$ and $\theta_s \in \mathcal{S}$ by definition of the SSHB_0 class. Hence $\theta \in \mathcal{N}$ by Lemma 5.2.4, part 4, and thus $\theta \in \hat{\mathcal{N}}$ since θ is not constant. Since P_s and Q_s do not have common zeros, it follows that P and Q do not have common nonzero zeros. In view of Lemma 11.1.1, part 1, ω satisfies (5.1.2) and does not have zeros in the open lower half-plane. Since the real entire functions P and Q do not have common nonzero real zeros, it follows that ω does not have nonzero real zeros. If $\omega(0) \neq 0$, we conclude that ω is of HB class and hence of SHB class since by assumption ω is symmetric with respect to the imaginary axis. If, however, $\omega(0) = 0$, then $P_s(0) = 0$ and $Q_s(0) \neq 0$, which implies that ω has a simple zero at 0. Writing $\omega(\lambda) = \lambda \omega_1(\lambda)$ with $\omega_1(\lambda) = \lambda^{-1}\tilde{P}(\lambda) + i\tilde{Q}(\lambda)$ it follows from $\frac{\mathrm{I}\,\omega_1}{\mathrm{R}\,\omega_1} = \theta$ that ω_1 is of HB class. Clearly, in this case $i\omega_1$ is symmetric with respect to the imaginary axis since ω has this property.

Conversely, if ω_1 is of HB class, then $\theta \in \mathcal{N}$. Hence $\theta_s \in \mathcal{S}$ by Lemma 5.2.4, part 4. With the arguments from the beginning of the proof of Theorem 5.1.7 we conclude that $P = \mathrm{R}\,\omega$ and $Q = \mathrm{I}\,\omega$ do not have common nonzero zeros. Hence P_s and Q_s do not have common nonzero zeros. Since $\omega_1(0) \neq 0$, 0 cannot be a common zero of P_s and Q_s. Because $Q \neq 0$ implies $Q_s \neq 0$, either θ_s is a positive constant or θ_s is positive on the negative imaginary axis by Lemma 5.2.2. Therefore ω is of SSHB_0 class.

2. Let ω be of $\overline{\mathrm{SSHB}}_0$ class. With the notation from Definition 5.2.7 we write $\omega = \hat{\omega}_2 \hat{\omega}_1$, where $\hat{\omega}_1(\lambda) = P_*(\lambda^2) + i\lambda Q_*(\lambda^2)$ and $\hat{\omega}_2(\lambda) = R_*(\lambda^2)$. Furthermore, $\hat{\omega}_1$ is of SHB_0 class and R_* has only nonnegative real zeros, if any. In view of part 1, $\hat{\omega}_1(\lambda) = \lambda^m \omega_1(\lambda)$, where ω_1 is of HB class. Defining $\omega_2(\lambda) = \lambda^m R_*(\lambda^2)$, it follows that ω_2 is a real entire function which has only real zero. Hence $\omega = \omega_1 \omega_2$ is a nontrivial function of $\overline{\mathrm{SHB}}$ class.

Conversely, let ω be a nontrivial function of $\overline{\mathrm{SHB}}$ class. Then its real zeros are symmetric with respect to the origin, and we can construct an even entire function ω_3 whose zeros are the nonzero real zeros of ω, counted with multiplicity. Let $n = 2k + m$ be the multiplicity of the zero 0 of ω, where $k \in \mathbb{N}_0$ and $m \in \{0, 1\}$. It follows that there is an entire function R_* such that $\lambda^{2k}\omega_3(\lambda) = R_*(\lambda^2)$. Since the function ω_2 defined by $\omega_2(\lambda) = \lambda^n \omega_3(\lambda)$ is a real entire function which accounts for all real zeros of ω and since ω is nontrivial, Theorem 5.1.9 gives a function ω_1 of HB class such that $\omega(\lambda) = \omega_2(\lambda)\omega_1(\lambda) = R_*(\lambda^2)\lambda^m \omega_1(\lambda)$. Clearly, $i^m \omega_1$ is symmetric with respect to the imaginary axis, and an application of part 1 completes the proof. $\qquad \square$

Recall that zeros of analytic functions will be counted with multiplicity.

Theorem 5.2.9. *Let $\omega \in \mathrm{SSHB}_\kappa$. Then*

1. *The zeros of ω lie in the open upper half-plane and on the imaginary axis.*
2. *The number of zeros of ω on $(-i\infty, 0)$ is equal to κ, the number of negative zeros of P_s. The zeros of ω on $(-i\infty, 0]$ are simple.*
3. *If $\kappa > 0$ and the zeros in the open lower half-plane are enumerated such that $\lambda_j = i\tau_j$, where $\tau_\kappa < \tau_{\kappa-1} < \cdots < \tau_1 < 0$, then $\omega(i|\tau_j|) \neq 0$ for $j = 1, \ldots, \kappa$.*
4. *If $\kappa > 1$, then each interval $(i|\tau_j|, i|\tau_{j+1}|)$, $j = 1, \ldots, \kappa - 1$, contains an odd number of zeros of ω, counted with multiplicity.*
5. *If $\kappa > 0$ and $\omega(0) \neq 0$, then the interval $(0, i|\tau_1|)$ contains an even number of zeros of ω, counted with multiplicity, or does not contain any zeros of ω.*
6. *If $\kappa > 0$ and $\omega(0) = 0$, then the interval $(0, i|\tau_1|)$ contains an odd number of zeros of ω, counted with multiplicity.*

Proof. 1. If λ is a nonzero real number, then $\omega(\lambda) = P_s(\lambda^2) + i\lambda Q_s(\lambda^2) \neq 0$ since P_s and Q_s are real on the real axis and do not have common zeros. Suppose now $\omega(\lambda_0) = 0$ where $\mathrm{Re}\,\lambda_0 \neq 0$ and $\mathrm{Im}\,\lambda_0 < 0$. Then $P_s(\lambda_0) \neq 0$ and

$$\frac{Q_s(\lambda_0^2)}{P_s(\lambda_0^2)} = i\lambda_0^{-1}. \tag{5.2.3}$$

Since $\frac{Q_s}{P_s} \in \mathcal{N}$ we have

$$0 \le \operatorname{Im} \lambda_0^2 \operatorname{Im}(i\lambda_0^{-1}) = 2 \operatorname{Re} \lambda_0 \operatorname{Im} \lambda_0 \frac{\operatorname{Re} \lambda_0}{|\lambda_0|^2} < 0,$$

which is impossible.

2. First let us prove that the zeros of ω on the negative semiaxis are simple. If $\frac{Q_s}{P_s}$ is constant, then Q_s is a real multiple of P_s, and P_s has no zeros. Hence ω has at most one zero. If $\frac{Q_s}{P_s}$ is not constant, then Lemma 11.1.1, part 1, shows that $P_s + iQ_s$ is of HB class, so that according to Theorem 5.1.13,

$$Q_s'(\lambda) P_s(\lambda) - Q_s(\lambda) P_s'(\lambda) > 0 \tag{5.2.4}$$

for all real λ.

Suppose that $\lambda = -i\tau$, $\tau > 0$, is a multiple zero of ω. Then

$$P_s(-\tau^2) + \tau Q_s(-\tau^2) = 0 \tag{5.2.5}$$

and

$$-2i\tau P_s'(-\tau^2) + iQ_s(-\tau^2) - 2i\tau^2 Q_s'(-\tau^2) = 0. \tag{5.2.6}$$

Since P_s and Q_s have no common zeros, (5.2.5) shows that $P_s(-\tau^2) \ne 0$. Multiplying (5.2.6) by $-iQ_s(-\tau^2)$ and using (5.2.5) we obtain

$$-2\tau[P_s'(-\tau^2)Q_s(-\tau^2) - P_s(-\tau^2)Q_s'(-\tau^2)] + Q_s^2(-\tau^2) = 0, \tag{5.2.7}$$

which is impossible in view of (5.2.4).

Suppose ω has a multiple zero at the origin. Then (5.2.5) and (5.2.6) with $\tau = 0$ implies $P_s(0) = Q_s(0) = 0$, which is impossible since P_s and Q_s have no common zeros.

Let $\lambda_k(\alpha)$ be a zero of $\omega(\lambda, \alpha) = P_s(\lambda^2) + i\lambda\alpha Q_s(\lambda^2)$, where we may assume in view of Theorem 9.1.1 that λ_k depends continuously on $\alpha \in [0, 1]$. Then

$$\lambda_k(\alpha) = i \frac{P_s(\lambda_k^2(\alpha))}{\alpha Q_s(\lambda_k^2(\alpha))}. \tag{5.2.8}$$

In the above reasoning on the simplicity of zeros we may replace Q_s with αQ_s for all $\alpha > 0$, and for $\alpha = 0$ we can argue with the simplicity of the zeros of P_s to conclude that all zeros on the negative imaginary semiaxis are simple for $\alpha \ge 0$. From Theorem 9.1.1 we therefore know that λ_k is differentiable, and we find

$$\lambda_k'(\alpha) = \frac{-i\lambda_k(\alpha)Q_s(\lambda_k^2(\alpha))}{2\lambda_k(\alpha)P_s'(\lambda_k^2(\alpha)) + i\alpha Q_s(\lambda_k^2(\alpha)) + 2i\alpha\lambda_k^2(\alpha)Q_s'(\lambda_k^2(\alpha))}. \tag{5.2.9}$$

Substituting (5.2.8) into (5.2.9) we obtain

$$\lambda_k'(\alpha) = -i \frac{\lambda_k(\alpha)Q_s^2(\lambda_k^2(\alpha))}{2\lambda_k(\alpha)[Q_s(\lambda_k^2(\alpha))P_s'(\lambda_k^2(\alpha)) - P_s(\lambda_k^2(\alpha))Q_s'(\lambda_k^2(\alpha))] + i\alpha Q_s^2(\lambda_k^2(\alpha))}. \tag{5.2.10}$$

For $\lambda_k(\alpha) = -i\tau(\alpha)$ with $\tau(\alpha) > 0$ and $\alpha \geq 0$, (5.2.10) and (5.2.4) imply $\tau'(\alpha) > 0$. This means that the zeros on the negative imaginary semiaxis move upwards. On the other hand they never cross the origin because a zero at the origin is independent of α and simple for all $\alpha > 0$. Since $\frac{Q_s}{P_s}$ belong to $\mathcal{N}_+^{\mathrm{ep}}$, there is $\gamma < 0$ such that

$$\frac{Q_s(-\tau^2)}{P_s(-\tau^2)} > 0 \text{ if } -\tau^2 < \gamma.$$

Hence, for $\tau > |\gamma|^{\frac{1}{2}}$, $P_s(-\tau^2) \neq 0$ and

$$\frac{w(-i\tau, \alpha)}{P_s(-\tau^2)} = 1 + \tau\alpha \frac{Q_s(-\tau^2)}{P_s(-\tau^2)} > 1,$$

so that $w(\cdot, \alpha)$ does not have any zeros on $(-i\infty, -i|\gamma|^{\frac{1}{2}})$, and therefore no zeros of $w(\cdot, \alpha)$ can join the imaginary axis from $-i\infty$. Hence the number of zeros of $w(\cdot, \alpha)$ on the negative imaginary semiaxis is independent of $\alpha \geq 0$, and it equals κ for $\alpha = 0$, so that, taking $\alpha = 1$, we have proved that w has κ simple zeros on the negative imaginary semiaxis.

3. Suppose $w(i\tau) = w(-i\tau) = 0$ where $\tau \in \mathbb{R} \setminus \{0\}$. Then

$$P_s(-\tau^2) + \tau Q_s(-\tau^2) = P_s(-\tau^2) - \tau Q_s(-\tau^2) = 0$$

and, consequently, $P_s(-\tau^2) = Q_s(-\tau^2) = 0$, which is impossible.

4. Let $\lambda_k(0)$ be a pure imaginary zero of $w(\lambda, 0) = P_s(\lambda^2)$. Then (5.2.9), which is true for all pure imaginary zeros $\lambda_k(\alpha)$ of $w(\cdot, \alpha)$ with sufficiently small $\alpha > 0$, implies $\operatorname{Re} \lambda_k'(0) = 0$ and $\operatorname{Im} \lambda_k'(0) > 0$. Hence, statement 4 is true for small $\alpha > 0$. Because of

$$w(-\bar{\lambda}, \alpha) = P_s(\bar{\lambda}^2) - i\bar{\lambda}\alpha Q_s(\bar{\lambda}^2) = \overline{w(\lambda, \alpha)}$$

for all $\lambda \in \mathbb{C}$, all zeros of $w(\cdot, \alpha)$ are symmetric with respect to the imaginary axis. Thus, the zeros of $w(\cdot, \alpha)$ can only leave or join the imaginary axis in pairs as α increases, and therefore statement 4 remains true for all $\alpha \in (0, 1]$.

5. The proof of this statement is the same as the proof of statement 4 since no zero can join the positive imaginary semiaxis through 0.

6. Here we have to observe that 0 is a double zero of $w(\cdot, \alpha)$ for $\alpha = 0$ and a simple zero for $\alpha > 0$, as we know from statement 2 in the case $\alpha = 1$, which readily extends to any $\alpha > 0$. Hence one zero of $w(\cdot, \alpha)$ has to move away from 0 as α becomes positive. As we have seen in the proof of part 4, zeros can leave the imaginary axis only in pairs, so that this eigenvalues must stay on the imaginary axis and therefore must move upward as α becomes positive. A reasoning as in the proof of part 4 completes the proof.   □

Remark 5.2.10. By definition, $w \in \overline{\mathrm{SSHB}}$ can be written as $w(\lambda) = R_*(\lambda^2)w_1(\lambda)$, where $w_1 \in \mathrm{SSHB}$. A zero λ of w_1 will be called a zero of w of type II, and its type II multiplicity is the multiplicity of λ as a zero of w_1. A zero of $\lambda \mapsto R_*(\lambda^2)$

will be called a zero of ω of type I, and its type I multiplicity as a zero of ω is its multiplicity as a zero of $\lambda \mapsto R_*(\lambda^2)$.

Corollary 5.2.11. *Let* $\omega \in \overline{\text{SSHB}}$. *Then:*

1. *The zeros of* ω *which are of type I lie on the real and imaginary axes with at most a finite number on the imaginary axis. They are located symmetrically with respect to the origin.*

2. *The zeros of* ω *which are of type* II *satisfy the statements 1–6 of Theorem 5.2.9.*

3. *All zeros of* ω *which are both of type I and type II lie on the imaginary axis.*

4. *In the notation of Definition 5.2.7 let* ζ_0 *and* ζ_1 *be the smallest zero of* $R_* P_*$ *and* P_*, *respectively. Put* $\gamma_j = \min\{\zeta_j, 0\}$, $j = 0, 1$. *If* $-i\tau$ *is a zero of* ω *with* $\tau > 0$, *then* $\tau^2 \le -\gamma_0$, *and* $\tau^2 < -\gamma_1$ *if* $-i\tau$ *is a zero of type* II.

Proof. Statement 1, 2 and 3 follow from Remark 5.2.10, Definition 5.2.7 and Theorem 5.2.9.

4. Since ω_1 is of SSHB class, the function $Q_* P_*^{-1}$ belongs to $\mathcal{N}_+^{\text{ep}}$ by Definition 5.2.6. Hence we have in view of Corollary 5.2.3 that $P_*(\zeta_1) = 0$, $Q_*(\zeta_1) \ne 0$, and $P_*(z)$ an $Q_*(z)$ are different from zero and have the same sign for $z < \zeta_1$. Then it is easy to see that $\omega_1(-i\tau) = P_*(-\tau^2) + \tau Q_*(-\tau^2) \ne 0$ for $\tau > 0$ with $\tau^2 \ge -\gamma_1$. This proves the statement for the zeros of type II. Since $-i\tau$ is a zero of ω of type I if and only if $-\tau^2$ is a zero of R_*, the proof is complete. $\qquad\square$

We will make use of the following simple

Lemma 5.2.12. *If* $\alpha \in \mathbb{C}^+$ *and* $\beta \in \overline{\mathbb{C}^+}$, *then*

$$\frac{\alpha + \beta}{1 - \alpha\beta} \in \mathbb{C}^+.$$

Proof. Since $0 < \arg\alpha < \pi$ and $0 \le \arg\beta \le \pi$, we have $0 < \arg\alpha\beta < 2\pi$, so that $\alpha\beta$ is not a positive real number, and the fraction is well defined. Then

$$\frac{\alpha + \beta}{1 - \alpha\beta} = \frac{(\alpha + \beta)(1 - \overline{\alpha}\overline{\beta})}{|1 - \alpha\beta|^2},$$

and the proof will be complete if we observe that

$$\text{Im}\left((\alpha + \beta)(1 - \overline{\alpha}\overline{\beta})\right) = \text{Im}(\alpha + \beta - |\alpha|^2\overline{\beta} - |\beta|^2\overline{\alpha})$$
$$= \text{Im}\,\alpha + \text{Im}\,\beta + |\alpha|^2\,\text{Im}\,\beta + |\beta|^2\,\text{Im}\,\alpha. \qquad\square$$

Lemma 5.2.13. *Let* $\theta_0 \in \mathcal{M}$ *be such that there are* $r_1, r_2 \ge 0$ *such that* $\theta_0(\lambda) \in \mathbb{C}^+$ *for all* $\lambda \in \mathbb{C}^+$ *with* $|\lambda| > r_1$ *and* $\lambda^{\frac{1}{2}}\theta_0(\lambda) \in \mathbb{C}^+$ *for all* $\lambda \in \mathbb{C}^+$ *with* $|\lambda| > r_2$, *where* $\lambda^{\frac{1}{2}}$ *is the square root in the upper half-plane. Let*

$$p(\lambda) = -\lambda + \xi^2 + \zeta^2, \quad q(\lambda) = 2\xi, \quad \xi, \zeta \in \mathbb{R}, \ \xi > 0,$$

and define

$$\theta(\lambda) = \frac{\theta_0(\lambda)p(\lambda) + q(\lambda)}{p(\lambda) - \lambda\theta_0(\lambda)q(\lambda)}.$$

Then $\theta \in \mathcal{M}$, $\theta(\lambda) \in \mathbb{C}^+$ for all $\lambda \in \mathbb{C}^+$ with $|\lambda| > r_1$, and $\lambda^{\frac{1}{2}}\theta(\lambda) \in \mathbb{C}^+$ for all $\lambda \in \mathbb{C}^+$ with $|\lambda| > r_2$. If θ_0 is real, then also θ is real.

Proof. We have

$$\theta(\lambda) = \frac{(-\lambda + \xi^2 + \zeta^2)\theta_0(\lambda) + 2\xi}{-\lambda + \xi^2 + \zeta^2 - \lambda\theta_0(\lambda)2\xi},$$

which gives in particular that θ is real if θ_0 has this property. Note that the denominator is not identically zero since for $\lambda \in \mathbb{C}^+$, $|\lambda| > r_1$, the complex number $\lambda(1 + \theta_0(\lambda)2\xi)$ is a product of two numbers in $\mathbb{C}^+$ and therefore cannot be a positive real number. Hence $\theta \in \mathcal{M}$. Assume there is $\lambda_0 \in \mathbb{C}^+$, $|\lambda_0| > r_1$, such that $\operatorname{Im} \theta(\lambda_0) \leq 0$. Putting

$$\psi(\eta) = \frac{(-\lambda_0 + \xi^2 + \zeta^2)\eta\theta_0(\lambda_0) + 2\xi}{-\lambda_0 + \xi^2 + \zeta^2 - \lambda_0\eta\theta_0(\lambda_0)2\xi}$$

for $\eta \in [0,1]$, the same reasoning as above, applied to $\lambda_0(1 + \eta\theta_0(\lambda_0)2\xi)$, gives that the denominator of ψ is never zero, so that ψ is continuous on $[0,1]$. We have $\psi(1) = \theta(\lambda_0) \notin \mathbb{C}^+$ and, since

$$\theta_1(\lambda) = \frac{q(\lambda)}{p(\lambda)} = \frac{2\xi}{-\lambda + \xi^2 + \zeta^2}$$

defines a Möbius transformation θ_1 mapping $\mathbb{C}^+$ onto itself, $\psi(0) = \theta_1(\lambda_0) \in \mathbb{C}^+$. By continuity, there is $\eta_0 \in (0,1]$ such that $\psi(\eta_0) \in \mathbb{R}$. Then

$$\eta_0\theta_0(\lambda_0) = \frac{(-\lambda_0 + \xi^2 + \zeta^2)\psi(\eta_0) - 2\xi}{-\lambda_0 + \xi^2 + \zeta^2 + \lambda_0\psi(\eta_0)2\xi}. \tag{5.2.11}$$

But in view of

$$-\psi(\eta_0)(\xi^2 + \zeta^2) - \left[(\xi^2 + \zeta^2)\psi(\eta_0) - 2\xi\right]\left[-1 + \psi(\eta_0)2\xi\right]$$
$$= -2\xi\left[(\xi^2 + \zeta^2)\psi^2(\eta_0) + 1 - 2\xi\psi(\eta_0)\right]$$
$$= -2\xi\left[(\xi\psi(\eta_0) - 1)^2 + \zeta^2\psi^2(\eta_0)\right] \leq 0,$$

the map

$$\lambda \mapsto \frac{(-\lambda + \xi^2 + \zeta^2)\psi(\eta_0) - 2\xi}{-\lambda + \xi^2 + \zeta^2 + \lambda\psi(\eta_0)2\xi}$$

is either a real constant function or a Möbius transformation which maps $\mathbb{C}^+$ onto $\mathbb{C}^-$, and we arrive at the contradiction that the left-hand side of (5.2.11) belongs to $\mathbb{C}^+$, whereas the right-hand side does not.

Since also

$$\lambda\theta_1(\lambda) = \frac{2\xi\lambda}{-\lambda + \xi^2 + \zeta^2}$$

defines a Möbius transformation mapping $\mathbb{C}^+$ onto itself, it is clear that

$$0 < \arg(\theta_1(\lambda)) < \arg\left(\lambda^{\frac{1}{2}}\theta_1(\lambda)\right) < \arg\left(\lambda\theta_1(\lambda)\right) < \pi \quad \text{for } \lambda \in \mathbb{C}^+,$$

so that $\lambda^{\frac{1}{2}}\theta_1(\lambda) \in \mathbb{C}^+$ if $\lambda \in \mathbb{C}^+$. Then

$$\lambda^{\frac{1}{2}}\theta(\lambda) = \frac{\lambda^{\frac{1}{2}}\theta_0(\lambda) + \lambda^{\frac{1}{2}}\theta_1(\lambda)}{1 - \lambda^{\frac{1}{2}}\theta_0(\lambda)\lambda^{\frac{1}{2}}\theta_1(\lambda)} \in \mathbb{C}^+ \quad \text{if } \lambda \in \mathbb{C}^+,\ |\lambda| > r_2,$$

by Lemma 5.2.12. $\square$

Lemma 5.2.14. *Let $\theta_0 \in \mathcal{M}$ and assume that there are $r_1 \geq 0$ and $r_2 \geq 0$ such that $\theta_0(\lambda) \in \mathbb{C}^+$ for all $\lambda \in \mathbb{C}^+$ with $|\lambda| > r_1$ and $\lambda^{\frac{1}{2}}\theta_0(\lambda) \in \mathbb{C}^+$ for all $\lambda \in \mathbb{C}^+$ with $|\lambda| > r_2$. Let*

$$p(\lambda) = -\lambda + \xi\zeta, \quad q(\lambda) = \xi + \zeta, \quad \xi, \zeta \in \mathbb{R},\ \xi + \zeta > 0,$$

and define

$$\theta(\lambda) = \frac{\theta_0(\lambda)p(\lambda) + q(\lambda)}{p(\lambda) - \lambda\theta_0(\lambda)q(\lambda)}.$$

Then $\theta \in \mathcal{M}$, θ has no zeros in $\{\lambda \in \mathbb{C}^+ : |\lambda| > r_1\}$ and no poles in $\{\lambda \in \mathbb{C}^+ : |\lambda| > r_3\}$, where $r_3 = \max\{r_1, r_2, -\xi\zeta\}$, and $\lambda^{\frac{1}{2}}\theta(\lambda) \in \mathbb{C}^+$ for all $\lambda \in \mathbb{C}^+$ with $|\lambda| > r_4 = \max\{r_2, -\xi\zeta\}$. If θ_0 is real, then also θ is real.

Proof. We have

$$\theta(\lambda) = \frac{(-\lambda + \xi\zeta)\theta_0(\lambda) + \xi + \zeta}{-\lambda + \xi\zeta - \lambda\theta_0(\lambda)(\xi + \zeta)},$$

which gives in particular that θ is real if θ_0 has this property. We are going to show that the denominator is not zero for all $\lambda \in \mathbb{C}^+$ with $|\lambda| > r_3$. Indeed, assume that

$$-\lambda + \xi\zeta - \lambda\theta_0(\lambda)(\xi + \zeta) = 0$$

for some such λ. Then

$$\theta_0(\lambda) = \frac{1}{\xi + \zeta}\left(\frac{\xi\zeta}{\lambda} - 1\right)$$

and thus

$$\lambda^{\frac{1}{2}}\theta_0(\lambda) = \frac{1}{\xi + \zeta}\left(\frac{\xi\zeta}{\lambda^{\frac{1}{2}}} - \lambda^{\frac{1}{2}}\right)$$

$$= \frac{1}{(\xi + \zeta)|\lambda|}\left(\xi\zeta\overline{\lambda}^{\frac{1}{2}} - |\lambda|\lambda^{\frac{1}{2}}\right),$$

so that

$$\text{Im}\left(\lambda^{\frac{1}{2}}\theta_0(\lambda)\right) = -\text{Im}\,\lambda^{\frac{1}{2}}\frac{\xi\zeta + |\lambda|}{(\xi + \zeta)|\lambda|}.$$

The right-hand side is negative for $\lambda \in \mathbb{C}^+$ with $|\lambda| > -\xi\zeta$, whereas the left-hand side is positive for $|\lambda| > r_2$ by assumption. This contradiction shows that θ has no poles in $\{\lambda \in \mathbb{C}^+ : |\lambda| > r_3\}$.

Assume there is $\lambda_0 \in \mathbb{C}^+$, $|\lambda_0| > r_1$, such that $\theta(\lambda_0) = 0$. Then

$$\theta_0(\lambda_0) = \frac{\xi + \zeta}{\lambda_0 - \xi\zeta},$$

where the right-hand side is a Möbius transformation mapping $\mathbb{C}^+$ onto $\mathbb{C}^-$, thus arriving at a contradiction since the left-hand side belongs to $\mathbb{C}^+$.

We define

$$\theta_1(\lambda) = \frac{q(\lambda)}{p(\lambda)} = \frac{\xi + \zeta}{-\lambda + \xi\zeta}$$

and observe that

$$\lambda^{\frac{1}{2}}\theta_1(\lambda) = \frac{\lambda^{\frac{1}{2}}(\xi + \zeta)}{-\lambda + \xi\zeta} = \frac{\lambda^{\frac{1}{2}}(\xi + \zeta)(-\bar{\lambda} + \xi\zeta)}{|-\lambda + \xi\zeta|^2}$$

and

$$\text{Im}\left(\lambda^{\frac{1}{2}}(-\bar{\lambda} + \xi\zeta)\right) = \text{Im}\,\lambda^{\frac{1}{2}}(|\lambda| + \xi\zeta)$$

show that $\lambda^{\frac{1}{2}}\theta_1(\lambda) \in \mathbb{C}^+$ if $\lambda \in \mathbb{C}^+$, $|\lambda| > -\xi\zeta$. Then

$$\lambda^{\frac{1}{2}}\theta(\lambda) = \frac{\lambda^{\frac{1}{2}}\theta_0(\lambda) + \lambda^{\frac{1}{2}}\theta_1(\lambda)}{1 - \lambda^{\frac{1}{2}}\theta_0(\lambda)\lambda^{\frac{1}{2}}\theta_1(\lambda)} \in \mathbb{C}^+ \quad \text{if } \lambda \in \mathbb{C}^+, \, |\lambda| > r_4,$$

by Lemma 5.2.12. $\qquad\square$

Lemma 5.2.15. *Let $\theta \in \mathcal{M} \cap \mathcal{N}_+^{\mathrm{ep}}$. Then there is $r \geq 0$ such that $\lambda^{\frac{1}{2}}\theta(\lambda) \in \mathbb{C}^+$ if $\lambda \in \mathbb{C}^+$ with $|\lambda| > r$.*

Proof. The case that θ is constant is trivial, so that we may assume now that θ is not constant and therefore $\theta \in \hat{\mathcal{N}}$. Since $\theta \in \mathcal{N}_+^{\mathrm{ep}}$, there is $r \geq 0$ such that θ is positive on $(-\infty, -r]$. By Corollary 5.2.3, θ has at least one pole, and the smallest pole is smaller than all zeros of θ, if any. Then the zeros and poles of ψ defined by $\psi(\lambda) = (\lambda + r)\theta(\lambda)$ interlace. We observe that

$$\psi'(\lambda) = \theta(\lambda) + (\lambda + r)\theta'(\lambda),$$

so that $\psi'(-r) = \theta(-r) > 0$, and it follows from Theorem 11.1.6 and Remark 11.1.7 that $\psi \in \hat{\mathcal{N}}$. For $\lambda \in \mathbb{C}^+$ we write

$$\lambda + r = \lambda^{\frac{1}{2}}\left(\lambda^{\frac{1}{2}} + \frac{r}{|\lambda|}\bar{\lambda}^{\frac{1}{2}}\right)$$

and observe that

$$\operatorname{Im}\left(\lambda^{\frac{1}{2}}+\frac{r}{|\lambda|}\overline{\lambda}^{\frac{1}{2}}\right)=\operatorname{Im}\lambda^{\frac{1}{2}}\left(1-\frac{r}{|\lambda|}\right)>0 \text{ if } \lambda\in\mathbb{C}^{+}, |\lambda|>r.$$

Hence, for $\lambda\in\mathbb{C}^{+}$, $|\lambda|>r$, we have

$$\pi>\arg(\lambda+r)=\arg\lambda^{\frac{1}{2}}+\arg\left(\lambda^{\frac{1}{2}}+\frac{r}{|\lambda|}\overline{\lambda}^{\frac{1}{2}}\right)>\arg\lambda^{\frac{1}{2}}>0$$

and therefore

$$\pi>\arg\psi(\lambda)=\arg((\lambda+r)\theta(\lambda))>\arg(\lambda^{\frac{1}{2}}\theta(\lambda))>\arg\theta(\lambda)>0,$$

whence $\lambda^{\frac{1}{2}}\theta(\lambda)\in\mathbb{C}^{+}$ if $\lambda\in\mathbb{C}^{+}$, $|\lambda|>r$.   □

The following theorem is the converse of Theorem 5.2.9.

Theorem 5.2.16. *Consider the entire function*

$$\omega(\lambda)=\lambda^{m}e^{u(\lambda)+i(\nu\lambda+\delta)}\prod_{k=1}^{(\infty)}\left(1-\frac{\lambda}{\alpha_{k}}\right)e^{\mathrm{R}P_{k}\left(\frac{\lambda}{\alpha_{k}}\right)},$$

where $m\in\mathbb{N}_{0}$, $\nu\geq0$, $\delta\in\mathbb{R}$, $u(\lambda)$ is a real entire function and

$$\sum_{k=1}^{(\infty)}\left|\operatorname{Im}\frac{1}{\alpha_{k}}\right|<\infty.$$

Assume that ω is symmetric with respect to the imaginary axis, i. e., $\omega(-\overline{\lambda})=\overline{\omega(\lambda)}$ for all $\lambda\in\mathbb{C}$, and assume that the zeros of ω satisfy the following conditions:

1. *The zeros of ω lie in the open upper half-plane and on the imaginary axis.*
2. *There are at most finitely many zeros of ω on $(-i\infty,0)$, these zeros are simple, and their number is denoted by κ.*
3. *If $\kappa>0$ and the zeros in the open lower half-plane are enumerated such that $\lambda_{j}=i\tau_{j}$, where $\tau_{\kappa}<\tau_{\kappa-1}<\cdots<\tau_{1}<0$, then $\omega(i|\tau_{j}|)\neq0$ for $j=1,\ldots,\kappa$.*
4. *If $\kappa>1$, then each interval $(i|\tau_{j}|,i|\tau_{j+1}|)$, $j=1,\ldots,\kappa-1$, contains an odd number of zeros of ω, counted with multiplicity.*
5. *If $\kappa>0$ and $\omega(0)\neq0$, then the interval $(0,i|\tau_{1}|)$ contains an even number of zeros of ω, counted with multiplicity, or does not contain any zeros of ω.*
6. *If $\omega(0)=0$, then this zero is simple, and if additionally $\kappa>0$, then the interval $(0,i|\tau_{1}|)$ contains an odd number of zeros of ω, counted with multiplicity.*
7. *If $\kappa>0$, then the interval $(i|\tau_{\kappa}|,i\infty)$ contains at least one zeros of ω.*

Then $\omega\in\mathrm{SSHB}_{\kappa}$ if ω has at least one nonzero zero.

Proof. Let P_s, Q_s, $\tilde{P}$, and $\tilde{Q}$ be as in (5.1.23) and (5.1.24). We are going to show that P_s and Q_s do not have common zeros. Indeed, assume there is $\lambda \in \mathbb{C} \setminus \{0\}$ such that $P_s(\lambda^2) = Q_s(\lambda^2) = 0$. Then $\omega(\pm\lambda) = P_s(\lambda^2) \pm i\lambda Q_s(\lambda^2) = 0$, which is impossible because in view of conditions 1 and 3 there are no nonzero zeros of ω which are symmetric with respect to the origin. Also, since 0 is at most a simple zero of ω and since $\tilde{P}$ and $\tilde{Q}$ are even functions, at least one of $P_s(0)$ and $Q_s(0)$ must be different from zero.

Next assume that ω has at least one nonzero zero. Then the zeros of ω are not symmetric with respect to the origin, but the zeros of $P_s(\lambda^2)$ and $i\lambda Q_s(\lambda^2)$ are. Hence neither P_s nor Q_s can be identically zero.

We are going to prove the statement of the theorem by induction on κ, where we are also going to prove and use that there is $r \geq 0$ such that

$$\lambda^{\frac{1}{2}} \frac{Q_s(\lambda)}{P_s(\lambda)} \in \mathbb{C}^+ \quad \text{if } \lambda \in \mathbb{C}^+, \ |\lambda| > r. \tag{5.2.12}$$

If $\kappa = 0$, then ω is an entire function without zeros in the closed lower half-plane, except possibly at 0. Due to Theorem 5.1.11, ω is of $\overline{\text{HB}}$ class. We have shown that $P_s \neq 0$ and $Q_s \neq 0$, so that

$$\theta(\lambda) = \frac{\lambda \tilde{Q}(\lambda)}{\tilde{P}(\lambda)}, \quad \theta_s = \frac{Q_s}{P_s},$$

are well-defined nonzero meromorphic functions. In view of Lemma 11.1.1, applied to (5.1.23), $\theta \in \mathcal{N}$, and then $\theta(\lambda) = \lambda \theta_s(\lambda^2)$ shows by Lemma 5.2.4, part 4, that $\theta_s \in \mathcal{S}$, so that $\omega \in \text{SSHB}_0$. Then Proposition 5.2.8 gives $\theta \in \hat{\mathcal{N}}$, and (5.2.12) follows from Lemma 5.2.15.

Now assume that $\kappa > 0$ and that the statement of this theorem and (5.2.12) are true for $\kappa - 1$. We are going to construct a family of entire functions $\omega(\cdot, \eta)$, $\eta \in [0,1]$, such that $\omega = \omega(\cdot, 0)$, the zeros of $\omega(\cdot, \eta)$ depend continuously on η, $\omega(\cdot, \eta)$ satisfies the same assumptions as ω for all $\eta > 0$, and $\omega(\cdot, 1)$ has $\kappa - 1$ zeros on $(-i\infty, 0)$. We start with writing ω in a form which is more convenient for our purposes. Let $k_0 \in \mathbb{N}$ be such that $k_0 - 2$ is the number of zeros of ω in $[0, i|\tau_1|)$. It will be convenient to choose a certain indexing for finitely many zeros of ω, to include 0 into this indexing if $\omega(0) = 0$, and to rewrite the corresponding factors in the product representation of ω in the following way. Let $\alpha_1 = i\tau_1$, let α_k, $k = 2, \ldots, k_0 - 1$, be all zeros of ω in $[0, i|\tau_1|)$, if any, where we may assume that $\operatorname{Im} \alpha_k \leq \operatorname{Im} \alpha_{k+1}$ for $k = 2, \ldots, k_0 - 2$, and let α_{k_0} be the zero on $(i|\tau_1|, i\infty)$ with the smallest absolute value, which exists by assumptions 4 and 7. For each α_k with $1 \leq k \leq k_0$ and $\alpha_k \neq 0$, $e^{\operatorname{R} P_k\left(\frac{\lambda}{\alpha_k}\right)}$ can be moved into $u(\lambda)$, possibly changing u to a different real entire function. Secondly, we will write

$$1 - \frac{\lambda}{\alpha_k} = -\alpha_k^{-1}(\lambda - \alpha_k)$$

for these k. The constants $-\alpha_k^{-1}$ can be absorbed into the exponential term, possibly changing u and δ, so that

$$\omega(\lambda) = \hat{\omega}(\lambda)\omega_0(\lambda),$$

where $\hat{\omega} = \hat{\omega}_1\hat{\omega}_2$,

$$\hat{\omega}_1(\lambda) = \prod_{k=2}^{k_0-1} [i(\lambda - \alpha_k)], \quad \hat{\omega}_2(\lambda) = [i(\lambda - i\tau_1)][i(\lambda - \alpha_{k_0})],$$

$$\omega_0(\lambda) = e^{\hat{u}(\lambda)+i(\nu\lambda+\hat{\delta})} \prod_{k=k_0+1}^{(\infty)} \left(1 - \frac{\lambda}{\alpha_k}\right) e^{R\,P_k\left(\frac{\lambda}{\alpha_k}\right)},$$

$\hat{u}$ is a real entire function and $\hat{\delta} \in \mathbb{R}$. In our construction, at most two linear factors from $\hat{\omega}$ of the form $i(\lambda - \alpha_k(\eta))$ will change with η at a time, continuously depending on the parameter η, so that the resulting functions $\hat{\omega}_1(\cdot, \eta)$ and $\hat{\omega}_2(\cdot, \eta)$ will be polynomials which depend continuously on η, uniformly on compact subsets.

Now we describe how to move $i\tau_1$ from the open lower half-plane into the open upper half-plane. According to the conditions 2, 5 and 6, the interval $[0, i|\tau_1|)$ contains an even number of zeros of ω and thus of $\hat{\omega}_1$. Let us arrange them in adjacent pairs $(\alpha_2, \alpha_3), \ldots, (\alpha_{k_0-2}, \alpha_{k_0-1})$, if any. Then we move elements of each pair along the interval $(0, i|\tau_1|)$, continuously as functions of the parameter η in $[0, \frac{1}{2}]$, to collide and after the collision we move them into the complex plane symmetrically with respect to the imaginary axis. In this process all other zeros are kept fixed. Now, since there are no zeros of $\hat{\omega}(\cdot, \eta)$ in the interval $(0, i|\tau_1|)$ for $\eta = \frac{1}{2}$, we move the zero $i\tau_1$ up along the imaginary axis until it appears in the upper half-plane in the interval $(0, i|\tau_1|)$ for $\eta = 1$. Here, this zero is paired with the zero α_{k_0}, which may stay fixed. Clearly, in this process all of the properties 1–7 are satisfied for $\hat{\omega}_1(\cdot, \eta)$, $\hat{\omega}_2(\cdot, \eta)$, ω_0 and any of their products with $\eta \in [0, 1]$.

Observe that

$$i\left(-\overline{\lambda} + \overline{\beta}\right) = \overline{i(\lambda - \beta)}, \quad \lambda, \beta \in \mathbb{C},$$

so that $\psi(-\overline{\lambda}) = \overline{\psi(\lambda)}$ if

$$\psi(\lambda) = i(\lambda - \beta_1(\mu)) \quad \text{or} \quad \psi(\lambda) = [i(\lambda - \beta_2(\mu))][i(\lambda + \overline{\beta_2(\mu)})],$$

where $\beta_1(\eta) \in i\mathbb{R}$ and $\beta_2(\mu) \in \mathbb{C}$. This shows $\hat{\omega}_1(-\overline{\lambda}, \eta) = \overline{\hat{\omega}_1(\lambda, \eta)}$, $\hat{\omega}_2(-\overline{\lambda}, \eta) = \overline{\hat{\omega}_2(\lambda, \eta)}$ and thus $\omega_0(-\overline{\lambda}, \eta) = \overline{\omega_0(\lambda, \eta)}$ for all $\lambda \in \mathbb{C}$ and $\eta \in [0, 1]$.

Therefore, $\omega(\cdot, 1)$ and $\hat{\omega}_1(\cdot, \eta)\omega_0$ satisfy all assumptions of this theorem, but with κ replaced with $\kappa - 1$. By induction hypothesis, $\omega(\cdot, 1) \in \mathrm{SSHB}_{\kappa-1}$.

Next we note that the symmetry proved above shows in view of Proposition 5.2.8 that there are real entire functions $\hat{P}_0(\cdot, \eta)$ and $\hat{Q}_0(\cdot, \eta)$ such that

$$\hat{\omega}_1(\lambda, \eta)\omega_0(\lambda) = \hat{P}_0(\lambda^2, \eta) + i\lambda\hat{Q}_0(\lambda^2, \eta).$$

If $\hat{\omega}_1(\cdot, \eta)\omega_0$ has at least one nonzero zero, then this function belongs to $\text{SSHB}_{\kappa-1}$ by induction hypothesis, so that

$$\hat{\theta}_0(\cdot, \eta) = \frac{\hat{Q}_0(\cdot, \eta)}{\hat{P}_0(\cdot, \eta)} \in \mathcal{N}.$$

Otherwise, if $\hat{\omega}_1(\cdot, \eta)\omega_0$ has no nonzero zero, then $\kappa = 1$, and $\hat{\omega}_1(0, \eta)\omega_0(0) \neq 0$ by condition 6 since the interval $(0, i|\tau_1|]$ contains no zeros of $\hat{\omega}_1(\cdot, \eta)\omega_0$, so that $\hat{\omega}_1(\cdot, \eta)\omega_0$ has no zeros at all. Then $\hat{\omega}_1(\cdot, \eta) = 1$ as it is an empty product. Hence $\hat{\omega}_1(\cdot, \eta)\omega_0 = \omega_0$ is independent of η, and so are $\hat{P}_0(\cdot, \eta)$ and $\hat{Q}_0(\cdot, \eta)$. In this case, $\hat{P}_0(0, \eta) = \omega_0(0) \neq 0$, so that $\hat{P}_0(\cdot, \eta) \neq 0$. However, $\hat{Q}_0(\cdot, \eta) = 0$ is possible as the trivial case $\omega_0 = 1$ demonstrates. But if $\hat{Q}_0(\cdot, \eta) \neq 0$, then $\hat{\omega}_1(\cdot, \eta)\omega_0$ satisfies the assumptions of this theorem with $\kappa = 0$, and the proof of that case shows that $\hat{\theta}_0(\cdot, \eta) \in \hat{\mathcal{N}}$, that $\hat{\omega}_1(\cdot, \eta)\omega_0 \in \text{SSHB}_0$, and that (5.2.12) holds for $\lambda \mapsto \lambda^{\frac{1}{2}} \frac{\hat{Q}_0(\lambda, \eta)}{\hat{P}_0(\lambda, \eta)}$.

Summarizing the above result, we have either

$$\hat{\theta}_0(\cdot, \eta) \in \mathcal{N}_+^{\text{ep}} \text{ and } \lambda \mapsto \lambda^{\frac{1}{2}} \frac{\hat{Q}_0(\lambda, \eta)}{\hat{P}_0(\lambda, \eta)} \text{ satisfies (5.2.12) for all } \eta \in [0, 1] \quad (5.2.13)$$

or

$$\hat{\theta}_0(\cdot, \eta) = 0 \text{ for all } \eta \in [0, 1]. \quad (5.2.14)$$

All of the pairs of zeros in $\hat{\omega}_1(\cdot, \eta)$ and $\hat{\omega}_2(\cdot, \eta)$ are of the form $(i\xi, i\zeta)$ with $\xi + \zeta > 0$ or $(i\xi + \zeta, i\xi - \zeta)$ with $\xi > 0$, where $\xi, \zeta \in \mathbb{R}$. In the first case,

$$[i(\lambda - i\xi)][i(\lambda - i\zeta)] = -\lambda^2 + i\lambda(\xi + \zeta) + \xi\zeta =: p(\lambda^2) + i\lambda q(\lambda^2),$$

with

$$p(\lambda) = -\lambda + \xi\zeta, \quad q(\lambda) = \xi + \zeta. \quad (5.2.15)$$

In the second case,

$$[i(\lambda - (i\xi + \zeta))][i(\lambda - (i\xi - \zeta))] = -\lambda^2 + i\lambda 2\xi + \xi^2 + \zeta^2 =: p(\lambda^2) + i\lambda q(\lambda^2),$$

with

$$p(\lambda) = -\lambda + \xi^2 + \zeta^2, \quad q(\lambda) = 2\xi. \quad (5.2.16)$$

In particular, with the real-imaginary parts decomposition

$$\hat{\omega}_2(\lambda, \eta) =: \hat{p}_2(\lambda^2, \eta) + i\lambda \hat{q}_2(\lambda^2, \eta)$$

it follows that

$$\begin{aligned}
\omega(\lambda, \eta) &= [\hat{\omega}_1(\lambda, \eta)\omega_0(\lambda)]\hat{\omega}_2(\lambda, \eta) \\
&= \left[\hat{P}_0(\lambda^2, \eta) + i\lambda \hat{Q}_0(\lambda^2, \eta)\right] [\hat{p}_2(\lambda^2, \eta) + i\lambda \hat{q}_2(\lambda^2, \eta)] \\
&= \hat{P}_0(\lambda^2, \eta)\hat{p}_2(\lambda^2, \eta) - \lambda^2 \hat{Q}_0(\lambda^2, \eta)\hat{q}_2(\lambda^2, \eta) \\
&\quad + i\lambda \left(\hat{Q}_0(\lambda^2, \eta)\hat{p}_2(\lambda^2, \eta) + \hat{P}_0(\lambda^2, \eta)\hat{q}_2(\lambda^2, \eta)\right),
\end{aligned}$$

so that

$$\omega(\lambda, \eta) = P_s(\lambda^2, \eta) + i\lambda Q_s(\lambda^2, \eta)$$

with

$$P_s(\lambda, \eta) = \hat{P}_0(\lambda, \eta)\hat{p}_2(\lambda, \eta) - \lambda\hat{Q}_0(\lambda, \eta)\hat{q}_2(\lambda, \eta),$$
$$Q_s(\lambda, \eta) = \hat{Q}_0(\lambda, \eta)\hat{p}_2(\lambda, \eta) + \hat{P}_0(\lambda, \eta)\hat{q}_2(\lambda, \eta).$$

Therefore

$$\theta_s(\lambda, \eta) = \frac{Q_s(\lambda, \eta)}{P_s(\lambda, \eta)} = \frac{\hat{\theta}_0(\lambda, \eta)\hat{p}_2(\lambda, \eta) + \hat{q}_2(\lambda, \eta)}{\hat{p}_2(\lambda, \eta) - \lambda\hat{\theta}_0(\lambda, \eta)\hat{q}_2(\lambda, \eta)} . \tag{5.2.17}$$

From the alternative (5.2.13), (5.2.14) we know that $\hat{\theta}_0(\cdot, \eta) = 0$ for some $\eta \in [0, 1]$ implies that $\hat{\theta}_0(\cdot, \eta) = 0$ for all $\eta \in [0, 1]$, so that in this case

$$\theta_s(\lambda) = \theta_s(\lambda, 0) = \frac{\hat{q}_2(\lambda, 0)}{\hat{p}_2(\lambda, 0)} = \frac{|\alpha_{k_0}| + \tau_1}{-\lambda + |\alpha_{k_0}|\tau_1} .$$

Clearly, θ is a Möbius transformation which belongs to $SSHB_1$, and θ satisfies (5.2.12) in view of Lemma 5.2.15.

Now let $\hat{Q}_0(\cdot, \eta) \neq 0$ for $\eta \in [0, 1]$. Then $\hat{\theta}_0(\cdot, \eta) \in \mathcal{N}_+^{\text{ep}}$ with $\kappa - 1$ negative poles, see (5.2.13). Writing $\omega_0(\lambda) = p_0(\lambda^2) + i\lambda q_0(\lambda^2)$ and $\theta_0(\lambda) = \frac{q_0(\lambda)}{p_0(\lambda)}$, we have by induction hypothesis that there is $r_2 \geq 0$ such that $\lambda^{\frac{1}{2}}\theta_0(\lambda) \in \mathbb{C}^+$ for all $\lambda \in \mathbb{C}^+$ with $|\lambda| > r_2$. Adding pairs of factors from $\hat{\omega}_1$ recursively, Lemmas 5.2.13 and 5.2.14 show that the corresponding estimates remain true with the same r_2, and therefore, eventually, $\lambda^{\frac{1}{2}}\hat{\theta}_0(\lambda, \eta) \in \mathbb{C}^+$ for all $\lambda \in \mathbb{C}^+$ with $|\lambda| > r_2$.

Also, starting with θ_0, recursively adding pairs of factors from $\hat{\omega}_1$ to ω_0 gives meromorphic functions

$$\theta_j(\lambda, \eta) = \frac{\theta_{j-1}(\lambda, \eta) + \psi_j(\lambda, \eta)}{1 - \lambda\theta_{j-1}(\lambda, \eta)\psi_j(\lambda, \eta)},$$

where either

$$\psi_j(\lambda, \eta) = \frac{\xi_j(\eta) + \zeta_j(\eta)}{-\lambda + \xi_j(\eta)\zeta_j(\eta)}$$

or

$$\psi_j(\lambda, \eta) = \frac{2\xi_j(\eta)}{-\lambda + \xi_j^2(\eta) + \zeta_j^2(\eta)}$$

with continuous functions $\xi_j : [0, 1] \to (0, \infty)$ and $\zeta_j : [0, 1] \to [0, \infty)$. Since there is $\gamma \in \mathbb{R}$ such that $\theta_0(\lambda) > 0$ for $\lambda \in (-\infty, \gamma)$, it follows in each step that also $\theta_j(\lambda, \eta) > 0$ for each $\lambda \in (-\infty, \gamma)$ and $\eta \in [0, 1]$. Because $\hat{\theta}(\lambda, \eta)$ is the final of these $\theta_j(\infty, \eta)$, also $\hat{\theta}(\lambda, \eta) > 0$ for each $\lambda \in (-\infty, \gamma)$ and $\eta \in [0, 1]$.

Since, by construction,

$$\hat{p}_2(\lambda, \eta) = -\lambda + \xi(\eta)\zeta(\eta), \quad \hat{q}_2(\lambda, \eta) = \xi(\eta) + \zeta(\eta)$$

for $\eta \in [0, 1]$ with $\xi(\eta) = |\alpha_{k_0}|$, $\zeta(\eta) = \tau_1$ if $\eta \in [0, \frac{1}{2}]$ and $\zeta(1) \in (0, |\tau_1|)$, where ζ increases with $\eta \in [\frac{1}{2}, 1]$, we have

$$-\xi(\eta)\zeta(\eta) \le |\alpha_{k_0}||\tau_1|, \quad \eta \in [0, 1].$$

By Lemma 5.2.14, there are $r_3, r_4 \ge 0$, independent of η, such that $\theta(\cdot, \eta)$ has no zeros in $\mathbb{C}^+$, $\theta(\cdot, \eta)$ has no poles in $\{\lambda \in \mathbb{C}^+ : |\lambda| > r_3\}$, and $\lambda^{\frac{1}{2}}\theta(\lambda, \eta) \in \mathbb{C}^+$ for $\lambda \in \mathbb{C}^+$, $|\lambda| > r_4$. In particular, (5.2.12) is satisfied.

Setting

$$\hat{\theta}_2(\lambda, \eta) = \frac{\hat{q}_2(\lambda, \eta)}{\hat{p}_2(\lambda, \eta)} = \frac{\xi(\eta) + \zeta(\eta)}{-\lambda + \xi(\eta)\zeta(\eta)},$$

(5.2.17) can be written as

$$\theta_s(\lambda, \eta) = \frac{\hat{\theta}_0(\lambda, \eta) + \hat{\theta}_2(\lambda, \eta)}{1 - \lambda\hat{\theta}_0(\lambda, \eta)\hat{\theta}_2(\lambda, \eta)},$$

which implies that $\theta_s(\lambda, \eta) > 0$ if $\lambda \in (-\infty, \hat{\gamma})$, where $\hat{\gamma} = \min\{\gamma, -|\alpha_{k_0}||\tau_1|\}$.

Next we are going to study the behaviour of the real zeros of $P(\cdot, \eta)$. Writing

$$\omega_0(\lambda) = P_0(\lambda^2) + i\lambda Q_0(\lambda^2),$$
$$\hat{\omega}(\lambda, \eta) = \hat{p}(\lambda^2, \eta) + i\lambda\hat{q}(\lambda^2, \eta),$$

it follows that

$$P_s(\lambda, \eta) = P_0(\lambda)\hat{p}(\lambda, \eta) - \lambda Q_0(\lambda)\hat{q}(\lambda, \eta),$$
$$Q_s(\lambda, \eta) = Q_0(\lambda)\hat{p}(\lambda, \eta) + P_0(\lambda)\hat{q}(\lambda, \eta).$$

Since $\hat{\omega}(\cdot, \eta)$ are polynomials, so are $\hat{p}(\cdot, \eta)$ and $\hat{q}(\cdot, \eta)$. Furthermore, since $\hat{p}(\cdot, \eta)$ and $\hat{q}(\cdot, \eta)$ can be constructed according to the above rules from building blocks of the form (5.2.15) and (5.2.16), it is easy to see that the corresponding polynomials $p_j(\cdot, \eta)$ and $q_j(\cdot, \eta)$ have the form $\hat{p}_j(\lambda, \eta) = (-\lambda)^j + O(\lambda^{j-1})$ and $\hat{q}_j(\lambda, \eta) = \sigma_j(\eta)(-\lambda)^{j-1} + O(\lambda^{j-2})$, where the σ_j are positive continuous functions and where the O-terms are uniform on $\eta \in [0, 1]$. Hence there is $\delta > r_3$ such that $\hat{p}(\cdot, \eta)$ and $\hat{q}(\cdot, \eta)$ do not have zeros on (δ, ∞).

First we consider the case that P_0 has infinitely many zeros, and therefore infinitely many positive zeros. We recall that $\theta_0 = \frac{Q_0}{P_0}$ is a Nevanlinna function. If λ_1 and λ_2 are two consecutive zeros of P_0 in (δ, ∞), then there is exactly one simple zero of Q_0 between λ_1 and λ_2, and therefore $Q_0(\lambda_1)$ and $Q_0(\lambda_2)$ have opposite signs. Then also

$$P_s(\lambda_1, \eta) = -\lambda_1 Q_0(\lambda_1)\hat{q}(\lambda_1, \eta) \quad \text{and} \quad P_s(\lambda_2, \eta) = -\lambda_2 Q_0(\lambda_2)\hat{q}(\lambda_2, \eta)$$

have opposite signs, so that between any two zeros of P_0 in (δ, ∞) there is a zero of $P_s(\cdot, \eta)$, for any $\eta \in [0, 1]$. Conversely, any such zero $\lambda(\eta)$, depending continuously on η, must stay in such an interval (λ_1, λ_2) since $P_s(\cdot, \eta)$ does not have nonreal zeros λ with $|\lambda| > r_3$.

On the other hand, if P_0 has finitely many zeros, then θ_0 is a rational function, and in view of Theorem 11.1.6 we may replace P_0 and Q_0 by real polynomials, thus replacing $P_s(\cdot, \eta)$ and $Q_s(\cdot, \eta)$ by polynomials whose leading coeffients are never zero. Hence also the zeros of $P_s(\cdot, \eta)$ must remain in a finite region.

In particular, since zeros λ of $P_s(\cdot, \eta)$ and $Q_s(\cdot, \eta)$ with $|\lambda| > r_3$ must be real, zeros of $P_s(\cdot, \eta)$ and $Q_s(\cdot, \eta)$ cannot leave or join the complex plane through ∞. Hence, if $(\lambda_j(1))_{j=1}^{(\infty)}$ denotes the zeros of $P_s(\cdot, 1)$ and $Q_s(\cdot, 1)$, then all zeros are real, may be arranged such that $\lambda_j(1) < \lambda_{j+1}(1)$ for all $j \in \mathbb{N}$, and they are alternatively zeros of $P_s(\cdot, 1)$ and of $Q_s(\cdot, 1)$. The above considerations show that there is a sequence of continuous functions $(\lambda_j)_{j=1}^{(\infty)}$ on $[0, 1]$ such that $(\lambda_j(\eta))_{j=1}^{(\infty)}$ is the sequence of the zeros of $P_s(\cdot, \eta)$ and $Q_s(\cdot, \eta)$ for all $\eta \in [0, 1]$. As η decreases from 1 to 0, the $\lambda_j(\eta)$ must remain distinct since $P_s(\cdot, \eta)$ and $Q_s(\cdot, \eta)$ do not have common zeros. Since real zeros of a real analytic functions can leave the real axis into the complex plane only in conjugate complex pairs, all zeros of $P_s(\cdot, \eta)$ must therefore stay on the real axis for all η, and it follows for all $\eta \in [0, 1]$ that the zeros of $P_s(\cdot, \eta)$ and $Q_s(\cdot, \eta)$ are real and interlace.

We now show that $\theta_s \in \hat{\mathcal{N}}$. The proof of this part is similar to the last paragraph of the proof of Theorem 11.1.6. To this end, we use that there is a product $\psi \in \hat{\mathcal{N}}$ of the form (11.1.2), with $C = 1$, such that ψ has the same zeros and poles as θ_s. Then $f = \frac{\theta_s}{\psi}$ is a real meromorphic function without poles and zeros, that is, a real entire function without zeros. From what we have already shown for θ_s and from Lemma 5.2.15 applied to ψ we have that there is $r > 0$ such that $\lambda^{\frac{1}{2}} \theta_s(\lambda) \in \mathbb{C}^+$ and $\lambda^{\frac{1}{2}} \psi(\lambda) \in \mathbb{C}^+$ for all $\lambda \in \mathbb{C}^+$ with $|\lambda| > r$. Then it follows for these λ that $f(\lambda) \notin (-\infty, 0)$. Also, there is $m > 0$ such that $|f(\lambda)| \le m$ for all $\lambda \in \mathbb{C}$ with $|\lambda| \le r$. It follows that $f(\mathbb{C}^+) \cap (-\infty, -m) = \emptyset$, and since f is real, also $f(\mathbb{C}^-) \cap (-\infty, -m) = \emptyset$. Continuing as in the last paragraph of the proof of Theorem 11.1.6, it follows that f is a positive constant, and then $\psi \in \hat{\mathcal{N}}$ gives $\theta_s \in \hat{\mathcal{N}}$.

Finally, since we already know that $\theta_s(\lambda) > 0$ for $\lambda \in (-\infty, \gamma)$, it follows that $\omega \in \mathrm{SSHB}_{\kappa'}$ for some $\kappa' \in \mathbb{N}_0$. In view of Theorem 5.2.9, ω has κ' zeros in $(0, -i\infty)$, so that $\kappa' = \kappa$. $\square$

The following simple example shows that Theorem 5.2.16 may be false or true without assumption 7.

Example 5.2.17. Let $0 < b < a$ and define $\omega(\lambda) = (i\lambda - a)(i\lambda + b)^2$. Then ω satisfies all assumptions of Theorem 5.2.16 except assumption 7, which does not hold. We can write $\omega(\lambda) = P_s(\lambda^2) + i\lambda Q_s(\lambda^2)$ with $P_s(\lambda) = (a - 2b)\lambda - ab^2$ and $Q_s(\lambda) = -\lambda - b(2a - b)$. Now it is easy to see that $\frac{Q_s}{P_s} \in \mathcal{N}_+^{\mathrm{ep}}$ if and only if $a < 2b$. Hence ω is of SSHB_1 class if $0 < b < a < 2b$, but ω does not belong to SSHB if $0 < 2b \le a$.

5.3 Notes

Hermite–Biehler polynomials were introduced and investigated in [107]. In [30], Biehler considered a class of polynomials for which the imaginary parts of its zeros have the same sign and showed that the zeros of the real and imaginary parts of these polynomials have only real simple zeros which interlace. By the change of parameter $\lambda \mapsto -i\lambda$ we obtain Hurwitz polynomials [120]. For a detailed review of these and related results we refer the reader to [113].

Hermite–Biehler entire functions and generalized Hermite–Biehler entire functions were described in [173]. We cite the following statements about the history of generalizations of Hermite–Biehler polynomials to Hermite–Biehler functions from [173, Chapter VII, pp. 306, 307].

For the solution of certain questions of the theory of automation one needs effective criteria for all the roots of the function

$$F(z) = \sum_{k,j=0}^{n,m} a_{kj} z^j e^{\lambda_k z} \qquad (5.3.1)$$

to lie in the left half-plane. N.G. Čebotarëv called functions of type (5.3.1) quasipolynomials. Quasipolynomials depend only on a finite number of parameters, and therefore it is natural to suppose that for them there exists an effective method of solving this problem.

In 1942 N.G. Čebotarëv found such an effective criterion for a very special case of polynomials [51]. In another paper N.G. Čebotarëv [50], generalizing the Sturm algorithm to quasipolynomials, gave a general principle for solving this problem for arbitrary quasipolynomials. However the application of this general principle required a generalization of the Hermite–Biehler theorem to quasipolynomials.

L.S. Pontryagin [232] in 1942 generalized the Hermite–Biehler theorem to quasipolynomials of the type $P(z, e^z)$, where $P(z, u)$ is a polynomial in two variables.

In carrying over the Hermite–Biehler criterion to arbitrary entire functions, an essential role is played by a particular class of entire functions. This class was introduced and studied by M.G. Kreĭn in his paper "On a class of entire and meromorphic functions", which was devoted to the extension of the Hurwitz criterion to entire functions. The definition of this class presented here is due to N.N. Meĭman [188]. It is equivalent to the definition given earlier by M.G. Kreĭn (see the book of N.I. Ahiezer and M.G. Kreĭn [7]).

For further results on Hermite–Biehler polynomials and related polynomials we refer to [110], [111], [45, P. 215] [112], [203], [253] and the review paper [113]. Different results connected with Hermite–Biehler functions can be found in [61].

Shifted Hermite–Biehler polynomials were introduced in [219]. For polynomials, an analogue of Theorem 5.2.9 was obtained in [219] and [264, Theorem 5.3].

The class of essentially positive Nevanlinna functions has been introduced by M. Kaltenbäck, H. Winkler and H. Woracek in [131].

In [227] the theory of de Branges Pontryagin spaces of entire functions was used to characterize zeros of shifted Hermite–Biehler functions. The main result there is [227, Theorem 3.1], which corresponds to Theorems 5.2.9 and 5.2.16. Our proof of Theorem 5.2.16 follows the geometric approach in [228].

Results on representing rational functions as continued fractions can be found in [270, Chapter IX], see also [264, Theorem 1.36] and [113, Sections 1.3–1.5].

Chapter 6

Applications of Shifted Hermite–Biehler Functions

In this chapter we revisit some of the applications in Chapter 2 and express their spectra as the zeros of entire functions.

Indeed, we are going to use shifted Hermite–Biehler functions and generalized shifted Hermite–Biehler functions to describe characteristic functions of those spectral problems from Chapter 2 which are described by quadratic operator pencils $L(\lambda) = \lambda^2 M - i\lambda K - A$ with the operator K having rank 1. The results in this chapter differ slightly form those obtained in Chapter 2, and this additional information by either method will be exploited in Chapters 7 and 8.

In Definition 1.5.2 we have defined eigenvalues of type I and type II, whereas in Remark 5.2.10 we have defined zeros of type I and type II. In Chapter 2 the operator approach allowed us to classify eigenvalues of our applications as being of type I and type II. In this chapter we will consider characteristic functions of our applications, and the zeros of these characteristic functions, which are of $\overline{\text{SSHB}}$ class, will be classified as being of type I and type II. But from Remark 1.5.8 we readily conclude that a nonzero eigenvalue is of type I if and only it is a zero of $\lambda \mapsto R_*(\lambda^2)$, with multiplicity, see Remark 5.2.10. Hence nonzero eigenvalues of type I and type II, respectively, are zeros of type I and type II, respectively, of the characteristic function.

Therefore the notions of type of an eigenvalue for an eigenvalue problem and type of a zero of the characteristic function of that eigenvalue problem coincide for nonzero numbers, and in this chapter we may henceforth write, e. g., "eigenvalue of type I" rather than "zero of type I of the characteristic function".

However, the situation may be different at 0. For example, if 0 is a simple eigenvalue of a pencil L as considered in Section 1.5, then it is an eigenvalue of type I by Remark 1.5.3, part 2. On the other hand, if this problem has a characteristic function of $\overline{\text{SSHB}}$ class, then 0 is a zero of type II since this zero is simple.

6.1   The generalized Regge problem

The Regge problem was considered in Section 2.1. Recall that s is the solution of the differential equation (2.1.1) which satisfies the initial conditions $s(\lambda, 0) = 0$, $s'(\lambda, 0) = 1$. Then the eigenvalues of the generalized Regge problem (2.1.1), (2.1.2), (2.1.5) are exactly the zeros of the entire function ϕ obtained by substituting s, the solution of (2.1.1), (2.1.2) satisfying $s'(0) = 1$, into (2.1.5):

$$\phi(\lambda) = s'(\lambda, a) + (i\lambda\alpha + \beta)s(\lambda, a), \qquad (6.1.1)$$

where $\alpha > 0$ and $\beta \in \mathbb{R}$.

Proposition 6.1.1. 1. *The function ϕ is an entire function of exponential type a.*
2. *The function ϕ is of sine type if and only if $\alpha \neq 1$.*

Proof. Substituting (12.2.22) and (12.2.23) with $n = 0$ into (6.1.1) we obtain

$$\phi(\lambda) = \frac{1 + \alpha}{2} e^{i\lambda a} + \frac{1 - \alpha}{2} e^{-i\lambda a} + o(e^{|\operatorname{Im}\lambda|a}). \qquad (6.1.2)$$

Clearly,

$$\phi(\lambda) = O\left(e^{|\operatorname{Im}\lambda|a}\right) \quad \text{and} \quad \phi(-i\tau) = \left(\frac{1 + \alpha}{2} + o(1)\right) e^{\tau a} \quad \text{for } \tau > 0, \quad (6.1.3)$$

which shows that ϕ is of exponential type a. Similarly,

$$\phi(i\tau) = \left(\frac{1 - \alpha}{2} + o(1)\right) e^{\tau a} \quad \text{for } \tau > 0, \qquad (6.1.4)$$

and therefore (6.1.3) and (6.1.4) show that ϕ is of exponential type both in the upper and lower half-planes if $\alpha \neq 1$. Furthermore, it immediately follows from (6.1.2) that there are positive real numbers h, m, M such that

$$me^{|\operatorname{Im}\lambda|a} \leq |\phi(\lambda)| \leq Me^{|\operatorname{Im}\lambda|a} \quad \text{for all } \lambda \in \mathbb{C} \text{ with } |\operatorname{Im}\lambda| \geq h$$

if $\alpha \neq 1$. Therefore ϕ is of sine type if $\alpha \neq 1$ by Definition 11.2.5. On the other hand, if $\alpha = 1$, then (6.1.2) gives $|\phi(\lambda)| = o(e^{\operatorname{Im}\lambda a})$ for all $\lambda \in \mathbb{C}_+$, and therefore ϕ is not a sine type function by Proposition 11.2.19. $\square$

Proposition 6.1.2. *The function ϕ defined in (6.1.1) belongs to the class* SSHB.

Proof. Since $s(\cdot, x)$ is an even function, there are unique entire functions f and g such that

$$f(\lambda^2) = s'(\lambda, a), \quad g(\lambda^2) = s(\lambda, a), \quad \lambda \in \mathbb{C}.$$

By Theorem 12.6.2, the sequence $(\mu_k^2)_{k=1}^\infty$ of the squares of the zeros of $s'(\cdot, a)$ interlaces with the sequence $(\nu_k^2)_{k=1}^\infty$ of the squares of the zeros of $s(\cdot, a)$:

$$\mu_1^2 < \nu_1^2 < \mu_2^2 < \nu_2^2 < \cdots \qquad (6.1.5)$$

That is, the zeros of f and g interlace. As in the proof of Proposition 6.1.1 it follows from (12.2.22) and (12.2.23) with $n = 0$ that $s'(\cdot, a)$ and $\lambda \mapsto \lambda s(\lambda, a)$ are sine type functions. In view of Lemma 11.2.29, there are nonzero complex numbers c and c' and integers $m, m' \in \{0, 1\}$ such that

$$\lambda s(\lambda, a) = c\lambda^{2m+1} \prod_{k=1}^{\infty}{}' \left(1 - \frac{\lambda^2}{\nu_k^2}\right),$$

$$s'(\lambda, a) = c'\lambda^{2m'} \prod_{k=1}^{\infty}{}' \left(1 - \frac{\lambda^2}{\mu_k^2}\right),$$

where $\prod'$ indicates that the factor, if any, for which $\nu_k = 0$ or $\mu_k = 0$, respectively, is omitted, and where $m + m' \leq 1$ in view of (6.1.5). It follows that

$$g(\lambda) = c\lambda^m \prod_{k=1}^{\infty}{}' \left(1 - \frac{\lambda}{\nu_k^2}\right),$$

$$f(\lambda) = c'\lambda^{m'} \prod_{k=1}^{\infty}{}' \left(1 - \frac{\lambda}{\mu_k^2}\right).$$

Since $s(\lambda, \cdot)$ is a real solution of (2.1.1) for real λ, f and g are real analytic. In particular, c and c' are real numbers. We conclude from Remark 11.1.7 that either $\frac{g}{f}$ or $\frac{f}{g}$ is a Nevanlinna function. Denote that function which is a Nevanlinna function by θ. Clearly, $\theta \in \mathcal{N}^{\mathrm{ep}}$ by Definition 5.1.26, part 1. Furthermore,

$$s(\lambda, a) \underset{\lambda \to \pm i\infty}{=} \frac{e^{|\lambda|a}}{2|\lambda|} + o\left(\frac{e^{|\lambda|a}}{|\lambda|}\right), \tag{6.1.6}$$

$$s'(\lambda, a) \underset{\lambda \to \pm i\infty}{=} \frac{e^{|\lambda|a}}{2} + o\left(e^{|\lambda|a}\right), \tag{6.1.7}$$

shows that $\theta(\lambda)$ is positive for $\lambda \to -\infty$, that is, $\theta \in \mathcal{N}_+^{\mathrm{ep}}$ by Definition 5.1.26, part 2. Since μ_1^2, the smallest zero of f, is smaller than the smallest zero of g, μ_1^2 must be a pole of θ by Corollary 5.2.3. It follows that $\theta = \frac{g}{f}$. Then $-\frac{f}{g} \in \mathcal{N}_-^{\mathrm{ep}}$ by Lemma 5.2.4, part 2, and therefore

$$-\frac{f + \beta g}{\alpha g} = -\frac{1}{\alpha}\frac{f}{g} - \frac{\beta}{\alpha} \in \mathcal{N}_-^{\mathrm{ep}}$$

if we observe that $\frac{f(\lambda)}{g(\lambda)} \to \infty$ as $\lambda \to -\infty$ by (6.1.6) and (6.1.7). We conclude, again from Lemma 5.2.4, part 2, that

$$\frac{\alpha g}{f + \beta g} \in \mathcal{N}_+^{\mathrm{ep}}.$$

Since $\phi(\lambda) = (f + \beta g)(\lambda^2) + i\alpha\lambda g(\lambda^2)$, the function ϕ belongs to the class SSHB by Definition 5.2.6. $\qquad\square$

Theorem 6.1.3. *The eigenvalues of the generalized Regge problem* (2.1.1), (2.1.2), (2.1.5) *have the properties:*

1. *Only a finite number of the eigenvalues lie in the closed lower half-plane.*
2. *All nonzero eigenvalues in the closed lower half-plane lie on the negative imaginary semiaxis and are simple. If their number κ is positive, they will be uniquely indexed as $\lambda_{-j} = -i|\lambda_{-j}|$, $j = 1, \ldots, \kappa$, satisfying $|\lambda_{-j}| < |\lambda_{-(j+1)}|$, $j = 1, \ldots, \kappa - 1$.*
3. *If $\kappa > 0$, then the numbers $i|\lambda_{-j}|$, $j = 1, \ldots, \kappa$, are not eigenvalues.*
4. *If $\kappa \geq 2$, then in each of the intervals $(i|\lambda_{-j}|, i|\lambda_{-(j+1)}|)$, $j = 1, \ldots, \kappa - 1$ the number of eigenvalues, counted with multiplicity, is odd.*
5. *If $\kappa > 0$ and 0 is not an eigenvalue, then the interval $(0, i|\lambda_{-1}|)$ contains an even number of eigenvalues, counted with multiplicity, or does not contain any eigenvalues.*
6. *If $\kappa > 0$ and 0 is an eigenvalue, then the interval $(0, i|\lambda_{-1}|)$ contains an odd number of eigenvalues, counted with multiplicity.*
7. *If $\alpha \neq 1$, then the generalized Regge problem has infinitely many eigenvalues, which lie in a horizontal strip of the complex plane.*

Proof. Parts 1–6 follow from Proposition 6.1.2 and Theorem 5.2.9.

Part 7 immediately follows from Proposition 11.2.8, part 2, since ϕ is a sine type function by Proposition 6.1.1. $\square$

Remark 6.1.4. The statement of part 7 of Theorem 6.1.3 may fail if $\alpha = 1$. For example, if $q = 0$ and $\beta = 0$, then

$$s(\lambda, x) = \frac{\sin(\lambda x)}{\lambda}, \ \lambda \in \mathbb{C} \setminus \{0\},$$

and therefore

$$\phi(\lambda) = \cos(\lambda a) + i\sin(\lambda a) = e^{i\lambda a},$$

which has no zeros at all.

Remark 6.1.5. We recall that Theorem 2.1.2 is concerned with the Regge problem, whereas Theorem 6.1.3 deals with the generalized Regge problem. Statements 1–4 are identical in both theorems, up to minor differences in formulation. Statement 5 in Theorem 2.1.2 distinguishes two cases determined by whether the operator A is injective. Since the operator A does not feature in this section, in statements 5 and 6 of Theorem 6.1.3 the condition on the nullspace of A is replaced by the equivalent condition on whether 0 is an eigenvalue. It is also important to note that infinitely many eigenvalues can only be guaranteed if $\alpha \neq 1$. Hence the condition $\alpha \neq 1$ will play a prominent role in Chapters 7 and 8. In Chapter 7 we will encounter sufficient conditions for the Regge problem to have infinitely many eigenvalues; however these eigenvalues will not lie in a horizontal strip but in a logarithmic strip of the complex plane.

6.2 Damped vibrations of Stieltjes strings

Let us recall problem (2.6.7)–(2.6.8) for the case with only one point of damping, which is derived from (2.6.3)–(2.6.6):

$$\frac{u_k^{(j)} - u_{k+1}^{(j)}}{l_k^{(j)}} + \frac{u_k^{(j)} - u_{k-1}^{(j)}}{l_{k-1}^{(j)}} - \lambda^2 m_k^{(j)} u_k^{(j)} = 0, \ k = 1, \ldots, n_j, \ j = 1, 2, \quad (6.2.1)$$

$$u_0^{(j)} = 0, \quad (6.2.2)$$

$$u_{n_1+1}^{(1)} = u_{n_2+1}^{(2)}, \quad (6.2.3)$$

$$\frac{u_{n_1+1}^{(1)} - u_{n_1}^{(1)}}{l_{n_1}^{(1)}} + \frac{u_{n_2+1}^{(2)} - u_{n_2}^{(2)}}{l_{n_2}^{(2)}} + i\lambda\nu u_{n_1+1}^{(1)} = 0. \quad (6.2.4)$$

Here $\nu > 0$ is the coefficient of damping at the interior point of damping and $m_k^{(j)} > 0$ for $k = 1, \ldots, n_j$ and $j = 1, 2$. Following [85, Supplement II, (16) and (17)], a recursive application of (6.2.1) leads to

$$u_k^{(j)}(\lambda) = R_{2k-2}^{(j)}(\lambda^2) u_1^{(j)}(\lambda), \ k = 2, \ldots, n_j + 1, \ j = 1, 2,$$

where the $R_{2k-2}^{(j)}$ are real polynomials of degree $k - 1$. For convenience we extend the above equations to $k = 1$ and $k = 0$ by putting $R_0^{(j)} = 1$ and $R_{-2}^{(j)} = 0$, the latter in view of (6.2.2). We further define

$$R_{2k-1}^{(j)} = \frac{R_{2k}^{(j)} - R_{2k-2}^{(j)}}{l_k^{(j)}}, \ k = 0, \ldots, n_j, \ j = 1, 2. \quad (6.2.5)$$

From (6.2.1) we obtain

$$\frac{R_{2k}^{(j)}(\lambda^2) - R_{2k-2}^{(j)}(\lambda^2)}{l_k^{(j)}} = \frac{R_{2k-2}^{(j)}(\lambda^2) - R_{2k-4}^{(j)}(\lambda^2)}{l_{k-1}^{(j)}} - \lambda^2 m_k^{(j)} R_{2k-2}^{(j)}(\lambda^2)$$

for $k = 1, \ldots, n_j$ and $j = 1, 2$. Substituting (6.2.5) into these equations, solving (6.2.5) for $R_{2k}^{(j)}$, and observing $R_0^{(j)} = 1$ and $R_{-2}^{(j)} = 0$, we obtain that the polynomials $R_k^{(j)}$ satisfy the recurrence relations and initial conditions

$$R_{2k-1}^{(j)}(\lambda^2) = -\lambda^2 m_k^{(j)} R_{2k-2}^{(j)}(\lambda^2) + R_{2k-3}^{(j)}(\lambda^2), \quad k = 1, \ldots, n_j, \ j = 1, 2, \quad (6.2.6)$$

$$R_{2k}^{(j)} = l_k^{(j)} R_{2k-1}^{(j)} + R_{2k-2}^{(j)}, \quad k = 1, \ldots, n_j, \ j = 1, 2, \quad (6.2.7)$$

$$R_{-1}^{(j)} = \frac{1}{l_0^{(j)}}, \ R_0^{(j)} = 1, \ R_{-2}^{(j)} = 0. \quad (6.2.8)$$

The conditions (6.2.3) and (6.2.4) at the joint of the two substrings give

$$R_{2n_1}^{(1)}(\lambda^2)u_1^{(1)}(\lambda) = R_{2n_2}^{(2)}(\lambda^2)u_1^{(2)}(\lambda), \tag{6.2.9}$$

$$R_{2n_1-1}^{(1)}(\lambda^2)u_1^{(1)}(\lambda) + R_{2n_2-1}^{(2)}(\lambda^2)u_1^{(2)}(\lambda) + i\nu\lambda R_{2n_1}^{(1)}(\lambda^2)u_1^{(1)}(\lambda) = 0. \tag{6.2.10}$$

The characteristic polynomial Φ of (6.2.1)–(6.2.3) is the determinant of the coefficient matrix of the linear system (6.2.9), (6.2.10):

$$\Phi(\lambda) = \phi(\lambda^2) + i\lambda\nu R_{2n_1}^{(1)}(\lambda^2)R_{2n_2}^{(2)}(\lambda^2), \tag{6.2.11}$$

where

$$\phi = R_{2n_1}^{(1)}R_{2n_2-1}^{(2)} + R_{2n_1-1}^{(1)}R_{2n_2}^{(2)}. \tag{6.2.12}$$

Proposition 6.2.1.

1. For $k = 1,\ldots,n_j$ and $j = 1,2$, the functions $\dfrac{R_{2k-2}^{(j)}}{R_{2k-1}^{(j)}}$ and $\dfrac{R_{2k}^{(j)}}{R_{2k-1}^{(j)}}$ are S-functions.

2. The function $\dfrac{\nu R_{2n_1}^{(1)}R_{2n_2}^{(2)}}{\phi}$ is an S-function.

3. For $m = 1,\ldots,2n_j$ and $j = 1,2$, all zeros of $R_m^{(j)}$ lie in $(0,\infty)$ and are simple, and $R_m^{(j)}$ is positive on $(-\infty,0]$. If $R_m^{(j)}(z) = 0$ for some $z \in \mathbb{C}$, then $R_{m-1}^{(j)}(z) \neq 0$.

4. If $R_{2n_1}^{(1)}(z) = R_{2n_2}^{(2)}(z) = 0$ for some $z \in \mathbb{C}$, then z is a simple zero of ϕ.

Proof. 1. The statement will be proved by induction on k. For convenience, we start with the case $k = 0$, which is obvious since $\dfrac{R_{-2}^{(j)}}{R_{-1}^{(j)}} = 0$ and $\dfrac{R_0^{(j)}}{R_{-1}^{(j)}} = l_0^{(j)}$ by (6.2.8). Now let $k \in \{1,\ldots,n_j\}$ and assume that the statement is true for $k - 1$. We observe that all $R_k^{(j)}$ for $k = 1,\ldots,n_j$ and $j = 1,2$ are nonzero polynomials. From (6.2.6) it follows that

$$\frac{R_{2k-1}^{(j)}(z)}{R_{2k-2}^{(j)}(z)} = -zm_k^{(j)} + \frac{R_{2k-3}^{(j)}(z)}{R_{2k-2}^{(j)}(z)}. \tag{6.2.13}$$

By induction hypothesis and Lemma 5.2.4, part 3, we obtain $-\dfrac{R_{2k-3}^{(j)}}{R_{2k-2}^{(j)}} \in S^{-1}$. Then $-\dfrac{R_{2k-1}^{(j)}}{R_{2k-2}^{(j)}} \in S^{-1}$ by (6.2.13), and $\dfrac{R_{2k-2}^{(j)}}{R_{2k-1}^{(j)}} \in S$ by Lemma 5.2.4, part 3. Hence (6.2.7) gives

$$\frac{R_{2k}^{(j)}}{R_{2k-1}^{(j)}} = l_k^{(j)} + \frac{R_{2k-2}^{(j)}}{R_{2k-1}^{(j)}} \in S. \tag{6.2.14}$$

2. The same reasoning as in the proof of part 1 shows that

$$-\frac{\phi}{R_{2n_1}^{(1)}R_{2n_2}^{(2)}} = -\frac{R_{2n_1-1}^{(1)}}{R_{2n_1}^{(1)}} - \frac{R_{2n_2-1}^{(2)}}{R_{2n_2}^{(2)}} \in S^{-1} \quad \text{and} \quad \frac{\nu R_{2n_1}^{(1)}R_{2n_2}^{(2)}}{\phi} \in S. \tag{6.2.15}$$

3. If $R_m^{(j)}(z) = 0$ and $R_{m-1}^{(j)}(z) = 0$, then a recursive substitution into (6.2.6) and (6.2.7) would lead to $R_p^{(j)}(z) = 0$ for all $p = m, m-1, \ldots, 0$, which is impossible due to (6.2.8). Hence $\dfrac{R_m^{(j)}}{R_{m-1}^{(j)}}$ does not have zero-pole cancellations, and in view of part 1, all zeros of $R_m^{(j)}$ lie in $[0, \infty)$ and are simple by Lemma 11.1.3. The recurrence relations (6.2.6)–(6.2.8) show that $R_m^{(j)}(0) > 0$, and since the real polynomials $R_m^{(j)}$ have no zeros on $(-\infty, 0)$, they must also be positive there.

4. By part 3, $z > 0$, and from the proof of part 3 we know that $-\dfrac{R_{2n_j-1}^{(j)}}{R_{2n_j}^{(j)}} \in \mathcal{S}$. Hence the function on the left in (6.2.15) belongs to $\mathcal{S}^{-1}$ and therefore Lemma 11.1.3 shows that it has a simple pole at z. Since its denominator has a double zero at z, its numerator ϕ must have a simple zero at z. □

Proposition 6.2.2. *The function Φ defined in (6.2.11) is a nontrivial function of* $\overline{\text{SHB}}$ *class with $\Phi(0) > 0$.*

Proof. Clearly, the functions ϕ and $\nu R_{2n_1}^{(1)} R_{2n_2}^{(2)}$ are real polynomials and therefore Φ is real on the imaginary axis, that is, Φ is symmetric with respect to the imaginary axis. Let z be a common zero of ϕ and $\nu R_{2n_1}^{(1)} R_{2n_2}^{(2)}$. Without loss of generality, we may assume that $R_{2n_1}^{(1)}(z) = 0$, and it follows from Proposition 6.2.1, part 3, that $z \in (0, \infty)$. Hence the function $\dfrac{\nu R_{2n_1}^{(1)} R_{2n_2}^{(2)}}{\phi}$ does not have zero-pole cancellations on $(-\infty, 0)$. By Proposition 6.2.1, part 2, the rational function $\dfrac{\nu R_{2n_1}^{(1)} R_{2n_2}^{(2)}}{\phi}$ has no poles on $(-\infty, 0)$. Altogether it follows that ϕ has no negative real zeros. Letting R_* be a real polynomial whose zeros are the common zeros of ϕ and $\nu R_{2n_1}^{(1)} R_{2n_2}^{(2)}$, it now easily follows that Φ can be written as in Definition 5.2.7, and therefore Φ is of $\overline{\text{SHB}}_0$ class. An application of Proposition 5.2.8 gives that Φ is a nontrivial function of SHB class. In view of Proposition 6.2.1, part 3, we have

$$\Phi(0) = \phi(0) = R_{2n_1}^{(1)}(0)R_{2n_2-1}^{(2)}(0) + R_{2n_1-1}^{(1)}(0)R_{2n_2}^{(2)}(0) > 0. \qquad \square$$

Theorem 6.2.3. *The eigenvalues $(\lambda_k)_{k=-(n_1+n_2)}^{n_1+n_2}$ of problem (6.2.1)–(6.2.4) can be indexed in such a way that they have the following properties:*

1. *$\operatorname{Im} \lambda_k \geq 0$ for $k = 0, \pm 1, \cdots \pm (n_1 + n_2)$;*
2. *$\lambda_{-k} = -\overline{\lambda_k}$ for not pure imaginary λ_{-k};*
3. *all real eigenvalues, if any, are simple and nonzero;*
4. *for each real eigenvalue λ_k, $\operatorname{Im} \Phi'(\lambda_k) = 0$;*
5. *the number of real eigenvalues does not exceed $2 \min\{n_1, n_2\}$.*

Proof. The eigenvalues of this problem are the zeros of the function Φ. Since the polynomials $R_{2n_j}^{(j)}$ and $R_{2n_j-1}^{(j)}$ have degree n_j, it follows that Φ has degree $2n_1 + 2n_2 + 1$, and the eigenvalues can be labelled as indicated.

An application of Proposition 6.2.2 and Corollary 5.2.11 proves part 1, and part 2 follows from the symmetry of Φ.

3. By Proposition 6.2.2, real eigenvalues are nonzero, and hence a real eigenvalue λ_k is a square root of a common zero of the real polynomials ϕ and $R_{2n_1}^{(1)} R_{2n_2}^{(2)}$. Assume without loss of generality that $R_{2n_1}^{(1)}(\lambda_k^2) = 0$. Then $R_{2n_1-1}^{(1)}(\lambda_k^2) \neq 0$ by Proposition 6.2.1, part 3, and

$$0 = \phi(\lambda_k^2) = R_{2n_1-1}^{(1)}(\lambda_k^2) R_{2n_2}^{(2)}(\lambda_k^2)$$

gives $R_{2n_2}^{(2)}(\lambda_k^2) = 0$. In view of Proposition 6.2.1, part 4, it follows that λ_k^2 is a simple zero of ϕ, which implies that λ_k is a simple zero of Φ.

4 and 5. For real λ, we have $\operatorname{Im} \Phi(\lambda) = \lambda R_{2n_1}^{(1)}(\lambda^2) R_{2n_2}^{(2)}(\lambda^2)$. From the proof of part 3 and from Proposition 6.2.1, part 3, we recall that real eigenvalues are square roots of common simple zeros of $R_{2n_1}^{(1)}$ and $R_{2n_2}^{(2)}$ and therefore double zeros of $\operatorname{Im} \Phi$. This immediately proves part 4. We have also shown that the number of real eigenvalues is bounded by $2n_1$ and $2n_2$, the degrees of $R_{2n_1}^{(1)}$ and $R_{2n_2}^{(2)}$, and part 5 follows. $\square$

6.3   Vibrations of smooth strings

Here we revisit problem (2.2.1)–(2.2.3). We recall that separation of variables and the Liouville transform lead to the boundary eigenvalue problem (2.2.4)–(2.2.6), which upon the assumption $\sigma(s) \equiv 2\varrho\rho(s)$ and the parameter transformation $\lambda = \pm\tau + i\varrho$, see the beginning of Subsection 2.2.2, takes the form

$$y'' - (q(x) - \varrho^2)y + \tau^2 y = 0, \tag{6.3.1}$$

$$y(\tau, 0) = 0, \tag{6.3.2}$$

$$y'(\tau, a) + (-\tilde{m}\tau^2 + i\tau|\tilde{\nu} - 2\tilde{m}\varrho| + \beta - \tilde{\nu}\varrho + \tilde{m}\varrho^2)y(\tau, a) = 0. \tag{6.3.3}$$

Recall that $s(\tau, \cdot)$ denotes the solution of (6.3.1) which satisfies the initial conditions $s(\tau, 0) = 0$, $s'(\tau, 0) = 1$. Then the spectrum of problem (6.3.1)–(6.3.3) coincides with the set of zeros of the function ϕ given by

$$\phi(\tau) = s'(\tau, a) + (-\tilde{m}\tau^2 + i\tau|\tilde{\nu} - 2\tilde{m}\varrho| + \beta - \tilde{\nu}\varrho + \tilde{m}\varrho^2)s(\tau, a). \tag{6.3.4}$$

Let us consider

$$\begin{aligned}
\theta_1(\tau^2) &= \frac{s(\tau, a)}{s'(\tau, a) + (-\tilde{m}\tau^2 + \beta - \tilde{\nu}\varrho + \tilde{m}\varrho^2)s(\tau, a)} \\
&= \left(\frac{s'(\tau, a)}{s(\tau, a)} + (-\tilde{m}\tau^2 + \beta - \tilde{\nu}\varrho + \tilde{m}\varrho^2) \right)^{-1}.
\end{aligned}$$

Since the above initial value problem is as considered in Section 6.1, we obtain as in the proof of Proposition 6.1.2 that θ as defined there satisfies $-\theta^{-1} \in \mathcal{N}_-^{\mathrm{ep}}$.

From

$$-\frac{1}{\theta_1(z)} = -\frac{1}{\theta(z)} + \tilde{m}z - \beta + \tilde{\nu}\varrho - \tilde{m}\varrho^2 \qquad (6.3.5)$$

we conclude that $-\theta_1^{-1} \in \mathcal{N}_-^{\mathrm{ep}}$, and Lemma 5.1.22 shows that $\theta_1 \in \mathcal{N}_+^{\mathrm{ep}}$. If the numerator and denominator in the definition of θ_1 would have common zeros, s would satisfy the initial conditions $s(\tau, a) = s'(\tau, a) = 0$ for some $\tau \in \mathbb{C}$, which would lead to the contradiction $s(\tau, \cdot) = 0$. Therefore ϕ is a shifted Hermite–Biehler function.

We can now apply Theorem 5.2.9 to the characteristic function ϕ of the problem (6.3.1)–(6.3.3), which leads to the following results if we recall that the zeros of ϕ are the eigenvalues of (6.3.1)–(6.3.3).

Theorem 6.3.1. *Let $\tilde{\nu} > 2\tilde{m}\varrho$. Then:*

1. *Only a finite number of the eigenvalues of problem (6.3.1)–(6.3.3) lie in the closed half-plane $\operatorname{Im}\lambda \le \varrho$.*

2. *All eigenvalues in the half-plane $\operatorname{Im}\lambda \le \varrho$ which are different from $i\varrho$ lie on $(0, i\varrho)$ and are simple. Their number will be denoted by κ. If $\kappa > 0$, they will be uniquely indexed as $\lambda_{-j} = i\varrho - i|\lambda_{-j} - i\varrho|$, $j = 1, \ldots, \kappa$, satisfying $|\lambda_{-j} - i\varrho| < |\lambda_{-(j+1)} - i\varrho|$, $j = 1, \ldots, \kappa - 1$.*

3. *If $\kappa > 0$, then the numbers $i\varrho + i|\lambda_{-j} - i\varrho|$, $j = 1, \ldots, \kappa$, are not eigenvalues.*

4. *If $\kappa \ge 2$, then in each of the intervals $(i\varrho + i|\lambda_{-j} - i\varrho|, i\varrho + i|\lambda_{-(j+1)} - i\varrho|)$, $j = 1, \ldots, \kappa - 1$, the number of eigenvalues, counted with multiplicity, is odd.*

5. *If $\kappa > 0$, then the interval $[i\varrho, i\varrho + i|\lambda_{-1} - i\varrho|)$ contains no or an even number of eigenvalues, counted with multiplicity. If $i\varrho$ is an eigenvalue, then it is simple.*

Theorem 6.3.2. *Let $\tilde{\nu} < 2\tilde{m}\varrho$. Then:*

1. *Only a finite number of the eigenvalues of problem (6.3.1)–(6.3.3) lie in the closed half-plane $\operatorname{Im}\lambda \ge \varrho$, and all other eigenvalues lie in the open strip $0 < \operatorname{Im}\lambda < \varrho$.*

2. *All eigenvalues in the half-plane $\operatorname{Im}\lambda \ge \varrho$ which are different from $i\varrho$ lie on $(i\varrho, i\infty)$ and are simple. Their number will be denoted by κ. If $\kappa > 0$, they will be uniquely indexed as $\lambda_{-j} = i\varrho + i|\lambda_{-j} - i\varrho|$, $j = 1, \ldots, \kappa$, satisfying $|\lambda_{-j}| < |\lambda_{-(j+1)}|$, $j = 1, \ldots, \kappa - 1$.*

3. *If $\kappa > 0$, then the numbers $i\varrho - i|\lambda_{-j} - i\varrho|$, $j = 1, \ldots, \kappa$, are not eigenvalues.*

4. *If $\kappa \ge 2$, then in each of the intervals $(i\varrho - i|\lambda_{-(j+1)} - i\varrho|, i\varrho - i|\lambda_{-j} - i\varrho|)$, $j = 1, \ldots, \kappa - 1$, the number of eigenvalues, counted with multiplicity, is odd.*

5. *If $\kappa > 0$, then the interval $(i\varrho - i|\lambda_{-1} - i\varrho|, i\varrho]$ contains no or an even number of eigenvalues, counted with multiplicity. If $i\varrho$ is an eigenvalue, then it is simple.*

It is evident that Theorems 6.3.1 and 6.3.2 are equivalent to Theorems 2.2.3 and 2.2.4.

We have seen for $\tilde{\nu} > 0$ and $\tilde{\nu} \ne 2\tilde{m}\varrho$ that the characteristic function ϕ is a shifted Hermite–Biehler function, say $\phi(\tau) = P_{\tilde{\nu}}(\tau^2) + i\lambda Q_{\tilde{\nu}}(\tau^2)$. Hence the zeros of $P_{\tilde{\nu}}$ and $Q_{\tilde{\nu}}$ are real, bounded below and interlace. But $Q_{2\tilde{m}\varrho} = 0$, whereas $P_{2\tilde{m}\varrho}$ has still infinitely many zeros which are real and bounded below. Hence, in case $\tilde{\nu} = 2\tilde{m}\varrho$, the characteristic function ϕ has infinitely many real zeros, at most finitely many pure imaginary zeros, and no other zeros. This confirms the result from Subsection 2.2.2.

6.4   Vibrations of star graphs

Here we revisit the problem (2.3.9)–(2.3.12). Applying the parameter transformation $\lambda = \pm\tau + i\varrho$ leads to

$$y_j''(\tau, x_j) + \tau^2 y_j(\tau, x_j) + (\varrho^2 - q_j(x))y_j(\tau, x_j) = 0, \quad j = 1, \ldots, p, \qquad (6.4.1)$$

$$y_j(\tau, 0) = 0, \quad j = 1, \ldots, p, \qquad (6.4.2)$$

$$v_1 y_1(\tau, a_1) = \cdots = v_p y_p(\tau, a_p), \qquad (6.4.3)$$

$$\sum_{j=1}^{p} \theta_j y_j'(\tau, a_j) + (-\tilde{m}\tau^2 + i|\tilde{\nu} - 2\tilde{m}\varrho|\tau + \tilde{m}\varrho^2 - \tilde{\nu}\varrho + \beta)v_1 y_1(\tau, a_1) = 0, \quad (6.4.4)$$

with positive constants v_j and θ_j, $j = 1, \ldots, p$.

For $j = 1, \ldots, p$ let us denote by $s_j(\tau, \cdot)$ the solutions of (6.4.1) which satisfy the initial conditions $s_j(\tau, 0) = 0$, $s_j'(\tau, 0) = 1$. The solutions of (6.4.1)–(6.4.2) are of the form $(y_j(\tau, \cdot))_{j=1}^{p} = (C_j s_j(\tau, \cdot))_{j=1}^{p}$. Then a complex number τ is an eigenvalue of (6.4.1)–(6.4.4) if and only if there is a nontrivial $(y_j(\tau, \cdot))_{j=1}^{p} = (C_j s_j(\tau, \cdot))_{j=1}^{p}$ which satisfies (6.4.3)–(6.4.4). Therefore we obtain a $p \times p$ system of linear algebraic equation with respect to the C_j's, which is singular if and only if τ is an eigenvalue of our problem. A coefficient matrix $B(\tau) = (b_{j,m}(\tau))_{j,m=1}^{p}$ for this system is given by $b_{j,j}(\tau) = v_j s_j(\tau, a_j)$ and $b_{j,j+1}(\tau) = -v_{j+1}s_{j+1}(\tau, a_{j+1})$ for $j = 1, \ldots, p-1$ from (6.4.3),

$$b_{p,1}(\tau) = \theta_1 s_1'(\tau, a_1) + (-\tilde{m}\tau^2 + i|\tilde{\nu} - 2\tilde{m}\varrho|\tau + \tilde{m}\varrho^2 - \tilde{\nu}\varrho + \beta)v_1 s_1(\tau, a_1),$$

$b_{p,j}(\tau) = \theta_j s_j'(\tau, a_j)$ for $j = 2, \ldots, p$ from (6.4.4), and $b_{j,k} = 0$ for all other pairs of indices j, k. If $\phi(\tau)$ denotes the determinant of $B(\tau)$, then the eigenvalues of (6.4.1)–(6.4.4) are the zeros of ϕ. An expansion of this determinant by its first column gives

$$\phi(\tau) = \sum_{j=1}^{p} \theta_j s_j'(\tau, a_j) \prod_{\substack{m=1 \\ m \neq j}}^{p} v_m s_m(\tau, a_m)$$

$$+ (-\tilde{m}\tau^2 + i|\tilde{\nu} - 2\tilde{m}\varrho|\tau + \tilde{m}\varrho^2 - \tilde{\nu}\varrho + \beta) \prod_{j=1}^{p} v_j s_j(\tau, a_j). \qquad (6.4.5)$$

We give now an alternative way to find ϕ, which involves some more abstract reasoning but less calculation at the end. An equivalent coefficient matrix can be written in 2×2 block matrix form as

$$\tilde{B}(\tau) = \begin{pmatrix} -B_{11}(\tau) & B_{12}(\tau) \\ B_{21}(\tau) & B_{22}(\tau) \end{pmatrix},$$

where $B_{11}(\tau)$ is the column vector of length $p-1$ with all entries equal $v_1 s_1(\tau, a_1)$, $B_{12} = \mathrm{diag}(v_2 s_2(\tau, a_2), \ldots, v_p s_p(\tau, a_p))$, $B_{21}(\tau) = b_{p,1}(\tau)$ from above, and $B_{2,2}(\tau)$ is the row vector $(\theta_2 s_2'(\tau, a_2), \ldots, \theta_p s_p'(\tau, a_2))$.

By Theorem 12.6.2, the zeros of $s_j(\cdot, a_j)$ are real and pure imaginary. Hence, for complex numbers τ which are not real or pure imaginary, we have the Schur factorization

$$\tilde{B}(\tau) = \begin{pmatrix} 1 & 0 \\ B_{22}(\tau)(B_{12}(\tau))^{-1} & I_{p-1} \end{pmatrix} \begin{pmatrix} -B_{11}(\tau) & B_{12}(\tau) \\ B_{22}(B_{12}(\tau))^{-1}B_{11}(\tau) + B_{21}(\tau) & 0 \end{pmatrix}.$$

From

$$B_{22}(B_{12}(\tau))^{-1}B_{11}(\tau) = \sum_{j=2}^{p} v_1 s_1(\tau, a_1) \frac{1}{v_j s_j(\tau, a_j)} \theta_j s_j'(\tau, a_j)$$

it easily follows that $\det \tilde{B}(\tau) = \pm\phi(\tau)$, which is true for all complex numbers τ.

Define the entire functions P and Q by

$$P(\tau^2) = \sum_{j=1}^{p} \theta_j s_j'(\tau, a_j) \prod_{\substack{m=1 \\ m \neq j}}^{p} v_m s_m(\tau, a_m)$$

$$+ (-\tilde{m}\tau^2 + \tilde{m}\varrho^2 - \tilde{\nu}\varrho + \beta) \prod_{j=1}^{p} v_j s_j(\tau, a_j) \qquad (6.4.6)$$

and

$$Q(\tau^2) = |\tilde{\nu} - 2\tilde{m}\varrho| \prod_{j=1}^{p} v_j s_j(\tau, a_j). \qquad (6.4.7)$$

In the remainder of this section we will assume that $\tilde{\nu} \neq 2\tilde{m}\varrho$. As we have seen above, all zeros of Q are real, and therefore common zeros of P and Q, if any, must be real. Furthermore,

$$-\frac{P(z)}{Q(z)} = \frac{1}{|\tilde{\nu} - 2\tilde{m}\varrho|} \left\{ -\sum_{j=1}^{p} \frac{\theta_j \tilde{s}_j'(z)}{v_j \tilde{s}_j(z)} + \tilde{m}z - \tilde{m}\varrho^2 + \tilde{\nu}\varrho - \beta \right\}, \qquad (6.4.8)$$

where $\tilde{s}_j(\tau^2) = s_j(\tau, a_j)$ and $\tilde{s}_j'(\tau^2) = s_j'(\tau, a_j)$. This equation is similar to (6.3.5), and as in Section 6.3 we conclude that $-\frac{P}{Q} \in \mathcal{N}_-^{\mathrm{ep}}$. An application of Lemma 5.1.22 gives $\frac{Q}{P} \in \mathcal{N}_+^{\mathrm{ep}}$. Since the zeros of Q are real and bounded below by Theorem 12.6.2, also common zeros of P and Q, if any, have this property. Hence ϕ is of $\overline{\mathrm{SSHB}}$ class.

Next we are going to show that zeros of type I and type II do not intersect. We start with $z \in \mathbb{R}$ such that $Q(z) = 0$. Without loss of generality we may assume that there is $1 \le q \le p$ such that $s_j(\tau, a_j) = 0$ for $j = 1, \ldots, q$ and $s_j(\tau, a_j) \ne 0$ for $j = q+1, \ldots, p$. It follows from Theorem 12.6.2 that the multiplicity of the zero z of Q is q, and in view of (6.4.8), the meromorphic function PQ^{-1} has either a simple pole or a removable singularity at z. We already know that $\psi_j := -\tilde{s}_j' \tilde{s}_j^{-1}$ is a Nevanlinna function with a simple pole at z for $j = 1, \ldots, q$. Since their derivatives ψ_j' are increasing in a deleted neighbourhood of z by Lemma 11.1.3, it follows that $x \mapsto \psi_j(x)(x - z) < 0$ for x in a sufficiently small deleted neighbourhood of z. Hence, for $j = 1, \ldots, q$, $\lim_{x \to z} \psi_j(x)(x - z) \le 0$. Since this limit is the residue of ψ_j at z and since ψ_j has a simple pole at z, it follows that this residue is negative. Summing up over $j = 1, \ldots, q$ in (6.4.8) shows that $-PQ^{-1}$ has a simple pole with negative residue at z, that is, z cannot be a removable singularity.

In particular, if z is also a zero of P, then its multiplicity as a zero of P is less than its multiplicity as a zero of Q. If now λ_* is a zero of type I of ω, then in the factorization $\omega(\lambda) = R_*(\lambda^2)(P_*(\lambda^2) + i\lambda Q_*(\lambda^2))$ we have $R_*(\lambda_*^2) = 0$, $P_*(\lambda_*^2) \ne 0$ and $Q_*(\lambda_*^2) = 0$, which shows that λ_* is not a zero of type II of ω. Since zeros of type I are symmetric, it follows that if $\lambda \ne 0$ is a zero of type II of ϕ, then $-\lambda$ is neither a zero of type I nor a zero of type II, that is, $\phi(-\lambda) \ne 0$.

An application of Theorem 5.2.9 and Corollary 5.2.11 gives the following results, which are equivalent to Theorems 2.3.9 and 2.3.10.

Theorem 6.4.1. *Let $\tilde{\nu} > 2\tilde{m}\varrho$. Then:*

1. *The sets of type I and type II eigenvectors do not intersect. Only a finite number, denoted by κ_2, of the eigenvalues of type II of problem (6.4.1)–(6.4.4) lie in the closed half-plane $\mathrm{Im}\,\lambda \le \varrho$.*

2. *All eigenvalues of type II in the closed half-plane $\mathrm{Im}\,\lambda \le \varrho$ lie on $(-i\infty, i\varrho)$ and their multiplicities are 1. If $\kappa_2 > 0$, they will be uniquely indexed as $\lambda_{-j} = i\varrho - i|\lambda_{-j} - i\varrho|$, $j = 1, \ldots, \kappa_2$, satisfying $|\lambda_{-j} - i\varrho| < |\lambda_{-(j+1)} - i\varrho|$, $j = 1, \ldots, \kappa_2 - 1$.*

3. *If $\kappa_2 > 0$, then the numbers $i\varrho + i|\lambda_{-j} - i\varrho|$, $j = 1, \ldots, \kappa_2$, are not eigenvalues.*

4. *If $\kappa_2 \ge 2$, then the intervals $(i\varrho + i|\lambda_{-j} - i\varrho|, i\varrho + i|\lambda_{-(j+1)} - i\varrho|)$, $j = 1, \ldots, \kappa_2 - 1$, contain an odd number of eigenvalues of type II, counted with multiplicity.*

5. *Let $\kappa_2 > 0$. Then the interval $(i\varrho, i\varrho + i|\lambda_{-1} - i\varrho|)$ contains no or an even number of eigenvalues of type II, counted with multiplicity, if $i\varrho$ is no eigenvalue or an eigenvalue of even multiplicity, and an odd number of eigenvalues of type II, counted with multiplicity, if $i\varrho$ is an eigenvalue of odd multiplicity.*

Theorem 6.4.2. *Let $\tilde{\nu} < 2\tilde{m}\varrho$. Then:*

1. *The sets of type I and type II eigenvectors do not intersect. Only a finite number, denoted by κ_2, of the eigenvalues of type II of problem (6.4.1)–(6.4.4) lie in the closed half-plane $\mathrm{Im}\,\lambda \ge \varrho$.*

2. *All eigenvalues of type II in the closed half-plane* $\operatorname{Im}\lambda \geq \varrho$ *lie on* $(i\varrho, i\infty)$ *and their multiplicities are 1. If* $\kappa_2 > 0$, *they will be uniquely indexed as* $\lambda_{-j} = i\varrho + i|\lambda_{-j} - i\varrho|$, $j = 1,\dots,\kappa_2$, *satisfying* $|\lambda_{-j}| < |\lambda_{-(j+1)}|$, $j = 1,\dots,\kappa_2 - 1$.
3. *If* $\kappa_2 > 0$, *then the numbers* $i\varrho - i|\lambda_{-j} - i\varrho|$, $j = 1,\dots,\kappa_2$, *are not eigenvalues.*
4. *If* $\kappa_2 \geq 2$, *then the intervals* $(i\varrho - i|\lambda_{-(j+1)} - i\varrho|, i\varrho - i|\lambda_{-j} - i\varrho|)$, $j = 1,\dots,\kappa_2 - 1$, *contain an odd number of eigenvalues of type II, counted with multiplicity.*
5. *Let* $\kappa_2 > 0$. *Then the interval* $(i\varrho - i|\lambda_{-1} - i\varrho|, i\varrho)$ *contains no or an even number of eigenvalues of type II, counted with multiplicity, if* $i\varrho$ *is no eigenvalue or an eigenvalue of even multiplicity, and an odd number of eigenvalues of type II, counted with multiplicity, if* $i\varrho$ *is an eigenvalue of odd multiplicity.*

6.5   Forked graphs

Here we revisit problem $(2.4.7)$–$(2.4.11)$, which is a particular case of problem $(6.4.1)$–$(6.4.4)$ with $p = 2$, $\tilde{m} = \varrho = \beta = 0$ and $\tilde{\nu} = \theta_j = \upsilon_j = 1$:

$$-y_j'' + q_j(x)y_j = \lambda^2 y_j, \quad x \in [0, a], \quad j = 1, 2, \tag{6.5.1}$$

$$y_1(\lambda, a) = y_2(\lambda, a), \tag{6.5.2}$$

$$y_1'(\lambda, a) + y_2'(\lambda, a) + i\lambda y_1(\lambda, a) = 0, \tag{6.5.3}$$

$$y_1(\lambda, 0) = 0, \tag{6.5.4}$$

$$y_2(\lambda, 0) = 0. \tag{6.5.5}$$

Therefore Theorem 6.4.1 leads to the following result.

Theorem 6.5.1. *The eigenvalues of* $(6.5.1)$–$(6.5.5)$ *have the following properties:*

1. *All eigenvalues of type I are located on the real and imaginary axes, are symmetric with respect to the origin, are not eigenvalues of type II, and at most finitely many of them lie on the imaginary axis.*
2. *Only a finite number, denoted by* κ_2, *of eigenvalues of type II lie in the closed lower half-plane.*
3. *All eigenvalues of type II in the closed lower half-plane lie on the negative imaginary semiaxis and their multiplicities are 1. If* $\kappa_2 > 0$, *they will be uniquely indexed as* $\lambda_{-j} = -i|\lambda_{-j}|$, $j = 1,\dots,\kappa_2$, *satisfying* $|\lambda_{-j}| < |\lambda_{-(j+1)}|$, $j = 1,\dots,\kappa_2 - 1$.
4. *If* $\kappa_2 > 0$, *then the numbers* $i|\lambda_{-j}|$, $j = 1,\dots,\kappa_2$, *are not eigenvalues.*
5. *If* $\kappa_2 \geq 2$, *then in each of the intervals* $(i|\lambda_{-j}|, i|\lambda_{-(j+1)}|)$, $j = 1,\dots,\kappa_2 - 1$, *the number of eigenvalues of type II, counted with multiplicity, is odd.*
6. *If 0 is an eigenvalue, then its multiplicity is 1 or 2.*
7. *Let* $\kappa_2 > 0$. *Then the interval* $(0, i|\lambda_{-1}|)$ *contains no or an even number of eigenvalues of type II, counted with multiplicity, if 0 is no eigenvalue or an*

eigenvalue of multiplicity 2, *and an odd number of eigenvalues of type II, counted with multiplicity, if* 0 *is a simple eigenvalue.*

Proof. We still have to prove statement 6. We observe that the characteristic function ϕ of (6.5.1)–(6.5.5) is a special case of (6.4.5) with $p = 2$. Since $\phi(\lambda) = P(\lambda^2) + i\lambda Q(\lambda^2)$ with P and Q as in (6.4.6) and (6.4.7), It follows that $P(0) = 0$. If $Q(0) \neq 0$, then 0 is a simple zero of ϕ. If, however, $Q(0) = 0$, then the multiplicity of Q can be at most 2 since $p = 2$. On the other hand, we have shown in Section 6.4 that for a common real zero z of P and Q, the multiplicity of z as a zero of Q exceeds the multiplicity of the zero of P by 1. Hence 0 must be a simple zero of P and a double zero of Q. Hence 0 is a zero of ϕ of multiplicity 2 in this case. $\quad\square$

Remark 6.5.2. Theorems 2.4.1 and 6.5.1 differ in that Theorem 2.4.1 has the additional statement 1 whereas Theorem 6.5.1 has the additional statement 6. The geometric simplicity of the eigenvalues cannot be proved with the methods of this subsection since the characteristic function does not contain any information on the geometric multiplicity of eigenvalues.

6.6   Lasso graphs

Here we revisit problem (2.5.5)–(2.5.7),

$$-y_1'' + q(x)y_1 = \lambda^2 y_1, \tag{6.6.1}$$

$$y_1(\lambda, 0) = y_1(\lambda, a), \tag{6.6.2}$$

$$y_1'(\lambda, 0) - y_1'(\lambda, a) - i\lambda y_1(\lambda, 0) = 0, \tag{6.6.3}$$

which differs from the previous applications in that solutions do not necessarily vanish at the initial points. With our usual notation $s(\lambda, \cdot)$ and $c(\lambda, \cdot)$ for the fundamental system of solutions of (6.6.1), see Theorem 12.2.9, we substitute the general solution

$$y(\lambda, \cdot) = As(\lambda, \cdot) + Bc(\lambda, \cdot)$$

of (6.6.1) into (6.6.2) and (6.6.3) and obtain

$$B = As(\lambda, a) + Bc(\lambda, a),$$
$$A - As'(\lambda, a) - Bc'(\lambda, a) - i\lambda B = 0.$$

The characteristic function, i.e., the determinant of the coefficients of the above linear system in the unknown parameters A and B, is

$$\phi(\lambda) = 2 - c(\lambda, a) - s'(\lambda, a) - i\lambda s(\lambda, a),$$

where we have used that the Wronskian $c(\lambda, \cdot)s'(\lambda, \cdot) - c'(\lambda, \cdot)s(\lambda, \cdot)$ is identically 1. It is well known and easily follows along the lines of the above calculations that

$P(\lambda^2) = 2 - c(\lambda, a) - s'(\lambda, a)$ is the characteristic function of the periodic problem (6.6.1), (6.6.2),

$$y_1'(\lambda, 0) - y_1'(\lambda, a) = 0, \tag{6.6.4}$$

while $Q(\lambda^2) = -s(\lambda, a)$ is the characteristic function of the Dirichlet problem (6.6.1),

$$y_1(\lambda, 0) = y_1(\lambda, a) = 0. \tag{6.6.5}$$

It is also well known, see, e.g., [271, Theorem 13.10], that the zeros $(\zeta_k)_{k=0}^\infty$ of P interlace with the zeros $(\nu_k)_{k=1}^\infty$ of Q:

$$\zeta_0 < \nu_1 < \zeta_1 \le \nu_2 \le \zeta_2 < \cdots. \tag{6.6.6}$$

Using (6.1.7), (6.1.6) and

$$c(\lambda, a) \underset{\lambda \to \pm i\infty}{=} \frac{e^{|\lambda|a}}{2} + o(e^{|\lambda|a}), \tag{6.6.7}$$

which follows from (12.2.24) with $n = 0$, we obtain that the real entire functions P and Q satisfy $P(z) < 0$ and $Q(z) < 0$ as $z \to -\infty$. We can conclude as in Proposition 6.1.1 that $\lambda \mapsto P(\lambda^2)$ and $\lambda \mapsto \lambda Q(\lambda^2)$ are sine type functions. Similar to the proof of Proposition 6.1.2 we then obtain with (6.6.6) and Corollary 5.2.3 that $QP^{-1} \in \mathcal{N}_+^{ep}$.

By (6.6.6), P and Q may have common zeros, which are all real. Hence we can write $\phi(\lambda) = R_*(\lambda^2)(P_*(\lambda^2) + i\lambda Q_*(\lambda^2))$, where R_* has only real zeros and where P_* and Q_* do not have common zeros. Hence ϕ is of $\overline{\text{SSHB}}$ class.

It follows from (6.6.6) that the zeros of $Q = R_* Q_*$ are simple, whereas the zeros of $P = R_* P_*$ have multiplicities 1 or 2. In particular, if 0 is a zero of ϕ, then its multiplicity is at most 3.

With the aid of Theorem 5.2.9 and Corollary 5.2.11 we obtain the following result, which is equivalent to the statements 2–7 of Theorem 2.5.1. We cannot prove statement 1 of Theorem 2.5.1 with the methods of this section since the characteristic function does not contain any information on the geometric multiplicity of eigenvalues.

Theorem 6.6.1. *The eigenvalues of problem* (6.6.1)–(6.6.3) *have the following properties.*

1. *All eigenvalues of type I are located on the real and imaginary axes, are symmetric with respect to the origin, and at most finitely many of them lie on the imaginary axis.*
2. *Only a finite number, denoted by κ_2, of eigenvalues of type II lie in the closed lower half-plane.*
3. *All eigenvalues of type II in the closed lower half-plane lie on the negative imaginary semiaxis and their type II multiplicities are 1. If $\kappa_2 > 0$, they will be uniquely indexed as $\lambda_{-j} = -i|\lambda_{-j}|$, $j = 1, \ldots, \kappa_2$, satisfying $|\lambda_{-j}| < |\lambda_{-(j+1)}|$, $j = 1, \ldots, \kappa_2 - 1$.*

4. If $\kappa_2 > 0$, then the numbers $i|\lambda_{-j}|$, $j = 1, \ldots, \kappa_2$, are not eigenvalues of type II.

5. If $\kappa_2 \geq 2$, then in each of the intervals $(i|\lambda_{-j}|, i|\lambda_{-(j+1)}|)$, $j = 1, \ldots, \kappa_2 - 1$, the number of eigenvalues of type II, counted with type II multiplicity, is odd.

6. If 0 is an eigenvalue, then its multiplicity is 1, 2 or 3.

7. Let $\kappa_2 > 0$. Then the interval $(0, i|\lambda_{-1}|)$ contains no or an even number of eigenvalues of type II, counted with type II multiplicity, if 0 is not an eigenvalue of type II, and an odd number of eigenvalues of type II, counted with type II multiplicity, if 0 is an eigenvalue of type II.

6.7 Beams with friction at one end

Here we revisit problem (2.7.2)–(2.7.6):

$$y^{(4)}(\lambda, x) - (gy')'(\lambda, x) = \lambda^2 y(\lambda, x), \tag{6.7.1}$$

$$y(\lambda, 0) = 0, \tag{6.7.2}$$

$$y''(\lambda, 0) = 0, \tag{6.7.3}$$

$$y(\lambda, a) = 0, \tag{6.7.4}$$

$$y''(\lambda, a) + i\alpha\lambda y'(\lambda, a) = 0. \tag{6.7.5}$$

Putting $z = \lambda^2$, there exists the canonical fundamental system of solutions $y_j(z, \cdot)$, $j = 1, \ldots, 4$, of (6.7.1) with $y_j^{(m)}(z, 0) = \delta_{j,m+1}$ for $m = 0, \ldots, 3$, which are real entire functions with respect to z, see, e.g., [189, Theorem 6.1.8]. The functions $y_2(\lambda^2, \cdot)$ and $y_4(\lambda^2, \cdot)$ form a basis of the solutions of the initial value problem (6.7.1)–(6.7.3). Substituting the generic linear combination of these two solutions into the boundary conditions (6.7.4)–(6.7.5), we obtain a 2×2 system of linear equations, and the determinant ϕ of its coefficient matrix is given by

$$\phi(\lambda) = f''(\lambda^2, a) + i\alpha\lambda f'(\lambda^2, a),$$

where

$$f(z, x) = y_2(z, a)y_4(z, x) - y_4(z, a)y_2(z, x).$$

Hence the eigenvalues of (6.7.1)–(6.7.5) are the zeros of ϕ. Similarly, we have for $k = 1, 2$ that the real entire functions $f^{(k)}(\cdot, a)$ are characteristic functions of the eigenvalue problems (6.7.1)–(6.7.4), $y^{(k)}(\lambda, a) = 0$ with $z = \lambda^2$. These two eigenvalue problems are realized by self-adjoint operators which are bounded below, which can be proved as in Proposition 2.7.1 by applying Theorems 10.3.5 and 10.3.8. Hence the zeros of $f^{(k)}(\cdot, a)$, $k = 1, 2$, are real and bounded below. Clearly, $f(z, \cdot)$ is a solution of (6.7.1)–(6.7.4), and therefore, in view of (6.7.1),

$$\int_0^a f^{(4)}(z, x)\overline{f(z, x)}\, dx - \int_0^a (gf'(z, \cdot))'(x)\overline{f(z, x)}\, dx = z\int_0^a |f(z, x)|^2\, dx.$$

Integrating by parts twice and taking (6.7.2)–(6.7.4) into account we obtain

$$-f''(z,a)\overline{f'(z,a)} + \int_0^a |f''(z,x)|^2\, dx + \int_0^a g(x)|f'(z,x)|^2\, dx = z \int_0^a |f(z,x)|^2\, dx.$$
(6.7.6)

Taking imaginary parts leads to

$$\mathrm{Im}(-f''(z,a)\overline{f'(z,a)}) = (\mathrm{Im}\, z) \int_0^a |f(\lambda,x)|^2\, dx.$$

For $\mathrm{Im}\, z > 0$ we therefore conclude that

$$\mathrm{Im}\, \frac{f'(z,a)}{f''(z,a)} = (\mathrm{Im}\, z) \frac{1}{|f''(z,a)|^2} \int_0^a |f(\lambda,x)|^2\, dx > 0.$$

Consequently, $\frac{f'(\cdot,a)}{f''(\cdot,a)} \in \mathcal{N}^{\mathrm{ep}}$. In view of Lemmas 10.3.6 and 10.3.7 it follows from (6.7.6) that there is a real number γ such that $f''(z,a)f'(z,a) > 0$ for $z < \gamma$. Hence $\frac{f'(\cdot,a)}{f''(\cdot,a)} \in \mathcal{N}_+^{\mathrm{ep}}$, which shows that ϕ is of $\overline{\mathrm{SSHB}}$ class.

An application of Theorem 5.2.9 and Corollary 5.2.11 gives the following result which is equivalent to Theorem 2.7.2 without the statements about the geometric multiplicity of eigenvalues.

Theorem 6.7.1. *The eigenvalues of (6.7.1)–(6.7.5) have the following properties:*

1. *All eigenvalues of type I are located on the imaginary and real axes and are symmetric with respect to the origin.*

3. *Only a finite number, denoted by κ_2, of the eigenvalues of type II lie in the closed lower half-plane.*

4. *All eigenvalues of type II in the closed lower half-plane lie on the negative imaginary semiaxis and their type II multiplicities are 1. If $\kappa_2 > 0$, they will be uniquely indexed as $\lambda_{-j} = -i|\lambda_{-j}|$, $j = 1, \ldots, \kappa_2$, satisfying $|\lambda_{-j}| < |\lambda_{-(j+1)}|$, $j = 1, \ldots, \kappa_2 - 1$.*

5. *If $\kappa_2 > 0$, then the numbers $i|\lambda_{-j}|$, $j = 1, \ldots, \kappa_2$, are not eigenvalues of type II.*

6. *If $\kappa_2 \geq 2$, then the number of eigenvalues of type II, counted with type II multiplicity, in each of the intervals $(i|\lambda_{-j}|, i|\lambda_{-(j+1)}|)$, $j = 1, \ldots, \kappa_2 - 1$, is odd.*

7. *Let $\kappa_2 > 0$. Then the interval $(0, i|\lambda_{-1}|)$ contains no or an even number of eigenvalues of type II, counted with type II multiplicity, if 0 is not an eigenvalue of type II, and an odd number of eigenvalues of type II, counted with type II multiplicity, if 0 is an eigenvalue of type II.*

6.8   Vibrations of damped beams

Here we reconsider problem (2.7.8)–(2.7.12). With the parameter transformation $\lambda = \pm\tau + i\varrho$ as in Subsection 2.7.2, the problem becomes

$$y^{(4)}(\tau, x) - (gy')'(\tau, x) - \varrho^2 y(\tau, x) = \tau^2 y(\tau, x), \qquad (6.8.1)$$
$$y(\tau, 0) = 0, \qquad (6.8.2)$$
$$y''(\tau, 0) = 0, \qquad (6.8.3)$$
$$y''(\tau, a) = 0, \qquad (6.8.4)$$

$$-y'''(\tau, a) + g(a)y'(\tau, a) + (m\varrho^2 - \alpha\varrho)y(\tau, a) + i\tau|\alpha - 2m\varrho|y(\tau, a) = \tau^2 my(\tau, a). \qquad (6.8.5)$$

Putting $z = \tau^2$, there exists the canonical fundamental system of solutions $y_j(z, \cdot)$, $j = 1, \ldots, 4$, of (6.8.1) with $y_j^{(m)}(z, 0) = \delta_{j, m+1}$ for $m = 0, \ldots, 3$, which are real entire functions with respect to z, see, e. g., [189, Theorem 6.1.8]. The functions $y_2(\tau^2, \cdot)$ and $y_4(\tau^2, \cdot)$ form a basis of the solutions of the initial value problem (6.8.1)–(6.8.3). Substituting the generic linear combination of these two solutions into the boundary conditions (6.8.4)–(6.8.5), we obtain a 2×2 system of linear equations, and the determinant ϕ of its coefficient matrix is given by

$$\phi(\tau) = -f'''(\tau^2, a) + g(a)f'(\tau^2, a) + (m\varrho^2 - \alpha\varrho)f(\tau^2, a) + i\tau|\alpha - 2m\varrho|f(\tau^2, a)$$
$$- \tau^2 mf(\tau^2, a),$$

where

$$f(z, x) = y_2''(z, a)y_4(z, x) - y_4''(z, a)y_2(z, x).$$

Hence the eigenvalues of (6.8.1)–(6.8.5) are the zeros of ϕ. Similarly, assuming now that $|\alpha - 2m\varrho| \neq 0$, we have that the real entire functions P and Q defined by

$$P(z) = -f'''(z, a) + g(a)f'(z, a) + (m\varrho^2 - \alpha\varrho)f(z, a) - zmf(z, a),$$
$$Q(z) = |\alpha - 2m\varrho|f(z, a),$$

are characteristic functions of the eigenvalue problems (6.8.1)–(6.8.4) with the fourth boundary condition being

$$-y'''(\tau, a) + g(a)y'(\tau, a) + (m\varrho^2 - \alpha\varrho)y(\tau, a)y(\tau, a) = \tau^2 my(\tau, a) \quad \text{for } P,$$
$$y(\tau, a) = 0 \quad \text{for } Q,$$

with $z = \tau^2$. With the reasoning from Section 6.7 if follows that the zeros of Q are real and bounded below. Similarly, with $\tilde{A}$ and $\tilde{M}$ from (2.7.14) and (2.7.16), the zeros of P are the eigenvalues of the self-adjoint operator $\tilde{M}^{-\frac{1}{2}}\tilde{A}\tilde{M}^{-\frac{1}{2}}$. Hence also the zeros of P are real and bounded below.

Clearly, $f(z, \cdot)$ is a solution of (6.8.1)–(6.8.4), and therefore, in view of (6.8.1),

$$\int_0^a f^{(4)}(z,x)\overline{f(z,x)}\,dx - \int_0^a (gf'(z,\cdot))'(x)\overline{f(z,x)}\,dx - \varrho^2 \int_0^a |f(z,x)|^2\,dx$$
$$= z\int_0^a |f(z,x)|^2\,dx.$$

Integrating by parts twice and taking (6.8.2)–(6.8.4) into account we obtain

$$f'''(z,a)\overline{f(z,a)} + \int_0^a |f''(z,x)|^2\,dx - g(a)f'(z,a)\overline{f(z,a)}$$
$$+ \int_0^a g(x)|f'(z,x)|^2\,dx - \varrho^2 \int_0^a |f(z,x)|^2\,dx = z\int_0^a |f(z,x)|^2\,dx. \quad (6.8.6)$$

Taking imaginary parts leads to

$$\mathrm{Im}(-P(z)\overline{Q(z)}) = (\mathrm{Im}\,z)|\alpha - 2m\varrho|\left[m|f(z,a)|^2 + \int_0^a |f(z,x)|^2\,dx\right].$$

For $\mathrm{Im}\,z > 0$ we therefore conclude that

$$\mathrm{Im}\,\frac{Q(z)}{P(z)} = (\mathrm{Im}\,z)\frac{|\alpha - 2m\varrho|}{|P(z)|^2}\left[m|f(z,a)|^2 + \int_0^a |f(z,x)|^2\,dx\right] > 0.$$

Consequently, $QP^{-1} \in \mathcal{N}^{\mathrm{ep}}$. From (6.8.6) we obtain for $z \in \mathbb{R}$ that

$$\frac{P(z)Q(z)}{|\alpha - 2m\varrho|} = [-f'''(z,a) + g(a)f'(z,a) + (m\varrho^2 - \alpha\varrho - zm)f(z,a)]\,f(z,a)$$
$$= \int_0^a |f''(z,x)|^2\,dx + \int_0^a g(x)|f'(z,x)|^2\,dx$$
$$- (z + \varrho^2)\int_0^a |f(z,x)|^2\,dx + (m\varrho^2 - \alpha\varrho - zm)|f(z,a)|^2.$$

In view of Lemmas 10.3.6 and 10.3.7 it follows that there is a real number γ such that $P(z)Q(z) > 0$ for $z < \gamma$. Hence $PQ^{-1} \in \mathcal{N}_+^{\mathrm{ep}}$, which shows that ϕ is of $\overline{\mathrm{SSHB}}$ class. An application of Theorem 5.2.9 and Corollary 5.2.11 gives the following results, which are slightly weaker than Theorems 2.7.5 and 2.7.6.

Theorem 6.8.1. *Assume that $\alpha > 2m\varrho$. Then:*

1. *Only a finite number, denoted by κ_2, of the eigenvalues of type II of problem (2.7.8)–(2.7.12) lie in the closed half-plane $\mathrm{Im}\,\lambda \le \varrho$.*
2. *All eigenvalues of type II in the closed half-plane $\mathrm{Im}\,\lambda \le \varrho$ which are different from $i\varrho$ lie on $(-i\infty, i\varrho)$. If $\kappa_2 > 0$, they will be uniquely indexed as $\lambda_{-j} = i\varrho - i|\lambda_{-j} - i\varrho|$, $j = 1, \ldots, \kappa_2$, satisfying $|\lambda_{-j} - i\varrho| < |\lambda_{-(j+1)} - i\varrho|$, $j = 1, \ldots, \kappa_2 - 1$.*

3. If $\kappa_2 > 0$, then the numbers $i\varrho + i|\lambda_{-j} - i\varrho|$, $j = 1, \ldots, \kappa_2$, are not eigenvalues of type II.

4. If $\kappa_2 \geq 2$, then in each of the intervals $(i\varrho + i|\lambda_{-j} - i\varrho|, i\varrho + i|\lambda_{-j-1} - i\varrho|)$, $j = 1, \ldots, \kappa_2 - 1$, the number of eigenvalues of type II, counted with type II multiplicity, is odd.

5. Let $\kappa_2 > 0$. Then the interval $(i\varrho, i\varrho + i|\lambda_{-1} - i\varrho|)$ contains no or an even number of eigenvalues of type II, counted with type II multiplicity, if 0 is not an eigenvalue of type II, and an odd number of eigenvalues of type II, counted with type II multiplicity, if 0 is an eigenvalue of type II.

Theorem 6.8.2. *Assume that $\alpha < 2mk$. Then:*

1. Only a finite number, denoted by κ_2, of the eigenvalues of type II of problem (2.7.8)–(2.7.12) lie in the closed half-plane $\operatorname{Im} \lambda \geq \varrho$.

2. All eigenvalues of type II in the closed half-plane $\operatorname{Im} \lambda \geq \varrho$ which are different from $i\varrho$ lie on $(i\varrho, i\infty)$. If $\kappa_2 > 0$, they will be uniquely enumerated as $\lambda_{-j} = i\varrho + i|\lambda_{-j} - i\varrho|$, $j = 1, \ldots, \kappa_2$, satisfying $|\lambda_{-j}| < |\lambda_{-(j+1)}|$, $j = 1, \ldots, \kappa_2 - 1$.

3. If $\kappa_2 > 0$, then the numbers $i\varrho - i|\lambda_{-j} - i\varrho|$, $j = 1, \ldots, \kappa_2$, are not eigenvalues of type II.

4. If $\kappa_2 \geq 2$, then in each of the intervals $(i\varrho - i|\lambda_{-(j+1)} - i\varrho|, i\varrho - i|\lambda_{-j} - i\varrho|)$, $j = 1, \ldots, \kappa_2 - 1$, the number of eigenvalues of type II, counted with type II multiplicity, is odd.

5. Let $\kappa_2 > 0$. Then the interval $(i\varrho - i|\lambda_{-1} - i\varrho|, i\varrho)$ contains no or an even number of eigenvalues of type II, counted with type II multiplicity, if 0 is not an eigenvalue of type II, and an odd number of eigenvalues of type II, counted with type II multiplicity, if 0 is an eigenvalue of type II.

6.9 Notes

The spectral properties of the generalized Regge problem in Theorem 6.1.3 were obtained in [223, Theorem 3.1].

Theorem 6.2.3 was proved in [33, Theorem 4.1].

Let us briefly discuss the physical meaning of the real eigenvalues of damped vibrations of Stieltjes strings. If no damping occurs, then the spectrum of the corresponding boundary problem, i.e., of problem (6.2.1)–(6.2.4) with $\nu = 0$, is real. Let $(u_1^{(k)}, u_2^{(k)}, \ldots, u_{n_1}^{(k)}, \tilde{u}_{n_2}^{(k)}, \ldots, \tilde{u}_2^{(k)}, \tilde{u}_1^{(k)})$ be the eigenvector corresponding to the eigenvalue λ_k with $1 < k \leq n_1 + n_2$. Here $u_j^{(k)}$ and $\tilde{u}_j^{(k)}$ are the amplitudes of vibrations of the point masses. Then the piecewise linear graph describing the threads between point masses, i.e., the amplitude function of vibrations of the kth frequency, possesses $k - 1$ nodes. If we now apply the one-dimensional damping and the point of damping is a node of this graph, then the real eigenvalue λ_k remains an eigenvalue for the damped string; otherwise it moves into the upper half-plane. The number of eigenvalues of the undamped string whose eigenvectors have nodes at the same point does not exceed $2 \min\{n_1, n_2\}$ since the corresponding piecewise linear graphs must be eigenfunctions of the Dirichlet problems on the intervals between the point masses. Thus we obtain statement 5 of Theorem 6.2.3 using just physical arguments.

It is shown in [229, Subsection 6.1] that sums of essentially positive Nevanlinna functions can be considered as ratios of Neumann and Dirichlet characteristic functions of a boundary problem for a star graph of strings.

In [194], K. Mochizuki and I. Trooshin consider a star graph with some finite rays and some infinite rays, which is more general than what we deal with in this book. The case were all rays are infinite was considered by M. Harmer in [106].

The first results on lasso graph (with zero potential) were obtained in [77].

Part III

Direct and Inverse Problems

Chapter 7

Eigenvalue Asymptotics

7.1 Asymptotics of the zeros of some classes of entire functions

The asymptotic behaviour of large eigenvalues provides important information about boundary value problems. Indeed, to be able to solve inverse problems, which will be accomplished in Chapter 8, we need to know the asymptotic behaviour of the eigenvalues of the underlying boundary eigenvalue problems.

The eigenvalue problems considered in this chapter will have a representation via quadratic operator pencils as considered in Chapter 1. Since the spectra of such problems are symmetric with respect to the imaginary axis by Lemma 1.1.11, we can index the eigenvalues taking this symmetry into account. A suitable indexing of the eigenvalues is of utmost importance for asymptotic expansion formulas which depend on the index of the eigenvalue as main parameter. We will use the notation $(\lambda_k)_{k=-\infty}^{\infty}$ for the sequence of the eigenvalues, counted with multiplicity, where the index k runs through all integers, with the possible exception of $k = 0$.

Definition 7.1.1. We call the indexing of the sequence of the eigenvalues proper if:

 (i) $\lambda_{-k} = -\overline{\lambda_k}$ for all λ_k which are not pure imaginary;
 (ii) $\operatorname{Re} \lambda_k \geq \operatorname{Re} \lambda_p$ for all $k > p > 0$;
 (iii) the multiplicities are taken into account;
 (iv) the index set is $\mathbb{Z}$ if the number of pure imaginary eigenvalues is odd and is $\mathbb{Z} \setminus \{0\}$ if the number of pure imaginary eigenvalues is even.

Whenever the eigenvalues are represented by the zeros of a characteristic function of the eigenvalue problem, the notation 'eigenvalues' in the above definition is synonymous with zeros of this characteristic function. If there is no upper bound for the real parts of the eigenvalues, then a proper indexing is only possible if each vertical strip in the complex plane contains only finitely many eigenvalues. In particular, there are only finitely many eigenvalues on the imaginary axis.

In this chapter we consider applications which were encountered in Chapters 2 and 6, and it will be shown that the respective eigenvalues can be indexed properly. In this section we will present results on asymptotic distributions of the zeros of some entire functions, which will be applied in the subsequent sections of this chapter.

Proposition 7.1.2. *Let σ, α be nonnegative real number with $\sigma \neq \alpha$ and let $a > 0$. Then the zeros of the entire function $\phi^{(1)}$ defined by*

$$\phi^{(1)}(\lambda) = \lambda(\sigma \cos \lambda a + i\alpha \sin \lambda a), \quad \lambda \in \mathbb{C}, \tag{7.1.1}$$

have the following asymptotic representations.

1. *For $\alpha > \sigma$, the properly indexed zeros $(\lambda_k^{(1)})_{k=-\infty, k\neq 0}^{\infty}$ of $\phi^{(1)}$ are $\lambda_{-1}^{(1)} = 0$,*

$$\lambda_k^{(1)} = \frac{\pi(|k| - 1)}{a} \operatorname{sgn} k + \frac{i}{2a} \log\left(\frac{\alpha + \sigma}{\alpha - \sigma}\right), \quad k \in \mathbb{Z} \setminus \{-1, 0\}. \tag{7.1.2}$$

2. *For $\alpha < \sigma$, the properly indexed zeros $(\lambda_k^{(1)})_{k=-\infty}^{\infty}$ of $\phi^{(1)}$ are $\lambda_0^{(1)} = 0$,*

$$\lambda_k^{(1)} = \frac{\pi}{a} \operatorname{sgn} k \left(|k| - \frac{1}{2}\right) + \frac{i}{2a} \log\left(\frac{\sigma + \alpha}{\sigma - \alpha}\right), \quad k \in \mathbb{Z} \setminus \{0\}. \tag{7.1.3}$$

Proof. With $\mu = \operatorname{Re} \lambda$ and $\nu = \operatorname{Im} \lambda$, the real and imaginary parts of $\frac{1}{\lambda}\phi^{(1)}(\lambda) = 0$ are

$$\cos \mu a (\sigma \cosh \nu a - \alpha \sinh \nu a) = 0,$$
$$\sin \mu a (\alpha \cosh \nu a - \sigma \sinh \nu a) = 0.$$

For all $\nu \in \mathbb{R}$, $\sigma \cosh \nu a - \alpha \sinh \nu a \neq 0$ if $\alpha < \sigma$ and $\alpha \cosh \nu a - \sigma \sinh \nu a \neq 0$ if $\alpha > \sigma$. Hence the stated results follow easily. $\qquad\square$

Lemma 7.1.3. *Let φ be an entire function which has the representation*

$$\varphi(\lambda) = \phi^{(1)}(\lambda) + M \sin \lambda a - iN \cos \lambda a + \psi(\lambda), \quad \lambda \in \mathbb{C}, \tag{7.1.4}$$

where $\phi^{(1)}$ is given by (7.1.1), $M, N \in \mathbb{C}$ and $\psi \in \mathcal{L}^a$.

1. *Let $0 \leq \sigma < \alpha$. Then the zeros $(\tilde{\lambda}_k)_{k=-\infty, k\neq 0}^{\infty}$ of φ behave asymptotically as follows:*

$$\tilde{\lambda}_k = \frac{\pi(|k| - 1)}{a} \operatorname{sgn} k + \frac{i}{2a} \ln\left(\frac{\alpha + \sigma}{\alpha - \sigma}\right) + \frac{P}{k} + \frac{\beta_k}{k},$$

    where

$$P = \frac{\alpha N - \sigma M}{\pi(\alpha^2 - \sigma^2)}$$

    and $(\beta_k)_{k=-\infty, k\neq 0}^{\infty} \in l_2$.

2. *Let* $0 \leq \alpha < \sigma$. *Then the zeros* $(\tilde{\lambda}_k)_{k=-\infty}^{\infty}$ *of* φ *behave asymptotically as follows:*

$$\tilde{\lambda}_k = \frac{\pi}{a} \operatorname{sgn} k \left(|k| - \frac{1}{2} \right) + \frac{i}{2a} \ln \left(\frac{\sigma + \alpha}{\sigma - \alpha} \right) + \frac{P}{k} + \frac{\beta_k}{k},$$

where P *and* $(\beta_k)_{k=-\infty, k \neq 0}^{\infty}$ *are as in part 1.*

Proof. Since $\sigma \neq \alpha$, we conclude as in the proof of Proposition 6.1.1 that the function $\phi^{(0)}$ defined by

$$\phi^{(0)}(\lambda) = \sigma \cos \lambda a + i\alpha \sin \lambda a, \quad \lambda \in \mathbb{C}, \tag{7.1.5}$$

is a sine type function. Let $r \in (0, \frac{\pi}{2a})$. Recall that the set Λ_r defined in (11.2.31) associated with the sine type function $\phi^{(0)}$ is the complement in $\mathbb{C}$ of the union of the open discs $C_{k,r}$ of radii r with the centres at the zeros $\lambda_k^{(0)}$ of $\phi^{(0)}$. For convenience, we will index zeros in such a way that they coincide with the zeros $\lambda_k^{(1)}$ of $\phi^{(1)}$ when $|k| \geq 2$. By Remark 11.2.21 we find $d > 0$ such that

$$|\phi^{(0)}(\lambda)| \geq d e^{a|\operatorname{Im} \lambda|} \text{ for all } \lambda \in \mathbb{C} \setminus \Lambda_r. \tag{7.1.6}$$

For all indices k, $|k| \geq 2$, and all $\lambda \in \mathbb{C}$ with $|\lambda - \lambda_k^{(1)}| \leq \frac{\pi}{2a}$ we have

$$\phi^{(0)}(\lambda) = \int_{\lambda_k^{(1)}}^{\lambda} (\phi^{(0)})'(z) \, dz$$

$$= (\phi^{(0)})'(\lambda_k^{(1)})(\lambda - \lambda_k^{(1)}) + \int_{\lambda_k^{(1)}}^{\lambda} \int_{\lambda_k^{(1)}}^{z} (\phi^{(0)})''(w) \, dw \, dz$$

$$= (\phi^{(0)})'(\lambda_k^{(1)})(\lambda - \lambda_k^{(1)}) + O(|\lambda - \lambda_k^{(1)}|^2).$$

The sum of arguments formula for $\sin \lambda_k^{(1)} a$ and $\cos \lambda_k^{(1)} a$, applied to the representation (7.1.2) and (7.1.3), shows for $|k| \geq 2$ that

$$\sin \lambda_k^{(1)} a = \sigma \varepsilon_k, \tag{7.1.7}$$

$$\cos \lambda_k^{(1)} a = -i\alpha \varepsilon_k, \tag{7.1.8}$$

where

$$\varepsilon_k = \begin{cases} i(-1)^{k-1}(\alpha^2 - \sigma^2)^{-\frac{1}{2}} & \text{if } \alpha > \sigma, \ |k| \geq 2, \\ \operatorname{sgn} k (-1)^{k-1}(\sigma^2 - \alpha^2)^{-\frac{1}{2}} & \text{if } \alpha < \sigma, \ |k| \geq 2. \end{cases} \tag{7.1.9}$$

Hence

$$(\phi^{(0)})'(\lambda_k^{(1)}) = -\sigma a \sin \lambda_k^{(1)} a + i\alpha a \cos \lambda_k^{(1)} a$$
$$= a(\alpha^2 - \sigma^2)\varepsilon_k.$$

Therefore there is $C > 0$ such that

$$|\phi^{(0)}(\lambda)| \geq a|\alpha|^2 - \sigma^2|\tfrac{1}{2}|\lambda - \lambda_k^{(1)}| - C|\lambda - \lambda_k^{(1)}|^2$$

for all indices k and all $\lambda \in \mathbb{C}$ with $|\lambda - \lambda_k^{(1)}| \leq \frac{\pi}{2a}$. We conclude that there are numbers $r_0 \in (0, \frac{\pi}{2a})$ and $c > 0$ such that $cr_0 \leq d$ and

$$|\phi^{(0)}(\lambda)| \geq c|\lambda - \lambda_k^{(1)}|$$

for all indices k and $\lambda \in C_{k,r_0}$. Hence we can choose $d = cr$ for $r \in (0, r_0)$ in (7.1.6). Since $\varphi - \phi^{(1)}$ is a sum of two sine type functions of type a and a function in $\mathcal{L}^a$, an application of Lemmas 11.2.6 and 12.2.4 gives a number $K > 0$ such that

$$|\varphi(\lambda) - \phi^{(1)}(\lambda)| < Ke^{a|\operatorname{Im}\lambda|} \text{ for all } \lambda \in \mathbb{C}.$$

Consequently, it follows for all $\lambda \in \mathbb{C} \setminus \Lambda_r$ with $|\lambda| > \frac{K}{d}$ that

$$|\phi^{(1)}(\lambda)| = |\lambda|\,|\phi^{(0)}(\lambda)| \geq |\lambda|de^{a|\operatorname{Im}\lambda|} > Ke^{a|\operatorname{Im}\lambda|} > |\varphi(\lambda) - \phi^{(1)}(\lambda)|. \quad (7.1.10)$$

In particular, $\varphi(\lambda) \neq 0$ for these λ. We can choose $k_0 \in \mathbb{N}$ and $R > \frac{K}{d}$ such that $\lambda_{k_0}^{(1)} + r < R < \lambda_{k_0+1}^{(1)} - r$. Then all discs $C_{k,r}$ with $|k| \leq k_0$ lie inside the circle with centre 0 and radius R, whereas all discs $C_{k,r}$ with $|k| > k_0$ lie outside the circle with centre 0 and radius R. In view of (7.1.10) we can apply Rouché's theorem to the disc with centre 0 and radius R and to the discs $C_{k,r}$ with $|k| > k_0$. Hence $\phi^{(1)}$ and φ have the same number of zeros, counted with multiplicity, inside the disc with centre 0 and radius R, whereas outside this disc, all zeros of φ are simple with exactly one zero inside each $C_{k,r}$ with $|k| > k_0$. This proves that the zeros $\tilde{\lambda}_k$ of φ satisfy

$$\tilde{\lambda}_k = \lambda_k^{(0)} + \delta_k, \quad\quad\quad (7.1.11)$$

with the same index set as the zeros of $\phi^{(1)}$ and with $|\delta_k| < r$ for $|k| > |k_0|$. For sufficiently large $|k|$, we can choose $r \in (0, r_0)$ and R such that $R > \frac{\pi|k|}{2a}$ and $R < \frac{2K}{d} = \frac{2K}{cr}$. Hence

$$r < \frac{2K}{cR} < \frac{4Ka}{c\pi|k|},$$

which proves

$$\delta_k \underset{|k| \to \infty}{=} O(|k|^{-1}). \quad\quad\quad (7.1.12)$$

Now we substitute (7.1.11) into $\varphi(\tilde{\lambda}_k) = 0$. To this end we observe that $(\psi(\tilde{\lambda}_k))_{k=-\infty}^{\infty} \in l_2$ in view of (7.1.12) and Lemmas 12.2.4 and 12.2.1. Hence, making use of (7.1.7) and (7.1.8), we calculate

$$
\begin{aligned}
0 = \varphi(\tilde{\lambda}_k) &= (\sigma\tilde{\lambda}_k - iN)\cos\tilde{\lambda}_k a + (i\alpha\tilde{\lambda}_k + M)\sin\tilde{\lambda}_k a + \psi(\tilde{\lambda}_k)\\
&= (\sigma\tilde{\lambda}_k - iN)\left[\cos\lambda_k^{(0)} a\cos\delta_k a - \sin\lambda_k^{(0)} a\sin\delta_k a\right]\\
&\quad + (i\alpha\tilde{\lambda}_k + M)\left[\sin\lambda_k^{(0)} a\cos\delta_k a + \cos\lambda_k^{(0)} a\sin\delta_k a\right] + \psi(\tilde{\lambda}_k)\\
&= (\sigma\tilde{\lambda}_k - iN)\varepsilon_k\left[-i\alpha\cos\delta_k a - \sigma\sin\delta_k a\right]\\
&\quad + (i\alpha\tilde{\lambda}_k + M)\varepsilon_k\left[\sigma\cos\delta_k a - i\alpha\sin\delta_k a\right] + \psi(\tilde{\lambda}_k)\\
&= \varepsilon_k\left[(\sigma M - \alpha N)\cos\delta_k a + \left((\alpha^2 - \sigma^2)\tilde{\lambda}_k + i(\sigma N - \alpha M)\right)\sin\delta_k a\right] + \psi(\tilde{\lambda}_k).
\end{aligned}
$$

Observing that $\tilde{\lambda}_k = O(|k|)$, $\cos\delta_k a = 1 + O(|k|^{-2})$, $\sin\delta_k a = \delta_k a + O(|k|^{-3})$, $\tilde{\lambda}_k = \frac{\pi}{a}k + O(1)$ and that any sequence satisfying $O(|k|^{-1})$ belongs to l_2, the above identity and (7.1.7) lead to

$$
0 = (\sigma M - \alpha N) + (\alpha^2 - \sigma^2)\pi k\delta_k + \gamma_k,
$$

where $(\gamma_k)_{k=-\infty}^{\infty} \in l_2$. Solving this equation with respect to δ_k gives

$$
\delta_k = \frac{P}{k} + \frac{\beta_k}{k},
$$

where $\{\beta_k\}_{-\infty}^{\infty} \in l_2$. □

Remark 7.1.4. If we assume in Lemma 7.1.3 that M and N are real and that ψ is pure imaginary on the imaginary axis, then φ is pure imaginary on the imaginary axis, and the asymptotic representations in Lemma 7.1.3 describe the properly indexed zeros of φ.

Lemma 7.1.5. *Let* $n \in \mathbb{N}_0$ *and let* $\tilde{\chi}$ *be an entire function of exponential type* $\leq a$ *having the form*

$$
\tilde{\chi}(\mu) = B_0\left(\left(\mu + \sum_{j=1}^{n+1}\varepsilon_j B_j\mu^{-j+1}\right)\sin\mu a + \left(\sum_{j=1}^{n+1}\varepsilon_{j+1}A_j\mu^{-j+1}\right)\cos\mu a\right)
$$
$$
+ \Psi_n(\mu)\mu^{-n},
$$

where $\varepsilon_j = 1$ *if* j *is even and* $\varepsilon_j = i$ *if* j *is odd,* $A_k \in \mathbb{R}$ *for* $k = 1,\ldots,n+1$, $B_k \in \mathbb{R}$ *for* $k = 0,\ldots,n+1$, $B_0 \neq 0$, $\Psi_n \in \mathcal{L}^a$, *and* $\Psi_n(\mu)\mu^{-n}$ *is real for pure imaginary* μ.

Then the properly indexed zeros $(\mu_k)_{k=-\infty,k\neq0}^{\infty}$ *of* $\tilde{\chi}$ *have the following asymptotic behaviour:*

$$
\mu_k \underset{n\to\infty}{=} \frac{\pi(k-1)}{a} + \sum_{j=1}^{n+1}\frac{\varepsilon_{j+1}P_j}{(k-1)^j} + \frac{b_k^{(n)}}{k^{n+1}}, \tag{7.1.13}
$$

where $(b_k^{(n)})_{k=2}^{\infty} \in l_2$.

The explicit values of p_1, p_2, p_3 are

$$p_1 = -A_1 \pi^{-1}, \tag{7.1.14}$$

$$p_2 = a\pi^{-2}(B_1 A_1 - A_2), \tag{7.1.15}$$

$$p_3 = a^2 \pi^{-3}[-A_3 + A_1 B_2 - A_2 B_1 + A_1(B_1^2 + A_1^2/3) - a^{-1}A_1^2]. \tag{7.1.16}$$

Proof. Without loss of generality we may assume that $B_0 = 1$. We define

$$\varphi(\mu) = i\tilde{\chi}(\mu), \ \mu \in \mathbb{C}.$$

Then φ has the representation

$$\varphi(\mu) = i\mu \sin \mu a - B_1 \sin \mu a + iA_1 \cos \mu a + \psi(\mu), \ \mu \in \mathbb{C}, \tag{7.1.17}$$

with $\psi \in \mathcal{L}^a$. Clearly, φ is as in Lemma 7.1.3 with $\sigma = 0$, $\alpha = 1$, $M = -B_1$ and $N = -A_1$, and ψ is pure imaginary on the imaginary axis. Hence it follows from Lemma 7.1.3 and Remark 7.1.4 that the zeros of $\tilde{\chi}$ can be indexed properly, and that the properly indexed zeros satisfy the asymptotic representation

$$\mu_k \underset{n\to\infty}{=} \frac{\pi(k-1)}{a} + \frac{p_1}{k} + \frac{\beta_k}{k}, \tag{7.1.18}$$

where $(\beta_k)_{k=2}^\infty \in l_2$. This proves (7.1.13) for $n = 0$ with $b_k^{(0)} = \beta_k - (k-1)^{-1}p_1$.

In order to prove (7.1.13) for $n > 0$, we write

$$\mu_k = \frac{\pi(k-1)}{a} + \delta_k, \ k \geq 2, \tag{7.1.19}$$

and assume that (7.1.13) holds for $n - 1$. By definition of $\tilde{\chi}$ there are real polynomials $q_{n,1}$ and $q_{n,2}$ of degree at most n such that

$$\tilde{\chi}(\mu) = -i \left[i\mu + q_{n,1}((i\mu)^{-1})\right] \sin \mu a + q_{n,2}((i\mu)^{-1}) \cos \mu a + \Psi_n(\mu)\mu^{-n}.$$

We already know from that (7.1.18) that $\delta_k \to 0$ as $k \to \infty$, and we can therefore choose $k_0 \geq 2$ such that $|\delta_k a| < \frac{\pi}{2}$ and $2|\mu_k^{-1}q_{n,1}((i\mu_k)^{-1})| < 1$ for $k \geq k_0$. For $k \geq k_0$ we substitute $\mu = \mu_k$ into the above equation. From $\tilde{\chi}(\mu_k) = 0$, $\sin \mu_k a = (-1)^{k-1} \sin \delta_k a$ and $\cos \mu_k a = (-1)^{k-1} \cos \delta_k$ we conclude that

$$i \tan \delta_k a = (i\mu_k)^{-1} \frac{q_{n,2}((i\mu_k)^{-1})}{1 + (i\mu_k)^{-1}q_{n,1}((i\mu_k)^{-1})} + \frac{h_{n,k}}{k^{n+1}}, \tag{7.1.20}$$

where $(h_{n,k})_{k=k_0}^\infty \in l_2$. We define

$$\tilde{\delta}_k = \sum_{j=1}^n \frac{\varepsilon_{j+1}p_j}{(k-1)^j}, \quad \tilde{\mu}_k = \frac{\pi(k-1)}{a} + \tilde{\delta}_k, \tag{7.1.21}$$

and observe that

$$\delta_k = \tilde{\delta}_k + \frac{b_k^{(n-1)}}{k^n}, \quad \mu_k = \tilde{\mu}_k + \frac{b_k^{(n-1)}}{k^n}. \tag{7.1.22}$$

Substituting this representation of μ_k into the right-hand side of (7.1.20) and observing that

$$\frac{1}{\mu_k} - \frac{1}{\tilde{\mu}_k} = \frac{\tilde{\mu}_k - \mu_k}{\mu_k \tilde{\mu}_k} = -\frac{b_k^{(n-1)}}{k^n \mu_k \tilde{\mu}_k},$$

we obtain

$$\tan \delta_k a = -i(i\tilde{\mu}_k)^{-1} \frac{q_{n,2}((i\tilde{\mu}_k)^{-1})}{1 + (i\tilde{\mu}_k)^{-1} q_{n,1}((i\tilde{\mu}_k)^{-1})} + \frac{\tilde{h}_{n,k}}{k^{n+1}}, \tag{7.1.23}$$

where $(\tilde{h}_{n,k})_{k=k_0}^\infty \in l_2$. Expanding the first summand on the right-hand side into its Taylor series in powers of $(i(k-1))^{-1}$, we obtain a real polynomial $q_{n,3}$ of degree at most $n+1$ with $q_{n,3}(0) = 0$ and a bounded sequence $(c_{n,k})_{k=k_0}^\infty$ such that

$$\tan \delta_k a = -i q_{n,3} \left[(i(k-1))^{-1} \right] + \frac{\tilde{h}_{n,k}}{k^{n+1}} + \frac{c_{n,k}}{k^{n+2}}. \tag{7.1.24}$$

Applying the arctan on both sides, we obtain a real polynomial $q_{n,4}$ of degree at most $n+1$ with $q_{n,4}(0) = 0$ such that

$$\delta_k = \frac{i}{a} q_{n,4} \left[(i(k-1))^{-1} \right] + \frac{b_k^{(n)}}{k^{n+1}}, \tag{7.1.25}$$

where $(b_k^{(n)})_{k=k_0}^\infty \in l_2$. This proves the representation (7.1.13), and since the representation (7.1.25) of δ_k must coincide with (7.1.22), it follows that the numbers p_j are independent of n.

To find p_2 and p_3, we need to find $q_{2,4}$. We have

$$q_{2,1}(z) = -B_1 - B_2 z + B_3 z^2, \quad q_{2,2}(z) = A_1 - A_2 z - A_3 z^2,$$

and therefore

$$\frac{z q_{2,2}(z)}{1 + z q_{2,1}(z)} = A_1 z + (A_1 B_1 - A_2) z^2 + (A_1 B_2 - A_2 B_1 - A_3 + A_1 B_1^2) z^3 + O(z^4).$$

We further calculate

$$z = (i\mu_k)^{-1} = \frac{a}{i\pi(k-1)} \left(1 + \frac{a p_1}{\pi(k-1)^2} + O\left(k^{-3}\right) \right)^{-1}$$

$$= \frac{a}{\pi} w \left(1 + \frac{a p_1}{\pi} w^2 \right) + O\left(w^4\right),$$

where $w = (i(k-1))^{-1}$. Substituting this value for z into the above equation and taking the Taylor polynomial in w of order 3 we find

$$q_{2,3}(w) = \frac{a}{\pi} A_1 w + \frac{a^2}{\pi^2}(A_1 B_1 - A_2)w^2$$
$$+ \left[\frac{a^3}{\pi^3}(A_1 B_2 - A_2 B_1 - A_3 + A_1 B_1^2) + \frac{a^2 p_1}{\pi^2} A_1\right] w^3.$$

From $\arctan(-i\zeta) = -i\zeta - \frac{i}{3}\zeta^3 + O(\zeta^5)$ we finally conclude that

$$\frac{i}{a} q_{2,4}(w) = -\frac{1}{\pi} A_1(iw) + i\frac{a}{\pi^2}(A_1 B_1 - A_2)(iw)^2$$
$$+ \left[\frac{a^2}{\pi^3}(A_1 B_2 - A_2 B_1 - A_3 + A_1 B_1^2) + \frac{a p_1}{\pi^2} A_1 + \frac{a^2}{3\pi^3} A_1^3\right](iw)^3.$$

Observing that $iw = (k-1)^{-1}$, we arrive at (7.1.15) and (7.1.16).   □

Lemma 7.1.6. *Let $\rho > 0$, $\eta \in \mathbb{R}$ and define*

$$\phi_1(z) = \rho e^{i\eta} e^z + z. \tag{7.1.26}$$

Then ϕ_1 has infinitely many zeros $(z_k)_{k=-\infty}^{\infty}$, counted with multiplicity, which satisfy the asymptotics

$$\operatorname{Re} z_k \underset{|k|\to\infty}{=} \log|k| + \log(2\pi) - \log \rho + O(k^{-1}), \tag{7.1.27}$$

$$\operatorname{Im} z_k \underset{|k|\to\infty}{=} -\left(2k - \frac{1}{2}\operatorname{sgn} k\right)\pi - \eta + O(|k|^{-1}\log|k|). \tag{7.1.28}$$

Proof. We put

$$z = \log u - i(x + \eta), \tag{7.1.29}$$

where $x \in \mathbb{R}$ and $u > 0$. Then the zeros of ϕ_1 are given by

$$\rho u e^{-ix} = -\log u + i(x + \eta).$$

Taking real and imaginary parts leads to the equations

$$\rho u \cos x = -\log u, \tag{7.1.30}$$
$$\rho u \sin x = -(x + \eta). \tag{7.1.31}$$

Note that $\sin x = 0$ implies $x + \eta = 0$, and thus the two equations (7.1.30) and (7.1.31) lead to

$$\rho^2 u^2 = \log^2 u + (x + \eta)^2, \tag{7.1.32}$$
$$(x + \eta)\cot x = \log u, \tag{7.1.33}$$

unless $x = -\eta = j\pi$ for some $j \in \mathbb{Z}$. Substituting $\log u$ from (7.1.33) into (7.1.32) shows that x must satisfy $f(x) = 0$, where

$$f(x) = \rho^2 e^{2(x+\eta)\cot x} - (x+\eta)^2(1+\cot^2 x).$$

We calculate

$$\frac{1}{2}f'(x) = \rho^2 e^{2(x+\eta)\cot x}\left(\cot x - (x+\eta)(1+\cot^2 x)\right)$$
$$- (x+\eta)(1+\cot^2 x) + (x+\eta)^2 \cot x(1+\cot^2 x).$$

For any zero x of f we find a unique u from (7.1.33), which then satisfies (7.1.32) and therefore

$$\rho^2 u^2 = (x+\eta)^2(1+\cot^2 x). \tag{7.1.34}$$

It follows that whenever $x + \eta \neq 0$ and $f(x) = 0$, then

$$\frac{1}{2}f'(x) = -\frac{\rho^2 u^2}{x+\eta}\left((x+\eta)^2 + ((x+\eta)\cot x - 1)^2\right). \tag{7.1.35}$$

We observe that $x + \eta > 0$ for all x in the interval $(k\pi, (k+1)\pi)$ if and only if $k \geq -\eta\pi^{-1}$, and if we also include the endpoints of the intervals, then $k > -\eta\pi^{-1}$ is required. Similarly, for $x+\eta < 0$ to hold for all x on such an interval, the condition $k \leq -\eta\pi^{-1} - 1$ or $k < -\eta\pi^{-1} - 1$, respectively, is necessary and sufficient. Thus f is decreasing at each zero x of f on each interval $(k\pi, (k+1)\pi)$ when $k \geq -\eta\pi^{-1}$, which means that f can have at most one zero on this interval. On the other hand, if $k > -\eta\pi^{-1}$, $(x+\eta)\cot x \to \infty$ as $x \searrow k\pi$ shows that $f(x) \to \infty$ as $x \searrow k\pi$, whereas $(x+\eta)\cot x \to -\infty$ as $x \nearrow (k+1)\pi$ shows that $f(x) \to -\infty$ as $x \nearrow (k+1)\pi$. Hence f has exactly one simple zero in the interval $(k\pi, (k+1)\pi)$ for $k > -\eta\pi^{-1}$. Similarly, it follows that f has exactly one simple zero in the interval $(k\pi, (k+1)\pi)$ for $k < -\eta\pi^{-1} - 1$.

By construction, every solution of (7.1.30), (7.1.31) is also a solutions of (7.1.32), (7.1.33), but not vice versa. Indeed, by (7.1.34), any solution of (7.1.32), (7.1.33) satisfies $\rho^2 u^2 \sin^2 x = (x+\eta)^2$, and in order for (7.1.31) to hold it is necessary and sufficient that $\sin x$ and $(x+\eta)$ have opposite signs, that is, k must be odd in case $k > -\eta\pi^{-1}$ and k must be even in case $k < -\eta\pi^{-1} - 1$.

Denoting the smallest integer k with $2k > -\eta\pi^{-1} + 1$ by k_+ and the largest integer k with $2k < -\eta\pi^{-1} - 1$ by k_-, we conclude that all solutions of (7.1.30), (7.1.31) with $x \geq (2k_+ - 1)\pi$ or $x \leq (2k_- + 1)\pi$ can be indexed as (x_k, u_k), where $x_k \in ((2k-1)\pi, 2k\pi)$ if $k \geq k_+$ and where $x_k \in (2k\pi, (2k+1)\pi)$ if $k \leq k_-$.

From (7.1.32) it follows that

$$\frac{u_k^2}{k^2} \geq \frac{(x_k+\eta)^2}{\rho^2 k^2} \underset{|k|\to\infty}{=} \rho^{-2}\left(2\pi + O(|k|^{-1})\right)^2,$$

so that

$$u_k^2 - \rho^{-2} \log^2 u_k = u_k^2 \left(1 - \frac{\log^2 u_k}{\rho^2 u_k^2}\right) \underset{|k|\to\infty}{=} u_k^2 \left(1 + O(|u_k|^{-1})\right)$$
$$\underset{|k|\to\infty}{=} u_k^2 \left(1 + O(|k|^{-1})\right)$$

and

$$\frac{\rho^2 u_k^2 - \log^2 u_k}{k^2} = \frac{(x_k + \eta)^2}{k^2} \underset{|k|\to\infty}{=} \left(2\pi + O(|k|^{-1})\right)^2$$

lead to

$$u_k \underset{|k|\to\infty}{=} 2\pi\rho^{-1}|k|(1 + O(|k|^{-1})).$$

Therefore

$$\log u_k \underset{|k|\to\infty}{=} \log|k| + \log(2\pi) - \log\rho + O(|k|^{-1}). \tag{7.1.36}$$

Together with (7.1.33) we conclude that

$$\cot x_k = \frac{\log u_k}{x_k + \eta} \underset{|k|\to\infty}{=} O(|k|^{-1}\log|k|),$$

which finally shows that

$$x_k \underset{|k|\to\infty}{=} \left(2k - \frac{1}{2}\operatorname{sgn} k\right)\pi + O(|k|^{-1}\log|k|). \tag{7.1.37}$$

We still have to account for those solutions of (7.1.30), (7.1.31) for which $x \in ((2k_- +1)\pi, (2k_+ -1)\pi)$. Since the values $x = x(\eta)$ of these solutions are uniformly bounded when η varies in a bounded interval, the corresponding solutions $\log u$ of (7.1.33) must be uniformly bounded by (7.1.32). Thus also the corresponding zeros z of ϕ_1, given by (7.1.29), are uniformly bounded. We recall that the zeros z_k which are already accounted for satisfy $|\operatorname{Im} z_k + 2k\pi + \eta| = |-(x_k + \eta) + 2k\pi + \eta| < \pi$ and depend continuously on η. It follows that for any $\eta_0 \in \mathbb{R}$, we can find a sufficiently large circle in $\mathbb{C}$ such that each zero of ϕ_1 already accounted for stays either inside or outside this circle for η close to η_0, and all zeros not yet accounted for stay inside this circle. Since ϕ_1 depends continuously on η, the argument principle, see [55, V.3.4], shows that the total number of zeros of ϕ_1 inside this circle is locally constant. Hence, with the indexing (7.1.27), (7.1.28) for large zeros, it follows that the index set for the zeros is independent of η, and it suffices to consider one value of η. Indeed, we choose $\eta = 0$, in which case $k_+ = 1$ and $k_- = -1$. This means that we have to solve (7.1.31), (7.1.32) for $x \in (-\pi, \pi)$. But we already know from (7.1.31) that there is no such solution if $x \neq 0$. However, if $x = 0$, then (7.1.31) holds, and the increasing function $u \mapsto \rho u + \log u$ has exactly one (simple) zero. Therefore (7.1.30) has exactly one solution u in case $x = 0$, which leads to one more zero z_0 of ϕ_1. Hence the sequence of zeros of ϕ_1 is indexed by $\mathbb{Z}$. □

Corollary 7.1.7. *There are $\delta_0 > 0$ and $k_0 > 0$ such that the function ϕ_1 defined in Lemma 7.1.6 satisfies*

$$|\phi_1(z)| \geq \frac{1}{2}|z_k||z - z_k| \quad \text{for} \quad |k| \geq k_0 \quad \text{and} \quad |z - z_k| \leq \delta_0. \qquad (7.1.38)$$

Proof. From $\rho e^{i\eta} = -z_k e^{-z_k}$ it follows that

$$\phi_1(z) = z - z_k e^{z - z_k}.$$

Then

$$\phi_1'(z) = 1 - z_k e^{z - z_k}, \quad \phi_1''(z) = -z_k e^{z - z_k},$$

and Taylor's theorem show that

$$|\phi_1(z)| \geq |1 - z_k||z - z_k| - \frac{1}{2}|z_k|e^{|z - z_k|}|z - z_k|^2.$$

Choosing k_0 such that $|z_k| > 4$ for $|k| \geq k_0$ and $\delta_0 > 0$ such that $\delta_0 e^{\delta_0} < \frac{1}{2}$ completes the proof. $\qquad \square$

Corollary 7.1.8. *Let ϕ_1 be defined as in Lemma 7.1.6. Let $\delta > 0$, $r > 0$ and let $U_{\delta,r}$ be the set of $z \in \mathbb{C}$ such that $|z| > r$ and $|z_k - z| \geq \delta$. Then for each $\delta > 0$ there are $r_0 > 0$ and $\gamma > 0$ such that*

$$|\phi_1(z)| \geq |z|\gamma, \qquad (7.1.39)$$

$$|\phi_1(z)| \geq \gamma e^{\mathrm{Re}\, z}, \qquad (7.1.40)$$

for all $z \in U_{\delta,r_0}$.

Proof. Defining $D_+(z) = z^{-1}\phi_1(z)$ and $D_-(z) = e^{-z}\phi_1(z)$, we have

$$D_+(z) = \rho e^{i\eta} z^{-1} e^z + 1, \quad D_-(z) = \rho e^{i\eta} + z e^{-z}.$$

From [189, Proposition A.2.6] we know that there is $l \in \mathbb{N}$ such that for all $\delta > 0$ there are $K(\delta) > 0$ and $g(\delta) > 0$ such that for all $R > K(\delta)$ there are at most l discs of radius $\frac{1}{2}\delta$ such that $|D_+(z)| \geq g(\delta)$ and $|D_-(z)| \geq g(\delta)$ for all z outside these discs and $R \leq |z| \leq R + 1$. Choosing $\delta < l^{-1}$, we see that components of the union of these discs consist of at most $2l$ discs. Each of these components must contain a zero of ϕ_1 because otherwise it can be removed since then $|D_+(z)| \geq g(\delta)$ and $|D_-(z)| \geq g(\delta)$ for all z inside these components by the minimum principle. We may assume that $g(\delta) < \frac{1}{2}\delta \min\{1, \rho e^{-\delta_0}\}$ and $\delta < \delta_0$, where δ_0 is from Corollary 7.1.7. In view of Corollary 7.1.7 we can apply the same reasoning to this component with all discs with centre z_k and radius δ removed, whenever z_k is in this component. Hence we have shown that each exceptional disc must contain one zero of ϕ_1, and thus is contained in a disc with centre z_k and radius δ for some k. For z in the boundary of the union of these discs we have $|z - r_k| = \delta$ for some k, and for sufficiently large $|z|$ and hence k, Corollary 7.1.7 shows that $|\phi_1(z)| \geq \frac{1}{2}|z_k|\delta \geq g(\delta)|z|$, and therefore $|D_+(z)| \geq g(\delta) =: \gamma$ for all $z \in U_{\delta,r_0}$ for a suitable $r_0 > 0$. This proves the estimate (7.1.39), and the proof of the estimate (7.1.40) is similar. $\qquad \square$

7.2 Eigenvalue asymptotics of the generalized Regge problem

We recall the generalized Regge problem, which we already encountered in Sections 2.1 and 6.1:

$$-y'' + q(x)y = \lambda^2 y, \tag{7.2.1}$$

$$y(\lambda, 0) = 0, \tag{7.2.2}$$

$$y'(\lambda, a) + (i\lambda\alpha + \beta)\, y(\lambda, a) = 0, \tag{7.2.3}$$

with a real-valued function $q \in L_2(0, a)$, $\alpha > 0$ and $\beta \in \mathbb{R}$. By (6.1.1), a characteristic function ϕ of (7.2.1)–(7.2.3) is given by

$$\phi(\lambda) = s'(\lambda, a) + (i\lambda\alpha + \beta)s(\lambda, a), \tag{7.2.4}$$

and the zeros of ϕ are exactly the eigenvalues of this eigenvalue problem. To find the eigenvalue asymptotics we will use the asymptotic representations (6.1.2) and (6.1.3) of ϕ.

Theorem 7.2.1.

1. *If $\alpha \in (1, \infty)$, then the properly indexed eigenvalues $(\lambda_k)_{k=-\infty}^{\infty}$ of problem (7.2.1)–(7.2.3) have the asymptotic behaviour*

$$\lambda_k = \frac{\pi k}{a} + \frac{i}{2a} \log\left(\frac{\alpha+1}{\alpha-1}\right) + \frac{P}{k} + \frac{\beta_k}{k}, \quad k \in \mathbb{Z} \setminus \{0\}, \tag{7.2.5}$$

   where

$$P = \frac{1}{2\pi}\left(\int_0^a q(x)\, dx - \frac{2\beta}{\alpha^2 - 1}\right) \tag{7.2.6}$$

   and $(\beta_k)_{k=-\infty}^{\infty} \in l_2$. In particular, the total algebraic multiplicity of the pure imaginary eigenvalues is odd.

2. *If $\alpha \in (0, 1)$, then the properly indexed eigenvalues $(\lambda_k)_{k=-\infty, k\neq 0}^{\infty}$ of problem (7.2.1)–(7.2.3) have the asymptotic behaviour*

$$\lambda_k = \frac{\pi\left(|k| - \frac{1}{2}\right)}{a}\operatorname{sgn} k + \frac{i}{2a}\log\left(\frac{\alpha+1}{1-\alpha}\right) + \frac{P}{k} + \frac{\tilde{\beta}_k}{k}, \quad k \in \mathbb{Z} \setminus \{0\}, \tag{7.2.7}$$

   where P is given by (7.2.6) and $\{\tilde{\beta}_k\}_{k=-\infty, k\neq 0}^{\infty} \in l_2$. In particular, the total algebraic multiplicity of the pure imaginary eigenvalues is even.

Proof. In view of Corollary 12.2.10 and Lemma 12.2.4,

$$s(\lambda, a) = \frac{\sin \lambda a}{\lambda} - K(a, a)\frac{\cos \lambda a}{\lambda^2} + \frac{\psi_1(\lambda)}{\lambda^2},$$

$$s'(\lambda, a) = \cos \lambda a + K(a, a)\frac{\sin \lambda a}{\lambda} + \frac{\psi_2(\lambda)}{\lambda},$$

where $\psi_1, \psi_2 \in \mathcal{L}^a$. Hence the characteristic function ϕ of the eigenvalue problem (7.2.1)–(7.2.3), see (7.2.4), has the representation

$$\phi(\lambda) = \phi^{(0)}(\lambda) + (K(a,a) + \beta)\frac{\sin \lambda a}{\lambda} - i\alpha K(a,a)\frac{\cos \lambda a}{\lambda} + \frac{\psi(\lambda)}{\lambda},$$

where $\phi^{(0)}(\lambda) = \cos \lambda a + i\alpha \sin \lambda a$, see (7.1.5),

$$\psi(\lambda) = -\beta K(a,a)\frac{\cos \lambda a}{\lambda} + \psi_2(\lambda) + \frac{i\lambda\alpha + \beta}{\lambda}\psi_1(\lambda).$$

Multiplying the equation for ϕ above by λ we see that ψ is an entire function. Clearly, ψ is an L_2-function on the real axis. Hence $\lambda \mapsto \lambda\phi(\lambda)$ is of the form φ as considered in Lemma 7.1.3 with $\sigma = 1$, $M = K(a,a) + \beta$ and $N = \alpha K(a,a)$. Since the multiplication with λ introduces one more zero, we have to remove this zero from the indexing in Lemma 7.1.3. In part 2 we simply remove the index 0, whereas in part 1 we remove the index 1 and shift all remaining indices towards 0 by 1. Here we observe that in part 1, the resulting remainder becomes

$$\frac{\hat{\beta}_k}{k \pm 1} = \frac{\hat{\beta}_k}{k}\left(1 - \frac{\pm 1}{k \pm 1}\right) = \frac{\hat{\beta}_k}{k},$$

where $(\hat{\beta}_k) \in l_2$ implies $(\beta_k) \in l_2$. Taking into account that $K(a,a) = \frac{1}{2}\int_0^a q(x)\,dx$, see (12.2.19), the asymptotic representation of the eigenvalues follows from Lemma 7.1.3. Finally, since $s(\cdot,a)$ and $s'(\cdot,a)$ are even real entire functions, we have $\phi(-\lambda) = \overline{\phi(\lambda)}$ for all $\lambda \in \mathbb{C}$, which shows that the indexing of the eigenvalues is proper. $\qquad\square$

Now we turn our attention to the case $\alpha = 1$.

Theorem 7.2.2. *If $\alpha = 1$ and $\beta \neq 0$, then the properly indexed eigenvalues of problem (7.2.1)–(7.2.3) behave asymptotically as*

$$\lambda_k \underset{|k|\to\infty}{=} \frac{\pi}{a}\left(k - \frac{1}{4}\operatorname{sgn}(k\beta)\right) + \frac{i}{2a}\left(\log|k| + \log(2\pi) - \log(|\beta|a)\right) + o(1), \quad (7.2.8)$$

where $k = 0$ belongs to the index set if and only if $\beta < 0$.

Proof. Substituting the representations (12.2.22) and (12.2.23) of $s(\lambda,a)$ and $s'(\lambda,a)$ for $n = 0$ into (7.2.4), we obtain that the characteristic function ϕ is

$$\phi(\lambda) = e^{i\lambda a} + \left(\frac{\beta + 2K(a,a)}{2i\lambda} - \beta\frac{K(a,a)}{2\lambda^2}\right)e^{i\lambda a} - \frac{\beta}{2\lambda}\left(-i + \frac{K(a,a)}{\lambda}\right)e^{-i\lambda a}$$

$$+ \frac{1}{\lambda}\int_0^a K_x(a,t)\sin \lambda t\,dt + \frac{i\lambda + \beta}{\lambda^2}\int_0^a K_t(a,t)\cos \lambda t\,dt. \quad (7.2.9)$$

In view of $|\sin \lambda t| \le e^{|\operatorname{Im}\lambda| t}$ for $t \in \mathbb{R}$, Hölder's inequality gives

$$\left| \int_0^a K_x(a,t) \sin \lambda t \, dt \right| \le \|K_x(a,\cdot)\|_2 \left(\frac{e^{2|\operatorname{Im}\lambda| a} - 1}{2|\operatorname{Im}\lambda|} \right)^{\frac{1}{2}}, \tag{7.2.10}$$

$$\left| \int_0^a K_t(a,t) \cos \lambda t \, dt \right| \le \|K_t(a,\cdot)\|_2 \left(\frac{e^{2|\operatorname{Im}\lambda| a} - 1}{2|\operatorname{Im}\lambda|} \right)^{\frac{1}{2}}, \tag{7.2.11}$$

for all nonreal λ, which extends to real λ if we take the limit as $|\operatorname{Im}\lambda| \to 0$. Putting $z = -2i\lambda a$, we can therefore write

$$-2i\lambda a e^{-i\lambda a} \phi(\lambda) =: \phi_0(z) = \phi_1(z) + p_1(z^{-1}) + \frac{p_2}{z} e^z + \psi(z), \tag{7.2.12}$$

where p_1 is a polynomial and $p_2 \in \mathbb{C}$,

$$\phi_1(z) = z + \beta a e^z, \quad \psi(z) = O\left(\left(\frac{e^{|\operatorname{Re}z|} - 1}{|\operatorname{Re}z|} e^{\operatorname{Re}z} \right)^{\frac{1}{2}} \right). \tag{7.2.13}$$

We observe that ϕ_1 is of the form (7.1.26) with

$$\rho e^{i\eta} = \beta a. \tag{7.2.14}$$

Choose $\delta > 0$, $r_0 > 0$ and $\gamma > 0$ according to Corollary 7.1.8. In view of (7.2.12) and (7.2.13) we can find positive constants $r_1 \ge r_0$ and C such that

$$|\phi_0(z) - \phi_1(z)| < C\left(1 + (1 + |\operatorname{Re}z|)^{-\frac{1}{2}} e^{\operatorname{Re}z} \right), \quad |z| > r_1. \tag{7.2.15}$$

In view of (7.1.39),

$$|\phi_0(z) - \phi_1(z)| < |\phi_1(z)| \text{ if } z \in U_{\delta,r_2}, \ \operatorname{Re}z \le 0, \tag{7.2.16}$$

where $r_2 = \max\{r_1, 2C\gamma^{-1}\}$. Similarly, (7.1.39) shows that

$$|\phi_0(z) - \phi_1(z)| < |\phi^{(1)}(z)| \text{ if } z \in U_{\delta,r_1}, \ \operatorname{Re}z > 0, \ |z| > 2C\gamma^{-1} e^{\operatorname{Re}z}. \tag{7.2.17}$$

From (7.2.15) and (7.1.40) it follows that

$$|\phi_0(z) - \phi_1(z)| \le |\phi_1(z)| C\gamma^{-1} \left(e^{-\operatorname{Re}z} + (1 + |\operatorname{Re}z|)^{-\frac{1}{2}} \right) \tag{7.2.18}$$

for $z \in U_{\delta,r_1}$. For $\operatorname{Re}z > 0$ and $|z| \le 2C\gamma^{-1} e^{\operatorname{Re}z}$ we conclude

$$e^{-\operatorname{Re}z} \le 2C\gamma^{-1} |z|^{-1} \tag{7.2.19}$$

and thus

$$\operatorname{Re}z \ge \log|z| + \log\gamma - \log(2C),$$

which gives

$$(1 + |\operatorname{Re} z|)^{-\frac{1}{2}} \le (C' + \log |z|)^{-\frac{1}{2}} \qquad (7.2.20)$$

for $C' = 1 + \log \gamma - \log(2C)$ and $|z| > e^{-C'}$. Substituting the estimates (7.2.19) and (7.2.20) into (7.2.18) and observing that the right-hand sides of (7.2.19) and (7.2.20) tend to 0 as $|z| \to \infty$, we conclude that there is $r_3 \ge r_2$ such that the right-hand side of (7.2.18) becomes less than $|\phi_1(z)|$ if additionally $|z| > r_3$. Hence (7.2.18) leads to

$$|\phi_0(z) - \phi_1(z)| < |\phi_1(z)| \text{ if } z \in U_{\delta, r_3}, \ \operatorname{Re} z > 0, \ |z| \le 2C\gamma^{-1} e^{\operatorname{Re} z}. \qquad (7.2.21)$$

The estimates (7.2.16), (7.2.17) and (7.2.21) give

$$|\phi_0(z) - \phi_1(z)| < |\phi_1(z)| \text{ if } z \in U_{\delta, r_3}. \qquad (7.2.22)$$

It now follows immediately that $\phi_0(z) \ne 0$ for all $z \in U_{\delta, r_3}$.

We can find an increasing sequence of positive numbers $(R_n)_{n=1}^{\infty}$ such that $R_n = n + O(1)$, $R_{n+1} - R_n < 2$ and $\{z \in \mathbb{C} : |z| = R_n\} \subset U_{\delta, r_3}$ for all $n \in \mathbb{N}$. Applying Rouché's theorem to the discs with radius R_n, $n \in \mathbb{N}$, and centre zero, we see that the numbers of zeros of ϕ_0 and ϕ_1 with modulus less than R_n are equal. In particular, ϕ_1 has an infinite number of zeros, which we denote by $(z_{1,k})_{k=-\infty}^{\infty}$. Then the above estimates show that we can apply Rouché's theorem to discs with centre z_k and radius δ, which gives $z_{1,k} - z_k = o(1)$ since δ can be made arbitrarily small. In the transition from ϕ to ϕ_0 we have introduced one additional zero, so that, with the exception of one value, the numbers $\tilde{\lambda}_k = \frac{iz_{1,k}}{2a}$ are the zeros of ϕ. By (7.2.14) we have that $\rho = |\beta|a$, $\eta = 0$ if $\beta > 0$ and $\eta = -\pi$ if $\beta < 0$. Hence (7.2.8) follows from (7.1.27) and (7.1.28) if we observe that some reindexing is needed to arrive at a properly indexed sequence. $\qquad \square$

In the case $\alpha = 1$, $\beta = 0$, where we face the classical Regge problem, see Section 2.1, the eigenvalue asymptotics are even more complicated. Indeed, we have seen in Remark 6.1.4 that there may be no eigenvalues at all. Thus conditions on the potential q are needed to assure that there are infinitely many eigenvalues. However, if we assume that there are infinitely many eigenvalues $(\lambda_k)_{k=1}^{\infty}$, then it follows from (7.2.9) with $\beta = 0$, (7.2.10) and (7.2.11) that

$$\left| 1 + \frac{K(a,a)}{i\lambda_k} \right| = O(|\lambda_k|^{-1}) e^{(|\operatorname{Im} \lambda_k| + \operatorname{Im} \lambda_k)a},$$

so that $\operatorname{Im} \lambda_k \to \infty$ as $|\lambda_k| \to \infty$. Hence we have shown

Proposition 7.2.3. *If $\alpha = 1$ and $\beta = 0$ and if problem (7.2.1)–(7.2.3) has infinitely many eigenvalues $(\lambda_k)_{k=1}^{\infty}$, then $\operatorname{Im} \lambda_k \to \infty$ as $k \to \infty$.*

Theorem 7.2.4. *Consider problem (7.2.1)–(7.2.3) with $\alpha = 1$ and $\beta = 0$ and assume that there is $p \in \mathbb{N}_0$ such that $q \in W_2^{p+1}(a,b)$, $q^{(j)}(a) = 0$ for $j = 0, \ldots, p-1$*

and $q^{(p)}(a) \neq 0$. Then the properly indexed eigenvalues of problem (7.2.1)–(7.2.3) have the asymptotic behavior

$$\lambda_k \underset{|k|\to\infty}{=} \frac{\pi}{a}\left(k + \frac{1}{4}\operatorname{sgn} k\left[p+1+(-1)^p \operatorname{sgn}(q^{(p)}(a))\right]\right)$$
$$+ \frac{i}{2a}\left((p+2)[\log|k|+\log(2\pi)-\log a]-\log|q^{(p)}(a)|\right)+o(1), \qquad (7.2.23)$$

and the index $k = 0$ is omitted if $(-1)^p \operatorname{sgn}(q^{(p)}(a)) < 0$.

Proof. Substituting (12.2.22) and (12.2.23) into (7.2.4) we obtain

$$\phi(\lambda) = s'(\lambda, a) + i\lambda s(\lambda, a)$$
$$= \cos\lambda a + i\sin\lambda a + \frac{1}{\lambda}K(a,a)(\sin\lambda a - i\cos\lambda a)$$
$$- \sum_{j=1}^{p+1}\frac{1}{\lambda^{j+1}}\left(\frac{\partial^{j-1}}{\partial t^{j-1}}K_x(a,a)\sin^{(j)}\lambda a + i\frac{\partial^j}{\partial t^j}K(a,a)\sin^{(j+1)}\lambda a\right)$$
$$+ \frac{1}{\lambda^{p+2}}\int_0^a\left(\frac{\partial^{p+1}}{\partial t^{p+1}}K_x(a,t)\sin^{(p+1)}\lambda t + i\frac{\partial^{p+2}}{\partial t^{p+2}}K(a,t)\sin^{(p+2)}\lambda t\right)dt.$$

Observing that $\sin^{(j)}(\tau) = \frac{1}{2}i^{j-1}e^{i\tau} + \frac{1}{2}(-i)^{j-1}e^{-i\tau}$, we can write

$$\phi(\lambda) = e^{i\lambda a} - \frac{1}{-i\lambda}K(a,a)e^{i\lambda a} + \sum_{j=1}^{p+1}\frac{A_j}{(-2i\lambda)^{j+1}}e^{i\lambda a}$$
$$+ \sum_{j=1}^{p+1}(-1)^j\frac{B_j}{(-2i\lambda)^{j+1}}e^{-i\lambda a} + \frac{1}{(-2i\lambda)^{p+2}}\psi_p(\lambda),$$

where $\psi_p(\lambda) = O\left(\left(\frac{e^{2|\operatorname{Im}\lambda|a}-1}{2|\operatorname{Im}\lambda|}\right)^{\frac{1}{2}}\right)$, see (7.2.10) and (7.2.11), where the A_j are real numbers, and where

$$B_j = 2^j\frac{\partial^{j-1}}{\partial t^{j-1}}[K_x(a,t)+K_t(a,t)]_{t=a}$$

for $j = 1,\ldots,p+1$. Taking partial derivatives in (12.2.18) shows that

$$\tilde{K}_x(x,t) = \frac{1}{4}q\left(\frac{x+t}{2}\right) + \frac{1}{2}\int_0^{\frac{x+t}{2}}q\left(s+\frac{x-t}{2}\right)\tilde{K}\left(s+\frac{x-t}{2}, s-\frac{x-t}{2}\right)ds$$
$$+ \frac{1}{2}\int_0^{\frac{x-t}{2}}q\left(\frac{x+t}{2}+u\right)\tilde{K}\left(\frac{x+t}{2}+u, \frac{x+t}{2}-u\right)du,$$

$$\tilde{K}_t(x,t) = \frac{1}{4}q\left(\frac{x+t}{2}\right) - \frac{1}{2}\int_0^{\frac{x+t}{2}} q\left(s+\frac{x-t}{2}\right)\tilde{K}\left(s+\frac{x-t}{2}, s-\frac{x-t}{2}\right)ds$$

$$+ \frac{1}{2}\int_0^{\frac{x-t}{2}} q\left(\frac{x+t}{2}+u\right)\tilde{K}\left(\frac{x+t}{2}+u, \frac{x+t}{2}-u\right)du,$$

and a straightforward calculation gives

$$K_x(a,t) + K_t(a,t) = \tilde{K}_x(a,t) - \tilde{K}_x(a,-t) + \tilde{K}_t(a,t) + \tilde{K}_t(a,-t)$$

$$= \frac{1}{2}q\left(\frac{a+t}{2}\right) + \int_0^{\frac{a-t}{2}} q\left(\frac{a+t}{2}+u\right)\tilde{K}\left(\frac{a+t}{2}+u, \frac{a+t}{2}-u\right)du$$

$$- \int_0^{\frac{a-t}{2}} q\left(s+\frac{a+t}{2}\right)\tilde{K}\left(s+\frac{a+t}{2}, s-\frac{a+t}{2}\right)ds.$$

It follows that

$$B_j = 2^j\frac{\partial^{j-1}}{\partial t^{j-1}}\left[K_x(a,t) + K_t(a,t)\right]_{t=a} = q^{(j-1)}(a) + \sum_{m=0}^{j-2} a_{jm}q^{(m)}(a)$$

for $j = 1,\ldots,p+1$, where the a_{jm} are real numbers. By assumption, $q^{(j)}(a) = 0$ for $j = 0,\ldots,p-1$ and $q^{(p)}(a) \neq 0$, and we therefore conclude that

$$\phi(\lambda) = e^{i\lambda a} + \sum_{j=0}^{p+1}\frac{A_j}{(-2i\lambda)^{j+1}}e^{i\lambda a} + (-1)^{p+1}\frac{q^{(p)}(a)}{(-2i\lambda)^{p+2}}e^{-i\lambda a} + \frac{1}{(-2i\lambda)^{p+2}}\psi_p(\lambda).$$

As in the proof of Theorem 7.2.2, we set $z = -2i\lambda a$ and obtain

$$(-2i\lambda a)^{p+2}e^{-i\lambda a}\phi(\lambda) =: \phi^{(2)}(z) = \phi_2(z) + \sum_{j=0}^{p+1}A'_j z^{p-j+1} + \psi(z),$$

where the A'_j are real numbers and where

$$\phi_2(z) = z^{p+2} - (-a)^{p+2}q^{(p)}(a)e^z, \quad \psi(z) = O\left(\left(\frac{e^{|\operatorname{Re}z|}-1}{|\operatorname{Re}z|}e^{\operatorname{Re}z}\right)^{\frac{1}{2}}\right).$$

Defining $\rho = a|q^{(p)}(a)|^{\frac{1}{p+2}}$ and putting $\eta = 0$ if $(-1)^p\operatorname{sgn}(q^{(p)}(a)) < 0$ and $\eta = \pi$ if $(-1)^p\operatorname{sgn}(q^{(p)}(a)) > 0$, we can write

$$\phi_2(z) = z^{p+2} + \rho^{p+2}e^{i\eta}e^z = (p+2)^{p+2}\prod_{j=1}^{p+2}\left(\frac{z}{p+2} + \frac{\rho}{p+2}e^{i\eta_j}e^{\frac{z}{p+2}}\right), \quad (7.2.24)$$

where

$$\eta_j = \frac{\eta - \pi + 2\pi j}{p+2} - \pi \quad \text{for } j = 1,\ldots,p+2.$$

In view of Lemma 7.1.6, the zeros of ϕ_2 can be indexed as $z_{j,\tilde{k}}$, $j = 1, \ldots, p+2$, $\tilde{k} \in \mathbb{Z}$, where

$$\operatorname{Re} z_{j,\tilde{k}} \underset{|\tilde{k}|\to\infty}{=} (p+2)\left(\log|\tilde{k}| + \log(2\pi) - \log\rho + \log(p+2) + O(|\tilde{k}|^{-1})\right), \quad (7.2.25)$$

$$\operatorname{Im} z_{j,\tilde{k}} \underset{|\tilde{k}|\to\infty}{=} -(p+2)\left(\left(2\tilde{k} - \frac{1}{2}\operatorname{sgn}\tilde{k}\right)\pi + \eta_j\right) + O(|\tilde{k}|^{-1}\log|\tilde{k}|). \quad (7.2.26)$$

It is straightforward to verify that Corollary 7.1.8, applied to each factor on the right-hand side of (7.2.24), also hold if ϕ_1 is replaced with ϕ_2, z^{-1} with $z^{-(p+2)}$, and if all zeros $z_{j,\tilde{k}}$, $j = 1, \ldots, p+2$, $\tilde{k} \in \mathbb{Z}$, are taken into account. As in the proof of Lemma 7.1.3 we obtain, mutatis mutandis, that the zeros of $\phi^{(2)}$ can be indexed as $\tilde{z}_{j,\tilde{k}}$, $j = 1, \ldots, p+2$, $\tilde{k} \in \mathbb{Z}$, such that $\tilde{z}_{j,\tilde{k}} = z_{j,\tilde{k}} + o(1)$ as $|\tilde{k}| \to \infty$. In the transition from ϕ to $\phi^{(2)}$ we have added a $(p+2)$-fold zero at 0. Hence, omitting the terms with index $\tilde{k} = 0$ from the above sequences and putting

$$\tilde{\lambda}_{(p+2)\tilde{k}+j} = \begin{cases} \frac{i}{2a}\tilde{z}_{j,\tilde{k}+1} & \text{if } \tilde{k} \geq 0, \\ \frac{i}{2a}\tilde{z}_{j,\tilde{k}} & \text{if } \tilde{k} < 0, \end{cases}$$

a straightforward calculation shows that $\lambda_k := \tilde{\lambda}_k$ for $k > 0$ has the representation (7.2.23) and that $-\overline{\tilde{\lambda}_k} = \tilde{\lambda}_{-k} + o(1)$ if $\eta = \pi$ and $-\overline{\tilde{\lambda}_k} = \tilde{\lambda}_{-k+1} + o(1)$ if $\eta = 0$. Hence proper indexing is achieved by putting $\lambda_k = \tilde{\lambda}_k$ for $k \leq 0$ in case $\eta = \pi$ and by putting $\lambda_k = \tilde{\lambda}_{k+1}$ for $k < 0$ in case $\eta = 0$. Finally, we observe that in the case $\eta = 0$, the index $k = 0$ is omitted. $\qquad\square$

7.3 Eigenvalue asymptotics of the damped string problem

In this section we will find eigenvalue asymptotics for problem (2.2.4)–(2.2.6),

$$y'' - 2i\lambda\varrho y - q(x)y + \lambda^2 y = 0, \quad (7.3.1)$$

$$y(\lambda, 0) = 0, \quad (7.3.2)$$

$$y'(\lambda, a) + (-\lambda^2 m + i\lambda\nu + \beta)y(\lambda, a) = 0, \quad (7.3.3)$$

with $q \in L_2(0,a)$, $\varrho > 0$, $m > 0$, $\nu > 0$ and $\beta \in \mathbb{R}$. Whereas in (2.2.4) ϱ is a function, we assume here that ϱ is a constant. Also, for simplicity of notation, we have replaced $\tilde{m}$ and $\tilde{\nu}$ with m and ν, respectively.

Theorem 7.3.1. *The properly indexed eigenvalues* $(\lambda_k)_{k=-\infty,k\neq 0}^{\infty}$ *of problem (7.3.1)–(7.3.3) have the following asymptotic behaviour:*

1. *if* $q \in L_2(0,a)$ *then*

$$\lambda_k \underset{k\to\infty}{=} \frac{\pi(k-1)}{a} + i\varrho + \frac{p_1}{k-1} + \frac{b_k^{(0)}}{k}, \quad (7.3.4)$$

2. *if $q \in W_2^1(0, a)$ then*

$$\lambda_k \underset{k \to \infty}{=} \frac{\pi(k-1)}{a} + i\varrho + \frac{p_1}{k-1} + \frac{ip_2}{(k-1)^2} + \frac{b_k^{(1)}}{k^2}, \qquad (7.3.5)$$

3. *if $q \in W_2^2(0, a)$ then*

$$\lambda_k \underset{k \to \infty}{=} \frac{\pi(k-1)}{a} + i\varrho + \frac{p_1}{k-1} + \frac{ip_2}{(k-1)^2} + \frac{p_3}{(k-1)^3} + \frac{b_k^{(2)}}{k^3}, \qquad (7.3.6)$$

where $(b_k^{(n)})_{n=1}^\infty \in l_2$ for $n = 0, 1, 2$,

$$p_1 = \frac{1}{\pi m} + \frac{K(a,a)}{\pi} - \frac{\varrho^2 a}{2\pi}, \qquad (7.3.7)$$

$$p_2 = \frac{a}{\pi^2 m}\left(\frac{\nu}{m} - 2\varrho\right), \qquad (7.3.8)$$

$$p_3 = a^2 \pi^{-3}\left(-A_3 + A_1 B_2 - A_2 B_1 + A_1\left(B_1^2 + \frac{A_1^2}{3}\right) - a^{-1}A_1^2\right), \qquad (7.3.9)$$

$$B_1 = 2\varrho - \frac{\nu}{m},$$

$$B_2 = -\frac{K(a,a)}{m} + K_t(a,a) - \frac{\beta}{m} - \frac{3}{2}\varrho^2 + \frac{\nu\varrho}{m} - \frac{1}{8}a^2\varrho^4$$
$$\qquad + \frac{1}{2}K(a,a)a\varrho^2 + \frac{a\varrho^2}{2m},$$

$$B_3 = \left(2\varrho - \frac{\nu}{m}\right)\left(K_t(a,a) - \frac{\varrho^2}{2} + \frac{a\varrho^2}{2}K(a,a) - \frac{a^2\varrho^4}{8}\right),$$

$$A_1 = \frac{1}{2}a\varrho^2 - \frac{1}{m} - K(a,a),$$

$$A_2 = \left(\frac{1}{2}a\varrho^2 - K(a,a)\right)\left(2\varrho - \frac{\nu}{m}\right),$$

$$A_3 = \frac{K_x(a,a)}{m} + K_{tt}(a,a) + K(a,a)\left(\frac{\beta}{m} + \varrho\left(2\varrho - \frac{\nu}{m}\right)\right) - \frac{a\varrho^2\beta}{2m} + \frac{a\nu\varrho^3}{2m}$$
$$\qquad - K(a,a)\frac{a\varrho^2}{2m} + K_t(a,a)\frac{a\varrho^2}{2} + \frac{a\varrho^4}{8}\left(\frac{a}{m} + aK(a,a) - 7\right) - \frac{a^3\varrho^6}{48}.$$

Proof. The eigenvalues of (7.3.1)–(7.3.3) coincide with the zeros of the entire function

$$\phi(\lambda) = s'(\tilde\tau(\lambda), a) + (-m\lambda^2 + i\lambda\nu + \beta)s(\tilde\tau(\lambda), a), \qquad (7.3.10)$$

where s is the function defined in Theorem 12.2.9 and $\tilde\tau(\lambda) = \sqrt{\lambda^2 - 2i\varrho\lambda}$. Here we have to observe that $s(\cdot, a)$ and $s'(\cdot, a)$ are even functions, and therefore $s(\tilde\tau(\lambda), a)$ and $s'(\tilde\tau(\lambda), a)$ are unambiguous. Letting $\mu = \lambda - i\varrho$, we have

$$\tau(\mu) := \tilde\tau(\lambda) = \sqrt{\mu^2 + \varrho^2}.$$

Below we will briefly write τ for $\tau(\mu)$. We substitute (12.2.22) and (12.2.23) into (7.3.10) and obtain for $n \in \mathbb{N}_0$ and $q \in W_2^n(0, a)$ that the entire function ϕ has the representation

$$
\phi(\lambda) = \cos \tau a + K(a, a) \frac{\sin \tau a}{\tau} - \sum_{j=1}^{n} \frac{\partial^{j-1}}{\partial t^{j-1}} K_x(a, a) \frac{\sin^{(j)} \tau a}{\tau^{j+1}} + \frac{\psi_{1,n}(\tau)}{\tau^{n+1}}
$$
$$
+ \left(-m\lambda^2 + i\nu\lambda + \beta \right) \left(\frac{\sin \tau a}{\tau} - \sum_{j=0}^{n} \frac{\partial^j}{\partial t^j} K(a, a) \frac{\sin^{(j+1)} \tau a}{\tau^{j+2}} + \frac{\psi_{2,n}(\tau)}{\tau^{n+2}} \right),
$$

$$(7.3.11)$$

where $\psi_{m,n} \in \mathcal{L}^a$ in view of Lemma 12.2.4.

We now consider the case $n = 2$. The cases $n = 0$ and $n = 1$ easily follow by adapting the proof below. Introducing the function $\tilde{\chi}$ defined by $\tilde{\chi}(\mu) = \phi(\lambda)$, (7.3.11) becomes

$$
\tilde{\chi}(\mu) = \left[u(\mu) \left(-\frac{K(a, a)}{\tau^2} + \frac{K_{tt}(a, a)}{\tau^4} \right) + 1 - \frac{K_x(a, a)}{\tau^2} \right] \cos \tau a
$$
$$
+ \left[u(\mu) \left(1 + \frac{K_t(a, a)}{\tau^2} \right) + K(a, a) + \frac{K_{xt}(a, a)}{\tau^2} \right] \frac{\sin \tau a}{\tau} + \frac{\Psi(\mu)}{\mu^2},
$$

where $\Psi \in \mathcal{L}^a$ and

$$
u(\mu) = -m\mu^2 + i(\nu - 2\varrho m)\mu + m\varrho^2 - \nu\varrho + \beta.
$$

In view of Corollary 12.3.2 with $b = 0$ and $c = -\varrho^2$, we can write $\tau^{-1} \sin \tau a$ and $\cos \tau a$ in terms of (12.3.8) and (12.3.9). Therefore

$$
\tilde{\chi}(\mu) = P_1(\mu) \sin \mu a + P_2(\mu) \cos \mu a + \psi_1(\mu)\mu^{-2},
$$

where $\psi_1 \in \mathcal{L}^a$ and

$$
P_1(\mu) \overset{2}{\simeq} \frac{1}{\mu} \left[u(\mu) \left(1 + \frac{K_t(a, a)}{\tau^2} \right) + K(a, a) + \frac{K_{xt}(a, a)}{\tau^2} \right] f_{1,1,3}(\mu^{-1})
$$
$$
+ i \left[u(\mu) \left(-\frac{K(a, a)}{\tau^2} + \frac{K_{tt}(a, a)}{\tau^4} \right) + 1 - \frac{K_x(a, a)}{\tau^2} \right] f_{2,2,2}(\mu^{-1}),
$$

$$
P_2(\mu) \overset{2}{\simeq} \left[u(\mu) \left(-\frac{K(a, a)}{\tau^2} + \frac{K_{tt}(a, a)}{\tau^4} \right) + 1 - \frac{K_x(a, a)}{\tau^2} \right] f_{2,1,2}(\mu^{-1})
$$
$$
+ \frac{i}{\mu} \left[u(\mu) \left(1 + \frac{K_t(a, a)}{\tau^2} \right) + K(a, a) + \frac{K_{x,t}(a, a)}{\tau^2} \right] f_{1,2,3}(\mu^{-1}),
$$

and where $f(\mu) \overset{n}{\simeq} g(\mu)$ means that

$$
f(\mu) - g(\mu) = \sum_{j=1}^{r} \psi_j(\mu) O(|\mu|^{-n}) + O(|\mu|^{-n-1})
$$

with $\psi_1, \ldots, \psi_r \in \mathcal{L}^a$. From $f_{1,1,3}(0) = f_{2,1,2}(0) = 1$, $f_{1,2,3}(0) = f_{2,2,2}(0) = 0$, and

$$\frac{1}{\tau^2} = \frac{1}{\mu^2} - \frac{\varrho^2}{\mu^4} + O(|\mu|^{-6})$$

we obtain constants $\tilde{B}_j$ and $\tilde{A}_j$, $j = 1, 2, 3$, such that

$$P_1(\mu) \overset{2}{\simeq} -m\mu + i\tilde{B}_1 + \tilde{B}_2\mu^{-1} + i\tilde{B}_3\mu^{-2} =: P_{1,2}(\mu),$$

$$P_2(\mu) \overset{2}{\simeq} \tilde{A}_1 + i\tilde{A}_2\mu^{-1} + \tilde{A}_3\mu^{-2} =: P_{2,2}(\mu).$$

We conclude that

$$\tilde{\chi}(\mu) = P_{1,2}(\mu) \sin \mu a + P_{2,2}(\mu) \cos \mu a + \Psi_2(\mu)\mu^{-2},$$

where $\Psi_2 \overset{0}{\simeq} 0$. Multiplying this identity with μ^2 we see that Ψ_2 is an entire function, and it follows that $\Psi_2 \in \mathcal{L}^a$. Hence $\tilde{\chi}$ is formally as considered in Lemma 7.1.5. In particular, $B_0 = -m$. From

$$\tilde{\chi}(-\overline{\mu}) = \phi\left(-\overline{(\mu + i\varrho)}\right) = \overline{\phi(\mu + i\varrho)} = \overline{\tilde{\chi}(\mu)}$$

we conclude that $\tilde{\chi}$ also satisfies the symmetry conditions as required in Lemma 7.1.5. Hence (7.3.6) follows from Lemma 7.1.5, and it remains to find the constants $B_j = \tilde{B}_j B_0^{-1}$ and $A_j = \tilde{A}_j B_0^{-1}$ for $j = 1, 2, 3$.

Putting $z = \mu^{-1}$, the functions in (12.3.4)–(12.3.7) are

$$h_1(z) = 1 + \frac{1}{2}\varrho^2 z^2 - \frac{1}{8}\varrho^4 z^4 + O(|\mu|^{-5}),$$

$$\frac{1}{h_1(z)} = 1 - \frac{1}{2}\varrho^2 z^2 + O(|\mu|^{-4}),$$

$$h_2(z) = \frac{h_1(z) - 1}{z} = \frac{1}{2}\varrho^2 z - \frac{1}{8}\varrho^4 z^3 + O(|\mu|^{-4}),$$

$$f_{2,1,2}(z) = f_{2,1,3}(z) \overset{3}{\simeq} \cos h_2(z) a \overset{3}{\simeq} 1 - \frac{1}{8}\varrho^4 a^2 z^2,$$

$$f_{1,1,3}(z) \overset{3}{\simeq} \frac{\cos h_2(z) a}{h_1(z)} \overset{3}{\simeq} 1 - \frac{1}{2}\varrho^2 z^2 - \frac{1}{8}\varrho^4 a^2 z^2,$$

$$f_{2,2,2}(z) \overset{2}{\simeq} i \sin h_2(z) a \overset{2}{\simeq} \frac{i}{2}\varrho^2 za,$$

$$f_{1,2,3}(z) \overset{3}{\simeq} -i\frac{\sin h_2(z) a}{h_1(z)} \overset{3}{\simeq} -\frac{i}{2}\varrho^2 za + \frac{3i}{8}\varrho^4 az^3 + \frac{i}{48}\varrho^6 a^3 z^3.$$

We further calculate

$$\frac{1}{\mu}\left[u(\mu)\left(1 + \frac{K_t(a,a)}{\tau^2}\right) + K(a,a) + \frac{K_{xt}(a,a)}{\tau^2}\right] = -m\mu + i(\nu - 2\varrho m)$$

$$+ \left(m\varrho^2 - \nu\varrho + \beta - mK_t(a,a) + K(a,a)\right)\mu^{-1} + i(\nu - 2\varrho m)K_t(a,a)\mu^{-2}$$

$$+ O(|\mu|^{-3}),$$

$$\left[u(\mu)\left(-\frac{K(a,a)}{\tau^2} + \frac{K_{tt}(a,a)}{\tau^4} \right) + 1 - \frac{K_x(a,a)}{\tau^2} \right] = 1 + mK(a,a)$$
$$- i(\nu - 2\varrho m)K(a,a)\mu^{-1} - K(a,a)\left(2m\varrho^2 - \nu\varrho + \beta \right)\mu^{-2} - mK_{tt}(a,a)\mu^{-2}$$
$$- K_x(a,a)\mu^{-2} + O(|\mu|^{-3}).$$

Evaluating now the coefficients of P_1 and P_2 with respect to powers of μ gives

$$\tilde{B}_1 = \nu - 2\varrho m,$$

$$\tilde{B}_2 = \frac{3m\varrho^2}{2} - \nu\varrho + \beta - mK_t(a,a) + K(a,a) + \frac{\varrho^4 a^2 m}{8} - \frac{\varrho^2 a}{2} - \frac{m}{2}\varrho^2 aK(a,a),$$

$$\tilde{B}_3 = (\nu - 2\varrho m)\left(K_t(a,a) - \frac{1}{2}\varrho^2 - \frac{1}{8}\varrho^4 a^2 + K(a,a)\frac{a\varrho^2}{2} \right),$$

$$\tilde{A}_1 = 1 + mK(a,a) - \frac{m}{2}\varrho^2 a,$$

$$\tilde{A}_2 = -(\nu - 2\varrho m)\left(K(a,a) - \frac{1}{2}\varrho^2 a \right),$$

$$\tilde{A}_3 = -mK_{tt}(a,a) - K_x(a,a) - \frac{1}{8}\varrho^4 a^2(1 + mK(a,a)) - \frac{1}{2}\nu\varrho^3 a + \frac{1}{2}\varrho^2 a\beta$$
$$- \frac{1}{2}m\varrho^2 aK_t(a,a) + \frac{1}{2}\varrho^2 aK(a,a) + \frac{7}{8}\varrho^4 am + \frac{1}{48}\varrho^6 a^3 m$$
$$- K(a,a)\left(2m\varrho^2 - \nu\rho + \beta \right).$$

This proves the representations of the B_j and A_j for $j = 1, 2, 3$. From Lemma 7.1.5 we now find p_1, p_2 and p_3.   □

7.4 Eigenvalue asymptotics of star graphs of strings

In this section we revisit problem (6.4.1)–(6.4.4), see also (2.3.9)–(2.3.12). For the sake of simplicity we suppose in what follows that $a_1 = a_2 = \cdots = a_p =: a$, $\varrho = \tilde{m} = 0$, $v_j = \theta_j = 1$ for all $j = 1, \ldots, p$. Thus, we deal with the problem

$$y_j'' + \lambda^2 y_j - q_j(x)y_j = 0, \quad j = 1, \ldots, p, \ x \in [0, a], \qquad (7.4.1)$$

$$y_j(\lambda, 0) = 0, \quad j = 1, \ldots, p, \qquad (7.4.2)$$

$$y_1(\lambda, a) = \cdots = y_p(\lambda, a), \qquad (7.4.3)$$

$$\sum_{j=1}^{p} y_j'(\lambda, a) + (i\lambda\tilde{\nu} + \beta)y_1(\lambda, a) = 0. \qquad (7.4.4)$$

We assume that $p \geq 2$, that $q_j, j = 1, \ldots, p$, are real-valued functions from $L_2(0, a)$, that $\tilde{\nu} \geq 0$ and that β is a real constant derived in Subsection 2.3.3. We recall from Section 6.4 that for $j = 1, \ldots, p$, $s_j(\lambda, \cdot)$ denotes the solution of (7.4.1) which satisfies the initial conditions $s_j(\lambda, 0) = 0$, $s_j'(\lambda, 0) = 1$ and that the characteristic function ϕ of the of problem (7.4.1)–(7.4.4) has been derived in (6.4.5). With our particular assumptions we have

$$\phi(\lambda) = \varphi(\lambda) + (i\lambda\tilde{\nu} + \beta)\chi(\lambda), \tag{7.4.5}$$

where

$$\varphi(\lambda) = \sum_{j=1}^{p} s'_j(\lambda, a) \prod_{\substack{m=1 \\ m \neq j}}^{p} s_m(\lambda, a), \tag{7.4.6}$$

$$\chi(\lambda) = \prod_{j=1}^{p} s_j(\lambda, a). \tag{7.4.7}$$

Lemma 7.4.1. *The function ϕ has the representation*

$$\phi(\lambda) = p\frac{\sin^{p-1} \lambda a}{\lambda^{p-1}} \cos \lambda a - (p-1) \sum_{j=1}^{p} B_j \pi \frac{\cos^2 \lambda a}{\lambda^p} \sin^{p-2} \lambda a + \sum_{j=1}^{p} B_j \pi \frac{\sin^p \lambda a}{\lambda^p}$$

$$+ \frac{v_1(\lambda)}{\lambda^p} + \left(i\tilde{\nu} + \frac{\beta}{\lambda}\right)\left(\frac{\sin^p \lambda a}{\lambda^{p-1}} - \sum_{j=1}^{p} B_j \pi \frac{\sin^{p-1} \lambda a}{\lambda^p} \cos \lambda a + \frac{v_2(\lambda)}{\lambda^p}\right), \tag{7.4.8}$$

where $v_1 \in \mathcal{L}^{pa}$, $v_2 \in \mathcal{L}^{pa}$ and

$$B_j = \frac{1}{2\pi} \int_0^a q_j(x)dx, \quad j = 1, \ldots, p. \tag{7.4.9}$$

Proof. In view of Theorem 12.2.9, Corollary 12.2.10, Lemma 12.2.4 and Remark 12.2.5 we have

$$s_j(\lambda, a) = \frac{\sin \lambda a}{\lambda} - B_j \pi \frac{\cos \lambda a}{\lambda^2} + \frac{\omega_j(\lambda)}{\lambda^2}, \quad j = 1, \ldots, p, \tag{7.4.10}$$

$$s'_j(\lambda, a) = \cos \lambda a + B_j \pi \frac{\sin \lambda a}{\lambda} + \frac{\tau_j(\lambda)}{\lambda}, \quad j = 1, \ldots, p, \tag{7.4.11}$$

where $\omega_j \in \mathcal{L}_e^a$, $\tau_j \in \mathcal{L}_o^a$. Substituting (7.4.10) and (7.4.11) into (7.4.6) and (7.4.7) we obtain

$$\varphi(\lambda) = p\frac{\sin^{p-1} \lambda a}{\lambda^{p-1}} \cos \lambda a - (p-1) \sum_{j=1}^{p} B_j \pi \frac{\cos^2 \lambda a}{\lambda^p} \sin^{p-2} \lambda a$$

$$+ \sum_{j=1}^{p} B_j \pi \frac{\sin^p \lambda a}{\lambda^p} + \frac{v_1(\lambda)}{\lambda^p}, \tag{7.4.12}$$

$$\chi(\lambda) = \frac{\sin^p \lambda a}{\lambda^p} - \sum_{j=1}^{p} B_j \pi \frac{\sin^{p-1} \lambda a}{\lambda^{p+1}} \cos \lambda a + \frac{v_2(\lambda)}{\lambda^{p+1}}. \tag{7.4.13}$$

From (7.4.12) and (7.4.13) it easily follows that v_1 and v_2 are entire functions of exponential type less or equal pa which are L_2-functions on the real line, i.e., $v_1, v_2 \in \mathcal{L}^{pa}$. Substituting (7.4.12) and (7.4.13) into (7.4.5) gives (7.4.8). $\square$

The "leading term" of ϕ is the entire function ϕ_0 given by

$$\phi_0\left(\lambda\right) = \frac{\sin^{p-1}\lambda a}{\lambda^{p-1}}\left(p\cos\lambda a + i\tilde{\nu}\sin\lambda a\right). \qquad (7.4.14)$$

Clearly, ϕ_0 has infinitely many zeros and $\phi_0(-\overline{\lambda}) = \overline{\phi_0(\lambda)}$. In view of Proposition 7.1.2, the sequence $(\lambda_k^{(0)})_{k=-\infty}^{\infty}$ of the zeros of ϕ_0 can be indexed properly. Therefore $\lambda_{-k}^{(0)} = -\overline{\lambda_k^{(0)}}$ for all $k \neq 0$, and the following result for nonnegative indices is obvious, see Proposition 7.1.2 and its proof.

Lemma 7.4.2.

1. *If $\tilde{\nu} > p$, then*

$$\lambda_0^{(0)} = \frac{i}{2a}\log\frac{\tilde{\nu}+p}{\tilde{\nu}-p}, \quad \lambda_{1+(k-1)p}^{(0)} = \frac{\pi k}{a} + \frac{i}{2a}\log\frac{\tilde{\nu}+p}{\tilde{\nu}-p} \quad \text{for } k \in \mathbb{N}. \quad (7.4.15)$$

$$\lambda_{j+(k-1)p}^{(0)} = \frac{\pi k}{a} \quad \text{for } j = 2,\ldots,p,\ k \in \mathbb{N}. \qquad (7.4.16)$$

2. *If $\tilde{\nu} < p$, then (7.4.16) holds, $\lambda_0^{(0)}$ is absent and instead of (7.4.15) we have*

$$\lambda_{1+(k-1)p}^{(0)} = \frac{\pi(k-\frac{1}{2})}{a} + \frac{i}{2a}\log\frac{p+\tilde{\nu}}{p-\tilde{\nu}} \quad \text{for } k \in \mathbb{N}. \qquad (7.4.17)$$

3. *If $p = \tilde{\nu}$, then the index 0 is absent and*

$$\lambda_{j+(k-1)(p-1)}^{(0)} = \frac{\pi k}{a} \quad \text{for } j = 1,\ldots,p-1,\ k \in \mathbb{N}. \qquad (7.4.18)$$

Lemma 7.4.3. *Let $\tilde{\nu} \neq p$. Then:*

1. *The function $\lambda \mapsto \lambda^{p-1}\phi(\lambda)$ is of sine type pa.*
2. *The zeros $(\lambda_k)_{k=-\infty}^{\infty}$ of ϕ can be indexed properly and satisfy*

$$\lambda_k \underset{|k|\to\infty}{=} \lambda_k^{(0)} + o(1), \qquad (7.4.19)$$

where the index 0 is absent in case $\tilde{\nu} < p$.

3. *There are a nonzero constant C and a nonnegative integer $m \leq p-1$ such that ϕ has the product representation*

$$\phi(\lambda) = C\lambda^m \lim_{n\to\infty}\prod_{k=-n}^{n}{}'\left(1 - \frac{\lambda}{\lambda_k}\right), \qquad (7.4.20)$$

where $\prod'$ means that m factors are omitted from the product in case $\tilde{\nu} > p$ and $m+1$ factors in case $\tilde{\nu} < p$.

Proof. 1. It is easy to see that the function $\lambda \mapsto \lambda^{p-1}\phi_0(\lambda)$ satisfies the conditions (i), (ii) and (iii) of Definition 11.2.5 with exponential type pa, that is, this function is of sine type pa. The representation (7.4.8) of ϕ easily shows that there is a positive constant c such that

$$|\phi(\lambda) - \phi_0(\lambda)| < \frac{c}{|\lambda|^p}e^{|\operatorname{Im}\lambda|pa}$$

for all $\lambda \in \mathbb{C} \setminus \{0\}$, and therefore also $\lambda \mapsto \lambda^{p-1}\phi(\lambda)$ is a sine type function of type pa.

2. Using Remark 11.2.21 or the periodicity of the entire function defined by $\lambda \mapsto \lambda^{p-1}\phi_0(\lambda) = \sin^{p-1}\lambda a(p\cos\lambda a + i\tilde{\nu}\sin\lambda a)$, we find for every $r > 0$ a constant $d > 0$ such that

$$|\phi_0(\lambda)| > \frac{d}{|\lambda|^{p-1}}e^{|\operatorname{Im}\lambda|pa}$$

for all $\lambda \in \mathbb{C} \setminus \bigcup_k C_k$ with $|\lambda| \geq r$, where the C_k are the open discs of radii r with centres at the points $\lambda_k^{(0)}$. Consequently, we have for all $\lambda \in \mathbb{C} \setminus \bigcup_k C_k$ with $|\lambda| > \frac{c}{d}$ that

$$|\phi_0(\lambda)| > |\phi(\lambda) - \phi_0(\lambda)|. \tag{7.4.21}$$

Due to Lemma 7.4.2, for each $k \in \mathbb{N}$, $C_{2+(k-1)p} = C_{3+(k-1)p} = \cdots = C_{kp}$ and hence the disc $C_{2+(k-1)p}$ contains exactly $p-1$ (equal) $\lambda_k^{(0)}$-s for all positive integers k. Taking r sufficiently small we obtain $C_{2+(k-1)p} \cap C_{1+(k'-1)p} = \emptyset$ for all $k, k' \in \mathbb{N}$. With different indexing, an analogous result holds for negative indices k. For sufficiently small $r > 0$ we apply Rouché's theorem to these small C_k for sufficiently large $|k|$ and to a large disc with radius larger than $\frac{c}{d}$ and centre 0 and obtain the assertion 2.

Assertion 3 is an immediate consequence of Lemma 11.2.29, where $m \leq p-1$ has been shown in Theorem 2.3.2. □

For convenience we state the following simple algebraic result, which can be easily proved by induction.

Lemma 7.4.4. *Let $w_1, \ldots, w_n$ and $z_1, \ldots, z_n$ be complex numbers. Then*

$$\prod_{j=1}^{n} w_j - \prod_{j=1}^{n} z_j = \sum_{m=1}^{n}(w_m - z_m)\prod_{j=1}^{m-1} w_j \prod_{j=m+1}^{n} z_j.$$

Lemma 7.4.5. *Let $(a_k)_{k=1}^{\infty}$, $(b_k)_{k=1}^{\infty}$ and $(z_k)_{k=1}^{\infty}$ be sequences of complex numbers such that $(b_k)_{k=1}^{\infty}$ is bounded, $z_k \to 0$ as $k \to \infty$, $z_k = a_k + b_k z_k^2$ and $A := \lim_{k\to\infty} ka_k$ exists. Then $z_k = a_k + O(k^{-2})$.*

Proof. If $a_k = 0$, then $z_k(1 - b_k z_k) = 0$ and therefore $z_k = 0$ except for at most finitely many indices k. Also, if $z_k = 0$, then $a_k = 0$. For these k, the statement of this lemma is trivial, and we may now assume that $z_k \neq 0$ and $a_k \neq 0$ for all k. Then

$$1 = \frac{a_k}{z_k} + b_k z_k$$

and $b_k z_k = o(1)$ imply that

$$\lim_{z \to \infty} k z_k = \lim_{k \to \infty} k a_k \lim_{z \to \infty} \frac{z_k}{a_k} = A.$$

Hence $z_k = O(k^{-1})$. Substituting this for z_k into the right-hand side of the identity $z_k = a_k + b_k z_k^2$ completes the proof. $\square$

Lemma 7.4.6. *Define the polynomial P by*

$$P(M) = \sum_{j=1}^{p} \prod_{\substack{m=1 \\ m \neq j}}^{p} (M - B_m), \qquad (7.4.22)$$

where the B_j, $j = 1, \ldots, p$ are given by (7.4.9). Then the zeros M_j, $j = 1, \ldots, p-1$, of the polynomial P defined by (7.4.22) are real but not necessarily different and will be indexed such that $M_j \leq M_{j+1}$ for $j = 1, \ldots, p - 2$.

Proof. Clearly, the polynomial P is the derivative of the polynomial P_1 defined by

$$P_1(M) = \prod_{m=1}^{p} (M - B_m). \qquad (7.4.23)$$

We assume that the B_m are indexed in such a way that $B_1 \leq B_2 \leq \cdots \leq B_p$. If M is a multiple zero of P_1 of multiplicity r, then M is a zero of multiplicity $r - 1$ of P. Hence the Mean Value Theorem shows that all $p - 1$ zeros $M_1, \ldots, M_{p-1}$ of P are real and can be indexed in such a way that $B_j \leq M_j \leq B_{j+1}$ for $j = 1, \ldots, p - 1$. If $B_j < B_{j+1}$, then $B_j < M_j < B_{j+1}$. $\square$

Theorem 7.4.7. *Let B_j, $j = 1, \ldots, p$ be given by (7.4.9) and let M_j, $j = 1, \ldots, p-1$, be the zeros of the polynomial P defined by (7.4.22). Then the properly indexed sequence $(\lambda_k)_{k=-\infty}^{\infty}$ of the eigenvalues of (7.4.1)–(7.4.4) with $\tilde{\nu} \neq p$ can be represented as the union of p properly indexed subsequences $\left(\rho_k^{(j)} \right)_{k=-\infty}^{\infty}$, $j = 1, \ldots, p$, where the index $k = 0$ is omitted unless $j = p$ and $\tilde{\nu} > p$, and where these subsequences have the following asymptotic expansions: For all $\tilde{\nu} \neq p$,*

$$\rho_k^{(j)} = \frac{\pi k}{a} + \frac{M_j}{k} + \frac{\beta_k^{(j)}}{k}, \qquad j = 1, \ldots, p-1, \ k \in \mathbb{N}. \qquad (7.4.24)$$

For $\tilde{\nu} > p$,

$$\rho_k^{(p)} = \frac{\pi k}{a} + \frac{i}{2a} \log \frac{\tilde{\nu} + p}{\tilde{\nu} - p} + \frac{1}{pk} \left(\sum_{j=1}^{p} B_j + \frac{p^2 \beta}{(p^2 - \tilde{\nu}^2)\pi} \right) + \frac{\beta_k^{(p)}}{k}, \qquad k \in \mathbb{N},$$

$$(7.4.25)$$

For $\tilde{\nu} < p$,

$$\rho_k^{(p)} = \frac{\pi(k-\frac{1}{2})}{a} + \frac{i}{2a}\log\frac{p+\tilde{\nu}}{p-\tilde{\nu}} + \frac{1}{pk}\left(\sum_{j=1}^{p} B_j + \frac{p^2\beta}{(p^2-\tilde{\nu}^2)\pi}\right) + \frac{\beta_k^{(p)}}{k}, \quad k \in \mathbb{N}.$$

(7.4.26)

Here $\left(\beta_k^{(j)}\right)_{k=-\infty,k\neq 0}^{\infty} \in l_2$ if $j = p$ or if $1 \leq j \leq p-1$ and M_j is a simple zero of P, and $\left(\beta_k^{(j)}\right)_{k=-\infty,k\neq 0}^{\infty} \in l_{2+\varepsilon}$ for any $\varepsilon > 0$ otherwise.

Proof. We are going to index $(B_j)_{j=1}^{p}$ and $(M_j)_{j=1}^{p-1}$ as in the proof of Lemma 7.4.6. Furthermore, p_j will denote the multiplicity of the zero M_j of P for each $j = 1, \ldots, p-1$.

We know from Lemmas 7.4.3 and 7.4.2 that all eigenvalues λ_k with $k > 0$ are represented by complex numbers of the form $\rho_k^{(j)} = \lambda_{j+1+(k-1)p}^{(0)} + o(1)$, $j = 1, \ldots, p-1$, $k \in \mathbb{N}$, and $\rho_k^{(p)} = \lambda_{1+(k-1)p}^{(0)} + o(1)$, $k \in \mathbb{N}$. From the representation of $\lambda_k^{(0)}$ it follows that there is $r_0 \in (0, \frac{\pi}{2a})$ such that the closed discs C_k^1 with centre $\lambda_{2+(k-1)p}^{(0)}$ and radius r_0 and the closed discs C_k^p with centre $\lambda_{1+(k-1)p}^{(0)}$ and radius r_0 and $k \in \mathbb{N}$ are mutually disjoint and that C_k^1 contains $p-1$ zeros while C_k^p contains one zero. Hence we still have to show for sufficiently large k that the zeros $\rho_k^{(j)}$, $j = 1, \ldots, p-1$, inside C_k^1 have the representation (7.4.24) and that the zero $\rho_k^{(p)}$ inside C_k^p has the representation (7.4.25) or (7.4.26), respectively.

We will prove (7.4.24) by starting with a sequence of simplified functions, depending on the index k, whose zeros near $\frac{\pi}{a}k$ can be found in terms of the zeros of a polynomial. These functions $\phi_{1,k}$ are defined as

$$\phi_{1,k}(\lambda) = \sum_{j=1}^{p}\prod_{\substack{m=1 \\ m\neq j}}^{p}\left(\tan\lambda a - \frac{aB_m}{k}\right) = \frac{a^{p-1}}{k^{p-1}}P\left(a^{-1}k\tan\lambda a\right),$$

where $k \in \mathbb{N}$ and $|\lambda - \frac{\pi}{a}k| < \frac{\pi}{2a}$.

Setting

$$\rho_k^{(0,m)} = \frac{\pi k}{a} + \frac{1}{a}\arctan\left(\frac{M_m a}{k}\right), \quad k \in \mathbb{N}, \ m = 1, \ldots, p-1,$$

(7.4.27)

we obtain

$$\tan\rho_k^{(0,m)}a = \tan\left(\pi k + \arctan\left(\frac{M_m a}{k}\right)\right) = \frac{M_m a}{k}$$

(7.4.28)

and therefore

$$\phi_{1,k}(\rho_k^{(0,m)}) = \frac{a^{p-1}}{k^{p-1}}P(M_m) = 0, \quad k \in \mathbb{N}, \ m = 1, \ldots, p-1.$$

We observe that the numbers $\rho_k^{(0,m)}$ are zeros of $\phi_{1,k}$ of multiplicity p_m since the derivative of tan has no zeros, so that there is a local one-to-one correspondence between the multiplicities of the zeros of $\phi_{1,k}$ and the zeros of P. Hence $\phi_{1,k}$ has exactly $p-1$ zeros inside C_k^1 for sufficiently large k, counted with multiplicity, and the numbers $\rho_k^{(0,m)}$, $m = 1,\dots,p-1$, defined in (7.4.28) are the zeros of $\phi_{1,k}$.

To prove (7.4.24) let M_ι be any of the zeros of P. In case of a multiple zero we may choose the index ι to be the smallest index j for which M_j equals this zero. We can find $r \in (0,r_0)$ such that

$$2r < \min\{|M_\iota - M_m| : m = 1,\dots,m,\ M_m \neq M_\iota\}.$$

Since M_ι is a zero of the polynomial P of multiplicity p_ι, we can find $\gamma_1 > 0$ such that

$$|P(M)| \geq \gamma_1 |M - M_\iota|^{p_\iota} \text{ if } |M - M_\iota| \leq 2r. \qquad (7.4.29)$$

For $|\lambda - \rho_k^{(0,\iota)}| < \frac{r}{k}$ we conclude in view of (7.4.28) and Taylor's Theorem that

$$|a^{-1}k\tan\lambda a - M_\iota| = a^{-1}k|\tan\lambda a - \tan\rho_k^{(0,\iota)}a| \leq a^{-1}k2a|\lambda - \rho_k^{(0,\iota)}| \leq 2r,$$

$$|a^{-1}k\tan\lambda a - M_\iota| \geq \frac{1}{2}k|\lambda - \rho_k^{(0,\iota)}|$$

if k is sufficiently large. In view of (7.4.29) it follows that there is $\gamma > 0$ such that

$$|\lambda^{p-1}\phi_{1,k}(\lambda)| = a^{p-1}\left(\frac{|\lambda|}{k}\right)^{p-1}|P(a^{-1}k\tan\lambda a)| \geq \gamma\left(k|\lambda - \rho_k^{(0,\iota)}|\right)^{p_\iota} \quad (7.4.30)$$

for $|\lambda - \rho_k^{(0,\iota)}| \leq \frac{r}{k}$ and sufficiently large $k \in \mathbb{N}$.

From (7.4.5)–(7.4.7) we obtain that

$$\frac{\lambda^{2(p-1)}}{\cos^p\lambda a}\phi(\lambda) = \sum_{j=1}^{p}\frac{1}{\cos\lambda a}\left(s_j'(\lambda,a) + \frac{1}{p}(i\lambda\tilde{\nu} + \beta)s_j(\lambda,a)\right)\prod_{\substack{m=1\\m\neq j}}^{p}\frac{\lambda^2}{\cos\lambda a}s_m(\lambda,a),$$

whereas

$$\lambda^{p-1}\phi_{1,k}(\lambda) = \sum_{j=1}^{p}\prod_{\substack{m=1\\m\neq j}}^{p}\left(\lambda\tan\lambda a - \frac{aB_m\lambda}{k}\right).$$

Taking the representations of $s_j(\lambda,a)$ and $s_j'(\lambda,a)$ in (7.4.10) and (7.4.11) into account, it easily follows that all factors of $\lambda^{2(p-1)}(\cos\lambda a)^{-p}\phi(\lambda)$ and $\lambda^{p-1}\phi_{1,k}(\lambda)$, which are of the form

$$\frac{1}{\cos\lambda a}\left(s_j'(\lambda,a) + \frac{1}{p}(i\lambda\tilde{\nu} + \beta)s_j(\lambda,a)\right),\quad \frac{\lambda^2}{\cos\lambda a}s_m(\lambda,a),\quad \lambda\tan\lambda a - \frac{aB_m\lambda}{k},$$

$$(7.4.31)$$

are uniformly bounded with respect to $j, m = 1, \ldots, p$, $|\lambda - \rho_k^{(0,\iota)}| < \frac{r}{k}$ and sufficiently large k. Note that Lemma 12.2.4 shows for $v \in \mathcal{L}^a$ that $|v(\lambda)| \to 0$ as $|\lambda| \to \infty$ subject to $|\lambda - \rho^{(0,\iota)}| \leq \frac{r}{k}$, $k \in \mathbb{N}$. Then the differences of corresponding factors in $\lambda^{2(p-1)}(\cos \lambda a)^{-p}\phi(\lambda)$ and $\lambda^{p-1}\phi_{1,k}(\lambda)$ satisfy

$$\frac{1}{\cos \lambda a}\left(s_j'(\lambda, a) + \frac{1}{p}(i\lambda\tilde{v} + \beta)s_j(\lambda, a)\right) - 1 = \frac{B_j\pi + p^{-1}(i\lambda\tilde{v} + \beta)}{\lambda}\tan \lambda a$$

$$-p^{-1}(i\lambda\tilde{v} + \beta)\frac{B_j\pi}{\lambda^2} + \frac{\lambda\tau_j(\lambda) + p^{-1}(i\lambda\tilde{v} + \beta)\omega_j(\lambda)}{\lambda^2 \cos \lambda a} = O(k^{-1}) \qquad (7.4.32)$$

and

$$\frac{\lambda^2}{\cos \lambda a}s_m(\lambda, a) - \left(\lambda \tan \lambda a - \frac{aB_m\lambda}{k}\right) = O(k^{-2}) + (-1)^k\omega_m(\lambda) = o(1) \quad (7.4.33)$$

for $|\lambda - \rho^{(0,\iota)}| \leq \frac{r}{k}$ and $k \in \mathbb{N}$.

In view of Lemma 7.4.4, the estimates (7.4.31), (7.4.32) and (7.4.33) show that there is a sequence of constant $c_k > 0$ such that $c_k \to 0$ as $k \to \infty$ and

$$\left|\frac{\lambda^{2(p-1)}}{\cos^p \lambda a}\phi(\lambda) - \lambda^{p-1}\phi_{1,k}(\lambda)\right| < c_k \qquad (7.4.34)$$

for $|\lambda - \rho^{(0,\iota)}| \leq \frac{r}{k}$ and $k \in \mathbb{N}$ sufficiently large. It follows from (7.4.30) and (7.4.34) that for sufficiently large k, say $k \geq k_0$, and all λ on the circle with centre $\rho_k^{(0,\iota)}$ and radius $\frac{1}{k}(\frac{c_k}{\gamma})^{\frac{1}{p_\iota}}$ the estimate

$$\left|\frac{\lambda^{p-1}}{\cos^p \lambda a}\phi(\lambda) - \phi_{1,k}(\lambda)\right| < |\phi_{1,k}(\lambda)|$$

holds, and an application of Rouché's Theorem shows that ϕ has p_ι zeros $\rho_k^{(\kappa)}$, $\kappa = \iota, \ldots, \iota + p_\iota - 1$, such that $|\rho_k^{(\kappa)} - \rho_k^{(0,\iota)}| < \frac{1}{k}(\frac{c_k}{\gamma})^{\frac{1}{p_\iota}}$. Hence we have shown that ϕ has zeros of the form

$$\rho_k^{(\kappa)} = \rho_k^{(0,\iota)} + o(k^{-1}), \quad \kappa = \iota, \ldots, \iota + p_\iota - 1,$$

and we can write

$$\rho_k^{(\kappa)} = \frac{\pi}{a}k + \frac{M_\kappa}{k} + \frac{\beta_k^{(\kappa)}}{k}, \quad k \geq k_0, \ \kappa = \iota, \ldots, \iota + p_\iota - 1,$$

where $\beta_k^{(\kappa)} \to 0$ as $k \to \infty$. For each k, these are $p - 1$ zeros, counted with multiplicity, of ϕ satisfying the asymptotics $\frac{\pi}{a}k + o(1)$. Hence, for large k, they account for all zeros of (7.4.1)–(7.4.4) inside the disc C_k^1, as required.

So far we have proved the representation (7.4.24) with $\beta_k^{(j)} \to 0$ as $k \to \infty$. In view of Lemma 12.2.1, applied to the sets of even and odd indices k separately, we have that the sequences $(\omega_j(\rho_k^{(\kappa)}))_{k=k_0}^\infty$ and $(\tau_j(\rho_k^{(\kappa)}))_{k=k_0}^\infty$ belong to l_2 for $j = 1, \ldots, p$. Substituting $\lambda = \rho_k^{(\kappa)}$ into (7.4.32) and (7.4.33) it follows that these terms belong to l_2 if considered as a sequence with index k. Then the sequence $(c_k)_{k=k_0}^\infty$ occurring in (7.4.34) belongs to l_2, and (7.4.34) and $\phi(\rho_k^{(\kappa)}) = 0$ imply

$$\left(\left(\rho_k^{(\kappa)} \right)^{p-1} \phi_{1,k}(\rho_k^{(\kappa)}) \right)_{k=k_0}^\infty \in l_2. \text{ Hence (7.4.30) gives}$$

$$\left(k(\rho_k^{(\kappa)} - \rho_k^{(0,\iota)}) \right)_{k=k_0}^\infty \in l_{2p_\iota} \quad \text{for } \kappa = \iota, \ldots, \iota + p_\iota - 1. \tag{7.4.35}$$

In view of $\frac{1}{a} \arctan\left(\frac{M_\kappa a}{k} \right) = \frac{M_\kappa}{k} + O(k^{-3})$, the representation (7.4.24) follows for all j for which M_j is a simple zero of P.

In order to find the estimate for the remainder term in case $p_\iota > 1$ we need a closer inspection of the terms $\rho_k^{(\kappa)} \tan \rho_k^{(\kappa)} a - \frac{a B_m \rho_k^{(\kappa)}}{k}$ with $\kappa = \iota, \ldots, p_\iota - 1$ and $B_m = M_\iota$. In these cases we have

$$\left| \rho_k^{(\kappa)} \tan \rho_k^{(\kappa)} a - \frac{a B_m \rho_k^{(\kappa)}}{k} \right| = \left| \rho_k^{(\kappa)} \right| \left| \tan \rho_k^{(\kappa)} a - \frac{a B_m}{k} \right|$$

$$= \left(\frac{\pi}{a} k + O(k^{-1}) \right) \left(\frac{\beta_k^{(\kappa)} a}{k} + O(k^{-2}) \right) = \pi \beta_k^{(\kappa)} + O(k^{-1}). \tag{7.4.36}$$

We define the sequence $(t_n)_{n=1}^\infty$ recursively by

$$t_1 = 2p_\iota, \quad \frac{1}{t_{n+1}} = \frac{1}{p_\iota}\left(\frac{1}{2} + \frac{p_\iota - 1}{t_n} \right) \quad \text{for } n \in \mathbb{N}.$$

We are going to show that for all $n \in \mathbb{N}$,

$$t_n > 2, \quad \left(\beta_k^{(\kappa)} \right)_{k=k_0}^\infty \in l_{t_n} \quad \text{for } \kappa = \iota, \ldots, \iota + p_\iota - 1. \tag{7.4.37}$$

Let $n \in \mathbb{N}$ and assume that (7.4.37) holds. From $t_n > 2$ we conclude

$$\frac{1}{t_{n+1}} < \frac{1}{p_\iota}\left(\frac{1}{2} + \frac{p_\iota - 1}{2} \right) = \frac{1}{2},$$

so that $t_{n+1} > 2$. It follows from (7.4.33) and (7.4.36) that

$$\left(\frac{\left(\rho_k^{(\kappa)} \right)^2}{\cos \rho_k^{(\kappa)} a} s_j(\rho_k^{(\kappa)}, a) \right)_{k=k_0}^\infty \in l_{t_n} \quad \text{for } \kappa, j = \iota, \ldots, \iota + p_\iota - 1. \tag{7.4.38}$$

Applying Lemma 7.4.4 to $\lambda^{2(p-1)}(\cos \lambda a)^{-p}\phi(\lambda) - \lambda^{p-1}\phi_{1,k}(\lambda)$ with $\lambda = \rho_k^{(\kappa)}$, we see that each summand is a product of p bounded factors, where one of them satisfies the estimate (7.4.32) or (7.4.33), whereas at least $p_\iota - 1$ of the other factors satisfy (7.4.36) or (7.4.38). It follows that the absolute value of any of these summands has an upper bound of the form $c_{1,k}c_{2,k}\cdots c_{p_\iota,k}$ with $(c_{1,k})_{k=k_0}^\infty \in l_2$ and $(c_{j,k})_{k=k_0}^\infty \in l_{t_n}$ for $j = 2, \ldots, p_\iota$. From

$$(c_{1,k}c_{2,k}\cdots c_{p_\iota,k})^{\frac{t_{n+1}}{p_\iota}} = \left(c_{1,k}^2\right)^{\frac{t_{n+1}}{2p_\iota}} \left(c_{2,k}^{t_n}\right)^{\frac{t_{n+1}}{t_n p_\iota}} \cdots \left(c_{p_\iota,k}^{t_n}\right)^{\frac{t_{n+1}}{t_n p_\iota}},$$

$$\left(c_{1,k}^2\right)_{k=k_0}^\infty, \left(c_{2,k}^{t_n}\right)_{k=k_0}^\infty, \ldots, \left(c_{p_\iota,k}^{t_n}\right)_{k=k_0}^\infty \in l_1, \quad \text{and} \quad \frac{t_{n+1}}{2p_\iota} + (p_\iota - 1)\frac{t_{n+1}}{t_n p_\iota} = 1$$

it follows in view of the generalized Hölder's inequality, see [108, (13.26)], that

$$(d_k)_{k=k_0}^\infty := \left(\left(c_{1,k}c_{2,k}\cdots c_{p_\iota-1,k}\right)^{\frac{1}{p_\iota}}\right)_{k=k_0}^\infty \in l_{t_{n+1}}.$$

Hence we find finitely many sequences $(d_{1,k})_{k=k_0}^\infty, \ldots, (d_{h,k})_{k=k_0}^\infty \in l_{t_{n+1}}$ with positive terms so that (7.4.34) for $\lambda = \rho_k^{(\kappa)}$ can be improved to

$$\left|\left(\rho_k^{(\kappa)}\right)^{p-1}\phi_{1,k}(\rho_k^{(\kappa)})\right| \leq \sum_{j=0}^h d_{j,k}^{p_\iota} \leq \left(\sum_{j=0}^h d_{j,k}\right)^{p_\iota} =: \tilde{d}_k$$

with $(\tilde{d}_k)_{k=k_0}^\infty \in l_{t_{n+1}}$. From (7.4.30) we conclude $\left(k(\rho_k^{(\kappa)} - \rho_k^{(0,\iota)})\right)_{k=k_0}^\infty \in l_{t_{n+1}}$ and therefore $\left(\beta_k^{(\kappa)}\right)_{k=k_0}^\infty \in l_{t_{n+1}}$. By the principle of mathematical induction, it follows that (7.4.37) holds for all $n \in \mathbb{N}$.

Hence the proof of the representation (7.4.24) will be complete if we show that $t_n \to 2$ as $n \to \infty$. From

$$\frac{t_n}{t_{n+1}} = \frac{1}{p_\iota}\left(\frac{t_n}{2} + p_\iota - 1\right) > 1$$

we see that $(t_n)_{n=1}^\infty$ is a decreasing sequence, which therefore has a limit $t \geq 2$. This limit t satisfies

$$\frac{1}{t} = \frac{1}{p_\iota}\left(\frac{1}{2} + \frac{p_\iota - 1}{t}\right),$$

which shows that $t = 2$.

Now we are going to prove (7.4.25) and (7.4.26). With the notation from Proposition 7.1.2 we can write $\lambda_{1+(k-1)p}^{(0)}$ as $\lambda_k^{(1)}$ with $\sigma = p$ and $\alpha = \tilde{\nu}$ and with an index shift in case $\tilde{\nu} > p$. We already know from Lemma 7.4.3 that $\rho_k^{(p)} - \lambda_k^{(1)} = o(1)$ and from (7.1.7) and (7.1.8) that

$$\sin\lambda_k^{(1)}a = p\varepsilon_k, \quad \cos\lambda_k^{(1)}a = -i\tilde{\nu}\varepsilon_k,$$

where $\varepsilon_k^2 \neq 0$ is independent of k.

By Taylor's theorem,

$$-\phi(\lambda_k^{(1)}) = \phi(\rho_k^{(p)}) - \phi(\lambda_k^{(1)}) = \phi'(\lambda_k^{(1)})(\rho_k^{(p)} - \lambda_k^{(1)}) + \gamma_k(\rho_k^{(p)} - \lambda_k^{(1)})^2, \quad (7.4.39)$$

where

$$|\gamma_k| \le \frac{1}{2}\sup\{|\phi''(\lambda)| : \lambda \in [\lambda_k^{(1)}, \rho_k^{(p)}]\}.$$

It is easy to see from Lemma 12.2.4 that the derivative of a function in $\mathcal{L}^a$ also belongs to $\mathcal{L}^a$. Hence it follows from (7.4.12) and (7.4.13) that

$$\gamma_k = O(k^{-(p-1)}).$$

Furthermore,

$$\varphi'(\lambda_k^{(1)}) = \frac{ap}{(\lambda_k^{(1)})^{p-1}}\left((p-1)\sin^{p-2}\lambda_k^{(1)}a\cos^2\lambda_k^{(1)}a - \sin^p\lambda_k^{(1)}a\right) + O((\lambda_k^{(1)})^{-p})$$

and

$$\chi'(\lambda_k^{(1)}) = \frac{ap}{(\lambda_k^{(1)})^p}\sin^{p-1}\lambda_k^{(1)}a\cos\lambda_k^{(1)}a + O((\lambda_k^{(1)})^{-(p+1)})$$

show that

$$\phi'(\lambda_k^{(1)}) = \frac{ap\sin^{p-2}\lambda_k^{(1)}a}{(\lambda_k^{(1)})^{p-1}}\left((p-1)\cos^2\lambda_k^{(1)}a - \sin^2\lambda_k^{(1)}a + i\tilde{\nu}\sin\lambda_k^{(1)}a\cos\lambda_k^{(1)}a\right)$$
$$+ O((\lambda_k^{(1)})^{-p})$$
$$= \frac{ap\sin^{p-2}\lambda_k^{(1)}a}{(\lambda_k^{(1)})^{p-1}}\varepsilon_k^2\left(\tilde{\nu}^2 - p^2\right) + O((\lambda_k^{(1)})^{-p}).$$

Next we calculate

$$\phi(\lambda_k^{(1)}) = \phi_0(\lambda_k^{(1)}) - (p-1)\sum_{j=1}^{p}B_j\pi\frac{\cos^2\lambda_k^{(1)}a\sin^{p-2}\lambda_k^{(1)}a}{(\lambda_k^{(1)})^p} + \sum_{j=1}^{p}B_j\pi\frac{\sin^p\lambda_k^{(1)}a}{(\lambda_k^{(1)})^p}$$
$$+ \beta\frac{\sin^p\lambda_k^{(1)}a}{(\lambda_k^{(1)})^p} - i\tilde{\nu}\sum_{j=1}^{p}B_j\pi\frac{\sin^{p-1}\lambda_k^{(1)}a\cos\lambda_k^{(1)}a}{(\lambda_k^{(1)})^p} + \frac{v_1(\lambda_k^{(1)}) + i\tilde{\nu}v_2(\lambda_k^{(1)})}{(\lambda_k^{(1)})^p}$$
$$+ O((\lambda_k^{(1)})^{-(p+1)})$$
$$= -\frac{\sin^{p-2}\lambda_k^{(1)}a}{(\lambda_k^{(1)})^p}\varepsilon_k^2\left[(\tilde{\nu}^2 - p^2)\sum_{j=1}^{p}B_j\pi - p^2\beta\right] + \frac{\alpha_k^{(p)}}{k},$$

where $(\alpha_k^{(p)})_{k=1}^{\infty} \in l_2$ in view of Lemma 12.2.1.

From the above calculations and with γ_k from (7.4.39) we infer that

$$b_k := -\frac{\gamma_k}{\phi'(\lambda_k^{(1)})} = O(1), \tag{7.4.40}$$

$$a_k := -\frac{\phi(\lambda_k^{(1)})}{\phi'(\lambda_k^{(1)})} = \frac{1}{ap\lambda_k^{(1)}}\left(\sum_{j=1}^p B_j \pi + \frac{p^2 \beta}{p^2 - \tilde{\nu}^2}\right) + \frac{\tilde{\beta}_k}{k}$$

$$= \frac{1}{pk}\left(\sum_{j=1}^p B_j + \frac{p^2 \beta}{(p^2 - \tilde{\nu}^2)\pi}\right) + \frac{\hat{\beta}_k}{k}, \tag{7.4.41}$$

where $(\tilde{\beta}_k)_{k=1}^\infty$ and $(\hat{\beta}_k)_{k=1}^\infty$ belong to l_2. Hence (7.4.39) can be written in the form $z_k = a_k + b_k z_k^2$ with $z_k = \rho_k^{(p)} - \lambda_k^{(1)}$, and Lemma 7.4.5 gives

$$\rho_k^{(p)} - \lambda_k^{(1)} = \frac{1}{pk}\left(\sum_{j=1}^p B_j + \frac{p^2 \beta}{(p^2 - \tilde{\nu}^2)\pi}\right) + \frac{\beta_k^{(p)}}{k}, \tag{7.4.42}$$

where $(\beta_k^{(p)})_{k=1}^\infty \in l_2$. This proves (7.4.25) and (7.4.26). $\qquad\square$

Corollary 7.4.8. *Let $\tilde{\nu} = 0$, let q_j belong to the Sobolev space $W_2^1(0,a)$ for $j = 1,\ldots,p$ and assume that the zeros of the polynomial P defined in Theorem 7.4.7 are simple. Then the subsequences in Theorem 7.4.7 have the asymptotic expansions*

$$\rho_k^{(j)} = \frac{\pi k}{a} + \frac{M_j}{k} + \frac{\beta_k^{(j)}}{k^2}, \quad j = 1,\ldots,p-1, \tag{7.4.43}$$

$$\rho_k^{(p)} = \frac{\pi(k - \frac{1}{2})}{a} + \frac{1}{p\left(k - \frac{1}{2}\right)}\left(\sum_{j=1}^p B_j + \frac{\beta}{\pi}\right) + \frac{\beta_k^{(p)}}{k^2}, \tag{7.4.44}$$

where $\left(\beta_k^{(j)}\right)_{k=-\infty, k \neq 0}^\infty \in l_2$.

Proof. Because of $\tilde{\nu} = 0$, the estimate (7.4.32) can be sharpened to $O(k^{-2})$. For (7.4.33) we simply observe that Corollary 12.2.10 shows that $\omega_j(\lambda)$ in (7.4.10) can be written as

$$\omega_j(\lambda) = D_j \frac{\sin \lambda a}{\lambda} + \frac{\omega_j^{(1)}}{\lambda}, \quad j = 1,\ldots,p,$$

where the D_j are real constants and $(\omega_j^{(1)})_{j=1}^\infty \in l_2$. Since $\sin \lambda a = O(k^{-1})$ for the λ considered in the part of the proof of Theorem 7.4.7 corresponding to (7.4.24), we see that the differences (7.4.33) are of the form $O(k^{-2}) + k^{-1}\omega(\lambda)$ with $\omega \in \mathcal{L}^a$. Then we can replace c_k with $k^{-1}c_k$ in (7.4.34), and (7.4.43) easily follows from this modification in the proof of Theorem 7.4.7.

To prove (7.4.44) we first observe that going one step further in the Taylor expansion of ϕ we find that γ_k defined in (7.4.39) can be written as

$$\gamma_k = \phi''(\lambda_k^{(1)}) + \gamma_k^{(1)}(\rho_k^{(p)} - \lambda_k^{(1)}),$$

where $\gamma_k^{(1)} = O((\lambda_k^{(1)})^{-(p-1)})$. Note that $\cos \lambda_k^{(1)} a = 0$ and $\sin \lambda_k^{(1)} a = (-1)^{k-1}$. It is easy to see that $\phi''(\lambda_k^{(1)}) = O((\lambda_k^{(1)})^{-p})$, and therefore $\gamma_k = O((\lambda_k^{(1)})^{-p})$. It follows that (7.4.40) can be improved to

$$b_k = O(k^{-1}).$$

Substituting the above representation of ω_j into (7.4.10) gives

$$s_j(\lambda_k^{(1)}, a) = \frac{(-1)^{k-1}}{\lambda_k^{(1)}} + O((\lambda_k^{(1)})^{-3}),$$

$$\frac{\partial}{\partial \lambda} s_j(\lambda_k^{(1)}, a) = O((\lambda_k^{(1)})^{-2}).$$

Similarly, $\tau_j(\lambda)$ in (7.4.11) can be written as

$$\tau_j(\lambda) = E_j \frac{\cos \lambda a}{\lambda} + \frac{\tau_j^{(1)}}{\lambda}, \ j = 1, \ldots, p,$$

where the E_j are real constants and $\tau_j^{(1)} \in l_2$. Thus we have

$$s_j'(\lambda_k^{(1)}, a) = B_j \pi \frac{(-1)^{k-1}}{\lambda_k^{(1)}} + \frac{\tau_j(\lambda_k^{(1)})}{(\lambda_k^{(1)})^2},$$

$$\frac{\partial}{\partial \lambda} s_j'(\lambda_k^{(1)}, a) = -(-1)^{k-1} a + O((\lambda_k^{(1)})^{-2}).$$

We calculate

$$\varphi(\lambda_k^{(1)}) = \frac{(-1)^{p(k-1)}}{(\lambda_k^{(1)})^p} \sum_{j=1}^{p} B_j \pi + \frac{\omega^{(1)}(\lambda_k^{(1)})}{(\lambda_k^{(1)})^{p+1}} + O((\lambda_k^{(1)})^{-(p+2)}),$$

$$\chi(\lambda_k^{(1)}) = \frac{(-1)^{p(k-1)}}{(\lambda_k^{(1)})^p} + O((\lambda_k^{(1)})^{-(p+2)}),$$

where $\omega^{(1)} \in \mathcal{L}^{pa}$. Therefore

$$\phi(\lambda_k^{(1)}) = \frac{(-1)^{p(k-1)}}{(\lambda_k^{(1)})^p} \left(\sum_{j=1}^{p} B_j \pi + \beta \right) + \frac{\omega^{(1)}(\lambda_k^{(1)})}{(\lambda_k^{(1)})^{p+1}} + O((\lambda_k^{(1)})^{-(p+2)}).$$

Furthermore,

$$\varphi'(\lambda_k^{(1)}) = -\frac{(-1)^{p(k-1)}}{(\lambda_k^{(1)})^{(p-1)}} pa + O((\lambda_k^{(1)})^{-(p+1)}),$$

$$\chi'(\lambda_k^{(1)}) = O((\lambda_k^{(1)})^{-(p+1)}),$$

gives

$$\phi'(\lambda_k^{(1)}) = -\frac{(-1)^{p(k-1)}}{(\lambda_k^{(1)})^{(p-1)}} pa + O((\lambda_k^{(1)})^{-(p+1)}).$$

Observing the definition of a_k in (7.4.41) we conclude that

$$\rho_k^{(p)} = \lambda_k^{(1)} + a_k + O(k^{-3}) = \lambda_k^{(1)} + \frac{1}{pa\lambda_k^{(1)}} \left(\sum_{j=1}^{p} B_j \pi + \beta \right) + \frac{\beta_k^{(p)}}{k^2}$$

$$= \lambda_k^{(1)} + \frac{1}{p(k-\frac{1}{2})} \left(\sum_{j=1}^{p} B_j + \frac{\beta}{\pi} \right) + \frac{\beta_k^{(p)}}{k^2},$$

where $\left(\beta_k^{(j)} \right)_{k=-\infty, k\neq 0}^{\infty} \in l_2$.   □

7.5 Notes

First results on asymptotics of eigenvalues of boundary value problems can be found in [257], [258]. Since then many results concerning eigenvalue asymptotics for concrete boundary value problems generated by the Sturm–Liouville equation and similar equations have been published, see, e. g., [17], [84] and the references therein.

   Asymptotics for the eigenvalues of the generalized Regge problem (7.2.5)–(7.2.7) with $\alpha \neq 1$ were obtained in [101]. However, asymptotics of the classical Regge problem, that is, the case when $\alpha = 1$ and $\beta = 0$, being investigated for a long time are not completely known. An analogue of formula (7.2.23) under the assumption that $q^{(j)}(a) = 0$ for $j = 1, \ldots, p-1$ and $q^{(p)}(a) \neq 0$ was obtained in the original paper by T. Regge [238, (19)], see also [150, (6)]. Under the above assumption on the behaviour of the potential at a, [150] also obtained two-fold series expansions into eigenfunctions and associated functions of the Regge problem for certain classes of sufficiently smooth functions. It was shown in [142] that the condition that the potential q satisfies $a \in \operatorname{supp} q$ is necessary for two-fold completeness in $L_2(0, a)$. Also sufficient conditions for two-fold completeness were given in [142]. More recently, it was shown in [248, Theorem 3] that under the assumption that $a \in \operatorname{supp} q$ and that all of its eigenvalues λ_n, $n \in \mathbb{Z}$ are simple with eigenfunctions y_n, then the system $\{(y_n, \lambda_n y_n) : n \in \mathbb{Z}\}$ is complete and minimal in $W_{2,U}^1(0, a) \times L_2(0, 1)$, where $W_{2,U}^1(0, a)$ is the space of functions y from

$W_2^1(0, a)$ satisfying $f(0) = 0$. For the case when $a \in \text{supp}\, q$ and $q^{(k)}(a) = 0$ for all $k \in \mathbb{N}_0$, asymptotics of eigenvalues are unknown in general. However, if additionally $|q(x)\,(\exp((a - x)^{-\alpha})|$ is bounded and bounded away from 0 as $x \to a - 0$ for some $\alpha > 0$, an asymptotic formula for the eigenvalues can be found in [243, III.4].

A double-sided Regge problem was considered in [234] and [287].

The undamped string problem is given by equation (2.8.2). For the main term of the asymptotics of the eigenvalues of boundary value problems generated by (2.8.2), the following formula was obtained by M.G. Kreĭn for a regular string in [152], see also [96, Chapter VI, (8.5)]:

$$\lim_{n \to \infty} \frac{n}{\lambda_n} = \frac{1}{\pi} \int_0^a \sqrt{\frac{dM}{dx}}\, dx. \tag{7.5.1}$$

This formula is independent of the boundary conditions, at least if they are self-adjoint and independent of the spectral parameter. We observe that if M is absolutely continuous with $M' \in L_2(0, a)$, then (2.8.2) is the undamped string equation (2.2.1) with $\sigma = 0$, i.e.,

$$v'' + \lambda^2 \rho v = 0. \tag{7.5.2}$$

For continuous ρ, formula (7.5.1) was known to J. Liouville [179, pp. 420, 426].

Another general result on asymptotics for the eigenvalues of string problems is the Barcilon formula

$$\rho(0) = \frac{1}{a^2 \mu_1^2} \prod_{k=1}^{\infty} \frac{\nu_k^4}{\mu_{k+1}^2 \mu_k^2}$$

associated with (7.5.2), where the string density ρ is continuous and bounded away from zero, $(\nu_k)_{k \in \mathbb{N}}$ are the eigenvalues of the Dirichlet problem and $(\mu_k)_{k \in \mathbb{N}}$ are the eigenvalues of the Neumann–Dirichlet problem. This formula was obtained in [22] by the method of Stieltjes continued fractions and proved by [244] using the Liouville transform under the assumption that the density of the string has a piecewise continuous derivative. For $\rho = 1$, the Barcilon formula reduces to Wallis' formula. In [129, Theorem 4.4] it was shown that this formula remains true even at least in some cases of singular strings if $\rho(0)$ is replaced by $\lim_{x \to +0} \frac{M(x)}{x}$.

For the free Laplacian on finite quantum graphs, that is $\rho = 1$ on each edge, the formula corresponding to (7.5.1) is

$$\lim_{n \to \infty} \frac{n}{\lambda_n} = \frac{1}{\pi} \sum_{k=1}^{g} l_k, \tag{7.5.3}$$

where l_k $(k = 1, 2, \ldots, g)$ are the lengths of the edges of the graph, see [83, Proposition 3.3].

Eigenvalue asymptotics for star graphs were found in [215, Lemma 1.3]. Theorem 7.4.7 on the eigenvalue asymptotics for problem (7.4.1)–(7.4.4) was obtained in [230, Theorem 2.7 and Remark 2.8] with a slightly weaker result on the remainder terms if not all of the M_j, $j = 1, \ldots, p-1$, are distinct.

Eigenvalue asymptotics for quantum star graphs can be found in [279, Theorems 2.1 and 2.2]. Eigenvalue asymptotics for boundary value problems generated by the Dirac equation on star graphs were obtained in [276, Theorems 2.1 and 2.2].

For finite connected graphs, the main term of the eigenvalue asymptotics can be obtained from (7.5.3). The next terms depend on complicated combinations of $\int_0^{l_j} q_j(x)\, dx$, $j = 1, \ldots, g$, where g is the number of edges. Some results can be found in [47, Theorem 5.4], [62, Theorem 7.9 and Corollary 7.10], [279, Theorems 2.1 and 2.2]. However, in case all edges have the same length and the same potentials on them, symmetric with respect to the midpoint of the edge, there is a connection between spectral theory of quantum graphs and the classical spectral theory of graphs; for the classical spectral theory on graphs see, e. g., [63], [53]. This connection was pointed out by [11] and was proved in [48], [76]. In this case it was shown in [225, p. 193] that the characteristic function for problems with Neumann boundary conditions at pendant vertices is of the form

$$\phi(\lambda) = s^{p-g}(\lambda, a) \prod_{r=1}^{p} (c(\lambda, a) - \alpha_r)$$

where α_r are the eigenvalues of the matrix $\tilde{A} = B^{-\frac{1}{2}} A B^{-\frac{1}{2}}$. Here A is the adjacency matrix of the graph, $B = \mathrm{diag}\{d(v_1), \ldots, d(v_p)\}$, $d(v_i)$ is the degree of the vertex v_i, g is the number of the edges and p is the number of vertices of the graph, $s(\cdot, a)$ is the characteristic function of the Dirichlet–Dirichlet problem on the edges while $c(\cdot, a)$ is the characteristic problem of the Neumann–Dirichlet problem on the edge. If $p > g$, then the zeros of the factor $s^{p-g}(\cdot, a)$ give the Dirichlet spectrum, that is, the set of eigenvalues of the Dirichlet–Dirichlet problem on an edge of this graph, each eigenvalue having multiplicity $p - g$.

For asymptotics of eigenvalues of boundary value problems generated by a fourth-order differential equations and spectral parameter dependent boundary conditions see [195], [197], [198], [199].

An interesting and important question is whether the spectrum of a boundary value problem lies in the open upper half-plane and is separated from the real axis. In other words, whether a positive constant ε exists such that $\mathrm{Im}\, \lambda_k \geq \varepsilon$ for all eigenvalues λ_k. It is important because if the spectrum is separated and the eigenvectors and associated vectors form a two-fold Riesz basis, then the solution of the corresponding initial-boundary-value problem exists and is decaying exponentially, see [272], [274]. It was shown in [104, Theorem 6.2] that the spectrum of a string with constant density and damping at both ends is separated from the real axis even if there is point mass at an interior point, while in [104, Theorem 6.3] it was shown that this is not true with one fixed end, one damped end and a point mass at an interior point. However, for a string with distributed damping the spectrum is separated from the real axis, see [60, Corollary 5.4]. In [213, Theorem

3.1] it was shown that if the string satisfies $\rho \in W_2^2(0, a)$ and if the string bears a point mass at the right end, then the spectrum is not separated from the real axis. See [249, Theorem 2.5] and [190, Theorem 2.1] for related problems.

The problem of optimal resonances, i. e., the problem of design of optical cavities with minimal rate of energy decay for a given frequency is closely connected with the distance between the spectrum and the real axis. For related results see [134, (3.2)], [135, Proposition 2.3].

Chapter 8

Inverse Problems

8.1 The inverse generalized Regge problem

In this section we will consider the inverse problem corresponding to the generalized Regge problem, which was introduced at the end of Section 2.1 and whose eigenvalues were discussed in Sections 6.1 and 7.2:

$$-y'' + q(x)y = \lambda^2 y, \qquad (8.1.1)$$
$$y(\lambda, 0) = 0, \qquad (8.1.2)$$
$$y'(\lambda, a) + (i\lambda\alpha + \beta)\, y(\lambda, a) = 0. \qquad (8.1.3)$$

Definition 8.1.1. The classes $\mathcal{B}$ ($\mathcal{B}^+$, $\mathcal{B}^-$) are the sets of 4-tuples (a, q, α, β), where $a > 0$, $q \in L_2(0, a)$ is real valued, $\beta \in \mathbb{R}$, $\alpha \in \mathbb{R}$ ($\alpha \in (1, \infty)$, $\alpha \in (0, 1)$).

Definition 8.1.2. Let $\kappa \in \mathbb{N}_0$. Then the properly indexed sequence $(\lambda_k)_{k=-\infty}^{\infty}$ (or $(\lambda_k)_{k=-\infty, k\neq 0}^{\infty}$) is said to have the SHB$_\kappa^+$ (respectively SHB$_\kappa^-$) property if the following conditions are satisfied, where *term* means term of this sequence.

 (i) All but κ terms lie in the open upper half-plane.
 (ii) All terms in the closed lower half-plane are pure imaginary and pairwise different. If $\kappa > 0$, we denote them as $\lambda_{-j} = -i|\lambda_{-j}|$, $j = 1, \ldots, \kappa$, and assume that $|\lambda_{-j}| < |\lambda_{-(j+1)}|$, $j = 1, \ldots, \kappa - 1$, if $\kappa > 1$.
 (iii) If $\kappa > 0$, the numbers $i|\lambda_{-j}|$, $j = 1, \ldots, \kappa$, with the exception of λ_{-1} if it is equal to zero, are not terms of the sequence.
 (iv) If $\kappa > 1$, then there is an odd number of terms in the interval $(i|\lambda_{-j}|, i|\lambda_{-(j+1)}|)$, $j = 1, \ldots, \kappa - 1$.
 (v) If $\kappa > 0$ and $|\lambda_{-1}| > 0$, then the interval $(0, i|\lambda_{-1}|)$ contains no terms at all or an even number of terms.
 (vi) If $\kappa > 0$, then the interval $(i|\lambda_{-\kappa}|, i\infty)$ contains a positive even number of terms in case SHB$_\kappa^+$ or an odd number of terms in case SHB$_\kappa^-$.

(vii) If $\kappa = 0$, then the sequence has an odd number of positive imaginary terms in case SHB_κ^+ or no or an even number of positive imaginary terms in case SHB_κ^-.

We start with a result which covers all $\alpha > 0$ but where the eigenvalues of (8.1.1)–(8.1.3) are only given implicitly as the zeros of an analytic function. Unfortunately, in the exceptional case $\alpha = 1$ we are unable to obtain an explicit form of the inverse problem. In case $\alpha \neq 1$, the explicit forms of the inverse problem are investigated in Theorems 8.1.4 and 8.1.5.

Theorem 8.1.3. *For the sequence* (λ_k)*, which can be infinite or finite and possibly empty, to be the sequence of the eigenvalues of problem* (8.1.1)–(8.1.3) *with* $(a, q, \alpha, \beta) \in \mathcal{B}$ *and with* $\alpha > 0$*, it is necessary and sufficient that* (λ_k) *be the sequence of the zeros of an entire function* χ *which belongs to the class* SSHB *and which is of the form*

$$\chi(\lambda) = \cos \lambda a + i\alpha \sin \lambda a + M \frac{\sin \lambda a}{\lambda} + iN \frac{\cos \lambda a}{\lambda} + \frac{\psi(\lambda)}{\lambda}, \ \lambda \in \mathbb{C} \setminus \{0\}, \quad (8.1.4)$$

where $M, N \in \mathbb{R}$ *and* $\psi \in \mathcal{L}^a$*.*

Proof. Let ϕ be the characteristic function of (8.1.1)–(8.1.3) given by (6.1.1). By Proposition 6.1.2, the function ϕ belongs to the class SSHB and satisfies (8.1.4) in view of Corollary 12.2.10.

Conversely, assume that χ belongs to the class SSHB and satisfies (8.1.4). The numbers $a > 0$ and $\alpha > 0$ are explicitly given in (8.1.4). We recall from Definition 5.2.6 that we can write the shifted Hermite–Biehler function χ in the form

$$\chi(\lambda) = \Phi_1(\lambda^2) + i\lambda \Phi_2(\lambda^2), \ \lambda \in \mathbb{C}, \quad (8.1.5)$$

with real entire functions Φ_1 and Φ_2. The representation (8.1.4) of χ shows that

$$\Phi_1(\lambda^2) = \cos \lambda a + M \frac{\sin \lambda a}{\lambda} + \frac{\tilde{\psi}_1(\lambda)}{\lambda}, \quad (8.1.6)$$

$$\Phi_2(\lambda^2) = \alpha \frac{\sin \lambda a}{\lambda} + N \frac{\cos \lambda a}{\lambda^2} + \frac{\tilde{\psi}_2(\lambda)}{\lambda^2}, \quad (8.1.7)$$

where $\tilde{\psi}_1 \in \mathcal{L}_o^a$ and $\tilde{\psi}_2 \in \mathcal{L}_e^a$. It follows from (8.1.6) and (8.1.7) that

$$\Phi_1(\lambda^2) \underset{\lambda \to \pm i\infty}{=} \frac{e^{|\lambda|a}}{2} + o\left(e^{|\lambda|a}\right), \quad \Phi_2(\lambda^2) \underset{\lambda \to \pm i\infty}{=} \alpha \frac{e^{|\lambda|a}}{2|\lambda|} + o\left(\frac{e^{|\lambda|a}}{|\lambda|}\right). \quad (8.1.8)$$

We shall prove that the sequence of the zeros of χ is the sequence of the eigenvalues of problem (8.1.1)–(8.1.3) with

$$\beta = M + \alpha^{-1}N, \quad (8.1.9)$$

and with some real-valued function $q \in L_2(0, a)$. To this end we consider the entire functions Ξ_1 and Ξ_2 defined by

$$\Xi_1 = \Phi_1 - \alpha^{-1}\beta\Phi_2, \tag{8.1.10}$$

$$\Xi_2 = \alpha^{-1}\Phi_2. \tag{8.1.11}$$

In view of (8.1.6), (8.1.7), (8.1.10) and (8.1.11) we have

$$\Xi_1(\lambda^2) = \cos \lambda a + (M - \beta)\frac{\sin \lambda a}{\lambda} + \frac{\psi_1(\lambda)}{\lambda}, \tag{8.1.12}$$

$$\Xi_2(\lambda^2) = \frac{\sin \lambda a}{\lambda} + \frac{N}{\alpha\lambda^2}\cos \lambda a + \frac{\psi_{2,0}(\lambda)}{\lambda^2}, \tag{8.1.13}$$

where $\psi_1 \in \mathcal{L}_o^a$ and $\psi_{2,0} \in \mathcal{L}_e^a$. We define

$$\psi_{2,1}(\lambda) = \frac{N}{\alpha}\cos \lambda a - \frac{4a^2}{\alpha}N\frac{\lambda^2 \cos \lambda a}{4\lambda^2 a^2 - \pi^2}.$$

Clearly, $\psi_{2,1}$ is an even entire function of exponential type $\leq a$, and for $x \in \mathbb{R}$, $|\psi_{2,1}(x)| = O(x^{-2})$ as $|x| \to \infty$. Therefore $\psi_{2,1} \in \mathcal{L}_e^a$, and we have

$$\Xi_2(\lambda^2) = \frac{\sin \lambda a}{\lambda} + \frac{4a^2}{\alpha}N\frac{\cos \lambda a}{4\lambda^2 a^2 - \pi^2} + \frac{\psi_2(\lambda)}{\lambda^2}, \tag{8.1.14}$$

where $\psi_2 = \psi_{2,0} + \psi_{2,1} \in \mathcal{L}_e^a$. Since Ξ_2 is an entire function, multiplication of (8.1.14) by λ^2 and evaluation at $\lambda = 0$ gives $\psi_2(0) = 0$.

Since χ belongs to the class SSHB, $\frac{\Phi_2}{\Phi_1}$ belongs to $\mathcal{N}_+^{ep}$ and Φ_1 and Φ_2 do not have common zeros by Definition 5.2.6. In view of Lemma 5.1.22, the function

$$\frac{\Xi_2}{\Xi_1} = \left(\frac{\Phi_1 - \alpha^{-1}\beta\Phi_2}{\alpha^{-1}\Phi_2}\right)^{-1} = \alpha^{-1}\left(\frac{\Phi_1}{\Phi_2} - \alpha^{-1}\beta\right)^{-1}$$

is a Nevanlinna function. By (8.1.8), $\frac{\Phi_1(\lambda)}{\Phi_2(\lambda)} \to \infty$ as $\lambda \to -\infty$, so that there is $\gamma \in \mathbb{R}$ such that $\frac{\Xi_2(\lambda)}{\Xi_1(\lambda)} > 0$ if $\lambda < \gamma$. This shows that the meromorphic function $\frac{\Xi_2}{\Xi_1}$ belongs to $\mathcal{N}_+^{ep}$.

There are various ways to ascertain that the functions Ξ_1 and Ξ_2 have infinitely many zeros. One way to do so is to observe that $\lambda \mapsto \Xi_1(\lambda^2)$ and $\lambda \mapsto \lambda\Xi_2(\lambda^2)$ are sine type functions and then to apply Proposition 11.2.8, 1. Since Φ_1 and Φ_2 do not have common zeros, also Ξ_1 and Ξ_2 have no common zeros. Because $\frac{\Xi_2}{\Xi_1}$ belongs to $\mathcal{N}_+^{ep}$, we know from Theorem 11.1.6 and Corollary 5.2.3 that the zeros $(\mu_k)_{k=1}^\infty$ of Ξ_1 interlace with the zeros $(\nu_k)_{k=1}^\infty$ of Ξ_2 in the following way:

$$\mu_1 < \nu_1 < \mu_2 < \nu_2 < \cdots. \tag{8.1.15}$$

By (8.1.12) and (8.1.14), the functions $\lambda \mapsto \Xi_1(\lambda^2)$ and $\lambda \mapsto \Xi_2(\lambda^2)$ are of the form (12.3.11) and (12.3.10), respectively, where

$$A = -\frac{aN}{\pi^2\alpha}, \quad B = \frac{(M - \beta)a}{\pi^2} = -\frac{aN}{\pi^2\alpha}.$$

Hence Lemma 12.3.3 shows that $\mu_k = v_k^2$ and $\nu_k = u_k^2$, $k \in \mathbb{N}$, with

$$v_k = \frac{\pi}{a}\left(k - \frac{1}{2}\right) - \frac{N}{\pi\alpha}\frac{1}{k} + \frac{\gamma_k^{(1)}}{k}, \qquad (8.1.16)$$

$$u_k = \frac{\pi}{a}k - \frac{N}{\pi\alpha}\frac{1}{k} + \frac{\gamma_k^{(2)}}{k}, \qquad (8.1.17)$$

where $(\gamma_k^{(j)})_{k=1}^{\infty} \in l_2$ for $j = 1, 2$. Thus the sequences $(\mu_k)_{k=1}^{\infty}$ and $(\nu_k)_{k=1}^{\infty}$ satisfy the assumptions of Theorem 12.6.2. Therefore there exists a real-valued $q \in L_2(0, a)$ such that $(\mu_k)_{k=1}^{\infty}$ is the spectrum of the Dirichlet–Neumann problem

$$y'' + (z - q(x))\,y = 0,$$
$$y(z, 0) = y'(z, a) = 0,$$

and $(\nu_k)_{k=1}^{\infty}$ is the spectrum of the corresponding Dirichlet problem

$$y'' + (z - q(x))\,y = 0,$$
$$y(z, 0) = y(z, a) = 0.$$

Putting $z = \lambda^2$, let $s(\lambda, \cdot)$ be the solution of $y'' + (\lambda^2 - q(x))y = 0$ with $y(\lambda, 0) = 0$, $y'(\lambda, 0) = 1$. From Corollary 12.2.10 we know that $s'(\cdot, a)$ is a sine type function, and since $s'(\cdot, a)$ is a characteristic function of the Dirichlet–Neumann problem with $z = \lambda^2$, it follows in view of Lemma 11.2.29 that there is a constant c such that

$$s'(\lambda, a) = c\lambda^m \prod_{k=1}^{\infty}{}' \left(1 - \frac{\lambda^2}{\mu_k}\right).$$

Here $\prod'$ indicates that the factor, if any, for which $\mu_k = 0$, is replaced by the term λ^m with $m = 2$, whereas $m = 0$ otherwise. On the other hand, it follows from (8.1.12) that $\lambda \mapsto \Xi_1(\lambda^2)$ is a sine type function and hence $\lambda \mapsto \Xi_1(\lambda^2)$ has the same product representation as $s'(\cdot, a)$ with the same constant c because both functions have the same leading term $\cos\lambda a$. Similarly, $\lambda \mapsto \lambda s(\lambda, a)$ and $\lambda \mapsto \lambda\Xi_2(\lambda^2)$ are sine type functions with the same zeros and the same leading terms. Hence it follows from (8.1.10), (8.1.11) and (8.1.5) that

$$\begin{aligned}
s'(\lambda, a) + (i\alpha\lambda + \beta)s(\lambda, a) &= \Xi_1(\lambda^2) + (i\alpha\lambda + \beta)\Xi_2(\lambda^2)\\
&= \Phi_1(\lambda^2) - \alpha^{-1}\beta\Phi_2(\lambda) + (i\alpha\lambda + \beta)\alpha^{-1}\Phi_2(\lambda^2)\\
&= \Phi_1(\lambda^2) + i\lambda\Phi_2(\lambda^2)\\
&= \chi(\lambda).
\end{aligned}$$

Therefore we have shown that the sequence of the zeros (λ_k) of χ is the sequence of the eigenvalues of the problem (8.1.1)–(8.1.3) with a, q, α and β as found above. $\square$

Theorem 8.1.4. *Let $\kappa \in \mathbb{N}_0$ and let $(\lambda_k)_{k=-\infty,k\neq0}^\infty$ be a properly indexed sequence which has the SHB_κ^- property and which satisfies the asymptotic representation*

$$\lambda_k = \left(k - \frac{1}{2}\right)b + ig + \frac{h}{k} + \frac{\gamma_k}{k}, \quad k \to \infty, \tag{8.1.18}$$

where $b > 0$, $g > 0$, $h \in \mathbb{R}$ and $(\gamma_k)_{k=-\infty,k\neq0}^\infty \in l_2$. Then there exists a unique $(a, q, \alpha, \beta) \in \mathcal{B}$ for which problem (8.1.1)–(8.1.3) has the spectrum $(\lambda_k)_{-\infty,k\neq0}^\infty$. Furthermore, $\alpha \in (0,1)$, i. e., $(a, q, \alpha, \beta) \in \mathcal{B}^-$.

Proof. We put

$$a = \frac{\pi}{b} > 0. \tag{8.1.19}$$

Since

$$\left(1 - \frac{\lambda}{\lambda_k}\right)\left(1 - \frac{\lambda}{\lambda_{-k}}\right) - 1 = \left(1 - \frac{\lambda}{\lambda_k}\right)\left(1 + \frac{\lambda}{\lambda_k}\right) - 1 = O(|\lambda|^2)O(k^{-2})$$

for $k > 1$, it follows for any $C_1 \in \mathbb{R} \setminus \{0\}$ that

$$\chi(\lambda) = C_1 \lim_{n\to\infty} (\lambda_{-1} - \lambda)(\lambda_1 - \lambda) \prod_{k=-n,\,|k|>1}^{n} \left(1 - \frac{\lambda}{\lambda_k}\right) \tag{8.1.20}$$

converges for all $\lambda \in \mathbb{C}$ and defines an entire function χ, see, e. g., [55, Theorem VII.5.9]. Here we have used that $\lambda_k \neq 0$ for $k \neq -1$ by definition of SHB_κ^-.

We now consider an auxiliary problem for $(a, 2\pi h a^{-1}, \alpha, 0) \in \mathcal{B}^-$, where $\alpha = \tanh ag$:

$$y'' + \left(\lambda^2 - 2\pi\frac{h}{a}\right)y = 0 \text{ on } (0, a), \tag{8.1.21}$$

$$y(\lambda, 0) = 0, \tag{8.1.22}$$

$$y'(\lambda, a) + i\lambda\alpha y(\lambda, a) = 0. \tag{8.1.23}$$

The differential equation (8.1.21) is of the form (12.3.1) with $b = 0$ and $c = 2\pi h a^{-1}$. We are going to apply Corollary 12.3.2 with $n = 1$ to find an asymptotic representation of the characteristic function χ_0 of problem (8.1.21)–(8.1.23). In the notation of Lemma 12.3.1 and observing (12.3.4) and (12.3.5) we conclude that $\tau(\lambda) = \sqrt{\lambda^2 - 2\pi h a^{-1}}$,

$$\chi_0(\lambda) = \cos\tau(\lambda)a + i\lambda\alpha\frac{\sin\tau(\lambda)a}{\tau(\lambda)},$$

$h_1(z) = \sqrt{1 - 2\pi h a^{-1}z^2} = 1 - \pi h a^{-1}z^2 + O(z^4)$ and $h_2(z) = z^{-1}(h_1(z) - 1) = -\pi h a^{-1}z + O(z^3)$. Therefore, by (12.3.6) and (12.3.7), $f_{2,1,1}(z) = f_{1,1,1}(z) = 1$ coincide with the Taylor polynomial about 0 of order 1 of $z \mapsto \cos h_2(z)a$, whereas

$-f_{1,2,1}(z) = f_{2,2,1}(z) = -i\pi hz$ is the Taylor polynomial about 0 of order 1 of $z \mapsto i \sin h_2(z)a$. From Corollary 12.3.2 we therefore conclude that

$$\chi_0(\lambda) = \cos \lambda a + i\alpha \sin \lambda a + \frac{\pi h}{\lambda}(\sin \lambda a - i\alpha \cos \lambda a) + \frac{\psi_0(\lambda)}{\lambda}, \qquad (8.1.24)$$

where $\psi_0 \in \mathcal{L}^a$ and $\psi_0(-\overline{\lambda}) = -\overline{\psi_0(\lambda)}$ for all $\lambda \in \mathbb{C}$.

We know from Proposition 6.1.1 that χ_0 is a sine type function of type a and from Theorem 7.2.1 that its sequence of zeros $(\zeta_k)_{k=-\infty,\ k\neq 0}^{\infty}$ can be properly indexed and satisfies

$$\zeta_k = \frac{\pi(k - \frac{1}{2})}{a} + ig + \frac{h}{k} + \frac{d_k}{k} \quad \text{for } k \geq 1, \quad (d_k)_1^{\infty} \in l_2. \qquad (8.1.25)$$

Comparing (8.1.18) with (8.1.25) implies

$$\lambda_k = \zeta_k + \frac{\tilde{b}_k}{\zeta_k}, \quad \left(\tilde{b}_k\right)_{k=-\infty,\ k\neq 0}^{\infty} \in l_2. \qquad (8.1.26)$$

Due to Lemma 11.3.15 and Remark 11.3.16 with $\omega = \chi_0$, $\tilde{\omega} = \chi$ with $C_1 = 1$, $a = 0$ and $n = 1$, we obtain

$$\chi(\lambda) = C_1 C_0 \left(1 + \frac{iT}{\lambda}\right) \chi_0(\lambda) + \frac{\psi(\lambda)}{\lambda},$$

where $C_0 \neq 0$ and T are constants and ψ belongs to the class $\mathcal{L}^a$. Since the sequence $(\lambda_k)_{k=-\infty,k\neq 0}^{\infty}$ is properly indexed, $\chi(-\overline{\lambda}) = \overline{\chi(\lambda)}$ for all $\lambda \in \mathbb{C}$. From $\tau^2(\lambda) = \lambda^2 - 2\pi h a^{-1}$ it follows that also $\chi_0(-\overline{\lambda}) = \overline{\chi_0(\lambda)}$ for all $\lambda \in \mathbb{C}$. Hence χ and χ_0 are real on the imaginary axis, and we conclude that $C_0 \in \mathbb{R} \setminus \{0\}$, $T \in \mathbb{R}$ and $\psi(-\overline{\lambda}) = -\overline{\psi(\lambda)}$ for all $\lambda \in \mathbb{C}$. We now choose $C_1 = C_0^{-1}$ and obtain

$$\chi(\lambda) = \left(1 + \frac{iT}{\lambda}\right) \chi_0(\lambda) + \frac{\psi(\lambda)}{\lambda}. \qquad (8.1.27)$$

Substituting (8.1.24) into (8.1.27) we obtain

$$\chi(\lambda) = \cos \lambda a + i\alpha \sin \lambda a + \frac{\pi h - T\alpha}{\lambda} \sin \lambda a + \frac{i(T - \pi h\alpha)}{\lambda} \cos \lambda a + \frac{\tilde{\psi}(\lambda)}{\lambda}, \qquad (8.1.28)$$

where $\tilde{\psi} \in \mathcal{L}^a$.

We note that in the product representation (5.1.3) for χ we can choose $P_k(\frac{\lambda}{\lambda_k}) = \frac{\lambda}{\lambda_k}$ if $\lambda_{-k} = -\overline{\lambda_k}$ and $P_k(\frac{\lambda}{\lambda_k}) = 1$ otherwise since

$$\sum_{k=-\infty,|k|>1}^{\infty} |\lambda_k|^{-2} < \infty.$$

In view of the definition of $\mathrm{R}\,P_k(\frac{\lambda}{\lambda_k})$ right before the formulation of Theorem 5.1.11, we have

$$\mathrm{R}\,\frac{\lambda}{\lambda_k} + \mathrm{R}\,\frac{\lambda}{\lambda_{-k}} = \frac{1}{2}\left(\frac{\lambda}{\lambda_k} + \frac{\lambda}{\overline{\lambda_k}}\right) + \frac{1}{2}\left(\frac{\lambda}{-\overline{\lambda_k}} + \frac{\lambda}{-\lambda_k}\right) = 0$$

whenever $\lambda_{-k} = -\overline{\lambda_k}$. Hence χ has the form of the function ω in Theorem 5.2.16, with u constant and $\nu = 0$. From $\lambda_k = O(|k|)$ as $|k| \to \infty$ and from the boundedness of the set $\{\operatorname{Im}\lambda_k : k \in \mathbb{Z}\}$ it follows that $\operatorname{Im}\frac{1}{\lambda_k} = O(k^{-2})$ as $|k| \to \infty$, and therefore

$$\sum_{\substack{k=-\infty \\ |k|>1}}^{\infty} \left|\operatorname{Im}\frac{1}{\lambda_k}\right| < \infty.$$

Since the zeros of χ belong to SHB_κ^-, it easily follows that χ satisfies the conditions 1–7 in Theorem 5.2.16. Thus Theorem 5.2.16 shows that $\chi \in \mathrm{SSHB}_{\kappa'}$, where $\kappa' = \kappa$ if $\lambda_{-1} \neq 0$ and where $\kappa' = \kappa - 1$ if $\lambda_{-1} = 0$.

Hence we have shown that χ belongs to the class SSHB and has the representation (8.1.28). In view of Theorem 8.1.3, there are $\beta \in \mathbb{R}$ and a real-valued function $q \in L_2(0, a)$ such that the sequence of the zeros $(\lambda_k)_{k=-\infty, k\neq 0}^{\infty}$ is the sequence of the eigenvalues of the problem (8.1.1)–(8.1.3) with $a > 0$, and $\alpha \in (0, 1)$ as found above.

Now we are going to show that $(a, q, \alpha, \beta) \in \mathcal{B}$ is uniquely determined by (8.1.18). Firstly, we are going to show that $\alpha \in (0, 1)$. By proof of contradiction, assume that $\alpha > 1$. Then the index set of the properly indexed eigenvalues would be $\mathbb{Z}$ by Theorem 7.2.1, 1, which is impossible since the index set of the properly indexed sequence of numbers in (8.1.18) is $\mathbb{Z} \setminus \{0\}$. Since $\{\operatorname{Im}\lambda_k : k \in \mathbb{Z}\}$ is a bounded set, it follows from Proposition 7.2.3 that $\alpha \neq 1$. If $\alpha = 0$, then the eigenvalues would have to be symmetric with respect to the origin, which is impossible since $g \neq 0$ in (8.1.18). Next assume that $\alpha < 0$. Then the substitution $\lambda \mapsto -\lambda$, $\alpha \mapsto -\alpha$ would show by Theorem 7.2.1 and Proposition 7.2.3 that the sequence of eigenvalues would have infinitely many terms with negative imaginary parts, contradicting $g > 0$ in the representation (8.1.18).

Hence we have established that $\alpha \in (0, 1)$, and therefore the representations (8.1.18) and (7.2.7) coincide. In particular, the first two terms must be equal, which shows that

$$b = \frac{\pi}{a} \quad \text{and} \quad g = \frac{1}{2a}\log\frac{\alpha+1}{1-\alpha}.$$

Thus we have proved that a and α are uniquely determined by b and g.

Let $q_1, q_2 \in L_2(0, a)$ and $\beta_1, \beta_2 \in \mathbb{R}$ such that $(a, q_1, \alpha, \beta_1)$ and $(a, q_2, \alpha, \beta_2)$ generate the same spectrum of problem (8.1.1)–(8.1.3), given by the sequence (8.1.18). For $j = 1, 2$, let s_j be the solution of (8.1.1) corresponding to $(a, q_j, \alpha, \beta_j)$ with $s_j(\lambda, 0) = 0$, $s_j'(\lambda, 0) = 1$ for all $\lambda \in \mathbb{C}$. Since the spectra of these two problems coincide, it follows again from Lemma 11.2.29 and the representations of $s_j(\cdot, a)$ and of $s'_j(\cdot, a)$ in Corollary 12.2.10 that

$$s'_1(\lambda, a) + (i\lambda\alpha + \beta_1)s_1(\lambda, a) = s'_2(\lambda, a) + (i\lambda\alpha + \beta_2)s_2(\lambda, a)$$

for all $\lambda \in \mathbb{C}$. Denoting this function by ϕ, we have seen in the proof of Theorem 7.2.1 that $\lambda \mapsto \lambda\phi(\lambda)$ is of the form (7.1.4) with $M = K_j(a, a) + \beta_j$ and $N = \alpha K_j(a, a)$, $j = 1, 2$, where $K_j(a, a) = \frac{1}{2}\int_0^a q_j(x)\,dx$ according to Theorem 12.2.9. Clearly, the numbers M and N are uniquely determined by ϕ, and therefore

$$\beta_1 = M - \alpha^{-1}N = \beta_2.$$

Hence β is uniquely determined by (8.1.18), and so also the functions Ξ_1 and Ξ_2 defined in (8.1.10) and (8.1.11). Consequently, the zeros of Ξ_1 and Ξ_2, considered in the proof of Theorem 8.1.3, are uniquely determined by (8.1.18). Thus also the potential q is uniquely determined in view of Theorem 12.6.2.   □

Theorem 8.1.5. *Let $\kappa \in \mathbb{N}_0$ and let $(\lambda_k)_{k=-\infty}^{\infty}$ be a properly indexed sequence which has the SHB_κ^+ property and which satisfies the asymptotic representation*

$$\lambda_k = kb + ig + \frac{h}{k} + \frac{\tilde{\gamma}_k}{k}, \quad k \to \infty, \tag{8.1.29}$$

where $b > 0$, $g > 0$, $h \in \mathbb{R}$ and $(\gamma_k)_{k=-\infty}^{\infty} \in l_2$. Then there exists a unique $(a, q, \alpha, \beta) \in \mathcal{B}$ for which problem (8.1.1)–(8.1.3) has the spectrum $(\lambda_k)_{-\infty}^{\infty}$. Furthermore, $\alpha \in (1, \infty)$, i. e., $(a, q, \alpha, \beta) \in \mathcal{B}^+$.

Proof. The proof is, mutatis mutandis, as in Theorem 8.1.4 if we observe that we have to put $\alpha = \coth ag$ in this case.   □

8.2   The inverse problem for damped Stieltjes strings

Theorem 6.2.3 states that conditions 1–5 there are necessary for a sequence of complex numbers to be the spectrum of the eigenvalue problem

$$\frac{u_k^{(j)} - u_{k+1}^{(j)}}{l_k^{(j)}} + \frac{u_k^{(j)} - u_{k-1}^{(j)}}{l_{k-1}^{(j)}} - m_k^{(j)}\lambda^2 u_k^{(j)} = 0, \quad k = 1, \ldots, n_j,\ j = 1, 2, \tag{8.2.1}$$

$$u_0^{(j)} = 0, \tag{8.2.2}$$

$$u_{n_1+1}^{(1)} = u_{n_2+1}^{(2)}, \tag{8.2.3}$$

$$\frac{u_{n_1+1}^{(1)} - u_{n_1}^{(1)}}{l_{n_1}^{(1)}} + \frac{u_{n_2+1}^{(2)} - u_{n_2}^{(2)}}{l_{n_2}^{(2)}} + i\lambda\nu u_{n_1+1}^{(1)} = 0. \tag{8.2.4}$$

Here we prove that these conditions are sufficient.

Theorem 8.2.1. *Let two positive numbers $l > 0$ and $\hat{l} \in (0, l)$ and two natural numbers n_1 and n_2 be given together with the sequence of complex numbers $(\lambda_k)_{-(n_1+n_2)}^{n_1+n_2}$ which satisfy*

(i) *$\operatorname{Im} \lambda_k \geq 0$ for $k = 0, \pm 1, \pm 2, \ldots, \pm(n_1 + n_2)$;*
(ii) *$\lambda_{-k} = -\overline{\lambda_k}$ for not pure imaginary λ_k;*
(iii) *all real terms of the sequence, if any, are simple and nonzero;*
(iv) *for each real term λ_k of the sequence, $\operatorname{Im} \Phi'(\lambda_k) = 0$ and $\operatorname{Im} \Phi''(\lambda_k) \neq 0$;*
(v) *the number of real terms of the sequence does not exceed $2 \min\{n_1, n_2\}$; i. e., conditions 1–5 of Theorem 6.2.3 are satisfied, where*

$$\Phi(\lambda) = \prod_{k=-(n_1+n_2)}^{n_1+n_2} \left(1 - \frac{\lambda}{\lambda_k} \right), \quad \lambda \in \mathbb{C}. \tag{8.2.5}$$

Then there exists a problem (8.2.1)–(8.2.4), i. e., a positive constant ν, sequences of positive numbers $(m_k^{(1)})_{k=1}^{n_1}$, $(m_k^{(2)})_{k=1}^{n_2}$, $(l_k^{(1)})_{k=0}^{n_1}$ and $(l_k^{(2)})_{k=0}^{n_2}$ with $\sum_{k=0}^{n_1} l_k^{(1)} = \hat{l}$ and $\sum_{k=0}^{n_2} l_k^{(2)} = l - \hat{l}$, the spectrum of which coincides with $(\lambda_k)_{-(n_1+n_2)}^{n_1+n_2}$.

Proof. Since the polynomial Φ satisfies the symmetry condition $\Phi(-\overline{\lambda}) = \overline{\Phi(\lambda)}$, $\lambda \in \mathbb{C}$, there are real polynomials P and Q such that

$$\Phi(\lambda) = P(\lambda^2) + i\lambda Q(\lambda^2), \quad \lambda \in \mathbb{C}, \tag{8.2.6}$$

see (5.1.23). We set

$$\nu = Q(0) \left(\hat{l}^{-1} + (l - \hat{l})^{-1} \right). \tag{8.2.7}$$

Using (8.2.5) and (8.2.6) we obtain

$$Q(0) = \frac{1}{i} \Phi'(0) = i \sum_{k=-(n_1+n_2)}^{n_1+n_2} \frac{1}{\lambda_k}.$$

Due to conditions (i), (ii) and (v), $Q(0) > 0$ and consequently $\nu > 0$.

If λ_k is a real zero of Φ, then by conditions (ii) and (iii), $\lambda_k \neq 0$ and $-\lambda_k$ are simple zeros of Φ, and since P and Q are real polynomials, it follows that λ_k^2 is a zero of both P and Q. For convenience, we let m be the number of positive real terms in the sequence $(\lambda_k)_{-(n_1+n_2)}^{n_1+n_2}$ and we assume that these positive real terms in the sequence have the indices $n_1 + n_2 - m + 1, \ldots, n_1 + n_2$. Introducing

$$R(z) = \prod_{k=n_1+n_2-m+1}^{n_1+n_2} \left(1 - \frac{z}{\lambda_k^2} \right), \quad z \in \mathbb{C}, \tag{8.2.8}$$

$$\tilde{\Phi}(\lambda) = \prod_{k=-(n_1+n_2-m)}^{n_1+n_2-m} \left(1 - \frac{\lambda}{\lambda_k} \right), \quad \lambda \in \mathbb{C}, \tag{8.2.9}$$

it follows that

$$\Phi(\lambda) = R(\lambda^2)\tilde{\Phi}(\lambda), \quad \lambda \in \mathbb{C}. \tag{8.2.10}$$

We can find again unique real polynomials $\tilde{P}$ and $\tilde{Q}$ such that

$$\tilde{\Phi}(\lambda) = \tilde{P}(\lambda^2) + i\lambda\tilde{Q}(\lambda^2), \quad \lambda \in \mathbb{C}. \tag{8.2.11}$$

By construction and condition (i), all zeros of $\tilde{\Phi}$ lie in the open upper half-plane. Hence, by Theorem 5.2.16 and Definitions 5.2.6, 5.1.24, 5.1.25, $\tilde{P}$ and $\tilde{Q}$ have no common zeros and the rational function $\frac{\tilde{Q}}{\tilde{P}}$ belongs to the class $\mathcal{S}_0$, where we have used that $\tilde{P}(0) = \tilde{\Phi}(0) \neq 0$. In view of Lemma 5.2.2, all zeros of $\tilde{P}$ and $\tilde{Q}$ are positive. From (8.2.10) it follows that

$$P = R\tilde{P} \quad \text{and} \quad Q = R\tilde{Q}. \tag{8.2.12}$$

From condition (iv) we obtain for $k = n_1 + n_2 - m + 1, \ldots, n_1 + n_2$ that

$$0 = \text{Im}\,\Phi'(\lambda_k) = Q(\lambda_k^2) + 2\lambda_k^2 Q'(\lambda_k^2),$$

which shows that $Q'(\lambda_k^2) = 0$. Therefore λ_k^2 is at least a double zero of Q. By definition of R, these values λ_k^2 are simple zeros of R. Hence the numbers λ_k^2 are zeros of $\tilde{Q}$ when $k = n_1 + n_2 - m + 1, \ldots, n_1 + n_2$. These zeros and all the remaining zeros of $\tilde{Q}$ are simple zeros of $\tilde{Q}$ in view of Theorem 11.1.6. It follows that λ_k^2 are double zeros of Q for $k = n_1 + n_2 - m + 1, \ldots, n_1 + n_2$, whereas the remaining zeros of Q are simple. Since Φ is a polynomial of odd degree $2n_1 + 2n_2 + 1$, the polynomial Q has degree $n_1 + n_2$. Because of condition (v), we can arrange the zeros of Q, counted with multiplicity, into two sequences $(\nu_k^{(1)})_{k=1}^{n_1}$ and $(\nu_k^{(2)})_{k=1}^{n_2}$ in such a way that $\nu_k^{(1)} = \nu_k^{(2)} = \lambda_{n_1+m_1-m+k}^2$ for $k = 1, \ldots, m$. In particular, the terms in each of these two sequences are mutually distinct, i.e., $\nu_k^{(j)} \neq \nu_h^{(j)}$ for $j = 1, 2$, $1 \leq k < h \leq n_j$.

By Lemma 5.2.2 and Theorem 11.1.6, the zeros of $\tilde{P}$ and $\tilde{Q}$ interlace and the smallest zero of $\tilde{P}$ is smaller than the smallest zero of $\tilde{Q}$. Hence the degree of $\tilde{Q}$ cannot exceed the degree of $\tilde{P}$. It follows that the degree of P is greater or equal the degree of Q, which is $n_1 + n_2$. On the other hand, the degree of P cannot be larger than $n_1 + n_2$ since the degree of Φ is $2n_1 + 2n_2 + 1$. Therefore the degree of P equals $n_1 + n_2$. Hence we have the partial fraction decomposition

$$\frac{P(z)}{Q(z)} = \frac{\tilde{P}(z)}{\tilde{Q}(z)} = \sum_{k=1}^{n_1} \frac{A_k^{(1)}}{z - \nu_k^{(1)}} + \sum_{k=1}^{n_2} \frac{A_k^{(2)}}{z - \nu_k^{(2)}} + B, \tag{8.2.13}$$

where we have to observe that for $k = 1, \ldots, m$, the denominators $z - \nu_k^{(1)} = z - \nu_k^{(2)}$ are duplicated, and therefore the numbers $A_k^{(1)} + A_k^{(2)}$ for $k = 1, \ldots, m$, $A_k^{(1)}$ for $k = m + 1, \ldots, n_1$, $A_k^{(2)}$ for $k = m + 1, \ldots, n_2$, and B are uniquely determined by

P and Q. We will show now they are all positive. In particular, we can also choose $A_k^{(1)}$ and $A_k^{(2)}$ for $k = 1, \ldots, m$ to be positive. Firstly, since the degrees of P and Q coincide, $B \neq 0$, and from $\frac{\tilde{Q}}{P} \in \mathcal{S}_0$ it follows that $B > 0$. Secondly, since $\frac{\tilde{Q}}{P}$ is a Nevanlinna function, we have $\operatorname{Im} \frac{\tilde{P}(z)}{\tilde{Q}(z)} < 0$ when $\operatorname{Im} z > 0$. Near each zero ν of $\tilde{Q}$, the dominant term of $\frac{\tilde{P}(z)}{\tilde{Q}(z)}$ is of the form $\frac{A_\nu}{z - \nu} = \frac{A_\nu(\bar{z} - \nu)}{|z - \nu|^2}$, whence $A_\nu > 0$.

Since $P(0) = \Phi(0) = 1$, we obtain from (8.2.13) and (8.2.7) that

$$-\sum_{k=1}^{n_1} \frac{A_k^{(1)}}{\nu_k^{(1)}} - \sum_{k=1}^{n_2} \frac{A_k^{(2)}}{\nu_k^{(2)}} + B = \frac{1}{\nu}\left(\frac{1}{\hat{l}} + \frac{1}{l - \hat{l}}\right). \tag{8.2.14}$$

We define

$$B_1 = \sum_{k=1}^{n_1} \frac{A_k^{(1)}}{\nu_k^{(1)}} + \frac{1}{\nu \hat{l}}, \tag{8.2.15}$$

$$B_2 = \sum_{k=1}^{n_2} \frac{A_k^{(2)}}{\nu_k^{(2)}} + \frac{1}{\nu(l - \hat{l})}, \tag{8.2.16}$$

and

$$\psi_j(z) = \sum_{k=1}^{n_j} \frac{A_k^{(j)}}{z - \nu_k^{(j)}} + B_j, \quad j = 1, 2. \tag{8.2.17}$$

Evidently, $B_1 > 0$, $B_2 > 0$, $B_1 + B_2 = B$ by (8.2.14), and $-\psi_1$ and $-\psi_2$ are Nevanlinna functions whose poles are positive. Hence ψ_1 and ψ_2 can have at most one simple zero in $(-\infty, 0]$. Then

$$\psi_1(0) = \nu^{-1}\hat{l}^{-1} > 0, \quad \psi_2(0) = \nu^{-1}(l - \hat{l})^{-1} > 0, \tag{8.2.18}$$

and $\lim_{z \to -\infty} \psi_j(z) = B_j > 0$, $j = 1, 2$, show that the functions ψ_j, $j = 1, 2$, are positive on $(-\infty, 0]$. Therefore the functions

$$\theta_j = \frac{1}{\nu \psi_j}, \quad j = 1, 2, \tag{8.2.19}$$

are $\mathcal{S}_0$-functions in view of Lemma 5.2.4, 3. In view of Proposition 5.2.5 we have the continued fractions expansions

$$\theta_j(z) = a_{n_j}^{(j)} + \cfrac{1}{-b_{n_j}^{(j)} z + \cfrac{1}{a_{n_j-1}^{(j)} + \cfrac{1}{-b_{n_j-1}^{(j)} z + \cdots + \cfrac{1}{a_1^{(j)} + \cfrac{1}{-b_1^{(j)} z + \cfrac{1}{a_0^{(j)}}}}}} \tag{8.2.20}$$

with $a_k^{(j)} > 0$ for $k = 0, \ldots, n_j$, $j = 1, 2$, and $b_k^{(j)} > 0$ for $k = 0, \ldots, n_j$, $j = 1, 2$. Here we have to observe that (5.2.1) gives

$$a_{n_j}^{(j)} = \lim_{|z| \to \infty} \theta_j(z) = \frac{1}{\nu} \lim_{|z| \to \infty} \frac{1}{\psi(z)} = \frac{1}{\nu B_j} > 0.$$

We identify the positive numbers $a_k^{(j)}$, $k = 0, \ldots, n_j$, $j = 1, 2$, with the lengths of subintervals $l_k^{(j)}$, and we identify the positive numbers $b_k^{(j)}$, $k = 1, \ldots, n_j$, $j = 1, 2$, with masses $m_k^{(j)}$. Defining the rational functions $\theta_{j,k}$, $k = 0, \ldots, n_j$, $j = 1, 2$, inductively by $\theta_{j,0}(z) = l_0^{(j)}$ and

$$\theta_{j,k}(z) = l_k^{(j)} + \cfrac{1}{-m_k^{(j)} z + \cfrac{1}{\theta_{j,k-1}(z)}}, \quad k = 1, \ldots, n_j, \ j = 1, 2, \tag{8.2.21}$$

it is clear that $\theta_j = \theta_{j,n_j}$ for $j = 1, 2$.

Let us prove that problem (8.2.1)–(8.2.4) generated by these masses and subintervals possesses the spectrum $(\lambda_k)_{k=-(n_1+n_2)}^{n_1+n_2}$. With the notations of Section 6.2 we deduce from (6.2.14), (6.2.13) and (6.2.8) that the functions $\dfrac{R_{2k}^{(j)}}{R_{2k-1}^{(j)}}$ satisfy the same recurrence relation (8.2.21) and the same initial condition as $\theta_{j,k}$, and we therefore conclude that $\theta_j = \dfrac{R_{2n_1}^{(j)}}{R_{2n_1-1}^{(j)}}$ for $j = 1, 2$. Using (8.2.13), (8.2.17), (8.2.19) and (6.2.12) we obtain

$$\frac{P}{Q} = \psi_1 + \psi_2 = \frac{1}{\nu \theta_1} + \frac{1}{\nu \theta_2} = \frac{R_{2n_1-1}^{(1)}}{\nu R_{2n_1}^{(1)}} + \frac{R_{2n_2-1}^{(2)}}{\nu R_{2n_2}^{(2)}} = \frac{\phi}{\nu R_{2n_1}^{(1)} R_{2n_2}^{(2)}}.$$

We know from the discussion at the beginning of Section 6.2 that $R_{2n_j}^{(j)}$ is a polynomial of degree n_j for $j = 1, 2$. Since θ_j has n_j simple zeros, which are the poles of ψ_j, these zeros coincide with the zeros of $R_{2n_j}^{(j)}$, $j = 1, 2$. Consequently, the zeros of Q and $R_{2n_1}^{(1)} R_{2n_2}^{(2)}$, counted with multiplicity, coincide, so that there is a nonzero constant C such that $R_{2n_1}^{(1)} R_{2n_2}^{(2)} = CQ$. We then also have $\phi = C\nu P$. Denoting the function Φ in (6.2.11) by $\hat{\Phi}$ in order to avoid confusion with the function defined in (8.2.5), we therefore have by (8.2.6) that

$$\hat{\Phi}(\lambda) = C\nu P(\lambda^2) + i\lambda \nu C Q(\lambda^2) = C\nu \Phi(\lambda), \quad \lambda \in \mathbb{C}.$$

This proves that Φ is the characteristic function of the eigenvalue problem (8.2.1)–(8.2.4), so that its eigenvalues coincide with $(\lambda_k)_{k=-(n_1+n_2)}^{n_1+n_2}$.

Using (8.2.18) and (8.2.20) or (8.2.21) as well as (8.2.18), we obtain

$$\sum_{k=0}^{n_j} l_k^{(j)} = \theta_j(0) = \frac{1}{\nu \psi_j(0)} = \begin{cases} \hat{l} & \text{if } j = 1, \\ l - \hat{l} & \text{if } j = 2. \end{cases} \qquad \Box$$

Remark 8.2.2. The solution of this inverse problem is not unique because of the ambiguity in choice of ψ_1 and ψ_2. It can be shown that the solution is unique if and only if $n_1 = 0$ or $n_2 = 0$.

8.3 The inverse problem for damped strings

In this section we continue the investigation of the eigenvalue problem for damped strings, considered in Sections 2.2, 6.3 and 7.3. We recall from (7.3.1)–(7.3.3) that this problem is given by

$$y'' - 2i\lambda \varrho y - q(x)y + \lambda^2 y = 0, \tag{8.3.1}$$

$$y(\lambda, 0) = 0, \tag{8.3.2}$$

$$y'(\lambda, a) + (-\lambda^2 m + i\lambda \nu + \beta)y(\lambda, a) = 0. \tag{8.3.3}$$

Lemma 8.3.1. *Let* $(\lambda_k)_{k=-\infty, k\neq 0}^{\infty}$ *be a properly indexed sequence satisfying the conditions*

(i) $\operatorname{Im} \lambda_k > 0$ *for all terms of the sequence;*

(ii) *the sequence has the asymptotic representation*

$$\lambda_k \underset{k\to\infty}{=} \frac{\pi(k-1)}{a} + i\varrho + \frac{p_1}{k-1} + \frac{i\,p_2}{(k-1)^2} + \frac{p_3}{(k-1)^3} + \frac{b_k}{k^3}, \tag{8.3.4}$$

where $a > 0$, $\varrho > 0$, $p_j \in \mathbb{R}$ *for* $j = 1, 2, 3$, $(b_k)_{k=1}^{\infty} \in l_2$.

Then the entire function χ *defined by*

$$\chi(\lambda) = \lim_{n\to\infty} \prod_{\substack{k=-n \\ k\neq 0}}^{n} \left(1 - \frac{\lambda}{\lambda_k}\right), \quad \lambda \in \mathbb{C}, \tag{8.3.5}$$

may be represented in the form

$$\chi(\lambda) = B_0 \left[(\mu + iB_1 + B_2\mu^{-1} + iB_3\mu^{-2}) \sin \mu a \right. $$
$$\left. + (A_1 + iA_2\mu^{-1} + A_3\mu^{-2}) \cos \mu a \right] + \Psi(\mu)\mu^{-2}, \tag{8.3.6}$$

where $\mu = \lambda - i\varrho$ *and* $A_k \in \mathbb{R}$, $k = 1, 2, 3$, $B_k \in \mathbb{R}$, $k = 0, 1, 2, 3$, $B_0 \neq 0$, $\Psi \in \mathcal{L}^a$ *and* $\Psi(-\overline{\mu}) = \overline{\Psi(\mu)}$.

Proof. Since

$$\left(1 - \frac{\lambda}{\lambda_k}\right)\left(1 - \frac{\lambda}{\lambda_{-k}}\right) - 1 = \left(1 - \frac{\lambda}{\lambda_k}\right)\left(1 + \frac{\lambda}{\lambda_k}\right) - 1 = O(|\lambda|^2)O(k^{-2})$$

for $k > 1$, it follows that χ defined in (8.3.5) is an entire function, see, e. g., [55, Theorem VII.5.9]. We now define the entire function $\tilde{\chi}$ by

$$\tilde{\chi}(\mu) := \chi(\mu + i\varrho) = \chi(\lambda).$$

Letting
$$\mu_k = \lambda_k - i\varrho, \ k \in \mathbb{Z} \setminus \{0\},$$

it follows that $(\mu_k)_{k=-\infty, k\neq 0}^{\infty}$ is the properly indexed sequence of zeros of $\tilde{\chi}$.

Next we consider an auxiliary problem of the form (8.3.1)–(8.3.3) whose eigenvalues have an asymptotic representation which may only differ in the remainder term from the asymptotic representation of the sequence $(\mu_k)_{k=-\infty, k\neq 0}^{\infty}$. To this end let d be a real number such that

$$d < \frac{2\pi}{a} p_1 + \varrho^2 - \frac{\overline{p_2}\,\pi^2}{a^2 \varrho}, \qquad (8.3.7)$$

where $\overline{p_2} = \max\{0, -p_2\}$. Then we consider problem (8.3.1)–(8.3.3) with $q = d$ and as yet unspecified constants $m > 0$, $\alpha > 0$, and $\beta \in \mathbb{R}$. We will use the upper index '(0)' for functions and constants related to this auxiliary problem.

The characteristic function $\chi^{(0)}$ is

$$\chi^{(0)}(\mu) = \cos\tau(\mu)a + \left(-m\,(\mu + i\varrho)^2 + i\nu\,(\mu + i\varrho) + \beta\right)\frac{\sin\tau(\mu)a}{\tau(\mu)}, \qquad (8.3.8)$$

where $\tau(\mu) = \sqrt{\mu^2 + \varrho^2 - d}$. In view of Theorem 7.3.1, part 3, the properly indexed zeros $(\mu_k^{(0)})_{k=-\infty, k\neq 0}^{\infty}$ of $\chi^{(0)}$ have the asymptotics

$$\mu_k^{(0)} = \lambda_k^{(0)} - i\varrho \underset{k\to\infty}{=} \frac{\pi(k-1)}{a} + \frac{p_1^{(0)}}{k-1} + \frac{ip_2^{(0)}}{(k-1)^2} + \frac{p_3^{(0)}}{(k-1)^3} + \frac{b_k^{(0)}}{k^3}, \qquad (8.3.9)$$

where $(b_k^{(0)})_{k=2}^{\infty} \in l_2$ and $p_1^{(0)}, p_2^{(0)}, p_3^{(0)}$ are given by (7.3.7)–(7.3.9), i.e.,

$$p_1^{(0)} = \frac{1}{\pi m} + \frac{K^{(0)}(a,a)}{\pi} - \frac{\varrho^2 a}{2\pi}, \qquad (8.3.10)$$

$$p_2^{(0)} = \frac{a}{\pi^2 m}\left(\frac{\nu}{m} - 2\varrho\right), \qquad (8.3.11)$$

$$p_3^{(0)} = C + \frac{a^2}{\pi^3}\frac{\beta}{m^2}, \qquad (8.3.12)$$

where C in (8.3.12) is a real number which is independent of β. From Theorem 12.2.9 we know that $K^{(0)}(a,a) = \frac{1}{2}ad$, and therefore

$$p_1 - \frac{K^{(0)}(a,a)}{\pi} + \frac{\varrho^2 a}{2\pi} = p_1 - \frac{a}{2\pi}(d - \varrho^2) > 0$$

in view of (8.3.7). Hence there is a unique m such that $p_1^{(0)} = p_1$, and this m is positive. From (8.3.11) we can now find a unique real number ν such that $p_2 = p_2^{(0)}$. Clearly, $\nu > 0$ if $p_2 \geq 0$. If $p_2 < 0$, then

$$p_2 + \frac{2a\varrho}{\pi^2 m} = p_2 + \frac{2a\varrho}{\pi}\left(p_1 - \frac{ad}{2\pi} + \frac{\varrho^2 a}{2\pi}\right) > 0$$

by (8.3.7), and therefore also $\nu > 0$ in this case. From (8.3.12) we finally can find a unique real number β such that $p_3 = p_3^{(0)}$.

We will now continue the investigation of $\chi^{(0)}$ with these values of m, ν and β. From the representation (8.3.8) of $\chi^{(0)}$ and from Corollary 12.3.2 with $b = 0$, with $n = 3$ in (12.3.8) and with $n = 2$ in (12.3.9) we know that there are real constants $B_j^{(0)}$, $j = 1, 2, 3$ and $A_j^{(0)}$, $j = 1, 2, 3$ and $\psi_j^{(0)} \in \mathcal{L}^a$ with $\psi_j^{(0)}(-\overline{\mu}) = \psi_j^{(0)}(\mu)$ such that

$$\chi^{(0)}(\mu) = B_0^{(0)} \left(\mu + iB_1^{(0)} + B_2^{(0)} \mu^{-1} + iB_3^{(0)} \mu^{-2} \right) \sin \mu a$$
$$+ \left(A_1^{(0)} + iA_2^{(0)} \mu^{-1} + A_3^{(0)} \mu^{-2} \right) \cos \mu a + \psi^{(0)}(\mu) \mu^{-2}. \qquad (8.3.13)$$

Here we took into account that $f_{1,2,3}(0) = 0$ by Corollary 12.3.2. Furthermore, Corollary 12.3.2 also shows that $B_0^{(0)} = -m f_{1,1,3}(0) = -m$.

We choose any index j such that $\mu_j^{(0)} \neq 0$ and $\mu_j \neq 0$ and we consider the entire function $\chi_1^{(0)}$ defined by

$$\chi_1^{(0)}(\mu) = \left(1 - \frac{\mu}{\mu_j^{(0)}} \right)^{-1} \chi^{(0)}(\mu). \qquad (8.3.14)$$

The leading term of $\chi_1^{(0)}$ as $|\mu| \to \infty$ is

$$\frac{\mu \mu_j^{(0)}}{\mu_j^{(0)} - \mu} B_0^{(0)} \sin \mu a.$$

Therefore it easily follows from Proposition 11.2.19 that $\chi_1^{(0)}$ is a sine type function of exponential type a. A comparison of (8.3.4) with (8.3.9) yields

$$\mu_k = \mu_k^{(0)} + \frac{\tilde{b}_k}{(\mu_k^{(0)})^3}, \quad k \in \mathbb{Z} \setminus \{0\},$$

where $(\tilde{b}_k)_{k=1}^\infty \in l_2$. An application of Lemma 11.3.15 and Remark 11.3.16 leads to

$$\tilde{\chi}(\mu) \left(1 - \frac{\mu}{\mu_j} \right)^{-1} = \chi_1^{(0)}(\mu) \left(\tilde{T}_0 + \frac{\tilde{T}_1}{\mu} + \frac{\tilde{T}_2}{\mu^2} + \frac{\tilde{T}_3}{\mu^3} \right) + \frac{\tilde{\varphi}(\mu)}{\mu^3},$$

where $\tilde{T}_k \in \mathbb{C}$, $k = 0, 1, 2, 3$, $\tilde{T}_0 \neq 0$, and $\tilde{\varphi}$ is an entire function belonging to $\mathcal{L}^a$. From

$$\left(1 - \frac{\mu}{\mu_j} \right) \left(1 - \frac{\mu}{\mu_j^{(0)}} \right)^{-1} = \frac{\mu_j^{(0)}}{\mu_j} \left(1 - \frac{\mu_j}{\mu} \right) \left(1 - \frac{\mu_j^{(0)}}{\mu} \right)^{-1}$$

$$= \frac{\mu_j^{(0)}}{\mu_j} \left(1 + \frac{\hat{T}_1}{\mu} + \frac{\hat{T}_2}{\mu^2} + \frac{\hat{T}_3}{\mu^3} + O(|\mu|^{-4}) \right)$$

with $\hat{T}_k \in \mathbb{C}$, $k = 1, 2, 3$ and (8.3.14) we can now conclude that

$$\tilde{\chi}(\mu) = \chi^{(0)}(\mu) \left(T_0 + i\frac{T_1}{\mu} + \frac{T_2}{\mu} + i\frac{T_3}{\mu^3} \right) + \frac{\varphi(\mu)}{\mu^3},$$

where $T_k \in \mathbb{C}$, $k = 0, 1, 2, 3$, $T_0 \neq 0$, and φ is an entire function belonging to $\mathcal{L}^a$. From the symmetry of $\tilde{\chi}$ and $\chi^{(0)}$ we immediately conclude that the numbers T_k, $k = 0, 1, 2, 3$, are real. Inserting the representation (8.3.13) for $\chi^{(0)}$ into the above identity leads to (8.3.6) with $B_0 = B_0^{(0)} T_0 \neq 0$.   □

Lemma 8.3.2. *Let $\varrho > 0$ and let $(\lambda_k)_{k=-\infty, k\neq 0}^\infty$ be a properly indexed sequence satisfying the conditions:*

(i) *Only a finite number, denoted by κ, of terms of the sequence lie in the closed half-plane $\operatorname{Im} \lambda \leq \varrho$.*

(ii) *All terms in the open half-plane $\operatorname{Im} \lambda \leq \varrho$ lie on $(0, i\varrho)$ and are pairwise different. If $\kappa > 0$, we denote them by $\lambda_{-j} = i\varrho - i|\lambda_{-j} - i\varrho|$, $j = 1, \ldots, \kappa$, satisfying $|\lambda_{-j} - i\varrho| < |\lambda_{-(j+1)} - i\varrho|$, $j = 1, \ldots, \kappa - 1$.*

(iii) *If $\kappa > 0$, then the numbers $i\varrho + i|\lambda_{-j} - i\varrho|$, $j = 1, \ldots, \kappa$, are not terms of the sequence $(\lambda_k)_{k=-\infty, k\neq 0}^\infty$.*

(iv) *If $\kappa \geq 2$, then in each interval $(i\varrho + i|\lambda_{-j} - i\varrho|, i\varrho + i|\lambda_{-(j+1)} - i\varrho|)$, $j = 1, \ldots, \kappa - 1$, the number of terms of the sequence $(\lambda_k)_{k=-\infty, k\neq 0}^\infty$ is odd.*

(v) *If $\kappa > 0$, then the interval $[i\varrho, i\varrho + i|\lambda_{-1} - i\varrho|)$ contains no or an even number of terms of the sequence $(\lambda_k)_{k=-\infty, k\neq 0}^\infty$.*

(vi) *The asymptotic representation (8.3.4) holds.*

Then χ belongs to the class SSHB and $\tilde{\chi} := \chi(\cdot + i\varrho)$ belongs to the class SSHB_κ.

Proof. Due to the asymptotics (8.3.4),

$$\sum_{k=-\infty, k\neq 0}^\infty \left| \operatorname{Im}(\lambda_k)^{-1} \right| < \infty \quad \text{and} \quad \sum_{k=-\infty, k\neq 0}^\infty \left| \operatorname{Im}\left(\lambda_k - \frac{i\varrho}{2}\right)^{-1} \right| < \infty.$$

Hence the statement of the present theorem follows from Theorem 5.2.16.   □

Denote by $\mathcal{M}^+$ ($\mathcal{M}^-$) the class of tuples $(a, \varrho, m, q, \nu, \beta)$, where $a > 0$, $\varrho > 0$, $m > 0$, $\nu > 2m\varrho$, $(0 < \nu < 2m\varrho)$, $\beta \in \mathbb{R}$ and $q \in L_2(0, a)$ is real valued.

Theorem 8.3.3. *Let $(\lambda_k)_{k=-\infty, k\neq 0}^\infty$ be a sequence of complex numbers satisfying the conditions of Lemma 8.3.2, and let $p_2 \neq 0$ in (8.3.4). Then there exists a unique $(a, \varrho, m, q, \nu, \beta)$ from $\mathcal{M}^+$ such that $(\lambda_k)_{k=-\infty, k\neq 0}^\infty$ is the spectrum of the eigenvalue problem (8.3.1)–(8.3.3) with $(a, \varrho, m, q, \nu, \beta)$.*

Proof. Since a and ϱ occur explicitly in (8.3.4), a and ϱ are uniquely determined by (8.3.4). Furthermore, replacing λ with $\lambda + i\varrho$ gives a one-to-one relation between the eigenvalue problem (8.3.1)–(8.3.3) and the eigenvalue problem

$$y'' - (q(x) - \varrho^2)y + \lambda^2 y = 0, \tag{8.3.15}$$

$$y(\lambda, 0) = 0, \tag{8.3.16}$$

$$y'(\lambda, a) + (-\lambda^2 m + i\lambda(\nu - 2m\varrho) + \beta - \nu\varrho + m\varrho^2)y(\lambda, a) = 0, \tag{8.3.17}$$

see (6.3.1)–(6.3.3). Hence, given any $\varrho > 0$ and a sequence of the form (8.3.4) with ϱ replaced by 0, i.e.,

$$\lambda_k \underset{k \to \infty}{=} \frac{\pi(k-1)}{a} + \frac{p_1}{k-1} + \frac{i\,p_2}{(k-1)^2} + \frac{p_3}{(k-1)^3} + \frac{b_k}{k^3}, \tag{8.3.18}$$

we have to show that there are unique m, q, ν, β, such that $(a, \varrho, m, q, \nu, \beta)$ belongs to $\mathcal{M}^+$ and such that (8.3.18) is the sequence of the eigenvalues of (8.3.15)–(8.3.17). Let $\tilde{\chi}$ be the entire function defined in Lemma 8.3.2 associated with the sequence given by (8.3.18). Then $\mu = \lambda$ in the notation of Lemma 8.3.2. Since $\tilde{\chi}$ is real on the imaginary axis and observing (8.3.6), we can write

$$B_0^{-1}\tilde{\chi}(\lambda) = \Phi_1(\lambda^2) + i\lambda\Phi_2(\lambda^2),$$

with real entire functions Φ_1 and Φ_2, where

$$\Phi_1(\lambda^2) = (\lambda + B_2\lambda^{-1})\sin \lambda a + (A_1 + A_3\lambda^{-2})\cos \lambda a + \Psi_1(\lambda^2)\lambda^{-2}, \tag{8.3.19}$$

$$\Phi_2(\lambda^2) = (B_1 + B_3\lambda^{-2})\frac{\sin \lambda a}{\lambda} + A_2\lambda^{-2}\cos \lambda a + \Psi_2(\lambda^2)\lambda^{-3}, \tag{8.3.20}$$

and where $\Psi_j \in \mathcal{L}^a$, $j = 1, 2$, are real entire functions, Ψ_1 is even and Ψ_2 is odd. Since $\tilde{\chi} \in \mathrm{SSHB}_\kappa$ by Lemma 8.3.2, it follows by Definition 5.2.6 that $\frac{\Phi_2}{\Phi_1} \in \mathcal{N}_+^{\mathrm{ep}}$ and that Φ_1 and Φ_2 have no common zeros. Hence we have that $\frac{\Phi_2(z)}{\Phi_1(z)} > 0$ as $z \to -\infty$. But for $\lambda = i\eta$ with $\eta \to \infty$ we have

$$\frac{\Phi_2(-\eta^2)}{\Phi_1(-\eta^2)} = -\frac{\eta^{-1}(B_1 + o(1))e^{\eta a}}{\eta(1 + o(1))e^{\eta a}} = (-B_1 + o(1))\eta^{-2},$$

so that $B_1 \leq 0$. Corollaries 5.2.3 and 11.1.8 imply that $B_1 \neq 0$. Altogether, it follows that

$$B_1 < 0. \tag{8.3.21}$$

From (7.1.15) we know that $A_1B_1 - A_2 = \frac{\pi^2}{a}p_2$. The assumption $p_2 \neq 0$ as well as the fact that only finitely many λ_k lie in the closed lower half-plane imply that $p_2 > 0$. Hence we have

$$A_1B_1 - A_2 > 0. \tag{8.3.22}$$

We now define

$$\Xi_2 = B_1^{-1}\Phi_2, \tag{8.3.23}$$

$$\Xi_{1,\delta}(\lambda) = \frac{B_1}{A_1 B_1 - A_2}\Phi_1(\lambda) - \frac{\delta\lambda}{A_1 B_1 - A_2}\Phi_2(\lambda)$$

$$+ \frac{1}{A_1 B_1 - A_2}\left(\frac{A_2^2}{B_1^2} - \frac{A_2 A_1}{B_1} + \frac{B_3}{B_1} - B_2\right)\Phi_2(\lambda), \quad \lambda, \delta \in \mathbb{C}, \tag{8.3.24}$$

$$\Xi_1 = \Xi_{1,1}. \tag{8.3.25}$$

It follows from (8.3.19), (8.3.20), (8.3.23) and (8.3.25) that

$$\Xi_1(\lambda^2) = \cos\lambda a - \frac{A_2}{B_1}\frac{\sin\lambda a}{\lambda} + \frac{\psi_2(\lambda)}{\lambda}, \tag{8.3.26}$$

$$\Xi_2(\lambda^2) = \frac{\sin\lambda a}{\lambda} + \frac{A_2}{B_1}\frac{\cos\lambda a}{\lambda^2} + \frac{B_3}{B_1}\frac{\sin\lambda a}{\lambda^3} + \frac{\psi_1(\lambda)}{\lambda^3} \tag{8.3.27}$$

where $\psi_j \in \mathcal{L}_o^a$ for $j = 1, 2$.

By (8.3.26) and (8.3.27), the functions $\lambda \mapsto \Xi_1(\lambda^2)$ and $\lambda \mapsto \Xi_2(\lambda^2)$ are of the form (12.3.11) and (12.3.10), respectively, if we observe that Ξ_2 is of the form (8.1.13) and can therefore be written in the form (8.1.14). Hence Lemma 12.3.3 shows that the zeros $(\zeta_k)_{k=1}^\infty$ of Ξ_1 and $(\xi_k)_{k=1}^\infty$ of Ξ_2 can be written in the form $\zeta_k = v_k^2$ and $\xi_k = u_k^2$, $k \in \mathbb{N}$, with

$$v_k = \frac{\pi}{a}\left(k - \frac{1}{2}\right) - \frac{A_2}{B_1\pi}\frac{1}{k} + \frac{\gamma_k^{(1)}}{k}, \tag{8.3.28}$$

$$u_k = \frac{\pi}{a}k - \frac{A_2}{B_1\pi}\frac{1}{k} + \frac{\gamma_k^{(2)}}{k}, \tag{8.3.29}$$

where $(\gamma_k^{(j)})_{k=-\infty, \ k\neq 0}^\infty \in l_2$ for $j = 1, 2$.

We are going to show that these zeros are real, bounded below, and interlace. To this end we observe that $\frac{\Phi_2}{\Phi_1} \in \mathcal{N}_+^{\mathrm{ep}}$ and the fact that Φ_1 and Φ_2 do not have common zeros imply that the zeros of Φ_2 are real and bounded below, see Definition 5.1.26 and Lemma 11.1.3. Since these zeros are the zeros $(\xi_k)_{k=1}^\infty$ of Ξ_2, we have shown that the sequence $(\xi_k)_{k=1}^\infty$ is real, bounded below, and increasing. Furthermore,

$$(A_1 B_1 - A_2)\frac{\Xi_{1,0}}{\Xi_2}(\lambda) = B_1^2\frac{\Phi_1}{\Phi_2}(\lambda) + \frac{A_2^2}{B_1} - A_2 A_1 + B_3 - B_1 B_2.$$

Since $\frac{\Phi_2}{\Phi_1}$ is a Nevanlinna function and since $A_1 B_1 - A_2 > 0$, we conclude with the aid of Lemma 5.1.22, in turn, that also $-\frac{\Phi_1}{\Phi_2}$, $-\frac{\Xi_{1,0}}{\Xi_2}$ and $\frac{\Xi_2}{\Xi_{1,0}}$ are Nevanlinna functions. The leading terms in $\Xi_{1,0}$ and Ξ_2 are positive on the negative real axis, i.e., for λ in (8.3.26) and (8.3.27) on the imaginary axis, and therefore also

$\frac{\Xi_2}{\Xi_{1,0}} \in \mathcal{N}_+^{\mathrm{ep}}$. Since Φ_1 and Φ_2 do not have common zeros, also $\Xi_{1,\delta}$ and Ξ_2 do not have common zeros for each $\delta \in \mathbb{C}$. From Corollary 5.2.3 and Lemma 11.1.3 we conclude that the zeros $(\zeta_{0,k})_{k=1}^\infty$ of $\Xi_{1,0}$ are real and interlace with the zeros of Ξ_2:

$$\zeta_{0,1} < \xi_1 < \zeta_{0,2} < \xi_2 < \cdots .$$

Let $k \geq 2$, let $\delta_0 \in \mathbb{R}$ and let η be a zero of Ξ_{1,δ_0} in (ξ_{k-1}, ξ_k), if any, of multiplicity l. In view of Theorem 11.1.1, $\Xi_{1,\delta}$ has exactly l continuous branches of zeros near η for δ near δ_0, counted with multiplicity. Since $\Xi_{1,\delta}$ is real analytic for $\delta \in \mathbb{R}$, nonreal zeros would appear in conjugate complex pairs, and therefore the parity of the number of the real zeros amongst these zeros is constant for real δ near δ_0. Because $\Xi_{1,\delta}$ and Ξ_2 do not have common zeros, no real zeros of $\Xi_{1,\delta}$ can enter the interval (ξ_{k-1}, ξ_k) through its endpoints. Altogether, it follows that the parity of real zeros of $\Xi_{1,\delta}$ in (ξ_{k-1}, ξ_k) is locally and therefore globally constant for $\delta \in \mathbb{R}$. Clearly, this parity is odd since $\zeta_{0,k}$ is the single zero of $\Xi_{1,0}$ in (ξ_{k-1}, ξ_k).

It follows in particular for $\delta = 1$ that $\Xi_1 = \Xi_{1,1}$ has an odd number of zeros in (ξ_{k-1}, ξ_k) for $k \geq 2$. From (8.3.28) and (8.3.29) we infer that

$$\zeta_k = \frac{\pi^2}{a^2} \left(k - \frac{1}{2} \right)^2 + O(1) \quad \text{and} \quad \xi_k = \frac{\pi^2}{a^2} k^2 + O(1).$$

Hence there is $k_0 \geq 2$ such that for $k > k_0$ we have that the interval (ξ_{k-1}, ξ_k) contains exactly one zero of Ξ_1 and that this zero is ζ_k. Each of the $k_0 - 1$ intervals (ξ_{k-1}, ξ_k), $k = 2, \ldots, k_0 - 1$, must contain at least one of the remaining k_0 zeros of Ξ_1. We are left with one zero of Ξ_1. Again, since Ξ_1 is real analytic, this zero must be real because otherwise there would be an additional conjugate complex zero. Also, this zero must be different from all ξ_k, $k \geq 1$, and cannot lie in any of the intervals (ξ_{k-1}, ξ_k), $k \geq 2$, because otherwise such an interval would have exactly two zeros of Ξ_1, contradicting the parity of the number of these zeros being odd. Hence this remaining zero must lie in $(-\infty, \xi_1)$, and we therefore have shown that

$$\zeta_1 < \xi_1 < \zeta_2 < \xi_2 < \cdots . \tag{8.3.30}$$

In view of (8.3.28), (8.3.29) and (8.3.30), the sequences $(\zeta_k)_{k=1}^\infty$ and $(\xi_k)_{k=1}^\infty$ satisfy the assumptions of Theorem 12.6.2, and therefore there exists a real-valued $q_0 \in L_2(0, a)$ such that $(\zeta_k)_{k=1}^\infty$ is the spectrum of the Dirichlet–Neumann problem

$$y'' + (z - q_0(x)) y = 0,$$
$$y(z, 0) = y'(z, a) = 0,$$

and $(\xi_k)_{k=1}^\infty$ is the spectrum of the corresponding Dirichlet problem

$$y'' + (z - q_0(x)) y = 0,$$
$$y(z, 0) = y(z, a) = 0.$$

Putting $z = \lambda^2$, let $s(\lambda, \cdot)$ be the solution of $y'' + (\lambda^2 - q_0(x))y = 0$ with $y(\lambda, 0) = 0$, $y'(\lambda, 0) = 1$. From Corollary 12.2.10 we know that $s'(\cdot, a)$ is a sine type function, and since $s'(\cdot, a)$ is a characteristic function of the Dirichlet–Neumann problem with $z = \lambda^2$, it follows in view of Lemma 11.2.29 that there is a constant c such that

$$s'(\lambda, a) = c\lambda^l \prod_{k=1}^{\infty}{}' \left(1 - \frac{\lambda^2}{\zeta_k}\right).$$

Here $\prod'$ indicates that the factor, if any, for which $\zeta_k = 0$, is replaced by the term λ^l with $l = 2$, whereas $l = 0$ otherwise. On the other hand, it follows from (8.3.26) that $\lambda \mapsto \Xi_1(\lambda^2)$ is a sine type function and hence $\lambda \mapsto \Xi_1(\lambda^2)$ has the same product representation as $s'(\cdot, a)$ with the same constant c because both functions have the same leading term $\cos \lambda a$. Therefore, $s'(\lambda, a) = \Xi_1(\lambda^2)$ for all $\lambda \in \mathbb{C}$. Similarly, $\lambda \mapsto \lambda s(\lambda, a)$ and $\lambda \mapsto \lambda\Xi_2(\lambda^2)$ are sine type functions with the same zeros and the same leading terms, and we conclude that $s(\lambda, a) = \Xi_2(\lambda^2)$ for all $\lambda \in \mathbb{C}$.

We set

$$m = \frac{B_1}{A_2 - A_1 B_1}, \quad \nu = m(2\varrho - B_1), \quad q = q_0 + \varrho^2,$$

$$\beta = \nu\varrho - m\varrho^2 - B_2 m + \frac{A_2}{B_1} + m\frac{B_3}{B_1}.$$

Then $s(\cdot, \lambda)$ is a solution of (8.3.15) satisfying the initial condition (8.3.16). From (8.3.21) and (8.3.22) we conclude that $m > 0$ and $\nu > 2m\varrho$. Hence it follows from (8.3.23) and (8.3.25) that

$$
\begin{aligned}
&s'(\lambda, a) + (-\lambda^2 m + i\lambda(\nu - 2m\varrho) + \beta - \nu\varrho + m\varrho^2)s(\lambda, a)\\
&= \Xi_1(\lambda^2) + (-\lambda^2 m + i\lambda(\nu - 2m\varrho) + \beta - \nu\varrho + m\varrho^2)\Xi_2(\lambda^2)\\
&= \frac{B_1}{A_1 B_1 - A_2}\Phi_1(\lambda^2) - \frac{\lambda^2}{A_1 B_1 - A_2}\Phi_2(\lambda^2)\\
&\quad + \frac{1}{A_1 B_1 - A_2}\left(\frac{A_2^2}{B_1^2} - \frac{A_2 A_1}{B_1} + \frac{B_3}{B_1} - B_2\right)\Phi_2(\lambda^2)\\
&\quad + (-\lambda^2 m + i\lambda(\nu - 2m\varrho) + \beta - \nu\varrho + m\varrho^2)B_1^{-1}\Phi_2(\lambda^2)\\
&= -m\Phi_1(\lambda^2) + \frac{m\lambda^2}{B_1}\Phi_2(\lambda^2) + \frac{1}{A_1 B_1 - A_2}\left(\frac{A_2^2}{B_1^2} - \frac{A_2 A_1}{B_1} + \frac{B_3}{B_1} - B_2\right)\Phi_2(\lambda^2)\\
&\quad + \left(-\frac{m\lambda^2}{B_1} + i\frac{\nu - 2m\varrho}{B_1}\lambda + \frac{\beta - \nu\varrho + m\varrho^2}{B_1}\right)\Phi_2(\lambda^2)\\
&= -m\tilde{\chi}(\lambda).
\end{aligned}
$$

Hence we have shown that the sequence (8.3.18) is the sequence of the eigenvalues of the problem (8.3.15)–(8.3.17) with m, q, α and β as found above and where $(a, \varrho, m, q, \nu, \beta)$ belongs to $\mathcal{M}^+$.

For the uniqueness we recall that we have already stated at the beginning of this proof that a and ϱ are unique. Hence let $(a, \rho, m, q, \nu, \beta)$ be the tuple constructed above and let $(a, \rho, m_1, q_1, \nu_1, \beta_1) \in \mathcal{M}^+$ for which problem (8.3.15)–(8.3.17) has the same sequence of eigenvalues $(\lambda_k)_{k=-\infty, k\neq 0}^{\infty}$. We are going to show that the two tuples are equal. Let s and s_1 be the solutions of (8.3.15), (8.3.16), $y'(\lambda, 0) = 1$ with q and q_1, respectively. The corresponding characteristic functions of (8.3.15)–(8.3.17) with respect to the parameter tuples $(a, \rho, m, q, \nu, \beta)$ and $(a, \rho, m_1, q_1, \nu_1, \beta_1)$ are

$$\chi(\lambda) = s'(\lambda, a) + (-\lambda^2 m + i\lambda(\nu - 2m\varrho) + \beta - \nu\varrho + m\varrho^2)s(\lambda, a)$$
$$\chi_1(\lambda) = s_1'(\lambda, a) + (-\lambda^2 m_1 + i\lambda(\nu_1 - 2m_1\varrho) + \beta_1 - \nu_1\varrho + m_1\varrho^2)s_1(\lambda, a),$$

respectively. From $m \neq 0$, $m_1 \neq 0$, and Corollary 12.2.10 we know that $\lambda \mapsto (\lambda - \lambda_1)^{-1}\chi(\lambda)$ and $\lambda \mapsto (\lambda - \lambda_1)^{-1}\chi_1(\lambda)$ are sine type functions. Hence it follows in view of Lemma 11.2.29 that the two characteristic functions are multiples of each other, i. e., there is $C \neq 0$ such that

$$\chi(\lambda) = C\chi_1(\lambda), \quad \lambda \in \mathbb{C}.$$

Since $s(\cdot, a)$, $s'(\cdot, a)$, $s_1(\cdot, a)$ and $s_1'(\cdot, a)$ are even entire functions, we have

$$\chi(\lambda) - \chi(-\lambda) = 2i\lambda(\nu - 2m\varrho)s(\lambda, a),$$
$$\chi_1(\lambda) - \chi_1(-\lambda) = 2i\lambda(\nu_1 - 2m_1\varrho)s_1(\lambda, a).$$

Hence $s(\cdot, a)$ and $s_1(\cdot, a)$ are multiples of each other. Since these two functions have the same leading term by Theorem 12.2.9, we conclude that $s(\cdot, a) = s_1(\cdot, a)$ and $\nu - 2m\varrho = C(\nu_1 - 2m_1\varrho)$. We now calculate

$$\chi(\lambda) + \chi(-\lambda) = 2s'(\lambda, a) + \left(-2\lambda^2 m - 2\varrho(\nu - m\varrho) + 2\beta\right) s(\lambda, a)$$
$$= C\left(2s_1'(\lambda, a) + \left(-2\lambda^2 m_1 - 2\varrho(\nu_1 - m_1\varrho) + 2\beta_1\right) s_1(\tau, a)\right)$$
$$= C\left(2s_1'(\lambda, a) + (-2\lambda^2 m_1 + 2\beta_1 + 2m_1\varrho^2)s(\lambda, a)\right) - 2\varrho(\nu - 2m\varrho)s(\lambda, a),$$

and obtain

$$s'(\lambda, a) + (-m\lambda^2 + \beta + m\varrho^2)s(\lambda, a) = Cs_1'(\lambda, a) + C(-m_1\lambda^2 + \beta_1 + m_1\varrho^2)s(\lambda, a).$$

Taking the asymptotic representations of $s(\cdot, a)$, $s'(\cdot, a)$, $s_1(\cdot, a)$ and $s_1'(\cdot, a)$ into account, see Corollary 12.2.10, we conclude that the coefficients of $\lambda^2 s(\lambda, a)$ must coincide, so that $m = Cm_1$. Then

$$s'(\lambda, a) + \beta s(\lambda, a) = C(s_1'(\lambda, a) + \beta_1 s(\lambda, a)).$$

Considering the leading terms again we obtain $C = 1$ and therefore $m = m_1$. Then $\nu - 2m\varrho = C(\nu_1 - 2m_1\varrho)$ implies $\nu = \nu_1$ and

$$s'(\lambda, a) + \beta s(\lambda, a) = s_1'(\lambda, a) + \beta_1 s(\lambda, a).$$

In view of (12.2.22) and (12.2.23) it follows from this latter identity that

$$\frac{K(a,a)+\beta}{\lambda}\sin\lambda a + o(\lambda^{-1}) = \frac{K_1(a,a)+\beta_1}{\lambda}\sin\lambda a + o(\lambda^{-1}), \quad \lambda \in \mathbb{R}\setminus\{0\},$$

and therefore $K(a,a)+\beta = K_1(a,a)+\beta_1$. Applying Theorem 7.3.1 to (8.3.15)–(8.3.17) with these two tuples of parameters, (7.3.7) gives that $K(a,a) = K_1(a,a)$. Consequently $\beta = \beta_1$ and $s'(\cdot,a) = s'_1(\cdot,a)$. Therefore $s_1(\cdot,a)$ and $s'_1(\cdot,a)$ have the same sequences of zeros as $s(\cdot,a)$ and $s'(\cdot,a)$. Using now the uniqueness statement in Theorem 12.6.2 proves that $q - \varrho^2 = q_1 - \varrho^2$, that is, $q = q_1$. $\qquad\square$

8.3.1   Recovering string parameters

In Theorem 8.3.3 we have given a solution for the inverse problem associated with the spectral problem (8.3.1)–(8.3.3), which is the particular case of the spectral problem (2.2.4)–(2.2.6) with $\sigma(s) \equiv 2\varrho\rho(s)$. However, (2.2.4)–(2.2.6) was obtained from the spectral problem (2.2.1)–(2.2.3) via the Liouville transform, and in this subsection we will address the question if and how the parameters in (2.2.1)–(2.2.3) can be recovered from a sequence satisfying the conditions of Lemma 8.3.2.

Theorem 8.3.4. *Let the sequence* $(\lambda_k)_{k=-\infty,k\neq0}^{\infty}$ *of complex numbers satisfy the conditions of Lemma 8.3.2 and let* $l > 0$. *Then there exists a unique string of length* l *with density* $\rho \in W_2^2(0,l)$, $\rho \gg 0$, *with a point mass* $m > 0$ *and a damping coefficient* $\nu > 0$ *at the right end, which generates problem* (2.2.1)–(2.2.3) *with* $\sigma(s) \equiv 2\varrho\rho(s)$ *and the spectrum of which coincides with* $(\lambda_k)_{k=-\infty,k\neq0}^{\infty}$.

Proof. By Theorem 8.3.3 there exists a unique tuple $(a,\varrho,\tilde{m},q,\tilde{\nu},\beta)$ from $\mathcal{M}^+$ such that the sequence $(\lambda_k)_{k=-\infty,k\neq0}^{\infty}$ represents the spectrum of the problem (8.3.1)–(8.3.3) with the parameters $(a,\varrho,\tilde{m},q,\tilde{\nu},\beta)$. For the operator pencil $\tilde{L}$ defined in Subsection 2.2.3 this means that the spectrum of the operator pencil $\lambda \mapsto \tilde{L}(\lambda - i\varrho)$ lies in the open upper half-plane. Writing $L(\lambda) = \lambda^2 M - i\lambda K - A$, $\lambda \in \mathbb{C}$, it follows that $M = \tilde{M}$, $K = \tilde{K} + 2\varrho\tilde{M}$, $A = \tilde{A} + \varrho^2\tilde{M} + \varrho\tilde{K}$, where $\tilde{M}$, $\tilde{K}$ and $\tilde{A}$ are as in Subsection 2.2.3. Therefore $M \gg 0$, $K \geq 0$ and hence the operator A is strictly positive by Theorem 1.3.3. This means that the lowest eigenvalue ν_1 of the operator A is positive. Since the operator A has the representation

$$D(A) = \left\{ \tilde{y} = \begin{pmatrix} y \\ c \end{pmatrix} : y \in W_2^2(0,a),\ y(0) = 0,\ c = y(a) \right\},$$

$$A\begin{pmatrix} y \\ c \end{pmatrix} = \begin{pmatrix} -y'' + qy \\ y'(a) + \beta y(a) \end{pmatrix},$$

it follows that

$$0 < \nu_1 = \min_{\tilde{y}\in D(A)\setminus\{0\}} \frac{(A\tilde{y},\tilde{y})}{\int_0^a |y(x)|^2\,dx + |y(a)|^2} \leq \min_{\tilde{y}\in D(A)\setminus\{0\}} \frac{(A\tilde{y},\tilde{y})}{\int_0^a |y(x)|^2\,dx}.$$

We define the auxiliary operator A_0 in $L_2(0, a)$ by

$$D(A_0) = \{y \in W_2^2(0, a) : y(0) = 0, y'(a) + \beta y(a) = 0\}, \quad A_0 y = -y'' + qy.$$

The operator A_0 is selfadjoint, see, e.g., [285, (4.2.1), (4.2.2)]. For all $y \in D(A_0)$ we have $\tilde{y} = \begin{pmatrix} y \\ y(a) \end{pmatrix} \in D(A)$ and $A\tilde{y} = \begin{pmatrix} A_0 y \\ 0 \end{pmatrix}$, and hence $(A\tilde{y}, \tilde{y}) = (A_0 y, y)$. It follows that

$$0 < \nu_1 \leq \min_{\tilde{y} \in D(A) \setminus \{0\}} \frac{(A\tilde{y}, \tilde{y})}{\int_0^a |y(x)|^2\, dx} \leq \min_{y \in D(A_0) \setminus \{0\}} \frac{(A_0 y, y)}{(y, y)}.$$

Therefore A_0 is bounded below with lower bound ν_1, and the lowest eigenvalue μ_1 of A_0 satisfies $\mu_1 \geq \nu_1 > 0$.

Let ϕ be the solution of the initial value problem

$$\phi'' - q\phi = 0, \quad \phi(a) = 1, \quad \phi'(a) + \beta\phi(a) = 0,$$

and let ψ be the solution of the initial value problem

$$\psi'' - q\psi + \mu_1\psi = 0, \quad \psi(a) = 1, \quad \psi'(a) + \beta\psi(a) = 0.$$

Since $\psi'(a) + \beta\psi(a) = 0$ and since μ_1 is an eigenvalue of A_0, ψ is an eigenfunction of A_0 with respect to its lowest eigenvalue μ_1. We conclude from the Sturm oscillation theorem, see, e.g., [16, Theorem 8.4.5], that ψ has no zeros on $(0, a)$. Both ϕ and ψ are real valued, and since $\mu_1 > 0$, it follows from the Sturm comparison theorem, see, e.g., [54, Chapter 8, Theorem 1.1] for continuous potentials and [271, Theorem 13.1] for integrable potentials, that also ϕ has no zeros on $(0, a)$. Furthermore, since 0 is not an eigenvalue of A_0 and since $\phi'(a) + \beta\phi(a) = 0$, it follows that $\phi(0) \neq 0$. Together with $\phi(a) = 1$ this implies that ϕ is strictly positive.

We recall from the very beginning of Subsection 2.2.3 that the density ρ of the string has to satisfy

$$q(x) = \rho^{-\frac{1}{4}}(s(x)) \frac{d^2}{dx^2} \rho^{\frac{1}{4}}(s(x)), \tag{8.3.31}$$

$$a = \int_0^l \rho^{\frac{1}{2}}(r)\, dr, \tag{8.3.32}$$

$$\beta = -\rho^{-\frac{1}{4}}(s(a)) \left. \frac{d\rho^{\frac{1}{4}}(s(x))}{dx} \right|_{x=a}, \tag{8.3.33}$$

where

$$x(s) = \int_0^s \rho^{\frac{1}{2}}(\tau)\, d\tau, \quad 0 \leq s \leq l. \tag{8.3.34}$$

and $x \mapsto s(x)$ is the inverse of $s \mapsto x(s)$. Any solution $\rho^{\frac{1}{4}} \circ s$ of (8.3.31), (8.3.33) is a multiple of ϕ, say $C\phi$, $C \in \mathbb{C}$. The equations (8.3.32) and (8.3.34) show that

$l = s(a)$ and

$$s'(x) = \frac{1}{x'(s(x))} = \frac{1}{\rho^{\frac{1}{2}}(s(x))} = \frac{1}{C^2 \phi^2(x)},$$

and therefore

$$l = s(a) = \int_0^a \frac{dx}{C^2 \phi^2(x)}.$$

Hence there is a unique strictly positive function $\rho \in W_2^2(0, l)$ satisfying (8.3.31)–(8.3.34), and this function is given by

$$\rho(s) = C\phi(x(s)), \quad 0 \le s \le l,$$

where

$$C = \left(\frac{1}{l} \int_0^a \frac{dx}{\phi^2(x)} \right)^{\frac{1}{2}}$$

and

$$s(x) = \int_0^x \frac{dt}{C^2 \phi^2(t)}, \quad 0 \le x \le a.$$

Finally, we know from Subsection 2.2.3 that $m = \rho^{\frac{1}{2}}(s(a))\tilde{m}$ and $\nu = \rho^{\frac{1}{2}}(s(a))\tilde{\nu}$.

Altogether, we have shown that there is a unique eigenvalue problem (2.2.1)–(2.2.3) with $\rho \in W_2^2(0, l)$ and $\sigma = 2\varrho\rho$ whose spectrum is the given sequence of complex numbers.   □

8.4 The inverse Sturm–Liouville problem on a star graph

We revisit the system (7.4.1)–(7.4.4) with $\tilde{\nu} = 0$:

$$y_j'' + \lambda^2 y_j - q_j(x)y_j = 0, \quad j = 1, \ldots, p, \ x \in (0, a), \tag{8.4.1}$$

$$y_j(\lambda, 0) = 0, \quad j = 1, \ldots, p, \tag{8.4.2}$$

$$y_1(\lambda, a) = \cdots = y_p(\lambda, a), \tag{8.4.3}$$

$$\sum_{j=1}^p y_j'(\lambda, a) + \beta y_1(\lambda, a) = 0. \tag{8.4.4}$$

In this section we deal with the inverse problem of recovering the potentials q_j, $j = 1, \ldots, p$, and the parameter β from spectra of (8.4.1)–(8.4.4) and the related problems

$$y_j'' + \lambda^2 y_j - q_j(x)y_j = 0, \quad j = 1, \ldots, p, \ x \in (0, a), \tag{8.4.5}$$

$$y_j(\lambda, 0) = y_j(\lambda, a) = 0, \quad j = 1, \ldots, p. \tag{8.4.6}$$

Denote by $\mathcal{Q}$ the class of tuples $((q_j)_{j=1}^p, \beta)$ where the real-valued functions q_j, $j = 1, \ldots, p$, belong to $L_2(0, a)$ and where $\beta \in \mathbb{R}$.

Theorem 8.4.1. Let $p + 1$ properly indexed sequences $(\nu_k^{(j)})_{k=-\infty, k\neq 0}^{\infty}$, $j = 1, \ldots, p$, and $(\zeta_k)_{k=-\infty, k\neq 0}^{\infty}$ of real numbers be given, satisfying the following conditions:

1. The sequences $(\nu_k^{(j)})_{k=-\infty, k\neq 0}^{\infty}$, $j = 1, \ldots, p$, are such that:

   (i) $\nu_1^{(j)} > 0$;

   (ii) $\nu_k^{(j)} \neq \nu_{k'}^{(j')}$ whenever $(k, j) \neq (k', j')$;

   (iii) $$\nu_k^{(j)} = \frac{\pi k}{a} + \frac{B_j}{k} + \frac{\delta_k^{(j)}}{k^2}, \quad j = 1, \ldots, p, \; k \in \mathbb{N}, \tag{8.4.7}$$

   where the B_j are real constants, $B_j \neq B_{j'}$ for $j \neq j'$, and where $(\delta_k^{(j)})_{k=1}^{\infty} \in l_2$ for $j = 1, \ldots, p$.

2. The sequence $(\zeta_k)_{k=-\infty, k\neq 0}^{\infty}$ can be represented as the union of p properly indexed subsequences $(\rho_k^{(j)})_{k=-\infty, k\neq 0}^{\infty}$, $k = 1, \ldots, p$, which have the asymptotic behavior

$$\rho_k^{(j)} = \frac{\pi k}{a} + \frac{M_j}{k} + \frac{\beta_k^{(j)}}{k^2}, \quad j = 1, \ldots, p-1, \; k \in \mathbb{N}, \tag{8.4.8}$$

$$\rho_k^{(p)} = \frac{\pi(k - \frac{1}{2})}{a} + \frac{B_0}{k} + \frac{\beta_k^{(n)}}{k^2}, \quad k \in \mathbb{N}, \tag{8.4.9}$$

where $(\beta_k^{(j)})_{k=-\infty, k\neq 0}^{\infty} \in l_2$ for $j = 1, \ldots, p$, $B_0 \in \mathbb{R}$ and M_j, $j = 1, \ldots, p-1$, are the roots of the polynomial P defined by (7.4.22).

3. The properly indexed sequences of real numbers $(\zeta_k)_{k=-\infty, k\neq 0}^{\infty}$ and $(\xi_k)_{k=-\infty}^{\infty}$ interlace, where $\xi_0 = 0$ and where the sequence $(\xi_k)_{k=-\infty, k\neq 0}^{\infty}$ is the union of the sequences $(\nu_k^{(j)})_{k=-\infty, k\neq 0}^{\infty}$, $j = 1, \ldots, p$:

$$\cdots < \xi_{-1} < \zeta_{-1} < \xi_0 < \zeta_1 < \xi_1 < \zeta_2 < \cdots . \tag{8.4.10}$$

Then there exists a unique $((q_j))_{j=1}^{p}, \beta)$ in Q such that the sequence $(\zeta_k)_{k=-\infty, k\neq 0}^{\infty}$ coincides with the spectrum of problem (8.4.1)–(8.4.4), where

$$\beta = \pi \left(pB_0 - \sum_{j=1}^{p} B_j \right), \tag{8.4.11}$$

and such that the sequences $(\nu_k^{(j)})_{k=-\infty, \; k\neq 0}^{\infty}$, $j = 1, \ldots, p$, coincide with the spectra of problems (8.4.5), (8.4.6).

Proof. First note that we know from the proof of Lemma 7.4.6 that the numbers M_j, $j = 1, \ldots, p - 1$, are mutually distinct since this is true for the numbers B_j for $j = 1, \ldots, p$.

According to Lemma 12.3.4, the entire functions defined by

$$\varphi_j(\lambda) = a \prod_{k=1}^{\infty} \left(\frac{a^2}{\pi^2 k^2} \left((\rho_k^{(j)})^2 - \lambda^2 \right) \right), \quad j = 1, \ldots, p-1, \qquad (8.4.12)$$

$$\varphi_p(\lambda) = \prod_{k=1}^{\infty} \left(\frac{a^2}{\pi^2 (k - \frac{1}{2})^2} \left((\rho_k^{(p)})^2 - \lambda^2 \right) \right), \qquad (8.4.13)$$

$$\Phi(\lambda) = p \prod_{j=1}^{p} \varphi_j(\lambda), \qquad (8.4.14)$$

$$s_j(\lambda) = a \prod_{k=1}^{\infty} \left(\frac{a^2}{\pi^2 k^2} \left((\nu_k^{(j)})^2 - \lambda^2 \right) \right), \quad j = 1, \ldots, p, \qquad (8.4.15)$$

satisfy the representations

$$\varphi_j(\lambda) = \frac{\sin \lambda a}{\lambda} - \frac{\pi M_j \cos \lambda a}{\lambda^2} + \frac{E_j \sin \lambda a}{\lambda^3} + \frac{f_j(\lambda)}{\lambda^3}, \quad j = 1, \ldots, p-1, \quad (8.4.16)$$

$$\varphi_p(\lambda) = \cos \lambda a + \frac{\pi B_0 \sin \lambda a}{\lambda} + \frac{E_p \cos \lambda a}{\lambda^2} + \frac{f_p(\lambda)}{\lambda^2}, \qquad (8.4.17)$$

$$s_j(\lambda) = \frac{\sin \lambda a}{\lambda} - \frac{\pi B_j \cos \lambda a}{\lambda^2} + D_j \frac{\sin \lambda a}{\lambda^3} + \frac{g_j(\lambda)}{\lambda^3}, \quad j = 1, \ldots, p, \quad (8.4.18)$$

where $E_j, D_j \in \mathbb{R}$ and $f_j, g_j \in \mathcal{L}^a$ for $j = 1, \ldots, p$. Here we have used that the function

$$\lambda \mapsto \cos \lambda a - \frac{4\lambda^2 a^2 \cos \lambda a}{4\lambda^2 a^2 - \pi^2}$$

is an entire function belonging to $\mathcal{L}^a$ which satisfies the estimate $O(\lambda^{-2})$ for $\lambda \in \mathbb{R}$ with $|\lambda| \to \infty$.

Substituting (8.4.7) into (8.4.16), (8.4.17) and (8.4.18) we obtain

$$\varphi_l(\nu_k^{(j)}) = (-1)^k \frac{a^2 (B_j - M_l)}{\pi k^2} + \frac{\delta_k^{(l,j)}}{k^3}, \quad l = 1, \ldots, n-1, \; j = 1, \ldots, n, \quad (8.4.19)$$

$$\varphi_p(\nu_k^{(j)}) = (-1)^k \left(1 - \frac{a^2 B_j^2}{2k^2} + \frac{a^2 B_j B_0}{k^2} + \frac{a^2 E_p}{\pi^2 k^2} \right) + \frac{\delta_k^{(p,j)}}{k^2}, \; j = 1, \ldots, p, \quad (8.4.20)$$

$$s_l(\nu_k^{(j)}) = (-1)^k \frac{a^2 (B_j - B_l)}{\pi k^2} + \frac{\tilde{\delta}_k^{(l,j)}}{k^3}, \quad j, l = 1, \ldots, p, \; j \neq l, \qquad (8.4.21)$$

where $(\delta_k^{(l,j)})_{k=-\infty, k\neq 0}^{\infty}, (\tilde{\delta}_k^{(l,j)})_{k=-\infty, k\neq 0}^{\infty} \in l_2$ for $j, l = 1, \ldots, p$.

In the following, j is any integer with $1 \leq j \leq p$. For $k \in \mathbb{Z} \setminus \{0\}$ define

$$X_k^{(j)} := \nu_k^{(j)} \left(\Phi(\nu_k^{(j)}) \prod_{\substack{l=1 \\ l \neq j}}^{p} \frac{1}{s_l(\nu_k^{(j)})} - \cos \nu_k^{(j)} a - \frac{\pi B_j \sin \nu_k^{(j)} a}{\nu_k^{(j)}} \right). \qquad (8.4.22)$$

Observe that the terms of a properly indexed sequence of real numbers satisfy $\alpha_{-k} = -\alpha_k$ for all $k \in \mathbb{N}$. Since the functions φ_l and s_l, $l = 1, \ldots, p$, are even functions, it is therefore clear that $X_{-k}^{(j)} = -X_k^{(j)}$. With P and P_1 as defined in (7.4.22) and (7.4.23) and taking into account that the M_l, $l = 1, \ldots, p-1$, are the roots of P, we have

$$\prod_{\substack{l=1 \\ l \neq j}}^{p} (B_j - B_l) = \lim_{t \to B_j} \frac{P_1(t)}{t - B_j} = P_1'(B_j) = P(B_j) = p \prod_{l=1}^{p-1} (B_j - M_l).$$

Substituting (8.4.19) and (8.4.20) into (8.4.14) with $\lambda = \nu_k^{(j)}$ we then see that

$$\Phi(\nu_k^{(j)}) \prod_{\substack{l=1 \\ l \neq j}}^{p} \frac{1}{s_l(\nu_k^{(j)})} = (-1)^k \frac{p \prod_{l=1}^{p-1} (B_j - M_l)}{p \prod_{\substack{l=1 \\ l \neq j}}^{p} (B_j - B_l)} + \frac{\tilde{\delta}_k^{(j)}}{k} = (-1)^k + \frac{\tilde{\delta}_k^{(j)}}{k}.$$

This together with the evident asymptotic representation

$$\cos \nu_k^{(j)} a + \frac{\pi B_j \sin \nu_k^{(j)} a}{\nu_k^{(j)}} \underset{k \to \infty}{=} (-1)^k + O(k^{-2})$$

shows that

$$(X_k^{(j)})_{k=-\infty, k \neq 0}^{\infty} \in l_2, \quad j = 1, \ldots, p. \tag{8.4.23}$$

We now apply Theorem 11.3.14 to the sine type function $s_{j,0}$ defined by $s_{j,0}(\lambda) = \lambda s_j(\lambda)$ and the sequence $(X_k^{(j)})_{k=-\infty}^{\infty}$ with $X_0^{(j)} = 0$. Hence the functions ε_j, $j = 1, \ldots, p$, defined by

$$\varepsilon_j(\lambda) = s_{j,0}(\lambda) \sum_{k=-\infty}^{\infty} \frac{X_k^{(j)}}{s_{j,0}'(\nu_k^{(j)})(\lambda - \nu_k^{(j)})} = \lambda s_j(\lambda) \sum_{\substack{k=-\infty \\ k \neq 0}}^{\infty} \frac{X_k^{(j)}}{s_j'(\nu_k^{(j)})\nu_k^{(j)}(\lambda - \nu_k^{(j)})} \tag{8.4.24}$$

converge uniformly on any compact subset of the complex plane and in the norm of $L_2(\mathbb{R})$ on $\mathbb{R}$ to an entire function ε_j which belongs to $\mathcal{L}_o^a$. Then the functions r_j, $j = 1, \ldots, n$, defined by

$$r_j(\lambda) = \cos \lambda a + \frac{\pi B_j \sin \lambda a}{\lambda} + \frac{\varepsilon_j(\lambda)}{\lambda}, \tag{8.4.25}$$

are even entire functions. Due to (8.4.24), $\varepsilon_j(\nu_k^{(j)}) = X_k^{(j)}$, and thus equation (8.4.22) implies

$$r_j(\nu_k^{(j)}) = \Phi(\nu_k^{(j)}) \prod_{\substack{l=1 \\ l \neq j}}^{p} \frac{1}{s_l(\nu_k^{(j)})}. \tag{8.4.26}$$

By Lemma 12.3.3, the zeros of the function r_j, $j = 1, \ldots, p$, are real with the possible exception of a finite number of pure imaginary zeros, and can be written as a properly indexed sequence $(\mu_k^{(j)})_{k=-\infty}^{\infty}$, $k \neq 0$. Hence we may assume that $(\mu_k^{(j)})^2 \leq (\mu_{k+1}^{(j)})^2$ for all $k \in \mathbb{N}$. Furthermore, it also follows from Lemma 12.3.3 that

$$\mu_k^{(j)} = \frac{\pi}{a}\left(k - \frac{1}{2}\right) + \frac{B_j}{k} + \frac{\gamma_k^{(j)}}{k}, \tag{8.4.27}$$

where $(\gamma_k^{(j)})_{k=-\infty}^{\infty}$, $k \neq 0 \in l_2$.

Next we are going to show that

$$\frac{\Phi(0)}{\prod\limits_{l=1}^{p} s_l(0)} > 0, \quad (-1)^k \Phi(\nu_k^{(j)}) \prod\limits_{\substack{l=1 \\ l \neq j}}^{p} \frac{1}{s_l(\nu_k^{(j)})} > 0, \; j = 1, \ldots, p, \; k \in \mathbb{N}. \tag{8.4.28}$$

The first inequality is obvious since $\Phi(0) > 0$ and $s_l(0) > 0$ by (8.4.12)–(8.4.15). Since the real entire function

$$\lambda \mapsto \Phi(\lambda) \prod\limits_{l=1}^{p} \frac{1}{s_l(\lambda)}$$

has simple poles and zeros by condition 3, it changes sign at its poles and zeros. Hence it follows from the first inequality in (8.4.28) and from (8.4.10) that for all $l \in \mathbb{N}$, this function is positive on the intervals (ξ_l, ζ_{l+1}) and negative on the intervals (ζ_l, ξ_l). Thus

$$\Phi(\nu_k^{(j)}) \prod\limits_{\substack{l=1 \\ l \neq j}}^{n} \frac{1}{s_l(\nu_k^{(j)})} \lim\limits_{\lambda \to \nu_k^{(j)}} \frac{(\lambda - \nu_k^{(j)})}{s_j(\lambda)} = \lim\limits_{\lambda \to \nu_k^{(j)}} \frac{\Phi(\lambda)(\lambda - \nu_k^{(j)})}{\prod\limits_{l=1}^{p} s_l(\lambda)} > 0. \tag{8.4.29}$$

With a similar reasoning we conclude that

$$\lim\limits_{\lambda \to \nu_k^{(j)}} \frac{(\lambda - \nu_k^{(j)})}{s_j(\lambda)}(-1)^k > 0, \; k \in \mathbb{N}. \tag{8.4.30}$$

The inequalities on the right-hand side of (8.4.28) now follow from (8.4.29) and (8.4.30).

Using (8.4.28) we obtain

$$(-1)^k r_j(\nu_k^{(j)}) > 0.$$

This means that between consecutive $\nu_k^{(j)}$ there is an odd number of $\mu_k^{(j)}$. From the asymptotic formulas (8.4.7) and (8.4.27) we can now conclude that

$$\nu_1^{(j)} < \mu_2^{(j)} < \nu_2^{(j)} < \mu_3^{(j)} < \cdots \quad \text{and} \quad (\mu_1^{(j)})^2 < (\nu_1^{(j)})^2. \tag{8.4.31}$$

Indeed, otherwise there would be some k and k' such that there are at least three distinct indices l for which $\mu_l^{(j)}$ lies between $\nu_k^{(j)}$ and $\nu_{k+1}^{(j)}$ and no index l such that $\mu_l^{(j)}$ lies between $\nu_{k'}^{(j)}$ and $\nu_{k'+1}^{(j)}$, which is impossible.

Due to (8.4.7), (8.4.27) and (8.4.31), the two sequences $((\nu_k^{(j)})^2)_{k=-\infty,k\neq 0}^{\infty}$ and $((\mu_k^{(j)})^2)_{-\infty,k\neq 0}^{\infty}$ satisfy the conditions of Theorem 12.6.2. Thus, there is a real function $q_j \in L_2(0,a)$ such that $(\nu_k^{(j)})_{k=-\infty,k\neq 0}^{\infty}$ is the sequence of eigenvalues of the Dirichlet–Dirichlet problem (8.4.5), (8.4.6) and $(\mu_k^{(j)})_{k=-\infty,k\neq 0}^{\infty}$ is the sequence of eigenvalues of the Dirichlet–Neumann problem

$$\begin{aligned} y_j'' + \lambda^2 y_j - q_j(x)y_j &= 0, \\ y_j(\lambda,0) = y_j'(\lambda,a) &= 0. \end{aligned} \tag{8.4.32}$$

Finally, we define β by (8.4.11).

We are going to prove that the sequence $(\zeta_k)_{k=-\infty,k\neq 0}^{\infty}$ is the spectrum of problem (8.4.1)–(8.4.4) with the above q_j, $j = 1,\ldots,p$, and β defined by (8.4.11). Indeed, for $j = 1,\ldots,p$, let $\tilde{s}_j(\lambda,\cdot)$ be the solution of (8.4.5) with the potential q_j which satisfies $\tilde{s}_j(\lambda,0) = 0$ and $\tilde{s}_j'(\lambda,0) = 1$. By (7.4.5) the characteristic function φ of problem (8.4.1)–(8.4.4) is given by

$$\phi(\lambda) = \sum_{m=1}^{p} \tilde{s}_m'(\lambda,a) \prod_{\substack{l=1 \\ l\neq m}}^{p} \tilde{s}_l(\lambda,a) + \beta \prod_{m=1}^{p} \tilde{s}_m(\lambda,a). \tag{8.4.33}$$

We already know that the sequence of the zeros of $\tilde{s}_j(\cdot,a)$ coincides with the sequence of the zeros of s_j and that the sequence of the zeros of $\tilde{s}_j'(\cdot,a)$ coincides with the sequence of the zeros of r_j. In view of (12.2.22), (12.2.23), (8.4.18) and (8.4.25), $\tilde{s}_j(\cdot,a)$ and s_j as well as $\tilde{s}_j'(\cdot,a)$ and r_j have the same leading terms. Since r_j and $\lambda \mapsto \lambda s_j(\lambda)$ are sine type functions, it follows from Lemma 11.2.29 that $\tilde{s}_j(\cdot,a) = s_j$ and $\tilde{s}_j'(\cdot,a) = r_j$. By (7.4.11),

$$\tilde{s}_j'(\lambda,a) = \cos\lambda a + \tilde{B}_j \pi \frac{\sin\lambda a}{\lambda} + \frac{\tilde{\tau}_j(\lambda)}{\lambda}$$

with $\tilde{B}_j \in \mathbb{R}$ and $\tilde{\tau}_j \in \mathcal{L}^a$. From the representation (8.4.27) of the zeros of $\tilde{s}_j(\cdot,a)$ and from Lemma 12.3.3 we conclude that $\tilde{B}_j = B_j$. Hence ϕ has the representation (7.4.5) with $\tilde{\nu} = 0$, (7.4.12), (7.4.13) and with the numbers B_j as given in (8.4.7). For each $k \in \mathbb{Z} \setminus \{0\}$ and $j = 1,\ldots,p$ we obtain with the aid of (8.4.26) that

$$\phi(\nu_k^{(j)}) = \sum_{m=1}^{p} r_m(\nu_k^{(j)}) \prod_{\substack{l=1 \\ l\neq m}}^{p} s_l(\nu_k^{(j)}) + \beta \prod_{m=1}^{p} s_m(\nu_k^{(j)})$$

$$= r_j(\nu_k^{(j)}) \prod_{\substack{l=1 \\ l\neq j}}^{p} s_l(\nu_k^{(j)}) = \Phi(\nu_k^{(j)}).$$

This implies that the entire function $\Delta := \phi - \Phi$ is zero at each $\nu_k^{(j)}$, $k \in \mathbb{Z} \setminus \{0\}$, $j = 1, \ldots, n$. Hence

$$\omega := \Delta \prod_{j=1}^{p} \frac{1}{s_j} \qquad (8.4.34)$$

is an entire function. Substituting (8.4.16) and (8.4.17) into (8.4.14) we obtain

$$\Phi(\lambda) = p \frac{\sin^{p-1} \lambda a}{\lambda^{p-1}} \cos \lambda a - p \frac{\sin^{p-2} \lambda a}{\lambda^p} \cos^2 \lambda a \sum_{j=1}^{p-1} \pi M_j + p B_0 \pi \frac{\sin^p \lambda a}{\lambda^p} + \frac{\psi(\lambda)}{\lambda^{p+1}},$$

$$(8.4.35)$$

where $\psi \in \mathcal{L}^{ap}$. We recall that for a polynomial $\lambda \mapsto d_0 \lambda^n + d_1 \lambda^{n-1} + \cdots$, the number $-d_0^{-1} d_1$ equals the sum of its zeros. Hence, writing

$$P_1(\lambda) = \lambda^p + \sum_{j=1}^{p} c_j \lambda^{p-j}$$

and observing that $P_1' = P$, we obtain

$$(p-1) \sum_{j=1}^{p} B_j = -(p-1)c_1 = p \sum_{j=1}^{p-1} M_j.$$

Comparing (7.4.8) with (8.4.35) and taking (8.4.11) into account, we obtain

$$\Delta(\lambda) = \frac{\tilde{\psi}(\lambda)}{\lambda^p},$$

where $\tilde{\psi} \in \mathcal{L}^{ap}$. Since $\lambda \mapsto \lambda^p \prod_{j=1}^{p} s_j(\lambda)$ is a sine type function of exponential type ap, it follows from Lemma 12.2.4 and Remark 11.2.21 that ω is bounded outside discs of radius $\delta < \frac{\pi}{4a}$ centred at the $\nu_k^{(j)}$ for $k \in \mathbb{Z} \setminus \{0\}$ and $j = 1, \ldots, p$ and that $|\omega(\lambda)| \to 0$ as $|\lambda| \to 0$ outside these discs. Hence the entire function ω is bounded by the Maximum Modulus Theorem. We conclude from Liouville's theorem that $\tilde{\psi} = 0$. Consequently, the sequence $(\zeta_k)_{k=-\infty, k \neq 0}^{\infty}$ coincides with the sequence of eigenvalues of problem (8.4.1)–(8.4.4) generated by the tuple $((q_j)_{j=1}^n, \beta)$ obtained in this proof.

By Theorem 12.6.2, the functions q_j, $j = 1, \ldots, p$, are uniquely determined by the sequence of the eigenvalues $(\nu_k^{(j)})_{k=-\infty, k \neq 0}^{\infty}$ of the Dirichlet–Dirichlet problem (8.4.5), (8.4.6) and the sequence of the eigenvalues $(\mu_k^{(j)})_{k=-\infty, k \neq 0}^{\infty}$ of the Dirichlet–Neumann problem (8.4.5), (8.4.32). But this latter sequence is the sequence of the zeros of the function r_j, and r_j is uniquely determined by the given sequences $(\nu_k^{(j)})_{k=-\infty, k \neq 0}^{\infty}$, $j = 1, \ldots, p$, and $(\zeta_k)_{k=-\infty, k \neq 0}^{\infty}$. Therefore the potentials q_j, $j = 1, \ldots, p$, are uniquely determined by these sequences. Finally, the numbers B_j, $j = 0, \ldots, p$, are uniquely determined by (8.4.7) and (8.4.9). Comparing (8.4.9) and (7.4.26) with $\tilde{\nu} = 0$, it follows that also β is uniquely determined. $\qquad \square$

8.5 Notes

The history of inverse problems generated by the Sturm–Liouville equation begins with V. Ambarzumian's theorem [12]. Ambarzumian considered the exceptional case of the Neumann–Neumann boundary value problem and stated that its spectrum uniquely determines the potential. A correct proof of this theorem was given by G. Borg [32], who also showed that in general two spectra of boundary value problems with self-adjoint separated boundary conditions uniquely determine the potential. For example one can take the spectra of Dirichlet–Dirichlet and Dirichlet–Neumann problems. We see from Theorem 8.1.4 that instead of two real spectra one can take one complex spectrum of problem (8.1.1)–(8.1.3). This was proved in [223, Theorem 4.7]. But the phenomenon has been already known since M.G. Kreĭn and A.A. Nudel′man's papers [156] and [157]. In [157] a class of strings was introduced which is more general than the class of regular strings.

A string on (l_1, a), $-\infty \le l_1 < a \le \infty$ with mass distribution function M is called regular at the right end a if $a < \infty$ and $\lim_{s \to a-0} M(s) < \infty$. The class $\mathfrak{S}$ denotes the set of all strings with regular right end and with finite momentum $\int_{l_1}^{a} (l - s) dM(s)$. It is easy to see that such a string has finite mass $\int_{l_1}^{a} dM(s)$ but its length can be infinite. General properties of such strings were investigated in [127], [125] and [71]. For a connection with canonical systems see [132].

One of the results of [157] is that the spectrum of a boundary value problem generated by (2.8.2) with the Neumann boundary condition

$$u'(0 + 0) = 0 \tag{8.5.1}$$

and a dissipative boundary condition linearly dependent on the spectral parameter

$$u(a) - i\lambda^{-1} u'(a - 0) = 0 \tag{8.5.2}$$

together with the total length of the string (if it is finite) uniquely determine the mass distribution on the string. They proved the following theorem.

Theorem 8.5.1 ([157, Theorem 3.1]). *Let $K = (\lambda_k)$ be a sequence of complex numbers. In order that K be the spectrum of problem (2.8.2), (8.5.1), (8.5.2) generated by a mass distribution $M \in \mathfrak{S}$ it is necessary and sufficient that the following conditions are satisfied:*

1. *The sequence K is symmetric with respect to the imaginary axis, and symmetrically located terms have the same multiplicity;*
2. *$\operatorname{Im} \lambda_k > 0$ for all $\lambda_k \in K$;*
3. *$\sum_j \operatorname{Im}(-\frac{1}{\lambda_j}) < \infty$;*
4. *$\sum_j |\lambda_j|^{-2} < \infty$.*

In the proof of this theorem the authors made essential use of the 'main' theorem on existence of a string corresponding to two spectra, which was proved in [35]–[41]. This 'main' theorem can already be found without proof in [127]; it is also presented in the monographs [71] and [42].

It should be mentioned that in implicit form a result of this kind has been obtained earlier by D.Z. Arov [15]. He proved the following theorem

Theorem 8.5.2 ([15, Theorem 4.1]). *For a set K of complex numbers located in the open upper half-plane symmetric with respect to the imaginary axis to be the spectrum of problem* (2.8.2), (8.5.1), (8.5.2) *generated by a regular string it is necessary and sufficient that K is the set of zeros of an entire function F of exponential type which satisfies the inequalities*

$$\int_{-\infty}^{\infty} (1+x^2)^{-1} |F(x)|^{-2}\, dx < \infty, \quad \int_{-\infty}^{\infty} (1+x^2)^{-1} \log^+ |F(x)|\, dx < \infty.$$

The last inequality means that F belongs to the Cartwright class, see [173, Chapter 5], where this class is called class A.

It was shown in [275] that two spectra of Neumann and Dirichlet boundary value problems uniquely determine the tension and the density of a string if the damping is a known constant function.

The problem of small vibrations of a smooth inhomogeneous string which is damped at one end is described by (8.1.1)–(8.1.3). Of course, from a physical point of view it is clear that the spectrum must lie in the open upper half-plane because the corresponding dynamical system, a damped string, is stable. In this case the operator $\tilde{A}$ defined in Subsection 2.2.3 is strictly positive. Hence it follows from Theorem 8.1.3 that for a sequence (λ_k), which can be countable or empty, to be the spectrum of problem (8.1.1)–(8.1.3) with $a > 0$, $\alpha > 0$, $\beta \in \mathbb{R}$, real-valued $q \in L_2(0, a)$ and strictly positive operator $\tilde{A}$ it is necessary and sufficient that (λ_k) be the set of zeros of an entire function χ of SHB class which has the representation (8.1.4). Numerical results for the generalized inverse Regge problem can be found in [240].

Since asymptotics of eigenvalues in the classical Regge problem are known only in special cases, e.g., if the potential q is continuous and $q(a) \neq 0$, the corresponding inverse problem is solved also in these particular cases [148].

In Section 8.2 we consider the case of a Stieltjes string damped at an interior point. The problem with damping at one of its ends was consider in [162]. A nice review containing experimental results can be found in [58]. A related matrix problem for a vibrational system with damping has been considered in [265], [266] and [163].

Certain results on the inverse Sturm–Liouville problem on a semiaxis and on an axis with the potential linearly dependent on the spectral parameter, which corresponds to a smooth inhomogeneous string vibration in an absorbing medium, were obtained in [123], [8], [9], [10]. The inverse problem for the diffusion equation on a finite interval was considered in [86] (without proof) and solved in [236], [119].

Theorem 8.2.1 was proved in [33, Theorem 4.1]. The theorem in [33] contains a misprint, namely condition (5) must be included.

Since this problem can be considered as a problem on a star graph with two edges, its undamped version was generalized in [34] to the case of a star graph with q edges. Namely, consider the problem

$$\frac{u_k^{(j)} - u_{k+1}^{(j)}}{l_k^{(j)}} + \frac{u_k^{(j)} - u_{k-1}^{(j)}}{l_{k-1}^{(j)}} - m_k^{(j)} \lambda^2 u_k^{(j)} = 0, \quad k = 1, \ldots, n_j, \ j = 1, \ldots, q,$$

(8.5.3)

$$u_0^{(j)} = 0, \quad j = 1, \ldots, q,$$

(8.5.4)

$$u_{n_1+1}^{(1)} = u_{n_2+1}^{(2)} = \cdots = u_{n_q+1}^q,$$

(8.5.5)

$$\sum_{j=1}^q \frac{u_{n_j+1}^{(j)} - u_{n_j}^{(1)}}{l_{n_j}} = 0,$$

(8.5.6)

which bears the Neumann condition at the interior vertex, together with the q Dirichlet problems

$$\frac{u_k^{(j)} - u_{k+1}^{(j)}}{l_k^{(j)}} + \frac{u_k^{(j)} - u_{k-1}^{(j)}}{l_{k-1}^{(j)}} - m_k^{(j)} \lambda^2 u_k^{(j)} = 0, \quad k = 1, \ldots, n_j,$$

(8.5.7)

$$u_0^{(j)} = u_{n_j+1}^{(j)} = 0,$$

(8.5.8)

$j = 1, \ldots, q$, on the edges. It was shown in [34, Theorem 3.1] that these $q + 1$ spectra, if they do not intersect, together with the total lengths of the edges uniquely determine the masses on the edges and the lengths of the subintervals between them. Moreover, [34, Theorems 2.2 and 3.1] give conditions which are necessary and sufficient for $q + 1$ sequences to be the spectra of problems (8.5.3)–(8.5.8). The proof of [34, Theorem 3.1] is constructive and allows to find the masses and the lengths of the subintervals explicitly.

In [224] a more complicated problem was considered: Neumann and Dirichlet conditions were imposed at a pendant vertex of a star graph of Stieltjes strings. The problem was completely solved, but the conditions on two sequences of real numbers to be the spectra of the Dirichlet and Neumann problems on a star graph of Stieltjes strings are given in implicit form in [224, Theorem 3.14]. In this paper and in [226], a connection was noticed with the algebraic problems of possible multiplicities of eigenvalues of the so-called tree patterned matrices, see [171], [202], [124].

Inverse problems for vibrations of tree graphs of Stieltjes strings were considered in [93] and [222]. In [222] it was shown that the two spectra corresponding to the Dirichlet and Neumann boundary conditions at the root of the tree and the length of the edge incident with the root uniquely determine the point masses and the lengths of the subintervals between the point masses of this edge. In this paper, expansion of a Nevanlinna function into a branching continued fraction was used to find the values of point masses and their distribution on a given metric tree. Branching continued fractions were studied in [252]. A similar results on the uniqueness of the potential for the Sturm–Liouville problem on tree graphs was obtained in [44] and [282].

Theorem 8.3.3 with $\varrho = 0$ was proved in [212, Theorem 2.2] and with $\varrho > 0$ in [213, Theorem 4.16]. In [190] an inverse problem for a smooth inhomogeneous string damped at one end and having massless interval at the damped end was solved.

The inverse Sturm–Liouville problem on a star graph with three edges was considered in [215]. For generalizations see [221], [269], [260], and for the case of non-local boundary conditions see [201]. Theorem 8.4.1 was proved in [221, Theorem 4.1]. Related numerical results can be found in [241].

If the sequences $(\nu_k^{(j)})_{k=-\infty,k\neq0}^{\infty}$ intersect, i. e., condition 1.(ii) of Theorem 8.4.1 is violated, and, consequently, also condition 3, then the solution of the inverse problem either is not unique or does not exist for the same reasons as in the case of the three spectra problem, see [91] and [214]. If the sequence $(\zeta_k)_{k=-\infty,k\neq0}^{\infty}$ and the properly indexed sequence $(\xi_k)_{k-\infty}^{\infty}$, which is defined as the union of (0) and the sequences $(\nu_k^{(j)})_{k=-\infty,k\neq0}^{\infty}$, $j-1,\ldots,p$, satisfy the statements of Corollary 5.2.11, then the solution of the inverse problem exists but it is not unique.

In quantum graph theory conditions (8.4.3) and (8.4.4) with $\beta = 0$ are sometimes called Neumann conditions at a interior vertex while conditions (8.4.6) are called Dirichlet conditions at this vertex. The spectrum of the problem with these Neumann conditions alone does not determine the potentials on the edges uniquely. However, an exceptional Ambarzumian like case where one spectrum uniquely determines the potentials exists for quantum graphs too. This was shown for a star graph in [218] and [277] and for trees in [46], [64, Section 5], [170]. Generalizations for Dirac system on graphs can be found in [278], [280].

Unlike in the case of inverse problems on a finite interval, in quantum graph theory another inverse problem arises. If the potential is zero on all edges, does the spectrum determine the form of the graph uniquely? The answer is negative in general, see [26], but positive if the edges are not commensurable, see [103].

In the case of equal lengths of the edges and of equal potentials, symmetric with respect to the middle of the edges, the problem of finding the shape of the graph becomes purely algebraic and can be reduced to the problem of finding the shape of the graph in the classical spectral graph theory [63]. It is known that the spectrum uniquely determines the shape of the graph if the number of its vertices does not exceed 4 for general graphs and 5 for connected graphs, see [63, p. 157 and pp. 272–274]. Counterexamples with 5 and 6 vertices, respectively, can be found in [63, p. 157].

In [216] the inverse problem of recovering the potential on the loop of a lasso graph was solved using the Jost function. In [159] the Titchmarsh–Weyl function was used to prove uniqueness for such problems.

Part IV

Background Material

Chapter 9

Spectral Dependence on a Parameter

9.1 Zeros of analytic functions of two variables

The following theorem on the representation of zeros of an analytic function in two variables is well known, however often formulated in slightly different forms. For the sake of completeness and to have it in exactly the form we need it, the theorem and its proof are given below. For slightly different formulations and proofs we refer the reader to [25, Appendix A 5.4, Theorem 3], [114, Section A.1, Lemma A.1.3] and [185, Section 45, Corollary, p. 303].

Theorem 9.1.1. *Let $\Phi \subset \mathbb{C}^2$ be an open set and let $f : \Phi \to \mathbb{C}$ be analytic. Let $(z_0, w_0) \in \Phi$ such that $f(z_0, w_0) = 0$ and such that $f(\cdot, w_0)$ is not identically zero in a neighbourhood of z_0. Let m be the multiplicity of the zero z_0 of $f(\cdot, w_0)$. Then there are numbers $\varepsilon > 0$, $\delta > 0$, and positive integers l, p_k and m_k, $k = 1, \ldots, l$, such that*

$$\sum_{k=1}^{l} p_k m_k = m$$

and such that for each $w \in \mathbb{C}$ with $|w - w_0| < \varepsilon$ the analytic function $f(\cdot, w)$ has exactly m zeros, counted with multiplicity, in the disc $\{z \in \mathbb{C} : |z - z_0| < \delta\}$. The zeros of $f(\cdot, w)$ can be organized into l groups and denoted by $z_{kj}(w)$, $k = 1, \ldots, l$, $j = 1, \ldots, p_k$, such that the z_{kj} are pairwise different when $0 < |w| < \varepsilon$, have multiplicity m_k, and are represented by Puiseux series

$$z_{kj}(w) = z_0 + \sum_{n=1}^{\infty} a_{kn}(((w - w_0)^{\frac{1}{p_k}})_j)^n, \tag{9.1.1}$$

where, for $j = 1, \ldots, p_k$,

$$((w - w_0)^{\frac{1}{p_k}})_j = |w - w_0|^{\frac{1}{p_k}} \exp\left(\frac{2\pi i(j - 1) + i \arg(w - w_0)}{p_k}\right) \tag{9.1.2}$$

and arg denotes the principal argument. The coefficients a_{kn} of (9.1.1) are complex numbers and the Puiseux series converges in the disc $\{w \in \mathbb{C} : |w - w_0| < \varepsilon\}$.

Proof. For simplicity of proof we may assume that $z_0 = 0 = w_0$. Since $f(\cdot, 0)$ is not identically zero, there is $\delta > 0$ such that $f(z, 0) \neq 0$ for $0 < |z| \leq \delta$. Since f is continuous in z and w, by a compactness argument we can choose $\varepsilon > 0$ such that $|f(z, w) - f(z, 0)| < |f(z, 0)|$ for $|z| = \delta$ and $|w| \leq \varepsilon$. By Rouché's theorem, for each w with $|w| \leq \varepsilon$, $f(\cdot, w)$ has exactly m zeros, say $\tilde{z}_1(w), \ldots, \tilde{z}_m(w)$, counted with multiplicity, in $\{z \in \mathbb{C} : |z| < \delta\}$. Here, for each w we take an arbitrary indexing, and the proof will be complete when we have shown that we can choose a particular indexing satisfying (9.1.1).

We observe that for $0 < \delta' < \delta$ there is $0 < \varepsilon' < \varepsilon$ such that $f(\cdot, w)$ has m zeros in $\{z \in \mathbb{C} : |z| < \delta'\}$ if $|w| < \varepsilon'$, and these roots are $\tilde{z}_1(w), \ldots, \tilde{z}_m(w)$. Hence

$$\max_{\iota=1}^{m} |\tilde{z}_\iota(w)| \to 0 \text{ as } w \to 0. \tag{9.1.3}$$

Now we fix w with $|w| \leq \varepsilon$ and consider one $\tilde{z}_\iota(w)$. This zero of $h = f(\cdot, w)$ has a certain multiplicity $0 < n \leq m$, which depends on w and ι, and we can write

$$h(z) = (z - \tilde{z}_\iota(w))^n g(z),$$

where g is analytic with $g(\tilde{z}_\iota(w)) \neq 0$. Then, for integers $k \geq 0$,

$$\operatorname{res}_{z = \tilde{z}_\iota(w)} \frac{z^k h'(z)}{h(z)} = \operatorname{res}_{z = \tilde{z}_\iota(w)} \frac{z^k n (z - \tilde{z}_\iota(w))^{n-1} g(z)}{(z - \tilde{z}_\iota(w))^n g(z)} = n \tilde{z}_\iota^k(w),$$

and the residue theorem gives that

$$s_k(w) := \frac{1}{2\pi i} \int_{|z|=\delta} \frac{z^k \frac{\partial}{\partial z} f(z, w)}{f(z, w)} \, dz = \sum_{\iota=1}^{m} \tilde{z}_\iota^k(w). \tag{9.1.4}$$

The functions s_k depend analytically on w for $|w| < \varepsilon$ since the above integrand depends continuously on z for $|z| = \delta$ and analytically on w for $|w| < \varepsilon$. We define

$$P(z, w) = \prod_{\iota=1}^{m} (z - \tilde{z}_\iota(w)). \tag{9.1.5}$$

Then P is a monic polynomial of degree m in z, and by Newton's identities, the coefficients of P are polynomials in $s_1, \ldots, s_m$ and hence analytic functions of w, see, e.g., [130] or [242, Theorem 4.3.7]. The discriminant of the polynomial P is defined by

$$\Delta(w) = \prod_{\iota=1}^{m-1} \prod_{k=\iota+1}^{m} (\tilde{z}_\iota(w) - \tilde{z}_k(w))^2, \quad |w| < \varepsilon. \tag{9.1.6}$$

Clearly, Δ is symmetric in $\tilde{z}_1, \ldots, \tilde{z}_m$, i.e., invariant under permutations, and by the fundamental theorem of symmetric polynomials and Newton's identities, see, e.g., [242, Theorem 4.3.7], Δ is a polynomial in $s_1, \ldots, s_m$ and hence analytic.

Observe that Δ is identically zero if and only if $P(\cdot, w)$ has at least one double zero for all $|w| < \varepsilon$, which in turn is true if and only if $P(\cdot, w)$ and $\dfrac{\partial}{\partial z} P(\cdot, w)$ have a common zero for each w with $|w| < \varepsilon$. If Δ is identically zero, let R be the greatest common divisor of P and $\dfrac{\partial}{\partial z} P$ in the polynomial ring (in z) over the field of meromorphic functions (in w) on $\{w \in \mathbb{C} : |w| < \varepsilon\}$. The function R can be found as the last nonzero remainder in the Euclidean algorithm applied to P and $\dfrac{\partial}{\partial z} P$. Since P and $\dfrac{\partial}{\partial z} P$ have a common zero $z(w)$, also each remainder in the Euclidean algorithm has this common zero $z(w)$ for each w which is not a pole of any of the coefficients. Therefore R must be a nonconstant polynomial in z. Dividing by the coefficient of the highest power in z, we may assume that R is monic. Since zeros z of R must also be zeros of P for those w which are not poles of the coefficients of R and P/R, we may assume with a suitable indexing of $\tilde{z}_\iota(w)$, $\iota = 1, \ldots, m$ that

$$R(z, w) = \prod_{\iota=1}^{n} (z - \tilde{z}_\iota(w)),$$

where n is the degree of the polynomial R. Since the $\tilde{z}_\iota(w)$ are bounded functions of w (recall that $|\tilde{z}_\iota(w)| < \delta$), so are the coefficients of R, which therefore must be analytic functions. Similarly,

$$\frac{P(z, w)}{R(z, w)} = \prod_{\iota=n+1}^{m} (z - \tilde{z}_\iota(w))$$

is analytic in w. This shows that P can be factored into nonconstant polynomials with respect to z if $\Delta = 0$. After a finite number of steps we have factored

$$P = \prod_{\kappa=1}^{\nu} P_\kappa$$

into polynomials P_κ in z whose discriminants Δ_κ are not identically zero (note that $\Delta_\kappa = 1$ if the degree of the polynomial is 1). Let $\kappa \in \{1, \ldots, \nu\}$. With a slight abuse of notation, we write $P_\kappa = P$, m for the degree of this P, and Δ for its discriminant. Choosing $\varepsilon > 0$ sufficiently small we may assume that Δ has no zeros in $\{w \in \mathbb{C} : 0 < |w| < \varepsilon\}$.

Clearly, if $P(0, w) = 0$ for all w, then we can write $P(z, w) = z\widetilde{P}(z, w)$, where $\widetilde{P}$ is a polynomial in z and $\widetilde{P}(0, w) \neq 0$ for $0 < |w| < \varepsilon$. Therefore we still have to consider the case that, for some $\varepsilon > 0$, $P(\cdot, w) = 0$ has m zeros which are mutually distinct and nonzero for $0 < |w| < \varepsilon$.

As in (9.1.4) we can now find local branches

$$\hat{z}_\iota(w) = \frac{1}{2\pi i} \int_{\Gamma_{\tilde{z}_\iota(w_0)}} \frac{z \dfrac{\partial}{\partial z} P(z, w)}{P}(z, w)\, dz, \quad \iota = 1, \ldots, m,$$

of zeros of $P(\cdot, w)$ which depend analytically on w near w_0, where $0 < |w_0| < \varepsilon$ and $\Gamma_{\tilde{z}_\iota(w_0)}$ is a contour about $\tilde{z}_\iota(w_0)$ with no other zero of $P(\cdot, w_0)$ inside or on that contour. Hence, by a standard compactness argument, along each curve on the Riemann surface of the logarithm over $\{z \in \mathbb{C} : 0 < |w| < \varepsilon\}$, for definiteness say starting at $\frac{\varepsilon}{2}$, there are unique analytic functions ζ_ι in a neighbourhood of that curve with $\zeta_\iota(\frac{\varepsilon}{2}, 0) = \tilde{z}_\iota(\frac{\varepsilon}{2})$, where (r, φ), $0 < r < \varepsilon$, $\varphi \in \mathbb{R}$, is a point on the Riemann surface over $re^{i\varphi}$. By the principle of analytic continuation, we therefore have well-defined analytic functions ζ_ι on this Riemann surface, $\iota = 1, \ldots, m$, and for each (r, φ), the m complex numbers $\zeta_\iota(r, \varphi)$, $\iota = 1, \ldots, m$, are the m distinct zeros of the polynomial $P(\cdot, re^{i\varphi})$.

Now fix ι, choose some $0 < r_0 < \varepsilon$ and consider the sequence

$$(\zeta_\iota(r_0, 2\pi p))_{p=-\infty}^{\infty}.$$

Since each term of the sequence is a zero of $P(\cdot, r_0)$, there are $p_0 \in \mathbb{Z}$ and $p \in \mathbb{N}$ such that

$$\zeta_\iota(r_0, 2\pi p_0) = \zeta_\iota(r_0, 2\pi(p_0 + p)).$$

Hence, by the uniqueness of the ζ_ι we have that since $\zeta_\iota(r, \varphi)$ and $\zeta_\iota(r, \varphi + 2\pi p)$ coincide in one point they must coincide everywhere, i. e., we have

$$\zeta_\iota(r, \varphi) = \zeta_\iota(r, \varphi + 2\pi p)$$

for all $0 < r < \varepsilon$ and $\varphi \in \mathbb{R}$. We may choose p to be minimal having this property. For $0 \leq p_1 < p_2 < p$ we have

$$\zeta_\iota(r, \varphi + 2\pi p_1) \neq \zeta_\iota(r, \varphi + 2\pi p_2) \quad \text{for all} \quad 0 < r < \varepsilon \quad \text{and} \quad \varphi \in \mathbb{R}$$

or

$$\zeta_\iota(r, \varphi + 2\pi p_1) = \zeta_\iota(r, \varphi + 2\pi p_2) \quad \text{for all} \quad 0 < r < \varepsilon \quad \text{and} \quad \varphi \in \mathbb{R}.$$

Since the latter case would contradict the minimality of p, we therefore have that

$$\zeta_\iota(r, \varphi + 2\pi q), \quad q = 0, \ldots, p - 1,$$

represents a group of p distinct zeros of $P(\cdot, re^{i\varphi})$ satisfying $\zeta_\iota(r, \varphi + 2\pi q + 2\pi p) = \zeta_\iota(r, \varphi + 2\pi q)$. In particular, for $v = (re^{i\varphi})^p$,

$$\xi_\iota(v) = \zeta_\iota(r^p, p\varphi)$$

is a uniquely defined analytic function on $\{v \in \mathbb{C} : 0 < |v| < \varepsilon^p\}$. By (9.1.3), $\xi_\iota(0) = 0$. Hence we have a power series expansion

$$\xi_\iota(v) = \sum_{n=1}^{\infty} a_n v^n.$$

Observing that

$$\zeta_\iota(r, \varphi + 2\pi(j-1)) = \xi_\iota((w^{\frac{1}{p}})_j) = \sum_{n=1}^\infty a_n((w^{\frac{1}{p}})_j)^n$$

for $j = 1, \ldots, p$ describes the set of zeros of $P(\cdot, w)$ for $w = re^{i\varphi}$ belonging to this group of p zeros, it follows that the Puiseux series representation (9.1.1) has been proved for this group of zeros.

Returning to P given by (9.1.5), it follows that there are m Puiseux series $z_{\kappa,k,j}$, $k = 1, \ldots, \nu$, $k = 1, \ldots, l_\kappa$, $j = 1, \ldots, p_{\kappa,k}$, of the form (9.1.1) which represent the zeros of $P(\cdot, w)$ for $|w| < \varepsilon$. We may assume that $\varepsilon > 0$ is so small that each two of these Puiseux series are either identical or are different for all w with $0 < |w| < \varepsilon$. By construction, $z_{\kappa,k,j} \neq z_{\kappa,k',j'}$ if $(k, j) \neq (k', j')$. Since the $z_{\kappa,k,j}(w)$, $j = 1, \ldots, p_{\kappa,k}$, are pairwise different, we have that $p_{\kappa,k}$ and the set of indices $n > 0$ with $a_{\kappa,k,n} \neq 0$ have no common multiple larger than 1. Therefore the number $p_{\kappa,k}$ is uniquely determined by the representation of the function $z_{\kappa,k,j}$ as a Puiseux series (9.1.1). Hence it follows that $p_{\kappa,k} = p_{\kappa',k'}$ if there are j and j' such that $z_{\kappa,k,j} = z_{\kappa',k',j'}$ and thus

$$\{z_{\kappa,k,j_1} : j_1 = 1, \ldots, p_{\kappa,k}\} = \{z_{\kappa',k',j_1} : j_1 = 1, \ldots, p_{\kappa,k}\}.$$

This proves that the zeros of $f(\cdot, w)$ can be indexed in such a way that they have the Puiseux series expansions (9.1.1) with the indices and multiplicities as stated in this theorem. □

Corollary 9.1.2. *Under the assumptions of Theorem 9.1.1, let*

$$\Phi_0 = \{z \in \mathbb{C} : |z| < \delta\} \times \{w \in \mathbb{C} : |w| < \varepsilon\}$$

and let

$$f_0(z, w) = \prod_{k=1}^l \prod_{j=1}^{p_k} (z - z_{kj}(w))^{m_k}, \quad (z, w) \in \Phi_0.$$

Then there is an analytic function f_1 without zeros on Φ_0 such that $f = f_0 f_1$.

Proof. If in (9.1.2) we allow the argument of $w - w_0$ to be any real number, then the values of $z_{kj}(w)$, $j = 1, \ldots, p_k$, are obtained from those with the principal value of the argument of η by a permutation of the indices $1, \ldots, p_k$. Hence the functions g_k defined by

$$g_k(z, w) = \prod_{j=1}^{p_k} (z - z_{kj}(w)), \quad (z, w) \in \Phi_0, \quad k = 1, \ldots, l,$$

are analytic on Φ_0, and then also $f_0 = \prod_{k=1}^l g_k$ is analytic on Φ_0. For each w with $|w| < \varepsilon$, the function $\frac{f(\cdot, w)}{f_0(\cdot, w)}$ has an analytic extension $f_1(\cdot, w)$ to $\{z \in \mathbb{C} : z < \delta\}$

without zeros. By Cauchy's theorem,

$$f_1(z, w) = \frac{1}{2\pi i} \int_{|\tau - z_0| = \gamma} \frac{f(\tau, w)}{f_0(\tau, w)} \frac{d\tau}{\tau - z},$$

where $|z_0| < \delta, 0 < \gamma < \delta - |z_0|, |z - z_0| < \gamma$, and γ is chosen so that $f_0(\tau, \omega) \neq 0$ for all τ with $|\tau - z_0| = \gamma$. This shows that f_1 is (locally) analytic in both variables. $\square$

9.2   Spectral dependence of analytic operator functions

For operator functions in Banach spaces which are of the form $T(\varepsilon) - \lambda I$ where T depends analytically on the complex parameter ε and where λ is the spectral parameter, T. Kato [136] has obtained a complete description of the dependence on ε of the eigenvalues and operators associated with the principal spaces for isolated eigenvalues of finite multiplicity. This has been generalized by V. Eni [73] to operator functions which are also analytic in the spectral parameter. T. Kato's proof depends on the analyticity of the eigenprojections $P(\varepsilon)$, and to be able to use this method V. Eni has linearized the operator function with respect to the spectral parameter λ. V. Eni also investigates analytic dependence of eigenvectors and associated vectors. H. Baumgärtel's monograph [25] presents a detailed and comprehensive treatment of analytic dependence of spectral data on the parameter. For an overview of the history of this problem we refer the reader to the introduction of [25].

The main result of this section is due to V. Eni [73] and its formulation is extracted from [25]. Eni's results and their proofs, published in [73] and [72] have a limited accessibility, whereas H. Baumgärtel does not state and prove this case explicitly (even for operator polynomials), although it follows relatively easily from other results and observations in [25], see [25, p. 370].

For the convenience of the reader we will present a full proof for operator functions which depend analytically on two parameters, one of which may be considered as the spectral parameter. Indeed, our first step is Lemma 9.2.1, which reduces the problem to a problem in finite-dimensional spaces and which would result, in general, in non-polynomial dependence of both parameters even if the original problem were linear in both parameters.

The proof of the following result is an adaptation of part of the proof of [189, Theorem 1.3.1] to our situation.

Lemma 9.2.1. *Let $\Phi \subset \mathbb{C}^2$ be open, let X and Y be Banach spaces and let $T : \Phi \to L(X, Y)$ be analytic and Fredholm operator valued. Furthermore, let $(\lambda_0, \eta_0) \in \Phi$ such that λ_0 is an isolated eigenvalue of $T(\cdot, \eta_0)$. Let $X_0 = N(T(\lambda_0, \eta_0))$ be the null space of $T(\lambda_0, \eta_0)$ and let $Y_1 = R(T(\lambda_0, \eta_0))$ be the range $T(\lambda_0, \eta_0)$. Let X_1 be a topological complement of X_0 in X and let Y_0 be a topological complement of Y_1 in Y. Then there are a neighbourhood $\Phi_0 \subset \Phi$ of (λ_0, η_0) and analytic operator functions $C : \Phi_0 \to L(Y, Y), D : \Phi_0 \to L(X, X), T_{11} : \Phi_0 \to L(X_1, Y_1)$ and*

$S : \Phi_0 \to L(X_0, Y_0)$ *such that* $C(\lambda, \eta)$, $D(\lambda, \eta)$ *and* $T_{11}(\lambda, \eta)$ *are invertible for all* $(\lambda, \eta) \in \Phi_0$ *and such that*

$$T = C \begin{pmatrix} T_{11} & 0 \\ 0 & S \end{pmatrix} D \ \text{on} \ \Phi_0, \tag{9.2.1}$$

where the operator matrix is taken with respect to the decompositions $X = X_1 \dotplus X_0$ *and* $Y = Y_1 \dotplus Y_0$.

Proof. Since $T(\lambda_0, \eta_0)$ is a Fredholm operator, X_0 is a finite-dimensional subspace of X and Y_1 is a closed finite-codimensional subspace of Y. Hence there are a finite-codimensional subspace $X_1 \subset X$ and a finite-dimensional subspace $Y_0 \subset Y$ such that

$$X = X_1 \dotplus X_0, \quad Y = Y_1 \dotplus Y_0$$

are topologically direct sums, see [259, p. 247]. With this decomposition of X and Y we have the operator matrix representation

$$T(\lambda, \eta) = \begin{pmatrix} T_{11}(\lambda, \eta) & T_{12}(\lambda, \eta) \\ T_{21}(\lambda, \eta) & T_{22}(\lambda, \eta) \end{pmatrix} : X_1 \dotplus X_0 \to Y_1 \dotplus Y_0 \tag{9.2.2}$$

for $(\lambda, \eta) \in \Phi$. The operator functions T_{ij}, $i, j = 1, 2$ are analytic in Φ.

It is clear that $N(T_{11}(\lambda_0, \eta_0)) = \{0\}$ and $R(T_{11}(\lambda_0, \eta_0)) = R(T(\lambda_0, \eta_0)) = Y_1$. Hence $T_{11}(\lambda_0, \eta_0)$ is invertible. Since T_{11} depends analytically and hence continuously on λ and η, the perturbation theory of invertible operators, see [137, Theorem IV.2.21], yields that there is an open neighbourhood Φ_0 of (λ_0, η_0) such that T_{11} is invertible on Φ_0. In Φ_0 we consider the Schur factorization

$$\begin{pmatrix} T_{11} & T_{12} \\ T_{21} & T_{22} \end{pmatrix} = \begin{pmatrix} I_{Y_1} & 0 \\ T_{21}T_{11}^{-1} & I_{Y_0} \end{pmatrix} \begin{pmatrix} T_{11} & 0 \\ 0 & T_{22} - T_{21}T_{11}^{-1}T_{12} \end{pmatrix} \begin{pmatrix} I_{X_1} & T_{11}^{-1}T_{12} \\ 0 & I_{X_0} \end{pmatrix}. \tag{9.2.3}$$

It is easy to see that the left-hand and right-hand factors

$$C = \begin{pmatrix} I_{Y_1} & 0 \\ T_{21}T_{11}^{-1} & I_{Y_0} \end{pmatrix} \quad \text{and} \quad D = \begin{pmatrix} I_{X_1} & T_{11}^{-1}T_{12} \\ 0 & I_{X_0} \end{pmatrix}$$

on the right-hand side of (9.2.3) are invertible on Φ_0. Putting

$$S := T_{22} - T_{21}T_{11}^{-1}T_{12}$$

on Φ_0 completes the proof. $\qquad\square$

Lemma 9.2.2. *Let* $\Omega \subset \mathbb{C}$ *be a domain, let* X *and* Y *be finite-dimensional spaces and let* $A : \Omega \to L(Y, X)$ *be meromorphic. Then there are complementary subspaces* X_0 *and* X_1 *of* X, *complementary subspaces* Y_0 *and* Y_1 *of* Y, *meromorphic operator* functions $C : \Omega \to L(X,X)$, $D : \Omega \to L(Y,Y)$ and $A_{11} : \Omega \to L(Y_1, X_1)$ and an open subset $\Omega_0 \subset \Omega$ such that the following is true. The set $\Omega \setminus \Omega_0$ is a discrete subset of Ω, the operators $C(\sigma)$, $D(\sigma)$ and $A_{11}(\sigma)$ are invertible for all $\sigma \in \Omega_0$, and

$$A = C \begin{pmatrix} A_{11} & 0 \\ 0 & 0 \end{pmatrix} D \quad on \ \Omega_0. \tag{9.2.4}$$

Proof. Let Ω' be the set of $\sigma \in \Omega$ for which $A(\sigma)$ is analytic, let m be the maximum of the rank of $A(\sigma)$, $\sigma \in \Omega'$, and choose $\sigma_0 \in \Omega$ such that $A(\sigma_0)$ has rank m. Similar to Lemma 9.2.1 we set $Y_0 = N(A(\sigma_0))$ and $X_1 = R(A(\sigma_0))$. Let X_0 and Y_1 be corresponding complementary subspaces. Then we can write

$$A = \begin{pmatrix} A_{11} & A_{12} \\ A_{21} & A_{22} \end{pmatrix} : Y_1 \dotplus Y_0 \to X_1 \dotplus X_0, \tag{9.2.5}$$

where the operator functions A_{ij}, $i,j = 1,2$, are meromorphic in Ω and analytic in Ω'.

For fixed bases of X_1 and Y_1, the function $\sigma \mapsto \det A_{11}(\sigma)$, $\sigma \in \Omega'$, is meromorphic on Ω and analytic on Ω'. By construction, $A_{11}(\sigma_0)$ is invertible, so that $\det A_{11}(\sigma_0) \neq 0$. Then $\Omega_0 = \{\sigma \in \Omega' : \det A_{11}(\sigma) \neq 0\}$ is an open subset of Ω such that $\Omega \setminus \Omega_0$ is a discrete subset of Ω since the poles and zeros of $\det A_{11}$ form a discrete subset of Ω. As in Lemma 9.2.1 we have a factorization

$$A = C \begin{pmatrix} A_{11} & 0 \\ 0 & S \end{pmatrix} D, \tag{9.2.6}$$

where C, D and S are meromorphic on Ω and C and D are analytic and invertible on Ω_0. Since rank $A(\sigma) \leq m = \operatorname{rank} A_{11}(\sigma_0)$ for all $\sigma \in \Omega'$, it follows that $S(\sigma) = 0$ since otherwise $m \geq \operatorname{rank} A(\sigma) > \operatorname{rank} A_{11}(\sigma) = m$. Therefore (9.2.6) leads to the meromorphic factorization (9.2.4). $\qquad\square$

Before proceeding with the statement and proof of the main results of this section, we need a few preparations for its proof.

We are going to use the following definitions only for finite-dimensional spaces, although they also hold for infinite-dimensional Banach spaces X and Y. The dual space X' is the (Banach) space of all linear functionals on X. For $v \in X$ and $u \in Y'$ we define the tensor product $v \otimes u$ by

$$(v \otimes u)(w) := \langle w, u \rangle v \quad (w \in Y),$$

where $\langle \cdot, \cdot \rangle$ is the canonical bilinear form on $Y \times Y'$. The tensor product $v \otimes u$ is a bounded linear operator from Y to X.

We will call vector functions $v_1, \ldots, v_k$ from a set U to a vector space V pointwise linearly independent if $v_1(\mu), \ldots, v_k(\mu)$ are linearly independent for each $\mu \in U$.

Proposition 9.2.3. *Let X and Y be finite-dimensional spaces, let Ω be a domain in $\mathbb{C}$, let $A : \Omega \to L(Y, X)$ be meromorphic and let $v_1, \ldots, v_r : \Omega \to X$ be meromorphic on Ω and pointwise linearly independent outside a discrete subset of Ω. Then there are meromorphic functions $v_k : \Omega \to X$, $k = r + 1, \ldots, n$, and $u_k : \Omega \to Y'$, $k = 1, \ldots, n$, such that*

$$A = \sum_{k=1}^{n} v_k \otimes u_k \qquad (9.2.7)$$

and such that $v_1, \ldots, v_n$ as well as $u_{r+1}, \ldots, u_n$ are pointwise linearly independent outside a discrete subset of Ω. There are meromorphic functions $y_k : \Omega \to Y$, $k = r + 1, \ldots, n$, such that $Ay_k = v_k$. The integer n is uniquely determined by these properties.

Proof. Let $\Omega' \subset \Omega$ be the set of $\sigma \in \Omega$ for which A and $v_1, \ldots, v_r$ are analytic, let $m = \max\{\operatorname{rank} A(\sigma) : \sigma \in \Omega'\}$ and let Ω_0 be the set of all $\sigma \in \Omega'$ with $\operatorname{rank} A(\sigma) = m$. Choose $\sigma_0 \in \Omega_0$ such that $R(A(\sigma)) \cap \operatorname{span}\{v_1(\sigma), \ldots, v_r(\sigma)\}$ has maximal rank, say s, for $\sigma = \sigma_0$. With the notation of Lemma 9.2.2 we put

$$\tilde{A}_{11} = \begin{pmatrix} A_{11} \\ 0 \end{pmatrix}$$

and $\tilde{v}_k = C^{-1} v_k$, $k = 1, \ldots, s$. Then $R(\tilde{A}_{11}(\sigma_0)) \cap \operatorname{span}\{\tilde{v}_1(\sigma_0), \ldots, \tilde{v}_r(\sigma_0)\}$ has rank s. Let $q = \dim X_0$. Since $\dim X_1 = m$, there are $m - s$ vectors $\tilde{x}_{s+1}, \ldots, \tilde{x}_m$ in X_1 and $q - r + s$ vectors $\tilde{x}_{r-s+1}^0, \ldots, \tilde{x}_q^0$ in X_0 such that $\tilde{v}_1(\sigma_0), \ldots, \tilde{v}_r(\sigma_0)$, $\tilde{x}_{s+1}, \ldots, \tilde{x}_m$ and $\tilde{x}_{r-s+1}^0, \ldots, \tilde{x}_q^0$ form a basis for X. Therefore the operator

$$V(\sigma) = \begin{pmatrix} \tilde{v}_1(\sigma) & \cdots & \tilde{v}_r(\sigma) & \tilde{x}_{s+1} & \cdots & \tilde{x}_m & \tilde{x}_{r-s+1}^0 & \cdots & \tilde{x}_q^0 \end{pmatrix} : \mathbb{C}^{m+q} \to X$$

depends meromorphically on σ in Ω and is invertible for $\sigma = \sigma_0$. Hence V^{-1} is meromorphic on Ω. Let Ω_1 be the set of $\sigma \in \Omega_0$ for which $V(\sigma)$ is invertible. Now choose $\tilde{x}_1, \ldots, \tilde{x}_s$ in X_1 such that $\tilde{x}_1, \ldots, \tilde{x}_m$ is a basis of X_1. For each $\sigma \in \Omega_1$, every $x \in X_1$ is a (unique) linear combination of $\tilde{v}_1(\sigma), \ldots, \tilde{v}_r(\sigma)$ and $\tilde{x}_{s+1}, \ldots, \tilde{x}_m$. Since $x = V(\sigma)V^{-1}(\sigma)x$ and $V^{-1}(\sigma)x$ depends meromorphically on σ for all $x \in X_1$, there are meromorphic functions α_{jk}, $j = 1, \ldots, s$, $k = 1, \ldots, r$, and $\beta_{j,k}$, $j = 1, \ldots, s$, $k = s + 1, \ldots, m$, on Ω such that

$$\tilde{x}_j = \sum_{k=1}^{r} \alpha_{jk}(\sigma) \tilde{v}_k(\sigma) + \sum_{k=s+1}^{m} \beta_{jk}(\sigma) \tilde{x}_k, \quad j = 1, \ldots, s, \quad \sigma \in \Omega_1. \qquad (9.2.8)$$

By the Hahn–Banach Theorem there are $\tilde{w}_k^0 \in X_1'$, $k = 1, \ldots, m$, such that

$$\langle \tilde{x}_j, \tilde{w}_k^0 \rangle = \delta_{jk}, \quad j, k = 1, \ldots, m. \qquad (9.2.9)$$

It is well known, and obvious for operator functions in finite-dimensional spaces, that with A_{11} also its adjoint $A'_{11} : \Omega \to L(X'_1, Y'_1)$ is meromorphic. For $k = 1, \ldots, m$ and $\sigma \in \Omega_0$ we have

$$
\left\{ \sum_{j=1}^{m} \tilde{x}_j \otimes \left[(A_{11}(\sigma))' \tilde{w}_j^0 \right] \right\} A_{11}^{-1}(\sigma) \tilde{x}_k = \sum_{j=1}^{m} \langle A_{11}^{-1}(\sigma) \tilde{x}_k, (A_{11}(\sigma))' \tilde{w}_j^0 \rangle \tilde{x}_j
$$

$$
= \sum_{j=1}^{m} \langle \tilde{x}_k, \tilde{w}_j^0 \rangle \tilde{x}_j
$$

$$
= \tilde{x}_k
$$

$$
= A_{11}(\sigma) A_{11}^{-1}(\sigma) \tilde{x}_k.
$$

Since $\tilde{x}_1, \ldots, \tilde{x}_m$ is a basis of X_1, we have that $A_{11}^{-1}(\sigma) \tilde{x}_1, \ldots, A_{11}^{-1}(\sigma) \tilde{x}_m$ is a basis of Y_1 for all $\sigma \in \Omega_0$, and it follows that

$$
A_{11}(\sigma) = \sum_{j=1}^{m} \tilde{x}_j \otimes \left[(A_{11}(\sigma))' \tilde{w}_j^0 \right], \quad \sigma \in \Omega_0. \tag{9.2.10}
$$

For $j = 1, \ldots, m$ and $\sigma \in \Omega_0$ let $\tilde{w}_j(\sigma)$ be the extension of $(A_{11}(\sigma))' \tilde{w}_j^0$ to Y' by zero on Y_0. Then, in view of (9.2.4), (9.2.8) and (9.2.10),

$$
A(\sigma) = C(\sigma) \left(\sum_{j=1}^{m} \tilde{x}_j \otimes \tilde{w}_j(\sigma) \right) D(\sigma)
$$

$$
= C(\sigma) \left(\sum_{k=1}^{r} \sum_{j=1}^{s} \alpha_{jk}(\sigma) \tilde{v}_k(\sigma) \otimes \tilde{w}_j(\sigma) \right) D(\sigma)
$$

$$
+ C(\sigma) \left(\sum_{k=s+1}^{m} \sum_{j=1}^{s} \beta_{jk}(\sigma) \tilde{x}_k \otimes \tilde{w}_j(\sigma) \right) D(\sigma)
$$

$$
+ C(\sigma) \left(\sum_{j=s+1}^{m} \tilde{x}_j \otimes \tilde{w}_j(\sigma) \right) D(\sigma)
$$

$$
= \sum_{k=1}^{r} v_k(\sigma) \otimes \left(\sum_{j=1}^{s} \alpha_{jk}(\sigma) D'(\sigma) \tilde{w}_j(\sigma) \right)
$$

$$
+ \sum_{k=s+1}^{m} C(\sigma) \tilde{x}_k \otimes \left(D'(\sigma) \tilde{w}_k(\sigma) + \sum_{j=1}^{s} \beta_{jk}(\sigma) D'(\sigma) \tilde{w}_j(\sigma) \right).
$$

Setting $n = r - s + m$, the representation (9.2.7) follows with

$$v_k = C\tilde{x}_{k+s-r}, \quad k = r+1, \ldots, n,$$

$$u_k = \sum_{j=1}^{s} \alpha_{jk} D' \tilde{w}_j, \quad k = 1, \ldots, r,$$

$$u_k = D' \tilde{w}_{k+s-r} + \sum_{j=1}^{s} \beta_{j,k+s-r} D' \tilde{w}_j, \quad k = r+1, \ldots, n.$$

For $k = r+1, \ldots, n$ define $\tilde{y}_k = A_{11}^{-1} \tilde{x}_{k+s-r}$ and $y_k = D^{-1} \tilde{y}_k$. Since $\tilde{y}_k \in Y_1$, (9.2.4) gives $A y_k = C A_{11} \tilde{y}_k = C \tilde{x}_{k-s-r} = v_k$.

From the pointwise linear independence of $\tilde{v}_1, \ldots, \tilde{v}_r, \tilde{x}_{s+1}, \ldots, \tilde{x}_m$ on Ω_1 it follows that their images under C, i.e., $v_1, \ldots, v_n$, are pointwise linearly independent on Ω_1 since $C(\sigma)$ is invertible for $\sigma \in \Omega_1$.

Since the $\tilde{w}_j^0$, $j = 1, \ldots, m$, are linearly independent by (9.2.9) and since $A_{11}'(\sigma)$ as well as $D'(\sigma)$ are invertible for $\sigma \in \Omega_1$, it follows that the vectors $D'(\sigma) \tilde{w}_j(\sigma)$, $j = 1, \ldots, m$, are linearly independent for $\sigma \in \Omega_1$. Hence it is clear from their definition that also the vectors $u_k(\sigma)$, $j = r+1, \ldots, n$, are linearly independent for all $\sigma \in \Omega_1$.

Finally, to show the uniqueness of n assume that we have a representation (9.2.7) with the stated properties. For a generic point $\sigma_0 \in \Omega$ where the operator and vector functions are analytic and the linear independence properties hold, we can find $\hat{y}_j \in Y$, $j = r+1, \ldots, n$, such that

$$\langle \hat{y}_j, u_k(\sigma_0) \rangle = \delta_{jk}, \quad j, k = r+1, \ldots, n.$$

Hence (9.2.7) gives

$$A(\sigma_0) \hat{y}_j = \sum_{k=1}^{r} \langle \hat{y}_j, u_k(\sigma_0) \rangle v_k(\sigma_0) + v_j(\sigma_0), \quad j = r+1, \ldots, n.$$

This shows that

$$\text{span}\{v_{r+1}(\sigma_0), \ldots, v_n(\sigma_0)\} \subset \text{span}\{v_1(\sigma_0), \ldots, v_r(\sigma_0)\} + R(A(\sigma_0))$$

and therefore

$$\text{span}\{v_1(\sigma_0), \ldots, v_n(\sigma_0)\} \subset \text{span}\{v_1(\sigma_0), \ldots, v_r(\sigma_0)\} + R(A(\sigma_0)). \tag{9.2.11}$$

Again by (9.2.7) the reverse inclusion holds. For generic σ_0, $v_1(\sigma_0), \ldots, v_n(\sigma_0)$ are linearly independent and the dimension of the vector space on the right-hand side of (9.2.11) is independent of σ_0. Hence n is uniquely determined by $v_1, \ldots, v_r$ and A. $\qquad\square$

Theorem 9.2.4. *Let* $\Phi \subset \mathbb{C}^2$ *be open, let* X *and* Y *be Banach spaces and let* $T : \Phi \to L(X, Y)$ *be analytic and Fredholm operator valued. Furthermore, let* $(\lambda_0, \eta_0) \in \Phi$ *such that* λ_0 *is an isolated eigenvalue of* $T(\cdot, \eta_0)$. *Denote by* m *the algebraic multiplicity of the eigenvalue* λ_0. *Then there are numbers* $\varepsilon > 0$, $\delta > 0$, *and positive integers* l, p_k *and* m_k, $k = 1, \ldots, l$, *such that*

$$\sum_{k=1}^{l} p_k m_k = m$$

and such that the following assertions are true:

1. *For each* $\eta \in \mathbb{C}$ *with* $|\eta - \eta_0| < \varepsilon$, $T(\cdot, \eta)$ *has exactly* m *eigenvalues, counted with algebraic multiplicity, in the disc* $\{\lambda \in \mathbb{C} : |\lambda - \lambda_0| < \delta\}$. *The eigenvalues can be organized into* l *groups and denoted by* λ_{kj}, $k = 1, \ldots, l$, $j = 1, \ldots, p_k$, *such that the* λ_{kj} *are pairwise different when* $0 < |\eta - \eta_0| < \varepsilon$, *have algebraic multiplicity* m_k, *and are represented by Puiseux series*

$$\lambda_{kj}(\eta) = \lambda_0 + \sum_{n=1}^{\infty} a_{kn} (((\eta - \eta_0)^{\frac{1}{p_k}})_j)^n. \qquad (9.2.12)$$

2. *For* $k = 1, \ldots, l$ *there are positive integers* r_k *and* $q_{k1}, \ldots, q_{kr_k}$ *such that* r_k *is the geometric multiplicity of the eigenvalue* $\lambda_{kj}(\eta)$, $j = 1, \ldots, p_k$, *if* $0 < |\eta - \eta_0| < \varepsilon$,

$$\sum_{\iota=1}^{r_k} q_{k\iota} = m_k,$$

and there are vectors

$$y_{kj}^{\iota \kappa}(\eta) = \sum_{n=0}^{\infty} b_{kn}^{\iota \kappa} (((\eta - \eta_0)^{\frac{1}{p_k}})_j)^n, \quad \iota = 1, \ldots, r_k, \ \kappa = 0, \ldots, q_{k\iota} - 1, \quad (9.2.13)$$

in X *such that* $y_{kj}^{\iota 0}(\eta), \ldots, y_{kj}^{\iota q_{k\iota}-1}(\eta)$ *is a chain of an eigenvector and associated vectors of* $T(\cdot, \eta)$ *at* $\lambda_{kj}(\eta)$ *if* $0 < |\eta - \eta_0| < \varepsilon$ *and where* $\{y_{kj}^{\iota 0}(\eta) : \iota = 1, \ldots, r_k\}$ *is a basis of* $N(T(\lambda_{kj}(\eta), \eta))$ *for* $k = 1, \ldots, l$ *and* $j = 1, \ldots, p_k$.

Proof. Without loss of generality assume that $\lambda_0 = 0$ and $\eta_0 = 0$. First we will assume that X and Y are finite dimensional. Since 0 is an isolated eigenvalue of $T(\cdot, 0)$, it follows that X and Y have the same dimensions, and with chosen fixed bases we can write T as a matrix function which depends analytically on λ and η. Then its determinant is analytic in λ and η, and $T(\lambda, \eta)$ is invertible if and only if $t(\lambda, \eta) = \det T(\lambda, \eta) \neq 0$. Furthermore, it is well known that the algebraic multiplicity of the eigenvalue λ of $T(\cdot, \eta)$ equals the multiplicity of the zero λ of $t(\cdot, \eta)$, see [189, Proposition 1.8.5]. By Theorem 9.1.1, the zeros of $t(\cdot, \eta)$ and thus the eigenvalues of $T(\cdot, \eta)$ have the representations as stated in part 1. Hence part 1 has been proved.

Now we are going to prove part 2. By Cramer's rule, $t(\lambda, \eta)T^{-1}(\lambda, \eta)$ is analytic in λ and η. Putting $t_0 = \prod_{k=0}^{l} g_k$ with

$$g_k(\lambda, \eta) = \prod_{j=1}^{p_k} (\lambda - \lambda_{kj}(\eta))^{m_k}, \quad |\lambda| < \varepsilon, \ |\eta| < \delta, \ k = 1, \ldots, l,$$

it follows from Corollary 9.1.2 that

$$S := t_0 T^{-1} \tag{9.2.14}$$

is analytic on Φ_0. We now focus our attention on one group of eigenvalues, for some fixed k. We put $\sigma = (\eta^{\frac{1}{p_k}})_j$ for some fixed j and $0 < |\eta| < \varepsilon$. The map $\eta \mapsto \sigma$ depends on j, but $\hat{\lambda}_k(\sigma) = \lambda_{kj}(\eta)$ defines an analytic function $\hat{\lambda}_k$ which is independent of j, see (9.2.12). Replacing η with σ^{p_k} in the Taylor series expansion of $t(\cdot, \eta)$ about $\lambda_{kj}(\eta)$ shows that the multiplicity of the zero $\hat{\lambda}_k(\sigma)$ of $t(\cdot, \sigma^{p_k})$ remains m_k.

Since $\eta = \sigma^{p_k}$, it follows that $T^{-1}(\lambda, \sigma^{p_k})$ is analytic in λ and σ when $t(\lambda, \sigma^{p_k}) \neq 0$ and has a pole, as a function of λ for fixed σ, of order at most m_k at $\hat{\lambda}_k(\sigma)$. Hence, for $0 < |\sigma| < \varepsilon^{\frac{1}{p_k}}$ we have a Laurent series expansion

$$T^{-1}(\lambda, \sigma^{p_k}) = \sum_{n=-m_k}^{\infty} (\lambda - \hat{\lambda}_k(\sigma))^n A_{k,n}(\sigma) \tag{9.2.15}$$

in a punctured neighbourhood of $\hat{\lambda}_k(\sigma)$, and the $A_{k,n}$ are given by Cauchy's formula

$$A_{k,n}(\sigma) = \frac{1}{2\pi i} \oint_{\Gamma_{\hat{\lambda}_k(\sigma)}} (\lambda - \hat{\lambda}_k(\sigma))^{-n-1} T^{-1}(\lambda, \sigma^{p_k}) \, d\lambda, \tag{9.2.16}$$

where $\Gamma_{\hat{\lambda}_k(\sigma)}$ is a counterclockwise simply connected closed curve surrounding $\hat{\lambda}_k(\sigma)$ with no other zero of $t(\cdot, \sigma^{p_k})$ inside this curve. Since locally we may choose this curve to be independent of σ, it follows that the $A_{k,n}$ are analytic functions for $0 < |\sigma| < \varepsilon^{\frac{1}{p_k}}$. We also observe that there is an integer $p \geq 2$, e.g., the least common multiple of all p_ι, $\iota = 1, \ldots, l$, such that each $\lambda_{k'j'}(\eta)$ can be written as a power series in $(\eta^{\frac{1}{p}})_2$. Since the functions $\lambda_{k'j'}(\eta)$ are pairwise distinct, it follows that the differences $\lambda_{k'j'}(\eta) - \lambda_{k''j''}(\eta)$ for $(k', j') \neq (k'', j'')$ are analytic functions in $(\eta^{\frac{1}{p}})_2$ without zeros in $0 < |\eta| < \varepsilon$. Hence there is a positive integer α such that $|\lambda_{k'j'}(\eta) - \lambda_{k''j''}(\eta)||\eta|^{-\frac{\alpha}{p}}$ has a positive lower bound for $0 < |\eta| \leq \frac{\varepsilon}{2}$ and $(k', j') \neq (k'', j'')$. Thus there is $\gamma > 0$ so that the curve $\Gamma_{\hat{\lambda}_k(\sigma)}$ can be chosen such that

$$(\lambda - \hat{\lambda}_k(\sigma))^{-1} = O(\sigma^{-\gamma}) \quad \text{and} \quad (\lambda - \lambda_{k'j'}(\sigma^{p_k}))^{-1} = O(\sigma^{-\gamma})$$

for all (k', j') considered here, $\lambda \in \Gamma_{\hat{\lambda}_k(\sigma)}$ and σ close to 0. We can therefore find a positive integer γ_1 such that

$$\frac{1}{t_0(\lambda, \sigma^{p_k})} = O(\sigma^{-\gamma_1}).$$

In view of (9.2.14) it follows that $A_{k,n}$ has a pole at 0.

Below we are going to use the following notation: for $0 < \varepsilon_\kappa \le \varepsilon$ let $\Omega_\kappa = \{\eta \in \mathbb{C} : |\eta| < \varepsilon_\kappa\}$, let $\Lambda_\kappa = \{(\hat{\lambda}_k(\eta), \eta) : \eta \in \mathbb{C} : 0 < |\eta| < \varepsilon_\kappa\}$, and let $\Phi_\kappa \subset \Phi$ be a neighbourhood of Λ_κ.

Let s be the smallest positive integer such that $A_{k,-s} \ne 0$. We are going to prove

Claim 1. For $\kappa = 0, \dots, s$ there are $\varepsilon_\kappa > 0$, integers $r_k^{(\kappa)}$ and $q_{k1}, \dots, q_{kr_k^{(\kappa)}}$ and $v_\iota(\lambda, \sigma) \in X$ and $u_\iota^{(\kappa)}(\lambda, \sigma) \in Y'$, $(\lambda, \sigma) \in \Phi_\kappa$, $\iota = 1, \dots, r_k^{(\kappa)}$, which are analytic on Φ_κ and polynomials in λ of degree less than $q_{k\iota}$ whose coefficients are meromorphic in σ on Ω_κ, such that the $v_\iota(\hat{\lambda}_k(\sigma), \sigma)$, $\iota = 1, \dots, r_k^{(\kappa)}$, are linearly independent, $u_\iota^{(\kappa)}(\lambda_k(\sigma), \sigma) \ne 0$ for $\iota = 1, \dots, r_k^{(\kappa)}$, such that $T(\cdot, \sigma^{p_k})v_\iota(\cdot, \sigma)$ has a zero of order $\ge q_{k\iota}$ at $\hat{\lambda}_k(\sigma)$ for all $0 \le |\sigma| < \varepsilon_\kappa$, and such that

$$(\lambda - \hat{\lambda}_k(\sigma))^{s-\kappa} \left[T^{-1}(\lambda, \sigma^{p_k}) - \sum_{\iota=1}^{r_k^{(\kappa)}} (\lambda - \hat{\lambda}_k(\sigma))^{-q_{k\iota}} v_\iota(\lambda, \sigma) \otimes u_\iota^{(\kappa)}(\lambda, \sigma) \right] \quad (9.2.17)$$

is analytic on Φ_κ.

Claim 1 is trivial for $\kappa = 0$ where we can take $r_k^{(0)} = 0$.

Suppose that Claim 1 is true for some $0 \le \kappa < s$. We set

$$A(\lambda, \sigma) := \sum_{\iota=-s+\kappa}^{\infty} (\lambda - \hat{\lambda}_k(\sigma))^\iota A_\iota(\sigma) \quad (9.2.18)$$

$$:= T^{-1}(\lambda, \sigma^{p_k}) - \sum_{\iota=1}^{r_k^{(\kappa)}} (\lambda - \hat{\lambda}_k(\sigma))^{-q_{k\iota}} v_\iota(\lambda, \sigma) \otimes u_\iota^{(\kappa)}(\lambda, \sigma)$$

for $(\lambda, \sigma) \in \Phi_k$ in some punctured neighbourhood of $\hat{\lambda}_k(\sigma)$, where $A_\iota(\sigma) \in L(Y, X)$ and A_ι is meromorphic in Ω_κ. By Proposition 9.2.3 there are a number $r_k^{(\kappa+1)} \ge r_k^{(\kappa)}$ and $\tilde{y}_{r_k^{(\kappa)}+1}(\sigma), \dots, \tilde{y}_{r_k^{(\kappa+1)}}(\sigma) \in Y$, $\tilde{u}_1^{(\kappa+1)}(\sigma), \dots, \tilde{u}_{r_k^{(\kappa+1)}}^{(\kappa+1)}(\sigma) \in Y'$ depending meromorphically on $\sigma \in \Omega_{\kappa+1}$ such that

$$A_{-s+\kappa}(\sigma) = \sum_{\iota=1}^{r_k^{(\kappa)}} v_\iota(\hat{\lambda}_k(\sigma), \sigma) \otimes \tilde{u}_\iota^{(\kappa+1)}(\sigma) + \sum_{\iota=r_k^{(\kappa)}+1}^{r_k^{(\kappa+1)}} A_{-s+\kappa}(\sigma) \tilde{y}_\iota(\sigma) \otimes \tilde{u}_\iota^{(\kappa+1)}(\sigma),$$

where $\tilde{u}_{r_k^{(\kappa)}+1}^{(\kappa+1)}(\sigma), \dots, \tilde{u}_{r_k^{(\kappa+1)}}^{(\kappa+1)}(\sigma)$ as well as

$$v_1(\hat{\lambda}_k(\sigma), \sigma), \dots, v_{r_k^{(\kappa)}}(\hat{\lambda}_k(\sigma), \sigma), A_{-s+\kappa}(\sigma)\tilde{y}_{r_k^{(\kappa)}+1}(\sigma), \dots, A_{-s+\kappa}(\sigma)\tilde{y}_{r_k^{(\kappa+1)}}(\sigma)$$

are linearly independent. Here $\varepsilon_{\kappa+1} > 0$ is chosen in such a way that 0 is the only pole of all the above meromorphic functions in σ.

For $\iota = 1, \ldots, r_k^{(\kappa)}$ let

$$u_\iota^{(\kappa+1)}(\lambda, \sigma) = u_\iota^{(\kappa)}(\lambda, \sigma) + (\lambda - \hat{\lambda}_k(\sigma))^{q_{k\iota} - s + \kappa} \widetilde{u}_\iota^{(\kappa+1)}(\sigma),$$

and for $\iota = r_k^{(\kappa)} + 1, \ldots, r_k^{(\kappa+1)}$ let

$$q_{k\iota} = s - \kappa,$$
$$\widetilde{v}_\iota(\lambda, \sigma) = (\lambda - \hat{\lambda}_k(\sigma))^{q_{k\iota}} A(\lambda, \sigma) \widetilde{y}_\iota(\sigma),$$
$$u_\iota^{(\kappa+1)}(\lambda, \sigma) = \widetilde{u}_\iota^{(\kappa+1)}(\sigma).$$

Observe that these $\widetilde{v}_\iota$ for $\iota = r_k^{(\kappa)} + 1, \ldots, r_k^{(\kappa+1)}$ are analytic in λ at each $\hat{\lambda}_k(\sigma)$ since $(\lambda, \sigma) \mapsto (\lambda - \hat{\lambda}_k(\sigma))^{q_{k\iota}} A(\lambda, \sigma)$ has this property by assumption. For $\iota = r_k^{(\kappa)} + 1, \ldots, r_k^{(\kappa+1)}$ let v_ι be the Taylor polynomial of $\widetilde{v}_\iota$ in λ of order $q_{k\iota} - 1$ at $\hat{\lambda}_k(\sigma)$, i.e.,

$$v_\iota(\lambda, \sigma) = \sum_{\tau=0}^{q_{k\iota}-1} (\lambda - \hat{\lambda}_k(\sigma))^\tau \frac{1}{\tau!} \frac{\partial^\tau}{\partial \lambda^\tau} \widetilde{v}_\iota(\lambda, \sigma) \Big|_{\lambda = \hat{\lambda}_k(\sigma)}, \quad \iota = r_k^{(\kappa)} + 1, \ldots, r_k^{(\kappa+1)}.$$

Since

$$v_\iota(\hat{\lambda}_k(\sigma), \sigma) = \widetilde{v}_\iota(\hat{\lambda}_k(\sigma), \sigma) = A_{-s+\kappa}(\sigma) \widetilde{y}_\iota(\sigma), \quad \iota = r_k^{(\kappa)} + 1, \ldots, r_k^{(\kappa+1)},$$

it follows that the $v_\iota(\hat{\lambda}_k(\sigma), \sigma)$, $\iota = 1, \ldots, r_k^{(\kappa+1)}$ are linearly independent. By assumption and definition, the v_ι are polynomials of degree less than $q_{k\iota}$ in λ for $\iota = r_k^{(\kappa)} + 1, \ldots, r_k^{(\kappa+1)}$, and the same is true for the $u_\iota^{(\kappa+1)}$ since $s - \kappa > 0$ and $q_{k\iota} = s - \kappa$ for $\iota = r_k^{(\kappa)} + 1, \ldots, r_k^{(\kappa+1)}$. Also, $u_\iota^{(\kappa+1)}(\hat{\lambda}_k(\sigma), \sigma) = u_\iota^{(\kappa)}(\hat{\lambda}_k(\sigma), \sigma) \neq 0$ for $\iota = 1, \ldots, r_k^{(\kappa)}$, whereas the linear independence of $\widetilde{u}_{r_k^{(\kappa)}+1}^{(\kappa+1)}(\sigma), \ldots, \widetilde{u}_{r_k^{(\kappa+1)}}^{(\kappa+1)}(\sigma)$ gives $u_\iota^{(\kappa+1)}(\hat{\lambda}_k(\sigma), \sigma) \neq 0$ for $\iota = r_k^{(\kappa)} + 1, \ldots, r_k^{(\kappa+1)}$.

For $\iota = r_k^{(\kappa)} + 1, \ldots, r_k^{(\kappa+1)}$ the definition of $\widetilde{v}_\iota$ and (9.2.18) give

$$T(\lambda, \sigma^{p_k}) \widetilde{v}_\iota(\lambda, \sigma) = (\lambda - \hat{\lambda}_k(\sigma))^{s-\kappa} \Bigg(\widetilde{y}_\iota(\sigma)$$

$$- \sum_{\iota'=1}^{r_k^{(\kappa)}} \langle \widetilde{y}_\iota(\sigma), u_{\iota'}^{(\kappa)}(\lambda, \sigma) \rangle \left[(\lambda - \hat{\lambda}_k(\sigma))^{-q_{k\iota}} T(\lambda, \sigma^{p_k}) v_\iota(\lambda, \sigma) \right] \Bigg).$$

Since $(\lambda - \hat{\lambda}_k(\sigma))^{-q_{k\iota}} T(\lambda, \sigma^{p_k}) v_\iota(\lambda, \sigma)$, $\iota = 1, \ldots, r_k^{(\kappa)}$, is analytic at $(\hat{\lambda}_k(\sigma), \sigma)$ by assumption, it follows that $T(\cdot, \sigma^{p_k}) \widetilde{v}_\iota(\cdot, \sigma)$, $\iota = r_k^{(\kappa)} + 1, \ldots, r_k^{(\kappa+1)}$, has a zero of order $\geq q_{k\iota}$. Since v_ι is the Taylor polynomial of $\widetilde{v}_\iota$ of order $q_{k\iota}$, the same is true for $T(\cdot, \sigma^{p_k}) v_\iota(\cdot, \sigma)$. Finally, we have that

$$
T^{-1}(\lambda, \sigma^{p_k}) - \sum_{\iota=1}^{r_k^{(\kappa+1)}} (\lambda - \hat\lambda_k(\sigma))^{-q_{k\iota}} v_\iota(\lambda, \sigma) \otimes u_\iota^{(\kappa+1)}(\lambda, \sigma)
$$

$$
= A(\lambda, \sigma) - \sum_{\iota=1}^{r_k^{(\kappa)}} (\lambda - \hat\lambda_k(\sigma))^{-q_{k\iota}} v_\iota(\lambda, \sigma) \otimes \left(u_\iota^{(\kappa+1)}(\lambda, \sigma) - u_\iota^{(\kappa)}(\lambda, \sigma) \right)
$$

$$
- \sum_{\iota=r_k^{(\kappa)}+1}^{r_k^{(\kappa+1)}} (\lambda - \hat\lambda_k(\sigma))^{-q_{k\iota}} v_\iota(\lambda, \sigma) \otimes u_\iota^{(\kappa+1)}(\lambda)
$$

$$
= A(\lambda, \sigma) - \sum_{\iota=1}^{r_k^{(\kappa)}} (\lambda - \hat\lambda_k(\sigma))^{-s+\kappa} v_\iota(\lambda, \sigma) \otimes \tilde u_\iota^{(\kappa+1)}(\sigma)
$$

$$
- \sum_{\iota=r_k^{(\kappa)}+1}^{r_k^{(\kappa+1)}} (A(\lambda, \sigma) \tilde y_\iota(\sigma)) \otimes \tilde u_\iota^{(\kappa+1)}(\sigma)
$$

$$
- \sum_{\iota=r_k^{(\kappa)}+1}^{r_k^{(\kappa+1)}} (\lambda - \hat\lambda_k(\sigma))^{-q_{k\iota}} (v_\iota(\lambda, \sigma) - \tilde v_\iota(\lambda, \sigma)) \otimes \tilde u_\iota^{(\kappa+1)}(\sigma)
$$

$$
= \sum_{\iota=-s+\kappa+1}^{\infty} (\lambda - \hat\lambda(\sigma))^\iota A_\iota(\sigma)
$$

$$
- \sum_{\iota=1}^{r_k^{(\kappa)}} (\lambda - \hat\lambda_k(\sigma))^{-s+\kappa} \left(v_\iota(\lambda, \sigma) - v_\iota(\hat\lambda_k(\sigma), \sigma) \right) \otimes \tilde u_\iota^{(\kappa+1)}(\sigma)
$$

$$
- \sum_{\iota=r_k^{(\kappa)}+1}^{r_k^{(\kappa+1)}} \sum_{\iota'=-s+\kappa+1}^{\infty} (\lambda - \hat\lambda_k(\sigma))^{\iota'} (A_{\iota'}(\sigma) \tilde y_\iota(\sigma)) \otimes \tilde u_\iota^{(\kappa+1)}(\sigma)
$$

$$
- \sum_{\iota=r_k^{(\kappa)}+1}^{r_k^{(\kappa+1)}} (\lambda - \hat\lambda_k(\sigma))^{-q_{k\iota}} (v_\iota(\lambda, \sigma) - \tilde v_\iota(\lambda, \sigma)) \otimes \tilde u_\iota^{(\kappa+1)}(\sigma).
$$

Therefore Claim 1 is proved for $\kappa+1$ since the pole orders at $\lambda = \hat\lambda_k(\sigma)$ of the first three sums do not exceed $s - \kappa - 1$, whereas the last sum is analytic at $\lambda = \hat\lambda_k(\sigma)$.

Putting $r_k = r_k^{(s)}$ and $u_\iota = u_\iota^{(s)}$ for $\iota = 1, \ldots, r_k$ shows that

$$
E(\lambda, \sigma) := T^{-1}(\lambda, \sigma^{p_k}) - \sum_{\iota=1}^{r_k} (\lambda - \hat\lambda_k(\sigma))^{-q_{k\iota}} v_\iota(\lambda, \sigma) \otimes u_\iota(\lambda, \sigma) \tag{9.2.19}
$$

is analytic at $(\hat{\lambda}_k(\sigma), \sigma)$. Multiplying this equation by $T(\lambda, \sigma^{p_k})$ from the right we see that the identity operator on X can be written as

$$I_X = \sum_{\iota=1}^{r_k} (\lambda - \hat{\lambda}_k(\sigma))^{-q_{k\iota}} v_\iota(\lambda, \sigma) \otimes (T^*(\lambda, \sigma^{p_k}) u_\iota(\lambda, \sigma)) + E(\lambda, \sigma) T(\lambda, \sigma^{p_k}),$$

(9.2.20)

where $T^*(\lambda, \sigma^{p_k})$ is the adjoint operator of $T(\lambda, \sigma^{p_k})$. Hence

$$\sum_{\iota=1}^{r_k} (\lambda - \hat{\lambda}_k(\sigma))^{-q_{k\iota}} v_\iota(\lambda, \sigma) \otimes (T^*(\lambda, \sigma^{p_k}) u_\iota(\lambda, \sigma))$$

is analytic in λ at $\hat{\lambda}_k(\sigma)$. From [189, Proposition 1.5.3], we conclude that the functions $\lambda \mapsto (\lambda - \hat{\lambda}_k(\sigma))^{-q_{k\iota}} T^*(\lambda, \sigma^{p_k}) u_\iota(\lambda, \sigma)$, $\iota = 1, \ldots, r_k$, are analytic at $\hat{\lambda}_k(\sigma)$ since the $v_\iota(\hat{\lambda}_k(\sigma), \sigma)$, $\iota = 1, \ldots, r_k$, are linearly independent. Therefore, the functions $T'(\cdot, \sigma^{p_k}) u_\iota(\cdot, \sigma)$ have a zero of order at least $q_{k\iota}$ at $\hat{\lambda}_k(\sigma)$.

Next we are going to show that

$$N(T(\hat{\lambda}_k(\sigma), \sigma)) = \text{span}\{v_\iota(\hat{\lambda}_k(\sigma), \sigma) : \iota = 1, \ldots, r_k\}.$$

It is clear that $v_\iota(\hat{\lambda}_k(\sigma), \sigma) \subset N(T(\hat{\lambda}_k(\sigma), \sigma))$ for $\iota = 1, \ldots, r_k$ since Tv_ι has a zero at $\lambda = \hat{\lambda}_k(\sigma)$ for $\iota = 1, \ldots, r_k$.

Conversely, let $x \in N(T(\hat{\lambda}_k(\sigma), \sigma))$. Then (9.2.20) leads to

$$x = \sum_{\iota=1}^{r_k} \left\langle x, (\lambda - \hat{\lambda}_k(\sigma))^{-q_{k\iota}} T^*(\lambda, \sigma^p) u_\iota(\lambda, \sigma) \right\rangle v_\iota(\lambda, \sigma) + E(\lambda, \sigma) T(\lambda, \sigma^p) x,$$

which shows that x is a linear combination of the $v_\iota(\hat{\lambda}_k(\sigma), \sigma)$, $\iota = 1, \ldots, r_k$.

We note that the above statements hold with analytic dependence on σ for $0 < |\sigma| < \varepsilon'$ and suitable positive ε'. Then, for these σ, the general assumptions as well as condition ii) of [189, Theorem 1.5.9] are satisfied since $E(\cdot, \sigma)$ is analytic at $\hat{\lambda}_k(\sigma)$. Therefore $v_1(\cdot, \sigma), \ldots, v_{r_k}(\cdot, \sigma)$ form a canonical system of root functions of $T(\cdot, \sigma^{p_k})$ at $\hat{\lambda}_k(\sigma)$, see [189, Definition 1.4.5] for notation. We conclude that the numbers $q_{k1}, \ldots, q_{kr_k}$ are the partial multiplicities of $T(\cdot, \sigma^p)$ at $\hat{\lambda}_k(\sigma)$ and that the number

$$\sum_{\iota=1}^{r_k} q_{k\iota} = m_k$$

is the algebraic multiplicity of the eigenvalue $\hat{\lambda}_k(\sigma)$ of $T(\cdot, \sigma^p)$, see [189, p. 15 and Proposition 1.8.5].

To complete the proof of the case that X and Y are finite dimensional we still have to show the representation (9.2.13). Since $\hat{\lambda}_k$ as well as σ^{p_k} are independent of the choice of $\sigma = (\eta^{\frac{1}{p_k}})_j$, it follows that (9.2.20) is valid for all j if it holds for one j, with the same vector functions v_ι and u_ι, $\iota = 1, \ldots, r_k$. Since v_ι is a polynomial of degree less than $q_{k\iota}$ we have

$$v_\iota(\lambda, \sigma) = \sum_{\kappa=0}^{q_{k\iota}-1} (\lambda - \hat\lambda_k(\sigma))^\kappa v_{\iota\kappa}(\sigma),$$

where $v_{\iota 0}(\sigma), \ldots, v_{\iota q_{k\iota}-1}(\sigma)$ is a chain of an eigenvector and associated vectors of $T(\cdot, \sigma^{p_k})$ at $\hat\lambda_k(\sigma)$, which depend meromorphically on σ. Letting ν be the maximum of the pole orders of these functions at 0 and observing that σ^ν does not depend on λ, we obtain that also $\sigma^\nu v_{\iota 0}(\sigma), \ldots, \sigma^\nu v_{\iota q_{k\iota}-1}(\sigma)$ is a chain of an eigenvector and associated vectors of $T(\cdot, \sigma^{p_k})$ at $\hat\lambda_k(\sigma)$ and that these function are analytic at 0. Substituting $\sigma = (\eta^{\frac{1}{p_k}})_j$ into the corresponding Taylor series expansions proves (9.2.13).

Finally, if X and Y are infinite dimensional, then we use the factorization (9.2.1) and consider S, which is an operator function in finite-dimensional spaces, where we may replace Φ with the neighbourhood Φ_0 from Lemma 9.2.1. Because of the invertibility of C, T_{11} and D, the eigenvalues of T and S coincide, together with their algebraic multiplicities. Hence the proof of part 1 is complete.

For the proof of part 2 let $\{\tilde{y}_{kj}^{\iota\kappa}(\eta)\}_{\kappa=0}^{q_{k\iota}}$ be the chain of an eigenvector and associated vectors for S as in (9.2.13). Then

$$y_{kj}^\iota(\lambda, \eta) := D^{-1}(\lambda, \eta) \sum_{\kappa=0}^{q_{k\iota}-1} (\lambda - \lambda_{k,j}(\eta))^\kappa \begin{pmatrix} 0 \\ \tilde{y}_{kj}^{\iota\kappa}(\eta) \end{pmatrix} \qquad (9.2.21)$$

is a root function of $C^{-1}(\cdot, \eta)T(\cdot, \eta)$ and thus of $T(\cdot, \eta)$ at $\lambda_{kj}(\eta)$, that is, the first $q_{k\iota}$ Taylor coefficients of (9.2.21) about $\lambda_{kj}(\eta)$ are a chain of an eigenvector and associated vectors which is of the form (9.2.13). $\qquad \square$

Chapter 10

Sobolev Spaces and Differential Operators

10.1   Sobolev spaces on intervals

For the convenience of the reader, we recall the main definitions and results on Sobolev spaces on intervals which are used in this monograph. For the general theory of Sobolev spaces, we refer the reader to [2]. However, in this monograph we are only concerned with Sobolev spaces on compact intervals, and therefore the particular results in [189, Chapter II] suffice, and it is some of those results which will be cited without proof here. Throughout this section we assume that a and b are real numbers with $a < b$.

Let $\mathcal{I} = (a, b)$ or $\mathcal{I} = \mathbb{R}$. First we recall the definition of the Lebesgue spaces $L_p(\mathcal{I})$ for $1 \le p \le \infty$. For Lebesgue measurable functions $f, g : \mathcal{I} \to \mathbb{C}$ we write $f = g$ ($f \le g$, etc.) if $f(x) = g(x)$ ($f(x) \le g(x)$, etc.) for almost all $x \in \mathcal{I}$,, i.e., if there is a Lebesgue measurable set $X \subset \mathcal{I}$ such that $\mathcal{I} \setminus X$ has Lebesgue measure zero and such that $f(x) = g(x)$ ($f(x) \le g(x)$, etc.) for all $x \in X$. The function f is called essentially bounded if there is such a Lebesgue measurable set X such that f is bounded on X. The relation $f = g$ in the above sense defines equivalence classes of Lebesgue measurable functions, and for $1 \le p < \infty$, $L_p(\mathcal{I})$ is defined as the set of equivalence classes of measurable functions f on $\mathcal{I}$ for which

$$\int_{\mathcal{I}} |f(x)|^p \, dx < \infty,$$

whereas $L_\infty(\mathcal{I})$ is the set of equivalence classes of essentially bounded functions on $\mathcal{I}$. As is customary, we will identify equivalence classes with any of the functions representing it. In particular, when such an equivalence class contains a continuous functions, then this continuous function is unique, and we will identify the equivalence class with its continuous representative. The sesquilinear form

$$(f, g) := \int_{\mathcal{I}} f(x) \overline{g(x)} \, dx, \quad f, g \in L_2(\mathcal{I}),$$

defines an inner product on $L_2(\mathcal{I})$ which makes $L_2(\mathcal{I})$ a Hilbert space.

A function $f \in C^\infty(\mathcal{I})$ is called a test function if its support is a compact subset of $\mathcal{I}$. The space of all test functions on $\mathcal{I}$ is denoted by $C_0^\infty(\mathcal{I})$. We identify $C_0^\infty(a, b) = C_0^\infty((a, b))$ with a subspace of $C_0^\infty(\mathbb{R})$ by setting $f = 0$ outside of (a, b) for each $f \in C_0^\infty(a, b)$.

A certain class of linear functionals on $C_0^\infty(a, b)$ is called the space of distributions on (a, b) and denoted by $\mathcal{D}'(a, b)$. For $u \in \mathcal{D}'(a, b)$ and $\varphi \in C_0^\infty(a, b)$ it is customary to write

$$(\varphi, u)_0 = u(\varphi).$$

Correspondingly, $\mathcal{D}'(\mathbb{R})$ denotes the space of distributions on $\mathbb{R}$, and we write $(\varphi, u)_{0,\mathbb{R}} = u(\varphi)$ for $u \in \mathcal{D}'(\mathbb{R})$ and $\varphi \in C_0^\infty(\mathbb{R})$.

It is not necessary to know the exact conditions for a linear functional on $C_0^\infty(a, b)$ to be a distribution; rather, it suffices to recall a few properties of distributions. One of these properties is that $L_2(a, b) \subset \mathcal{D}'(a, b)$ via

$$(\varphi, f)_0 := \int_a^b \varphi(x)\overline{f(x)}\, dx = (\varphi, f), \quad \varphi \in C_0^\infty(a, b), \ f \in L_2(a, b). \qquad (10.1.1)$$

We have chosen to write $(\cdot, \cdot)_0$ as a sesquilinear form rather than a bilinear form on the dual pair $(C_0^\infty(a, b), \mathcal{D}'(a, b))$ so that we do not have to resort to conjugate complex functions in any of the two forms occurring in (10.1.1).

A second property is that every $u \in \mathcal{D}'(a, b)$ has a derivative $u' \in \mathcal{D}'(a, b)$ which is defined by $(\varphi, u')_0 = -(\varphi', u)_0$ for $\varphi \in C_0^\infty(a, b)$. Hence every function $f \in L_2(a, b)$ has a derivative in the sense of distributions, and by the integration by parts formula, this derivative coincides with the classical derivative if f is continuously differentiable, or somewhat more general, if f is absolutely continuous.

Thirdly, for $u \in \mathcal{D}'(a, b)$ and $\psi \in C^\infty(a, b)$,

$$(\varphi, \psi u)_0 = (\overline{\psi}\varphi, u)_0, \quad \varphi \in C_0^\infty(a, b), \qquad (10.1.2)$$

defines a unique distribution ψu on (a, b).

Definition 10.1.1. Let $k \in \mathbb{N}_0$. The space

$$W_2^k(a, b) = \{f \in L_2(a, b) : \forall j \in \{1, \ldots, k\}, \ f^{(j)} \in L_2(a, b)\}$$

is called a Sobolev space. Here the derivatives $f^{(j)}$ are the derivatives in the sense of distributions. For $f \in W_2^k(a, b)$ we set

$$\|f\|_{2,k} = \left(\sum_{j=0}^k \|f^{(j)}\|^2\right)^{\frac{1}{2}},$$

where $\|\ \|$ is the norm on $L_2(a, b)$.

Note that $W_2^0(a, b) = L_2(a, b)$.

Proposition 10.1.2 ([189, Propositions 2.1.6 and 2.1.7]). *Let $k \in \mathbb{N}$. Then $W_2^k(a,b)$ is a Banach space with respect to the norm $\| \; \|_{2,k}$, $W_2^k(a,b) \subset C^{k-1}([a,b])$, and the corresponding inclusion map is continuous.*

Proposition 10.1.3 ([189, Proposition 2.3.2]). *Let $k \in \mathbb{N}_0$, $l \in \mathbb{N}$ and $k \leq l$. Then the multiplication operator from $W_2^k(a,b) \times W_2^l(a,b)$ to $W_2^k(a,b)$ is a continuous bilinear map.*

For $f \in L_2(a,b)$ let f_e be its extension by zero onto $\mathbb{R}$. Then $f_e \in \mathcal{D}'(\mathbb{R})$.

Proposition 10.1.4 ([189, Proposition 2.2.4 and Theorem 2.2.5]). *For $k \in \mathbb{N}$ set*

$$W_2^{-k}[a,b] = \left\{ u \in \mathcal{D}'(\mathbb{R}) : \exists \, (u_j)_{j=0}^k \in (L_2(a,b))^{k+1}, \; u = \sum_{j=0}^k (u_j)_e^{(j)} \right\}.$$

Then, for $u = \sum_{j=0}^k (u_j)_e^{(j)} \in W_2^{-k}[a,b]$ and $f \in W_2^k(a,b)$,

$$(f,u)_{2,k} := \sum_{j=0}^k \int_a^b (-1)^j f^{(j)}(x) \overline{u_j(x)} \, dx$$

does not depend on the representation of u. For $\varphi \in C_0^\infty(\mathbb{R})$ we have

$$(\varphi|_{(a,b)}, u)_{2,k} = (\varphi, u)_{0,\mathbb{R}}.$$

Equipped with a suitable norm, $W_2^{-k}[a,b]$ is a representation of the dual of the Banach space $W_2^k(a,b)$ with respect to the sesquilinear form $(\; , \;)_{2,k}$.

For $k \in \mathbb{N}$ denote the restriction of elements in $W_2^{-k}[a,b]$ to distributions on (a,b) by $W_2^{-k}(a,b)$. From [114, Theorem 3.1.4], [189, Proposition 2.1.5] and the definition of $W_p^{-k}[a,b]$ in Proposition 10.1.4 we immediately obtain

Proposition 10.1.5. *For every $k \in \mathbb{Z}$, $j \in \mathbb{N}$ and $u \in \mathcal{D}'(a,b)$ we have $u \in W_2^k(a,b)$ if and only if $u^{(j)} \in W^{k-j}(a,b)$.*

Remark 10.1.6. Let $k \in \mathbb{N}_0$, $l \in \mathbb{N}$, $k \leq l$, and $g \in W_2^k(a,b)$. By Proposition 10.1.3, $g\cdot$ is a continuous operator from $W_2^l(a,b)$ to $W_2^k(a,b)$. Propositions 10.1.3 and 10.1.4 yield that its adjoint $(g\cdot)^*$ is an operator from $W_2^{-k}[a,b]$ to $W_2^{-l}[a,b]$. For $v \in L_2(a,b)$ and $f \in W_p^l(a,b)$ it follows that $(g\cdot)^* v_e = (\overline{g}v)_e$, so that we will write

$$(g\cdot)^* u =: \overline{g} \cdot u =: \overline{g}u$$

for all $u \in W_2^{-k}[a,b]$.

Theorem 10.1.7 ([189, Lemma 2.4.1 and Theorem 2.4.2]). *For each $k \in \mathbb{N}$, the inclusion maps $W_2^k(a,b) \hookrightarrow C^{k-1}[a,b]$ and $W_p^k(a,b) \hookrightarrow W_p^{k-1}(a,b)$ are compact.*

10.2   Lagrange identity and Green's formula

In this section we recall some basic properties of differential operators on Sobolev spaces. Rather than using the most general formulation, we will restrict ourselves to assumptions which cover all examples and applications in this monograph. For proofs and more general assumptions, we refer the reader to [271, Section 2] and [189, Chapter VII].

Let $n = 2k$ where $k \in \mathbb{N}$. We consider nth-order differential expressions of the form

$$\ell y = \sum_{j=0}^{k} \left(g_j y^{(j)} \right)^{(j)} \tag{10.2.1}$$

on an interval $[a, b]$, where $g_j \in W_2^j(a, b)$, $j = 0, \ldots, k$, are real valued functions and $|g_k(x)| \geq \varepsilon$ for some $\varepsilon > 0$ and all $x \in [a, b]$. We will often consider applications where $g_k = 1$ or $g_k = -1$ and $g_j \in C^j[a, b]$. In any case, ℓy is well defined for $y \in W_2^n(a, b)$, in which case $\ell y \in L_2(a, b)$. The operator L_0 defined by

$$D(L_0) = W_2^n(a, b), \quad L_0 y = \ell y, \quad y \in W_2^n(a, b), \tag{10.2.2}$$

is called the maximal operator associated with the differential expression ℓ on $[a, b]$.

We will now prove the Lagrange identity and Green's formula. In general, one would need to define the adjoint differential expression. However, in our situation, the differential expression ℓ is formally self-adjoint, so that there is no need to introduce the adjoint differential expression here.

Definition 10.2.1. Let $y \in W_2^n(a, b)$. For $j = 0, \ldots, n$, the jth quasi-derivative of y, denoted by $y^{[j]}$, is recursively defined by

$$y^{[j]} = y^{(j)} \quad \text{for } j = 0, \ldots, k - 1,$$
$$y^{[k]} = g_k y^{(k)},$$
$$y^{[j]} = (y^{[j-1]})' + g_{n-j} y^{[n-j]} \quad \text{for } j = k + 1, \ldots, n.$$

Observe that quasi-derivatives depend on the differential expression (10.2.1). Quasi-derivates are convenient for the formulation of the Lagrange identity when dealing with differential operators which have fairly general coefficients, see, e. g., [271, Section 2] and [196, Section 2]. Note, however, that these definitions may differ from each other. In particular, our definition is not exactly the same as in [271, Theorem 2.2]. A straightforward proof by induction shows that

$$y^{[k+m]} = \sum_{j=k-m}^{k} \left(g_j y^{(j)} \right)^{(j+m-k)} \quad \text{for } m = 0, \ldots, k.$$

In particular, $m = k$ gives

$$\ell y = y^{[n]}.$$

Proposition 10.2.2. *For the differential expression ℓ and $y, z \in W_2^n(a,b)$, we have*

$$(\ell y)\bar{z} = \sum_{j=0}^{k} (-1)^j g_j y^{(j)} \overline{z^{(j)}} + \frac{d}{dx}[y,z]_1 \tag{10.2.3}$$

on $[a,b]$ almost everywhere, where

$$[y,z]_1 = \sum_{j=1}^{k} (-1)^{j-1} y^{[n-j]} \overline{z^{[j-1]}}, \tag{10.2.4}$$

and

$$(\ell y, z) = \sum_{j=0}^{k} (-1)^j (g_j y^{(j)}, z^{(j)}) + [y,z]_1(b) - [y,z]_1(a). \tag{10.2.5}$$

Proof. (10.2.5) follows from (10.2.3) by integration, so that we only have to prove the latter. A straightforward calculation for $j = 1, \dots, k$ gives

$$\frac{d}{dx}\left(y^{[n-j]}\overline{z^{[j-1]}}\right) = y^{[n-j]}\overline{z^{(j)}} + y^{[n-j+1]}\overline{z^{(j-1)}} - g_{j-1}y^{(j-1)}\overline{z^{(j-1)}}.$$

Therefore

$$\frac{d}{dx}[y,z]_1 = \sum_{j=1}^{k} (-1)^{j-1} y^{[n-j]}\overline{z^{(j)}} + \sum_{j=0}^{k-1} (-1)^j y^{[n-j]}\overline{z^{(j)}} - \sum_{j=0}^{k-1} (-1)^j g_j y^{(j)}\overline{z^{(j)}}$$

$$= y^{[n]}\bar{z} - \sum_{j=0}^{k} (-1)^j g_j y^{(j)}\overline{z^{(j)}}. \qquad \square$$

We can now formulate the Lagrange identity and Green's formula.

Theorem 10.2.3. *For the differential expression ℓ and $y, z \in W_2^n(a,b)$, the Lagrange identity*

$$(\ell y)\bar{z} - y(\overline{\ell z}) = \frac{d}{dx}[y,z] \tag{10.2.6}$$

holds on $[a,b]$ almost everywhere, where

$$[y,z] = \sum_{j=1}^{k} (-1)^j \left(y^{[j-1]}\overline{z^{[n-j]}} - y^{[n-j]}\overline{z^{[j-1]}}\right), \tag{10.2.7}$$

and Green's formula

$$(\ell y, z) - (y, \ell z) = [y,z](b) - [y,z](a) \tag{10.2.8}$$

is valid, where $(\cdot, \cdot)$ is the inner product in $L_2(a,b)$.

Proof. Green's formula follows from the Lagrange identity by integration, so that we only have to prove the latter. Observing that the functions g_j are real valued, we conclude from (10.2.3) that

$$
(\ell y)\bar{z} - y(\overline{\ell z}) = (\ell y)\bar{z} - \overline{(\ell z)\bar{y}}
$$
$$
= \frac{d}{dx}[y, z]_1 - \frac{d}{dx}\overline{[z, y]_1}
$$
$$
= \frac{d}{dx}[y, z].
$$

$\square$

10.3 Self-adjoint differential operators

In this section we will deduce criteria for self-adjointness for a class of differential operators. This class will cover all applications in this monograph. Readers interested in a particular problem may find it easier to verify the results in this section by considering just that particular case.

Let $p \in \mathbb{N}$ and for $j = 1, \ldots, p$ let ℓ_j be formally self-adjoint differential expressions of even order $n_j = 2k_j$ on the interval $[0, a_j]$ as defined in (10.2.1), i. e.,

$$
\ell_j y_j = \sum_{m=0}^{k_j} (g_{j,m} y_j^{(m)})^{(m)}, \tag{10.3.1}
$$

where $g_{j,m} \in W_2^m(0, a_j)$ for $j = 1, \ldots, p$, $m = 0, \ldots, k_j$, are real valued functions and $|g_{j,k_j}(x)| \geq \varepsilon$ for some $\varepsilon > 0$ and all $x \in [a, b]$. Let $q \in \mathbb{N}_0$. We are going to use the notation

$$
Y = \begin{pmatrix} y_1 \\ \vdots \\ y_p \\ c \end{pmatrix} = \begin{pmatrix} Y_0 \\ c \end{pmatrix}, \quad Z = \begin{pmatrix} z_1 \\ \vdots \\ z_p \\ d \end{pmatrix} = \begin{pmatrix} Z_0 \\ d \end{pmatrix}, \quad W = \begin{pmatrix} w_1 \\ \vdots \\ w_p \\ e \end{pmatrix} = \begin{pmatrix} W_0 \\ e \end{pmatrix}
$$

for elements in the Hilbert space

$$
H = \bigoplus_{j=1}^{p} L_2(0, a_j) \oplus \mathbb{C}^q.
$$

Furthermore, for $y_j \in W_2^{n_j}(0, a_j)$, $j = 1, \ldots, p$, we use the notations

$$
\hat{y}_j = (y_j(0), \ldots, y_j^{[n_j-1]}(0), y_j(a_j), \ldots, y_j^{[n_j-1]}(a_j)), \quad j = 1, \ldots, p,
$$
$$
\hat{Y} = \hat{Y}_0 = (\hat{y}_1, \ldots, \hat{y}_p)^\mathsf{T},
$$

with corresponding notations for $\hat{Z}$ and $\hat{W}$. Setting

$$
n = \sum_{j=1}^{p} n_j,
$$

it is clear that $\hat{Y} \in \mathbb{C}^{2n}$. Finally, let

$$r \in \mathbb{N}_0, \ U_1 \text{ an } r \times 2n \text{ matrix}, \ U_2 \text{ a } q \times 2n \text{ matrix}, \ V \text{ a } q \times 2n \text{ matrix}. \quad (10.3.2)$$

Then the operator A in H is defined by

$$AY = \begin{pmatrix} \ell_1 y_1 \\ \vdots \\ \ell_p y_p \\ V\hat{Y} \end{pmatrix}, \ D(A) = \left\{ Y \in \bigoplus_{j=1}^{p} W_2^{n_j}(0, a_j) \oplus \mathbb{C}^p : U_1 \hat{Y} = 0, \ c = U_2 \hat{Y} \right\}.$$

$$(10.3.3)$$

We have to observe that the Lagrange brackets introduced in (10.2.7) depend on the quasi-derivatives and hence on the differential equation ℓ. In order to avoid further notation, we will use the convention that an expression of the form $[y_j, z_j]$ means the Lagrange bracket with respect to the operator ℓ_j. As usual, V^* denotes the adjoint, i.e., the conjugate complex transpose, matrix of V, and $c \in \mathbb{C}^q$ is identified with a $q \times 1$ matrix.

In view of Remark 10.1.6 it is clear that the definition of the quasi-derivatives, Definition 10.2.1, can be extended to $z_j \in L_2(0, a_j)$ giving quasi-derivatives $z_j^{[m]} \in W_2^{-m}[0, a_j]$ in the sense of distributions for $m = 0, \ldots, n_j$. In particular,

$$z_j^{[n_j]} = \sum_{m=0}^{k_j} (g_{j,m} z_j^{(m)})^{(m)}. \quad (10.3.4)$$

Proposition 10.3.1. *Let* $Z, W \in H$ *such that* $(AY, Z) = (Y, W)$ *for all* $Y \in D(A)$. *Then* $z_j \in W_2^{n_j}(0, a_j)$ *and* $\ell_j z_j = w_j$ *for* $j = 1, \ldots, n$.

Proof. Fixing $j \in \{1, \ldots, p\}$, taking $y_j \in C_0^\infty(0, a_j)$, letting $c = 0$ and $y_{j'} = 0$ for all $j' \neq j$, it follows that $Y \in D(A)$ and

$$(y_j, w_j) = (Y, W) = (AY, Z) = (\ell_j y_j, z_j)$$

$$= \sum_{m=0}^{k_j} \left((g_{j,m} y_j^{(m)})^{(m)}, z_j \right)$$

$$= \sum_{m=0}^{k_j} (-1)^m \left(g_{j,m} y_j^{(m)}, z_j^{(m)} \right)_{2,m}$$

$$= \sum_{m=0}^{k_j} (-1)^m \left(y_j^{(m)}, g_{j,m} z_j^{(m)} \right)_{2,m}$$

$$= \sum_{m=0}^{k_j} \left(y_j, (g_{j,m} z_j^{(m)})^{(m)} \right)_0.$$

Since $C_0^\infty(0, a_j)$ is dense in $L_2(a, b)$, it follows that

$$\sum_{m=0}^{k_j}(g_{j,m}z_j^{(m)})^{(m)} = w_j \in L_2(0, a_j). \tag{10.3.5}$$

Assume that $z_j \notin W_2^{n_j}(0, a_j)$. Then there is an integer m_j, $0 \le m_j < n_j$, such that $z_j \in W_2^{m_j}(0, a_j)$ and $z_j \notin W_2^{m_j+1}(0, a_j)$. Then

$$(g_{j,k_j}z_j^{(k_j)})^{(k_j)} = w_j - \sum_{m=0}^{k_j-1}(g_{j,m}z_j^{(m)})^{(m)} \in W_2^{m_j-n_j+1}(0, a_j).$$

By Proposition 10.1.5 it follows that

$$g_{j,k_j}z_j^{(k_j)} \in W_2^{m_j-k_j+1}(0, a_j).$$

Since $\frac{1}{g_{j,k_j}} \in W_2^{k_j}(0, a_j) \subset W_2^{m_j-k_j+1}(0, a_j)$, see [189, Proposition 2.5.8], we conclude from Remark 10.1.6 that

$$z_j^{(k_j)} \in W_2^{m_j-k_j+1}(0, a_j).$$

Another application of Proposition 10.1.5 gives the contradiction $z_j \in W_2^{m_j}(0, a_j)$. Hence $z_j \in W_2^{n_j}(0, a_j)$, and (10.3.5) shows that $\ell_j z_j = w_j$. $\square$

Proposition 10.3.2. *Assume that* rank $\begin{pmatrix} U_1 \\ U_2 \end{pmatrix} = r + q$. *Then the operator A is densely defined, and*

$$D(A^*) \subset \bigoplus_{j=1}^{p} W_2^{n_j}(0, a_j) \oplus \mathbb{C}^q.$$

Proof. Let $W \in H$ be orthogonal to $D(A)$. For $Z = 0 \in H$ and all $Y \in D(A)$ it follows that

$$(AY, Z) = 0 = (Y, W),$$

and therefore $w_j = \ell_j z_j = 0$ by Proposition 10.3.1. Then

$$0 = (Y, W) = e^* c = e^* U_2 \hat{Y}. \tag{10.3.6}$$

It is well known, see, e. g., [196, Corollary 2.7], that the map $y_j \mapsto \hat{y}_j$ from $W_2^{n_j}(0, a_j)$ to C^{2n_j} is surjective. With respect to the decomposition $\mathbb{C}^{2n} = N(U_1)^\perp \oplus N(U_1)$ we write

$$\begin{pmatrix} U_1 \\ U_2 \end{pmatrix} = \begin{pmatrix} U_{11} & 0 \\ U_{21} & U_{22} \end{pmatrix},$$

where the U_{ij} may be represented as matrices in bases of $N(U_1)^\perp$ and $N(U_1)$, respectively. Since rank $\begin{pmatrix} U_1 \\ U_2 \end{pmatrix} = r + q$ and U_1 and U_2 are $r \times 2n$ and $q \times 2n$

matrices, respectively, it follows that U_{22} has rank q and is therefore surjective. Thus there is $Y_0 \in \bigoplus_{j=1}^{p} W_2^{n_j}(0, a_j)$ such that $\hat{Y}_0 \in N(U_1)$ and $U_{22}\hat{Y}_0 = e$. Putting

$$Y = \begin{pmatrix} Y_0 \\ V\hat{Y}_0 \end{pmatrix},$$

it follows $U_1\hat{Y} = 0$ and therefore $Y \in D(A)$. Then (10.3.6) gives

$$0 = e^*U_{22}\hat{Y} = e^*e,$$

which shows $e = 0$. Altogether, we have $W = 0$, which proves the denseness of $D(A)$.

Let $Z \in D(A^*)$. Then, by definition of the adjoint operator, there is $W \in H$ such that $(AY, Z) = (Y, W)$ for all $Y \in D(A)$, and the statement about $D(A^*)$ follows from Proposition 10.3.1. $\qquad \square$

Proposition 10.3.3. *Assume that rank* $\begin{pmatrix} U_1 \\ U_2 \end{pmatrix} = r + q$. *Then* $Z \in D(A^*)$ *if and only if* $Z \in \bigoplus_{j=1}^{p} W_2^{n_j}(0, a_j) \oplus \mathbb{C}^q$ *and there is* $e \in \mathbb{C}^q$ *such that*

$$\sum_{j=1}^{p} [y_j, z_j](a_j) - \sum_{j=1}^{p} [y_j, z_j](0) + d^*V\hat{Y} - e^*U_2\hat{Y} = 0 \tag{10.3.7}$$

for all $\hat{Y} \in N(U_1)$. *For* $Z \in D(A^*)$, *e is unique and*

$$A^*Z = \begin{pmatrix} \ell_1 z_1 \\ \vdots \\ \ell_p z_p \\ e \end{pmatrix}. \tag{10.3.8}$$

Proof. By definition of the adjoint, $Z \in D(A^*)$ if and only if there is $W \in H$ such that $(AY, Z) = (Y, W)$ for all $Y \in D(A)$. By Propositions 10.3.1 and 10.3.2, if $Z \in D(A^*)$, then $Z \in \bigoplus_{j=1}^{p} W_2^{n_j}(0, a_j) \oplus \mathbb{C}^q$ and $\ell_j z_j = w_j$ for $j = 1, \ldots, p$.

Therefore, for $Y \in D(A)$, $W \in H$, and $Z \in \bigoplus_{j=1}^{p} W_2^{n_j}(0, a_j) \oplus \mathbb{C}^q$ with $\ell_j z_j = w_j$ for $j = 1, \ldots, p$,

$$(AY, Z) - (Y, W) = \sum_{j=1}^{p} (\ell_j y_j, z_j) - \sum_{j=1}^{p} (y_j, \ell_j z_j) + d^*V\hat{Y} - e^*c.$$

In view of Green's formula, (10.2.8), we conclude that

$$(AY, Z) - (Y, W) = \sum_{j=1}^{p} [y_j, z_j](a_j) - \sum_{j=1}^{p} [y_j, z_j](0) + d^* V \hat{Y} - e^* U_2 \hat{Y}.$$

This and $A^* Z = W$ if $(AY, Z) - (Y, W) = 0$ for all $Y \in D(A)$ completes the proof if we observe that $Y \in D(A)$ if and only if $U_1 \hat{Y} = 0$ and $c = U_2 \hat{Y}$. $\square$

Since A is self-adjoint if and only if $D(A) = D(A^*)$ and $A^* Z = AZ$ for all $Z \in D(A) = D(A^*)$, we have the following characterization of self-adjointness of A.

Theorem 10.3.4. *Assume that* $\operatorname{rank} \begin{pmatrix} U_1 \\ U_2 \end{pmatrix} = r + q$. *Then A is self-adjoint if and only if the following holds for all* $Z \in \bigoplus_{j=1}^{p} W_2^{n_j}(0, a_j) \oplus \mathbb{C}^q$:

$Z \in D(A)$ *if and only if there is* $e \in \mathbb{C}^q$ *such that* $e = V \hat{Z}$ *and*

$$\sum_{j=1}^{p} [y_j, z_j](a_j) - \sum_{j=1}^{p} [y_j, z_j](0) + d^* V \hat{Y} - e^* U_2 \hat{Y} = 0 \tag{10.3.9}$$

for all $\hat{Y} \in N(U_1)$.

For $m \in \mathbb{N}$ define

$$J_{m,0} = \left((-1)^{s-1} \delta_{s,m+1-t} \right)_{s,t=1}^{m}, \quad J_{m,1} = \begin{pmatrix} 0 & J_{m,0} \\ -J_{m,0}^* & 0 \end{pmatrix}, \quad J_m = \begin{pmatrix} -J_{m,1} & 0 \\ 0 & J_{m,1} \end{pmatrix}. \tag{10.3.10}$$

Then it follows for $y_j, z_j \in W_2^{n_j}(0, a_j)$ that

$$[y_j, z_j]_1 = \begin{pmatrix} z_j^{[0]} \\ \vdots \\ z_j^{[k_j-1]} \end{pmatrix}^* J_{k_j,0} \begin{pmatrix} y_j^{[k_j]} \\ \vdots \\ y_j^{[n_j-1]} \end{pmatrix}, \quad [y_j, z_j] = \begin{pmatrix} z_j^{[0]} \\ \vdots \\ z_j^{[n_j-1]} \end{pmatrix}^* J_{k_j,1} \begin{pmatrix} y_j^{[0]} \\ \vdots \\ y_j^{[n_j-1]} \end{pmatrix}, \tag{10.3.11}$$

so that

$$[y_j, z_j](a_j) - [y_j, z_j](0) = (J_{k_j} \hat{y}_j, \hat{z}_j).$$

Finally, define

$$J = \bigoplus_{j=1}^{p} J_{k_j} \quad \text{and} \quad U_3 = \begin{pmatrix} J \\ V \\ -U_2 \end{pmatrix}. \tag{10.3.12}$$

Note that J is a $2n \times 2n$ matrix and that U_3 is a $(2n + 2q) \times 2n$ matrix. We also observe that (10.3.9) can be written as

$$\begin{pmatrix} \hat{Z} \\ d \\ e \end{pmatrix}^* U_3 \hat{Y} = 0$$

and that $Z \in D(A)$ and $e = V\hat{Z}$ is equivalent to

$$\begin{pmatrix} \hat{Z} \\ d \\ e \end{pmatrix} \in N(U),$$

where

$$U = \begin{pmatrix} U_1 & 0 & 0 \\ U_2 & -I & 0 \\ V & 0 & -I \end{pmatrix} \tag{10.3.13}$$

and where I is the $q \times q$ identity matrix.

Observing that $(N(U))^\perp = R(U^*)$, Theorem 10.3.4 can therefore be reformulated as

Theorem 10.3.5. *In the notation of* (10.3.2) *and* (10.3.3), *assume that*

$$\operatorname{rank} \begin{pmatrix} U_1 \\ U_2 \end{pmatrix} = r + q.$$

Then A is self-adjoint if and only if

$$U_3(N(U_1)) = R(U^*).$$

Theorem 10.3.5 may become quite unwieldy since one might have to deal with matrices of possibly quite large sizes. In many cases, it might therefore be easier to apply Theorem 10.3.4.

Lemma 10.3.6. *Let $j \in \mathbb{N}_0$, $g \in L_2(0, a)$ and $\eta > 0$. Then, for all $y \in W_2^{j+1}(0, a)$,*

$$|(gy^{(j)}, y^{(j)})| \le \eta \|y^{(j+1)}\|^2 + \left[\frac{(1 + 5\kappa)a}{4\eta} \|g\|^2 + \frac{2\kappa^{\frac{1}{2}}}{a^{\frac{1}{2}}} \|g\| \right] \|y^{(j)}\|^2,$$

where the estimate is true in general with $\kappa = 1$, whereas it holds with $\kappa = 0$ if additionally $y^{(j)}(0) = 0$. If additionally $\eta \le 2a$, then

$$|y^{(j)}(0)|^2, |y^{(j)}(a)|^2 \le \eta \|y^{(j+1)}\|^2 + \frac{4}{\eta} \|y^{(j)}\|^2.$$

Proof. It suffices to consider the case $j = 0$; the general case is then obtained by applying this special case to $y^{(j)}$. Also, the first estimate is trivial if $g = 0$, so that we will assume that $g \ne 0$.

By Hölder's inequality and using that $y(x) - y(0) = \int_0^x y'(t)\, dt$ we obtain for $x \in [0, a]$ that

$$|y(x) - y(0)| \le \left(\int_0^a |y'(t)|^2\, dt \right)^{\frac{1}{2}} \left(\int_0^a dt \right)^{\frac{1}{2}} = a^{\frac{1}{2}} \|y'\|. \tag{10.3.14}$$

Then $|y(0)| \leq |y(x)| + a^{\frac{1}{2}}\|y'\|$ leads to the estimate $|y(0)|^2 \leq 2|y(x)|^2 + 2a\|y'\|^2$. Integrating from 0 to a gives $a|y(0)|^2 \leq 2\|y\|^2 + 2a^2\|y'\|^2$, which is equivalent to

$$|y(0)|^2 \leq \frac{2}{a}\|y\|^2 + 2a\|y'\|^2. \qquad (10.3.15)$$

From (10.3.14) we have $|y(x)| \leq |y(0)| + a^{\frac{1}{2}}\|y'\|$, and making use of (10.3.15) we obtain the estimate

$$|y(x)|^2 \leq 2|y(0)|^2 + (1+\kappa)a\|y'\|^2 \leq (1+5\kappa)a\|y'\|^2 + \frac{4\kappa}{a}\|y\|^2, \qquad (10.3.16)$$

with $\kappa = 1$ in general and where we may choose $\kappa = 0$ if $y(0) = 0$. Again by Hölder's inequality,

$$\left| \int_0^a g(x)|y(x)|^2\, dx \right| \leq \left(\int_0^a |g(x)|^2 dx \right)^{\frac{1}{2}} \left(\int_0^a |y(x)|^4 dx \right)^{\frac{1}{2}}$$

$$\leq \|g\| \left(\int_0^a |y(x)|^2\, dx \left[(1+5\kappa)a\|y'\|^2 + \frac{4\kappa}{a}\|y\|^2 \right] \right)^{\frac{1}{2}}$$

$$\leq \|g\|\|y\| \left[(1+5\kappa)^{\frac{1}{2}} a^{\frac{1}{2}} \|y'\| + \frac{2\kappa^{\frac{1}{2}}}{a^{\frac{1}{2}}}\|y\| \right]$$

$$\leq \|g\| \left[(1+5\kappa)^{\frac{1}{2}} a^{\frac{1}{2}} \frac{b^{-1}\|y'\|^2 + b\|y\|^2}{2} + \frac{2\kappa^{\frac{1}{2}}}{a^{\frac{1}{2}}}\|y\|^2 \right]$$

for any positive b. Choosing $b = \dfrac{(1+5\kappa)^{\frac{1}{2}} a^{\frac{1}{2}}}{2\eta}\|g\|$, it follows that

$$|(gy, y)| \leq \eta\|y'\|^2 + \left[\frac{(1+5\kappa)a}{4\eta}\|g\|^2 + \frac{2\kappa^{\frac{1}{2}}}{a^{\frac{1}{2}}}\|g\| \right] \|y\|^2,$$

which proves the first estimate of this theorem. For the second estimate we just have to observe that (10.3.14) and (10.3.15) also hold on the interval $[0, \frac{\eta}{2}]$, and that the corresponding norms over this smaller interval in (10.3.15) can be replaced by the possibly larger norms on the original interval $[0, a]$. The estimate at a follows with the obvious change of notation. $\qquad \square$

Lemma 10.3.7. *Let $j \in \mathbb{N}$, $m \in \mathbb{N}_0$ such that $m < j$ and $\eta > 0$. Then there is $C > 0$ such that for all $y \in W_2^j(0, a)$,*

$$\|y^{(m)}\|^2 \leq \eta\|y^{(j)}\|^2 + C\|y\|^2. \qquad (10.3.17)$$

Proof. The statement is trivial for $m = 0$. Now let $j = 2$ and $m = 1$. Then the integration by parts formula

$$\|y'\|^2 = (y', y') = y'(a)\overline{y(a)} - y'(0)\overline{y(0)} - (y'', y),$$

Lemma 10.3.6 and Hölder's inequality show that for each $\iota = 1, \ldots, 6$ and $\eta_\iota > 0$ there is a constant $C_\iota > 0$, which is independent of y, such that

$$
\begin{aligned}
\|y'\|^2 &\leq 2(\eta_1\|y''\| + C_1\|y'\|)(\eta_2\|y''\| + C_2\|y\|) + \eta_3\|y''\|^2 + C_3\|y\|^2 \\
&\leq 2\eta_1\eta_2(\eta_4\|y''\|^2 + C_4\|y'\|^2) + 2\eta_2 C_1\|y'\|^2 + 2\eta_1 C_2(\eta_5\|y''\|^2 + C_5\|y\|^2) \\
&\quad + C_1 C_2(\eta_6\|y'\|^2 + C_6\|y\|^2) + \eta_3\|y''\|^2 + C_3\|y\|^2 \\
&\leq (2\eta_1\eta_2\eta_4 + 2\eta_1 C_2\eta_5 + \eta_3)\|y''\|^2 \\
&\quad + (2\eta_1\eta_2 C_4 + 2\eta_2 C_1 + C_1 C_2\eta_6)\|y'\|^2 \\
&\quad + (2\eta_1 C_2 C_5 + C_1 C_2 C_6 + C_3)\|y\|^2.
\end{aligned}
$$

Let $\eta' > 0$. Choosing suitable pairs (η_4, C_4), (η_1, C_1), (η_2, C_2), (η_3, C_3), (η_5, C_5), (η_6, C_6), in that order, we obtain a constant C_7, depending on the choice of the above pairs, such that

$$\|y'\|^2 \leq \frac{1}{2}\eta'\|y''\|^2 + \frac{1}{2}\|y'\|^2 + \frac{1}{2}C_7\|y\|^2,$$

which implies

$$\|y'\|^2 \leq \eta'\|y''\|^2 + C_7\|y\|^2.$$

This proves (10.3.17) for $j = 2$.

We are going to prove the general case of (10.3.17) by induction on j. Assume (10.3.17) is true for $1 \leq m < j \leq k$. First let $2 \leq m \leq k$. Then, applying the induction hypothesis to y', it follows that for each $\iota = 1, 2$ and each $\eta_\iota > 0$ there is $C_\iota > 0$ such that

$$
\begin{aligned}
\|y^{(m)}\|^2 = \|(y')^{(m-1)}\|^2 &\leq \eta_1\|(y')^{(k)}\|^2 + C_1\|y'\|^2 \\
&\leq \eta_1\|(y')^{(k)}\|^2 + C_1\eta_2\|y^{(m)}\|^2 + C_1 C_2\|y\|^2.
\end{aligned}
$$

Choosing $\eta_1 = \frac{\eta}{2}$ and then $\eta_2 > 0$ such that $C_1\eta_2 \leq \frac{1}{2}$, (10.3.17) follows for $j = k + 1$. Finally, if $m = 1$, then, with a notation as for the case $m \geq 2$, we obtain, making use of (10.3.17) for $m = 1$ and $j = 2$ as well as $m = 2$ and $j = k + 1$,

$$\|y'\|^2 \leq \eta_1\|y''\|^2 + C_1\|y\|^2 \leq \eta_1\eta_2\|y^{(k+1)}\|^2 + (\eta_1 C_2 + C_1)\|y\|^2.$$

Thus we have shown that (10.3.17) is true for $1 \leq m < j \leq k + 1$. $\square$

In order to give easily checkable conditions for A to be bounded below, we will use the following notation. For $k \in \mathbb{N}$ define the $2k \times 2k$ and $4k \times 4k$ matrices $P_{k,0}$ and P_k by

$$P_{k,0} = \begin{pmatrix} I_k & 0 \\ 0 & 0 \end{pmatrix}, \quad P_k = \begin{pmatrix} P_{k,0} & 0 \\ 0 & P_{k,0} \end{pmatrix}.$$

Furthermore, we define

$$P = \bigoplus_{j=1}^{p} P_{k_j}.$$

Theorem 10.3.8. *Assume that A is self-adjoint. Then A has a compact resolvent. Assume additionally that*

(i) $(-1)^{k_j} g_{j,k_j} > 0$ *for all $j = 1, \ldots, p$,*

(ii) *each component of $U_1 \hat{Y}$ either contains only quasiderivatives $y_j^{[m]}$ with $m < k_j$ or contains only quasiderivatives $y_j^{[m]}$ with $m \geq k_j$,*

(iii) *each component of $U_2 \hat{Y}$ either contains only quasiderivatives $y_j^{[m]}$ with $m < k_j$ or contains only quasiderivatives $y_j^{[m]}$ with $m \geq k_j$,*

(iv) *for each component of $U_2 \hat{Y}$ which only contains quasiderivatives $y_j^{[m]}$ with $m \geq k_j$, the corresponding component of $V\hat{Y}$ only contains quasiderivatives $y_j^{[m]}$ with $m < k_j$.*

Then A is bounded below.

Proof. Since A is self-adjoint, its spectrum is a subset of $\mathbb{R}$, and $A + iI$ is therefore invertible. If $D(A)$ is equipped with the norm of $\bigoplus_{j=1}^{p} W_2^{n_j}(0, a_j) \oplus \mathbb{C}^q$, then it follows by the closed graph theorem that $(A + iI)^{-1}$ is bounded from $\bigoplus_{j=1}^{p} L_2(0, a_j) \oplus \mathbb{C}^q$ to $D(A)$. Theorem 10.1.7 shows that the embedding from $D(A)$ into $\bigoplus_{j=1}^{p} L_2(0, a_j) \oplus \mathbb{C}^q$ is compact. Hence $(A + iI)^{-1}$ is compact.

Now let $Y = (y_1, \ldots, y_p, c)^\mathsf{T} \in D(A)$. Then it follows from (10.3.3) and (10.2.5) that

$$(AY, Y) = \sum_{j=1}^{p} (\ell_j y_j, y_j) + \hat{Y}^* U_2^* V \hat{Y} \tag{10.3.18}$$

$$= \sum_{j=1}^{p} \sum_{m=0}^{k_j} (-1)^m (g_{j,m} y_j^{(m)}, y_j^{(m)}) + \sum_{j=1}^{p} [y_j, y_j]_1(a_j) - \sum_{j=1}^{p} [y_j, y_j]_1(0) + \hat{Y}^* U_2^* V \hat{Y}.$$

In view of the general assumption and $(-1)^{k_j} g_{j,k_j} > 0$, there is $\varepsilon > 0$ such that $(-1)^{k_j} g_{j,k_j}(x) \geq \varepsilon$ for all $x \in [0, a]$. Therefore

$$\sum_{m=0}^{k_j} (-1)^j (g_{j,m} y_j^{(m)}, y_j^{(m)}) \geq \varepsilon (y_j^{(k_j)}, y_j^{(k_j)}) - \sum_{m=0}^{k_j-1} |(g_{j,m} y_j^{(m)}, y_j^{(m)})|.$$

An application of Lemmas 10.3.6 and 10.3.7 shows that there is $C > 0$ such that

$$\sum_{j=1}^{p} \sum_{m=0}^{k_j} (-1)^m (g_{j,m} y_j^{(m)}, y_j^{(m)}) \geq -C\|Y\|^2, \quad Y \in D(A). \tag{10.3.19}$$

From (10.3.10) and (10.3.11) we infer that

$$\sum_{j=1}^{p}[y_j,y_j]_1(a_j) - \sum_{j=1}^{p}[y_j,y_j]_1(0) + \hat{Y}^*U_2^*V\hat{Y} = \hat{Y}^*PJ\hat{Y} + \hat{Y}^*U_2^*V\hat{Y} \quad (10.3.20)$$

for $\hat{Y} \in \mathbb{C}^{4n}$. Now let $Y \in D(A)$, so that $\hat{Y} \in N(U_1)$. In view of Theorem 10.3.5, there is $\Phi = (\phi_1, \phi_2, \phi_3)^\mathsf{T} \in \mathbb{C}^{4n} \oplus \mathbb{C}^r \oplus \mathbb{C}^q$ such that

$$U_3\hat{Y} = U^*\Phi = \begin{pmatrix} U_1^*\phi_1 + U_2^*\phi_2 + V^*\phi_3 \\ -\phi_2 \\ -\phi_3 \end{pmatrix}.$$

By definition of U_3 in (10.3.12) it follows that $\phi_3 = U_2\hat{Y}$, $\phi_2 = -V\hat{Y}$, and then

$$J\hat{Y} = U_1^*\phi_1 - U_2^*V\hat{Y} + V^*U_2\hat{Y}.$$

We apply $\hat{Y}^*P$ on the left-hand side. Observing that P is Hermitian, the resulting equation can be written as

$$\hat{Y}^*PJ\hat{Y} = (U_1P\hat{Y})^*\phi_1 - \hat{Y}^*PU_2^*V\hat{Y} + \hat{Y}^*PV^*U_2\hat{Y}.$$

By assumption (ii), each row of U_1P is either zero or the corresponding row of U_1. Hence $U_1P\hat{Y} = 0$ for all $\hat{Y} \in N(U)$. Therefore

$$\hat{Y}^*PJ\hat{Y} + \hat{Y}^*U_2^*V\hat{Y} = (U_2(I-P)\hat{Y})^*V\hat{Y} + (VP\hat{Y})^*U_2\hat{Y} \quad (10.3.21)$$

for all $\hat{Y} \in N(U_1)$. Now $P\hat{Y}$ only contains quasi-derivatives $y_j^{[m]}$ at 0 and a_j with $m < k_j$, and in view of Lemma 10.3.6 there is a constant $C_1 > 0$ such that

$$\|VP\hat{Y}\| \leq C_1\|Y\|, \quad Y \in D(A). \quad (10.3.22)$$

Since $U_2\hat{Y}$ is the last block component of $Y \in D(A)$, we also have

$$\|U_2\hat{Y}\| \leq \|Y\|, \quad Y \in D(A). \quad (10.3.23)$$

Essentially the same reasoning applies to $(U_2(I-P)\hat{Y})^*V\hat{Y}$. By assumption (iii), each component of $U_2(I-P)\hat{Y}$ is either 0 or equals the corresponding component of $U_2\hat{Y}$. Since we only have to consider the components of $V\hat{Y}$ for which the corresponding component of $U_2(I-P)\hat{Y}$ is not zero, assumption (iv) and Lemma 10.3.6 show that

$$|(U_2(I-P)\hat{Y})^*V\hat{Y}| \leq C_1\|Y\|^2, \quad Y \in D(A). \quad (10.3.24)$$

Altogether, we have from (10.3.18)–(10.3.24) that

$$(AY,Y) \geq -(C + 2C_1)\|Y\|^2, \quad Y \in D(A). \qquad \square$$

Chapter 11

Analytic and Meromorphic Functions

11.1 Meromorphic functions mapping $\mathbb{C}^+$ into itself

For convenience, in this chapter we will provide full proofs of the results from [173] which are cited in Chapter 5. One of the gaps in B.Ja. Levin's results is that [173, Theorem 1, p. 308] only deals with the case that the meromorphic functions under consideration have infinitely many positive and negative zeros and poles. We therefore will provide more detailed proofs and will cover all possible cases. However, the layout of the proofs is as presented in [173, Chapter VII].

It is sometimes convenient to consider a meromorphic function f on $\Omega \subset \mathbb{C}$ as a function from Ω to $\overline{\mathbb{C}} = \mathbb{C} \cup \{\infty\}$ by putting $f(\lambda) = \infty$ for poles λ of f.

Lemma 11.1.1. *Let P and Q be real entire functions without common nonreal zeros. Consider the functions*

$$\omega = P + iQ, \quad \theta = \frac{Q}{P}, \quad F = \frac{\omega}{\overline{\omega}}.$$

1. *The following statements are equivalent:*
   (i) *$|F(\lambda)| < 1$ for all $\lambda \in \mathbb{C}^+$ such that $\overline{\omega}(\lambda) \neq 0$;*
   (ii) *ω has no zeros in the open lower half-plane, and $|F(\lambda)| < 1$ for all $\lambda \in \mathbb{C}^+$;*
   (iii) *$P \neq 0$ and $\operatorname{Im} \theta(\lambda) > 0$ for all $\lambda \in \mathbb{C}^+$ with $P(\lambda) \neq 0$;*
   (iv) *All zeros of P and Q are real, $\theta(\mathbb{C}^+) \subset \mathbb{C}^+$ and $\theta(\mathbb{C}^-) \subset \mathbb{C}^-$.*

2. *Assume that $P \neq 0$. Then the following statements are equivalent:*
   (i) *$|F(\lambda)| \leq 1$ for all $\lambda \in \mathbb{C}^+$ such that $\overline{\omega}(\lambda) \neq 0$;*
   (ii) *ω has no zeros in the open lower half-plane, and $|F(\lambda)| \leq 1$ for all $\lambda \in \mathbb{C}^+$;*
   (iii) *$\operatorname{Im} \theta(\lambda) \geq 0$ for all $\lambda \in \mathbb{C}^+$ with $P(\lambda) \neq 0$;*
   (iv) *All zeros of P are real, $Q = 0$ or all zeros of Q are real, $\theta(\mathbb{C}^+) \subset \overline{\mathbb{C}^+}$ and $\theta(\mathbb{C}^-) \subset \overline{\mathbb{C}^-}$.*

Proof. 1. (i)$\Rightarrow$(ii): From

$$\begin{pmatrix} \omega \\ \overline{\omega} \end{pmatrix} = \begin{pmatrix} 1 & i \\ 1 & -i \end{pmatrix} \begin{pmatrix} P \\ Q \end{pmatrix}$$

and the assumption that P and Q do not have common nonreal zeros it follows that also ω and $\overline{\omega}$ have no common nonreal zeros. In particular, $\omega \neq 0$. Since F is bounded in the open upper half-plane, F does not have a pole there, and hence $\overline{\omega}$ has no zeros in the open upper half-plane, which means that ω has no zeros in the open lower half-plane.

(ii)$\Rightarrow$(iii): From

$$\theta = \frac{Q}{P} = i\frac{\overline{\omega} - \omega}{\overline{\omega} + \omega} = i\frac{1 - F}{1 + F} \tag{11.1.1}$$

it is clear that $|F(\lambda)| < 1$ implies $P \neq 0$ and $\operatorname{Im}\theta(\lambda) > 0$ for $\lambda \in \mathbb{C}^+$ with $P(\lambda) \neq 0$.

(iii)$\Rightarrow$(iv): Since $\theta(\lambda) \neq 0$ for all $\lambda \in \mathbb{C}^+$ with $P(\lambda) \neq 0$, Q has no zeros in the open upper half-plane and therefore no zeros in the open lower half-plane since zeros of real analytic functions are symmetric with respect to the real axis. But then $\frac{1}{\theta} = PQ^{-1}$ maps $\mathbb{C}^+$ into $\mathbb{C}^- \cup \{0\}$, where 0 occurs if and only if P has a zero in the open upper half-plane. In this case, by the open mapping theorem, see [55, IV.7.5], PQ^{-1} would be constant, which leads to the contradiction $P = 0$. Therefore P has no zeros in $\mathbb{C}^+$ and as with Q we conclude that P has no nonreal zeros. By assumption (iii), $\theta(\mathbb{C}^+) \subset \mathbb{C}^+$, and since θ is real analytic, $\theta(\mathbb{C}^-) \subset \mathbb{C}^-$ follows.

(iv)$\Rightarrow$(i) is an immediate consequence of (11.1.1).

2. The difference to part 1 is that $|F(\lambda)| = 1$ and $\operatorname{Im}\theta(\lambda) = 0$ is possible for some $\lambda \in \mathbb{C}^+$. But by the open mapping theorem, this gives a constant function, and it is easy to see that for such functions, (i)–(iv) are equivalent. $\square$

For the sake of completeness, we recall the following well-known result.

Lemma 11.1.2. *Let f be a nonconstant meromorpic function on $\mathbb{C}$. Then $\overline{\mathbb{C}} \setminus f(\mathbb{C})$ has at most two elements.*

Proof. If $\mathbb{C} \subset f(\mathbb{C})$, nothing has to be shown. If there is $\beta \in \mathbb{C} \setminus f(\mathbb{C})$, consider the Möbius transformation $m(\lambda) = \dfrac{1}{\lambda - \beta}$ and define $g = m \circ f$. Then g is a nonconstant entire function, and by the Little Picard Theorem, see [55, XII.2.3], g assumes all complex numbers, with one possible exception, i.e., $\overline{\mathbb{C}} \setminus g(\mathbb{C})$ has at most two elements. Since m is bijective on $\overline{\mathbb{C}}$, also $\overline{\mathbb{C}} \setminus f(\mathbb{C})$ has at most two elements. $\square$

Lemma 11.1.3. *Let θ be a real meromorphic function on $\mathbb{C}$ which maps the open upper half-plane into the open upper half-plane. Then θ has at least one zero or pole, $\mathbb{C}^+ \setminus \theta(\mathbb{C}^+)$ is either the empty set or consists of one complex number, and all zeros and poles of θ lie on the real axis, are simple and interlace. For each $x \in \mathbb{R}$ which is not a pole of θ, the inequality $\theta'(x) > 0$ holds.*

Proof. Since θ maps the open upper half-plane into the open upper half-plane, θ has neither poles nor zeros in the open upper half-plane, and hence also not in the open lower half-plane since θ is real. Therefore all zeros and poles of θ are real.

Assume θ has no zero or pole. Then θ is an entire function which takes real values only on the real axis. Since θ is continuous and does not have zeros, θ must therefore be either positive or negative on the real axis. Hence θ does not assume any value on either the positive or the negative real semiaxis, and it follows by Lemma 11.1.2 that θ is constant. Since θ is real analytic, this constant would be real, contradicting the fact that θ maps $\mathbb{C}^+$ into itself. Therefore θ must have at least one pole or zero.

Now let us consider any circle centred on the real axis which does not meet zeros or poles of θ. Since θ maps $\mathbb{C}^+$ into itself, the argument of $\theta(\lambda)$ for λ on the upper half of the circle lies in $[0, \pi]$, so that the difference of the arguments of $\theta(\lambda)$ is π, 0, or $-\pi$ when λ moves along the upper semicircle. Using that θ is real, the same conclusion can be made when λ moves along the lower semi-circle. Hence the change of argument of the function θ along the full circle is 2π, 0, or -2π. By the argument principle, this means that the difference of the numbers of zeros and poles, counted with multiplicity, inside the circle equals 1, 0, or -1. In particular, if we choose circles enclosing just one zero or pole of θ, this shows that the zeros and poles are simple. If, however we choose two adjacent zeros or poles $x_1 < x_2$ and a circle which contains x_1 and x_2, but no zeros and poles x with $x < x_1$ or $x > x_2$, then it follows that between two zeros of θ there must be a pole of θ and between two poles of θ there must be a zero of θ, that is, zeros and poles interlace.

Now assume there is $\alpha \in \mathbb{C}^+$ such that $\theta(\lambda) \neq \alpha$ for all $\lambda \in \mathbb{C}^+$. Since θ is real analytic, it follows that $\alpha \notin \theta(\mathbb{C})$ and then $\overline{\alpha} \notin \theta(\mathbb{C})$, so that $\theta(\mathbb{C}) = \overline{\mathbb{C}} \setminus \{\alpha, \overline{\alpha}\}$ by Lemma 11.1.2. Hence $\theta(\mathbb{C}^+) = \theta(\mathbb{C}) \cap \mathbb{C}^+ = \mathbb{C}^+ \setminus \{\alpha\}$.

Finally, when $x \in \mathbb{R}$ is not a pole of θ, then also $\theta - \theta(x)$ is real analytic and maps $\mathbb{C}^+$ into itself. Hence, by what we have already proved, x is a simple zero of $\theta - \theta(x)$. Therefore

$$\theta(\lambda) = \theta(x) + (\lambda - x)h(\lambda),$$

where h is real analytic in a neighbourhood of x and $h(x) \neq 0$. Since

$$\operatorname{Im}\theta(x + i\varepsilon) > 0 \quad \text{for } \varepsilon > 0,$$

it follows that

$$\operatorname{Re} h(x + i\varepsilon) = \frac{1}{\varepsilon} \operatorname{Im}\theta(x + i\varepsilon) > 0.$$

By continuity,

$$\theta'(x) = h(x) = \lim_{\varepsilon \searrow 0} \operatorname{Re} h(x + i\varepsilon) \geq 0,$$

and $\theta'(x) > 0$ follows since $h(x) \neq 0$. $\qquad\square$

Remark 11.1.4. In Lemma 11.1.3 one may impose the weaker assumption that θ only maps the domain of θ in the open upper half-plane into the open upper half-plane, i. e., θ is allowed to have poles in $\mathbb{C}^+$. Then $-\frac{1}{\theta}$ maps the open upper half-plane into the closed upper half-plane, and a pole of θ becomes a zero of $-\frac{1}{\theta}$. However, since a nonconstant analytic function maps open sets into open sets, see, e. g., [55, IV.7.5], $-\frac{1}{\theta(\lambda)}$ would have negative imaginary part for some λ in the open upper half-plane, and so also $\theta(\lambda)$. This proves that θ cannot have poles in $\mathbb{C}^+$. By the same reason, allowing θ to map the open upper half-plane into the closed upper half-plane would imply that either $\theta(\mathbb{C}^+) \subset \mathbb{C}^+$ or θ is constant.

Lemma 11.1.5. *Let p and q be real polynomials with $\deg p = \deg q + 1$ and such that all zeros of p and q are real and simple and interlace. Then p and q is a real pair.*

Proof. By Definition 5.1.3 we have to show that all zeros of $\mu p + \vartheta q$ are real, where $\mu, \vartheta \in \mathbb{R}$ such that $\mu^2 + \vartheta^2 > 0$. This is trivial if $\mu = 0$ or $\vartheta = 0$ or $\deg p = 1$. Hence it is sufficient to consider $p + \vartheta q$ with $\vartheta \neq 0$ and $\deg p \geq 2$. Choose two adjacent zeros x_1 and x_2 of p. By assumption, q has exactly one zero between x_1 and x_2, this zero is simple, and $q(x_1) \neq 0$, $q(x_2) \neq 0$. Therefore $q(x_1)$ and $q(x_2)$ are different from zero and have opposite signs. From $p(x_j) + \vartheta q(x_j) = \vartheta q(x_j)$ for $j = 1, 2$ it then follows that $p + \vartheta q$ changes sign between x_1 and x_2. Since there are $\deg p - 1$ pairs of such numbers x_1 and x_2, $p + \vartheta q$ has at least $\deg p - 1$ real zeros. But $p + \vartheta q$ is real, so that nonreal zeros must occur in pairs of a number and its conjugate complex number, i. e., the number of nonreal zeros must be even, and it follows that $p + \vartheta q$ cannot have any nonreal zeros. $\square$

The following theorem clarifies the statement and proof of [173, Theorem 1, p. 308].

Theorem 11.1.6. *A real meromorphic function θ on $\mathbb{C}$ maps the open upper half-plane into itself if and only if θ is represented in the form*

$$\theta(\lambda) = C \frac{(\lambda - b_0)_-}{(\lambda - a_0)_+} \prod_{k \in I'_{a,b}} \left(1 - \frac{\lambda}{b_k} \right) \left(1 - \frac{\lambda}{a_k} \right)^{-1}, \qquad (11.1.2)$$

where $C > 0$, the sets $\mathfrak{a}$, $\mathfrak{b}$, $I_\mathfrak{a}$, $I_\mathfrak{b}$ satisfy the properties as layed out in Remark 5.1.6, in particular $\mathfrak{a} \cup \mathfrak{b} \neq \emptyset$, and where $k \in I'_{\mathfrak{a},\mathfrak{b}}$ means that the factors $\left(1 - \frac{\lambda}{a_k} \right)$ occur for $k \in I_\mathfrak{a} \setminus \{0\}$, the factors $\left(1 - \frac{\lambda}{b_k} \right)^{-1}$ occur for $k \in I_\mathfrak{b} \setminus \{0\}$, that is, if for some k only one factor occurs, then the non-occurring factor has to be replaced by 1, and, similarly, $\lambda - a_0$ and $\lambda - b_0$ have to be replaced by 1 if $0 \notin \mathfrak{a}$ or by -1 if $0 \notin \mathfrak{b}$, respectively.

Proof. Assume that θ is of the form (11.1.2). We will first consider the case that $I_\mathfrak{b} = I_\mathfrak{a}$. The series

$$\sum_{k \in I_\mathfrak{a} \setminus \{0\}} \left(\frac{1}{a_k} - \frac{1}{b_k} \right)$$

converges absolutely if it has infinitely many terms. For example,

$$\sum_{k\geq 1}\left(\frac{1}{a_k}-\frac{1}{b_k}\right)$$

is a series of differences of positive interlacing numbers which is therefore bounded above by $\frac{1}{a_1}$. Since the functions

$$\lambda \mapsto \left(1-\frac{\lambda}{a_k}\right)^{-1}$$

are uniformly bounded on compact subsets of $\mathbb{C}$ not containing the points a_k, with bound independent of k, it follows that the series

$$\sum_{k\in I_a\setminus\{0\}}\left[\left(1-\frac{\lambda}{b_k}\right)\left(1-\frac{\lambda}{a_k}\right)^{-1}-1\right]=\lambda\sum_{k\in I_a\setminus\{0\}}\left(\frac{1}{a_k}-\frac{1}{b_k}\right)\left(1-\frac{\lambda}{a_k}\right)^{-1}$$

converges uniformly on such sets.

For all $k\in I_a\setminus\{0\}$ we have $a_kb_k>0$ and therefore

$$\arg\frac{1-\dfrac{\lambda}{b_k}}{1-\dfrac{\lambda}{a_k}}=\arg(\lambda-b_k)-\arg(\lambda-a_k).$$

Denoting $\alpha=\arg(\lambda-a_k)$ and $\beta=\arg(\lambda-b_k)$, the sketch

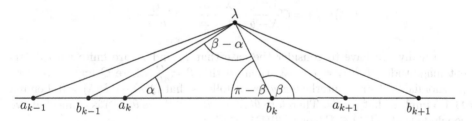

shows that for $\operatorname{Im}\lambda>0$ this argument is the angle subtended by the segment $[a_k,b_k]$ of the real axis from the point λ. Since

$$\arg\theta(\lambda)=\sum_{k\in I_a}[\arg(\lambda-b_k)-\arg(\lambda-a_k)],$$

it follows that $0<\arg\theta(\lambda)<\pi$ if $\operatorname{Im}\lambda>0$, i.e., $\theta(\mathbb{C}^+)\subset\mathbb{C}^+$.

Next we consider the case that $I_b=I_a-1$. Interchanging $\mathfrak{a}$ and $\mathfrak{b}$ we have an interlacing pair $\tilde{\mathfrak{a}}=\mathfrak{b}$, $\tilde{\mathfrak{b}}=\mathfrak{a}$ with $I_{\tilde{\mathfrak{b}}}=I_{\tilde{\mathfrak{a}}}$. Then

$$\psi(\lambda)=\frac{\lambda-\tilde{b}_0}{\lambda-\tilde{a}_0}\prod_{k\in I'_{\tilde{\mathfrak{a}},\tilde{\mathfrak{b}}}}\left(1-\frac{\lambda}{\tilde{b}_k}\right)\left(1-\frac{\lambda}{\tilde{a}_k}\right)^{-1}$$

defines a real meromorphic function which satisfies $\psi(\mathbb{C}^+) \subset \mathbb{C}^+$ by the first part of the proof, and thus $\frac{1}{\psi}(\mathbb{C}^+) \subset \mathbb{C}^-$. In order to relate θ to ψ we have to investigate the indexing of the zeros and poles of θ and ψ.

If $0 \notin I_a$, then $\tilde{b}_0 = a_1$, $\tilde{a}_0 = b_0$ shows that

$$\theta(\lambda)\psi(\lambda) = C\frac{\lambda - a_1}{1 - \frac{\lambda}{a_1}} = -Ca_1 < 0.$$

Hence $\theta(\mathbb{C}^+) \subset \mathbb{C}^+$.

If $0 \notin I_b$, then $\tilde{b}_0 = a_0$, $\tilde{a}_0 = b_{-1}$ shows that

$$\theta(\lambda)\psi(\lambda) = (-C)\frac{1 - \frac{\lambda}{b_{-1}}}{\lambda - b_{-1}} = \frac{C}{b_{-1}} < 0.$$

Hence $\theta(\mathbb{C}^+) \subset \mathbb{C}^+$.

If $0 \in I_a \cap I_b$, then we consider two subcases. The first subcase is that $\tilde{b}_0 = a_1$, $\tilde{a}_0 = b_0$, $\tilde{b}_{-1} = a_0$, so that

$$\theta(\lambda)\psi(\lambda) = C\frac{\lambda - a_1}{1 - \frac{\lambda}{a_1}}\frac{1 - \frac{\lambda}{\tilde{b}_{-1}}}{\lambda - \tilde{b}_{-1}} = C\frac{a_1}{\tilde{b}_{-1}} < 0.$$

The second subcase is that $\tilde{b}_0 = a_0$, $\tilde{a}_0 = b_{-1}$, $\tilde{a}_1 = b_0$, so that

$$\theta(\lambda)\psi(\lambda) = C\frac{1 - \frac{\lambda}{b_{-1}}}{\lambda - b_{-1}}\frac{\lambda - \tilde{a}_1}{1 - \frac{\lambda}{\tilde{a}_1}} = C\frac{\tilde{a}_1}{b_{-1}} < 0.$$

Finally, we have to consider the case that I_a and I_b are finite with different magnitude. If $\theta(\lambda) = \alpha \in \mathbb{R}$, then writing $\theta = \frac{Q}{P}$ where Q and P are real polynomials with real interlacing zeros, it follows that $Q(\lambda) - \alpha P(\lambda) = 0$. Lemma 11.1.5 implies that $\lambda \in \mathbb{R}$. Therefore $\theta(\mathbb{C}^+) \cap \mathbb{R} = \emptyset$. Since $\theta(\mathbb{C}^+)$ is connected, we conclude that $\theta(\mathbb{C}^+) \subset \mathbb{C}^+$ or $(-\theta)(\mathbb{C}^+) \subset \mathbb{C}^+$.

If θ has more zeros than poles, let λ_0 be the smallest zero and define

$$\psi(\lambda) = (\lambda - \lambda_0)\prod_{k \in I_a}\frac{\lambda - b_k}{\lambda - a_k}.$$

As for θ above, we conclude that $\psi(\mathbb{C}^+) \subset \mathbb{C}^+$ or $(-\psi)(\mathbb{C}^+) \subset \mathbb{C}^+$, and with the aid of

$$\psi'(\lambda_0) = \prod_{k \in I_a}\frac{\lambda_0 - b_k}{\lambda_0 - a_k} > 0$$

and Lemma 11.1.3 we conclude that $(-\psi)(\mathbb{C}^+) \subset \mathbb{C}^+$ is impossible, which shows that $\psi(\mathbb{C}^+) \subset \mathbb{C}^+$. Observe that $\lambda_0 = b_0$ if and only if $0 \notin I_a$.

In case $0 \notin I_a$ we therefore have

$$\theta(\lambda) = C(\lambda - \lambda_0) \prod_{k \in I_a} \frac{1 - \frac{\lambda}{b_k}}{1 - \frac{\lambda}{a_k}}$$

$$= C(\lambda - \lambda_0) \prod_{k \in I_a} \frac{a_k}{b_k} \frac{\lambda - b_k}{\lambda - a_k}$$

$$= C \prod_{k \in I_a} \frac{a_k}{b_k} \psi(\lambda),$$

so that $\theta(\mathbb{C}^+) \subset \mathbb{C}^+$.

If $0 \in I_a$, then $\lambda_0 < a_0$, so that $\lambda_0 = b_k$ for some $k < 0$ and therefore $\lambda_0 < 0$. Then

$$\theta(\lambda) = C\left(1 - \frac{\lambda}{\lambda_0}\right) \frac{\lambda - b_0}{\lambda - a_0} \prod_{k \in I_a \setminus \{0\}} \frac{1 - \frac{\lambda}{b_k}}{1 - \frac{\lambda}{a_k}}$$

$$= -\frac{C}{\lambda_0}(\lambda - \lambda_0) \frac{\lambda - b_0}{\lambda - a_0} \prod_{k \in I_a \setminus \{0\}} \frac{a_k}{b_k} \frac{\lambda - b_k}{\lambda - a_k}$$

$$= -\frac{C}{\lambda_0} \prod_{k \in I_a \setminus \{0\}} \frac{a_k}{b_k} \psi(\lambda),$$

so that $\theta(\mathbb{C}^+) \subset \mathbb{C}^+$.

If θ has more poles than zeros, let $\tilde{a} = -b$, $\tilde{b} = -a$, $I_{\tilde{a}} = -I_b$, $I_{\tilde{b}} = -I_a$, so that $\tilde{a}_{-k} = -b_k$ and $\tilde{b}_{-k} = -a_k$ for the relevant indices k. Note that this setting is consistent since $\tilde{b}_{-1} = -a_1 < 0 < -b_{-1} = a_1$. Let

$$\tilde{\theta}(\lambda) = \frac{(\lambda - \tilde{b}_0)_-}{(\lambda - \tilde{a}_0)_+} \prod_{k \in I'_{\tilde{a},\tilde{b}}} \left(1 - \frac{\lambda}{\tilde{b}_k}\right) \left(1 - \frac{\lambda}{\tilde{a}_k}\right)^{-1}$$

$$= \frac{(\lambda - \tilde{b}_0)_-}{(\lambda - \tilde{a}_0)_+} \prod_{k \in I'_{a,b}} \left(1 + \frac{\lambda}{a_k}\right) \left(1 + \frac{\lambda}{b_k}\right)^{-1}.$$

We note that $0 \in I_a$ because we have more poles than zeros. If also $0 \in I_b$, then

$$\frac{\lambda - \tilde{b}_0}{\lambda - \tilde{a}_0} = \frac{-\lambda - a_0}{-\lambda - b_0},$$

whereas if $0 \notin I_b$ and hence $0 \notin I_{\tilde{a}}$, then

$$\frac{(\lambda - \tilde{b}_0)_-}{(\lambda - \tilde{a}_0)_+} = \lambda - \tilde{b}_0 = -(-\lambda - a_0) = \frac{(-\lambda - a_0)_+}{(-\lambda - b_0)_-}.$$

Thus in either case

$$\theta(\lambda) = \frac{C}{\tilde{\theta}(-\lambda)}.$$

Since $\tilde{\theta}$ is real meromorphic with more zeros than poles, it follows from the previous case that $\tilde{\theta}(\mathbb{C}^+) \subset \mathbb{C}^+$ and then $\tilde{\theta}(\mathbb{C}^-) \subset \mathbb{C}^-$, so that $\theta(\mathbb{C}^+) \subset \mathbb{C}^+$.

Conversely, assume that θ is a real meromorphic function such that $\theta(\mathbb{C})^+ \subset \mathbb{C}^+$. By Lemma 11.1.3 the zeros and poles of θ are simple and interlace. Denoting the sets of zeros and poles by $\mathfrak{a}$ and $\mathfrak{b}$, respectively, we have $\mathfrak{a} \cup \mathfrak{b} \neq \emptyset$ by Lemma 11.1.3. It follows by the first part of the proof that the real meromorphic function ψ of the form (11.1.2) with $C = 1$ maps the open upper half-plane into itself. Then $f = \frac{\theta}{\psi}$ is a meromorphic function without zeros and poles, i.e., an entire function without zeros. Therefore there is an entire function h such that $f = h^2$, see [55, Theorem VIII.2.2]. We have to show that $f = C$ with a positive constant C. Since $\theta(\mathbb{C}^+) \subset \mathbb{C}^+$ and $\psi(\mathbb{C}^+) \subset \mathbb{C}^+$, it follows that $f(\lambda) \notin (-\infty, 0]$ for all $\lambda \in \mathbb{C}^+$, and similarly for all $\lambda \in \mathbb{C}^-$ because f is real analytic. Since f is real on the real line, either $f(\mathbb{R}) \subset (0, \infty)$ or $f(\mathbb{R}) \subset (-\infty, 0)$. Therefore $h(\mathbb{C} \setminus \mathbb{R}) \subset \mathbb{C} \setminus i\mathbb{R}$ and $h(\mathbb{R}) \subset \mathbb{R}$ or $h(\mathbb{R}) \subset i\mathbb{R}$. Since h is never zero, it follows that $h(\mathbb{R})$ is a subset of $(0, \infty)$, $(-\infty, 0)$, $(0, i\infty)$, or $(-i\infty, 0)$. In any case, one of $(0, i\infty)$ or $(-i\infty, 0)$ has no common points with $h(\mathbb{C})$. By Lemma 11.1.2, this implies that h is constant. Since this constant must belong to $h(\mathbb{C} \setminus \mathbb{R})$ and $h(\mathbb{R})$, this constant is real and nonzero, so that $f = C$ with a positive constant C. $\qquad\square$

Remark 11.1.7. If we only assume that $\mathfrak{a}$ and $\mathfrak{b}$ interlace without the requirement of the particular enumeration from Remark 5.1.6, with a possible gap for the index 0 in the index sets, and where $(\lambda - a_0)_-$ is possibly replaced by $(\lambda - a_0)_+$, where $0 \notin \mathfrak{a} \cup \mathfrak{b}$ is allowed and where we only require $C \neq 0$, then it is clear that this representation of θ is a nonzero constant multiple of that in Theorem 11.1.6. Hence $\theta(\mathbb{C}^+) \subset \mathbb{C}^+$ or $(-\theta)(\mathbb{C}^+) \subset \mathbb{C}^+$. By Lemma 11.1.3, this weaker assumption and $\theta'(x) > 0$ for some $x \in \mathbb{R}$ would be necessary and sufficient for θ to map the open upper half-plane into the open upper half-plane.

Corollary 11.1.8. *Let θ be a real meromorphic function which maps the upper half-plane into itself. Assume that the set of poles of θ is bounded below and that the smallest pole of θ is smaller than the smallest zero of θ. Then there are positive constants c_1, c_2, and γ such that*

$$c_1 |\lambda|^{-1} \leq |\theta(\lambda)| \leq c_2 \quad \text{for } \lambda < -\gamma.$$

Proof. If we take any two factor in (11.1.2) corresponding to a pole and a zero, the corresponding limit as $\lambda \to -\infty$ is a nonzero constant, and it is therefore sufficient to consider an infinite product

$$\prod_{k=k_0}^{\infty} \left(1 - \frac{\lambda}{b_k}\right) \left(1 - \frac{\lambda}{a_k}\right)^{-1}$$

with $0 < a_{k_0} < b_{k_0} < a_{k_0+1} < b_{k_0+1} < \cdots$. Then we conclude that

$$\left(1 - \frac{\lambda}{b_k}\right)\left(1 - \frac{\lambda}{a_k}\right)^{-1} = \frac{a_k b_k - a_k \lambda}{a_k b_k - b_k \lambda} < 1 \quad \text{for } \lambda < 0, \ k \geq k_0,$$

which proves that $|\theta(\lambda)| \leq c_2$ for suitable $c_2, \gamma > 0$ and $\lambda < \gamma$. Similarly,

$$\left(1 - \frac{\lambda}{b_k}\right)\left(1 - \frac{\lambda}{a_{k+1}}\right)^{-1} = \frac{a_{k+1} b_k - a_{k+1} \lambda}{a_{k+1} b_k - b_k \lambda} > 1 \quad \text{for } \lambda < 0, \ k \geq k_0,$$

and

$$\left(1 - \frac{\lambda}{a_{k_0}}\right)^{-1} = -\frac{a_{k_0}}{\lambda}(1 + o(1)) \quad \text{for } \lambda < 0$$

proves that $|\theta(\lambda)| \geq c_1 |\lambda|^{-1}$ for suitable $c_1, \gamma > 0$ and $\lambda < \gamma$. $\qquad \square$

The following lemma is a particular case of [173, Theorem 1, p. 224 and Theorem 2, p. 225]. For convenience of the reader we provide a proof, taking just those parts of the proofs in [173] which are needed for this special case.

Lemma 11.1.9. *Let ψ be analytic and bounded on $\overline{\mathbb{C}^+}$ and let $|\psi(\lambda)| = 1$ for all $\lambda \in \mathbb{R}$. Let $a_1, a_2, \ldots$ denote the zeros of ψ, counted with multiplicity. Then*

$$\sum_{k=1}^{(\infty)} \left| \mathrm{Im} \, \frac{1}{a_k} \right| < \infty.$$

Proof. The zeros may be indexed in such a way that $|a_j| \leq |a_{j+1}|$. Let $\varepsilon > 0$ such that $|a_j| > \varepsilon$ for all j and let $R > \varepsilon$ such that $|a_j| \neq R$ for all j. Let C be the positively oriented boundary of the region given by $\varepsilon < |\lambda| < R$, $\mathrm{Im} \, \lambda > 0$ and consider the contour integral

$$J = \frac{1}{2\pi i} \int_C \left(\frac{1}{R^2} - \frac{1}{\lambda^2} \right) \log \psi(\lambda) \, d\lambda.$$

Here we have to observe that J depends on the choice of the branch of $\log \psi$. We first choose a fixed branch on the semicircle in the upper half-plane with centre 0 and radius R. Then, for each such R, we extend this branch of $\log \psi$ to a continuous branch along C, say, starting and ending at ε. Then

$$J = \frac{1}{2\pi i} \int_\varepsilon^R \left(\frac{1}{R^2} - \frac{1}{x^2} \right) \log \psi(x) \, dx + \frac{1}{2\pi} \int_0^\pi \left(\frac{1}{R^2} - \frac{e^{-2i\theta}}{R^2} \right) \log \psi(Re^{i\theta}) Re^{i\theta} \, d\theta$$

$$+ \frac{1}{2\pi i} \int_{-R}^{-\varepsilon} \left(\frac{1}{R^2} - \frac{1}{x^2} \right) \log \psi(x) \, dx$$

$$+ \frac{1}{2\pi} \int_\pi^0 \left(\frac{1}{R^2} - \frac{e^{-2i\theta}}{\varepsilon^2} \right) \log \psi(\varepsilon e^{i\theta}) \varepsilon e^{i\theta} \, d\theta. \tag{11.1.3}$$

On the other hand, integration by parts yields

$$J = \frac{1}{2\pi i} \Phi_C \left[\left(\frac{\lambda}{R^2} + \frac{1}{\lambda} \right) \log \psi(\lambda) \right] - \frac{1}{2\pi i} \int_C \left(\frac{\lambda}{R^2} + \frac{1}{\lambda} \right) \frac{\psi'(\lambda)}{\psi(\lambda)} \, d\lambda, \qquad (11.1.4)$$

where Φ_C denotes the increase of the function along C, starting and ending at ε. By the argument principle, this increase equals $2\pi i(\varepsilon/R^2 + 1/\varepsilon)n(R)$, where $n(R)$ is the number of zeros of ψ with modulus less than R. We recall that the assumptions on ψ imply that all zeros $(a_k)_{k=1}^{(\infty)}$ of ψ lie in $\mathbb{C}^+$. The residue theorem shows that the second summand in (11.1.4) is

$$-\sum_{k=1}^{n(R)} \left(\frac{a_k}{R^2} + \frac{1}{a_k} \right).$$

Equating imaginary parts of (11.1.3) and (11.1.4) gives

$$\frac{1}{\pi R} \int_0^\pi \sin\theta \log|\psi(Re^{i\theta})| \, d\theta + A_\varepsilon(\psi, R) = -\sum_{k=1}^{n(R)} \operatorname{Im}\left(\frac{a_k}{R^2} + \frac{1}{a_k} \right), \qquad (11.1.5)$$

where

$$A_\varepsilon(\psi, R) = \operatorname{Im} \frac{1}{2\pi} \int_\pi^0 \left(\frac{1}{R^2} - \frac{e^{-2i\theta}}{\varepsilon^2} \right) \log \psi(\varepsilon e^{i\theta}) \varepsilon e^{i\theta} \, d\theta$$

$$- \frac{1}{2\pi} \int_\varepsilon^R \left(\frac{1}{R^2} - \frac{1}{x^2} \right) \log|\psi(x)\psi(-x)| \, dx.$$

Since ψ is bounded in the closed upper half-plane, $\log|\psi|$ is bounded above, and therefore the left-hand side of (11.1.5) is bounded above as $R \to \infty$. For $1 \le k \le n(R)$, the identities

$$\left| \operatorname{Im} \frac{1}{a_k} \right| = \left| \operatorname{Im} \frac{\overline{a_k}}{|a_k|^2} \right| = \operatorname{Im} \frac{a_k}{|a_k|^2} = \operatorname{Im} \left(\frac{a_k}{|a_k|^2} - \frac{a_k}{R^2} \right) \frac{R^2}{R^2 - |a_k|^2}$$

$$= \left(\frac{\operatorname{Im} a_k}{|a_k|^2} - \frac{\operatorname{Im} a_k}{R^2} \right) \frac{R^2}{R^2 - |a_k|^2} = \left(\operatorname{Im} \frac{1}{\overline{a_k}} - \frac{\operatorname{Im} a_k}{R^2} \right) \frac{R^2}{R^2 - |a_k|^2}$$

$$= -\operatorname{Im} \left(\frac{1}{a_k} + \frac{a_k}{R^2} \right) \frac{R^2}{R^2 - |a_k|^2}$$

show that the right-hand side of (11.1.5) is a sum of positive terms. Choosing now $r > \varepsilon$ and $R \ge 2r$ such that $|a_j| \notin \{r, R\}$ for all indices j, the above identity gives for $k = 1, \ldots, n(r)$ that

$$\left| \operatorname{Im} \frac{1}{a_k} \right| \le -\operatorname{Im} \left(\frac{1}{a_k} + \frac{a_k}{R^2} \right) \frac{4R^2}{4R^2 - R^2} = -\frac{4}{3} \operatorname{Im} \left(\frac{1}{a_k} + \frac{a_k}{R^2} \right),$$

so that

$$\sum_{j=1}^{n(r)} \left| \mathrm{Im}\, \frac{1}{a_k} \right| \leq -\frac{4}{3} \sum_{j=1}^{n(R)} \mathrm{Im}\left(\frac{1}{a_k} + \frac{a_k}{R^2} \right).$$

Then (11.1.5) and the fact that the left-hand side of (11.1.5) is bounded above as $R \to \infty$ complete the proof.   □

11.2 Sine type functions

Definition 11.2.1. An entire function w is said to be an *entire function of finite order* if there is $k \geq 0$ such that

$$|w(\lambda)| \leq e^{|\lambda|^k}$$

for all $\lambda \in \mathbb{C}$ with sufficiently large modulus. The largest lower bound of all k for which this estimate is true is denoted by ρ_w and is called the *order of w*.

For a different characterization, we introduce the following notation. For an entire function w and $r > 0$ let

$$M_w(r) = \max_{|\lambda|=r} |w(\lambda)|. \tag{11.2.1}$$

If w is not a constant function, then $M_w(r) \to \infty$ as $r \to \infty$ by Liouville's theorem. Then

$$\rho = \rho_w = \limsup_{r\to\infty} \frac{\log \log M_w(r)}{\log r} \tag{11.2.2}$$

is well defined with $\rho_w \in [0, \infty]$, w is of finite order if and only if $\rho_w < \infty$, and in this case ρ_w is the order of w, see, e. g., [173, p. 3].

For a nonconstant entire function w of order ρ one may fine-tune the defining estimate by considering the inequality

$$|M_w(r)| < e^{ar^\rho}.$$

Similar to (11.2.2) we now define, see [173, p. 3],

$$\sigma = \sigma_w = \limsup_{r\to\infty} \frac{\log M_w(r)}{r^{\rho_w}}. \tag{11.2.3}$$

Definition 11.2.2. An entire function w of finite order is said to be of *minimal type* if $\sigma = 0$, of *normal type* if $0 < \sigma < \infty$, and of *maximal type* if $\sigma = \infty$.

Definition 11.2.3. An entire function w is said to be of *exponential type* if

$$\sigma = \sigma_w = \limsup_{r\to\infty} \frac{\log M_w(r)}{r} < \infty, \tag{11.2.4}$$

and then the number σ is called the exponential type of w.

The definitions immediately show the following characterization.

Remark 11.2.4 ([173, p. 84]). An entire function ω is of exponential type if and only if

(i) ω is of finite order ρ less than 1, in which case its exponential type equals 0

or

(ii) ω is of finite order $\rho = 1$ and minimal or normal type, in which case its exponential type equals its type.

If in the above definitions we restrict the function ω to the upper or lower half-planes, then we obtain corresponding definitions of order, type, and exponential order in that half-plane.

Definition 11.2.5 ([176, §1]). An entire function ω of positive exponential type is said to be a sine type function if

 (i) there is $h > 0$ such that all zeros of ω lie in the strip $\{\lambda \in \mathbb{C} : |\operatorname{Im} \lambda| < h\}$,
 (ii) there are $h_1 \in \mathbb{R}$ and positive numbers $m < M$ such that $m \le |\omega(\lambda)| \le M$ holds for $\lambda \in \mathbb{C}$ with $\operatorname{Im} \lambda = h_1$,
(iii) the exponential type of ω in the lower half-plane coincides with the exponential type of ω in the upper half-plane.

For convenience, we introduce the functions ω_η, $\eta \in \mathbb{R}$, defined by

$$\omega_\eta(\lambda) = \omega(\lambda + i\eta), \ \lambda \in \mathbb{C}. \tag{11.2.5}$$

Let f be a function which is analytic on an angular domain

$$S = S_{\phi_1,\phi_2} = \{\lambda : \phi_1 < \arg \lambda < \phi_2\},$$

and continuous on $\overline{S}$. We will refer to such functions as functions on S. Define

$$M_{f,\phi_1,\phi_2}(r) = \max_{\phi \in [\phi_1,\phi_2]} |f(re^{i\phi})|, \ r > 0.$$

Lemma 11.2.6. Let ω be a function of exponential type σ and let $\eta \in \mathbb{R}$.

1. ω_η is a function of exponential type σ.
2. If there are $h_1 \in \mathbb{R}$ and a positive number M such that $|\omega(\lambda)| \le M$ holds for $\lambda \in \mathbb{C}$ with $\operatorname{Im} \lambda = h_1$, then there is $M_{\omega,\eta} > 0$ such that

$$|\omega_\eta(\lambda)| \le M_{\omega,\eta} e^{\sigma|\operatorname{Im} \lambda|}, \ \lambda \in \mathbb{C}. \tag{11.2.6}$$

3. If ω is a sine type function, then also ω_η is a sine type function.

Proof. 1. From the definition of the maximum modulus function (11.2.1) and the Maximum Modulus Theorem we have $M_\omega(r - |\eta|) \le M_{\omega_\eta}(r) \le M_\omega(r + |\eta|)$ for all $r > |\eta|$ and therefore

$$\sigma = \limsup_{r \to \infty} \frac{\log M_\omega(r - |\eta|)}{r} \le \limsup_{r \to \infty} \frac{\log M_{\omega_\eta}(r)}{r} \le \limsup_{r \to \infty} \frac{\log M_\omega(r + |\eta|)}{r} = \sigma,$$

which shows that ω_η is of exponential type σ.

2. First we are going to prove (11.2.6) for $\eta = h_1$. For $\varepsilon \geq 0$ define

$$\psi_\varepsilon(\lambda) = \omega_{h_1}(\lambda)e^{i(\sigma+\varepsilon)\lambda}, \ \lambda \in \mathbb{C}.$$

Then, in view of

$$|\psi_\varepsilon(\lambda)| = |\omega_{h_1}(\lambda)|e^{-(\sigma+\varepsilon)\operatorname{Im}\lambda}, \ \lambda \in \mathbb{C},$$

it follows for all $\varepsilon > 0$ that ψ_ε is bounded on the real axis as well as on the positive imaginary semiaxis since the type of ω_{h_1} is σ. Applying the Phragmén–Lindelöf principle in the form of [55, Corollary VI.4.2] to the first and second quadrants in the upper half-plane, it follows that ψ_ε is bounded on the closed upper half-plane for all $\varepsilon > 0$. Then [55, Corollary VI.4.4] shows that also ψ_0 is bounded on the closed upper half-plane. Hence there is $M_{\omega,h_1} > 0$ such that

$$|\omega_{h_1}(\lambda)|e^{-\sigma\operatorname{Im}\lambda} \leq M_{\omega,h_1}, \ \operatorname{Im}\lambda \geq 0.$$

Since also $\overline{\omega_{h_1}}$ is of exponential type σ and bounded on the real axis, the above estimate also applies to $\overline{\omega_{h_1}}$, and therefore

$$|\omega_{h_1}(\lambda)|e^{\sigma\operatorname{Im}\lambda} \leq M_{\omega,h_1}, \ \operatorname{Im}\lambda \leq 0,$$

where we can take the same constant in both estimates.

The estimate (11.2.6) for arbitrary η now easily follows from this special case:

$$|\omega_\eta(\lambda)| = |\omega_{h_1}(\lambda + i(\eta - h_1))| \leq M_{\omega,h_1}e^{\sigma|\operatorname{Im}\lambda + \eta - h_1|} \leq M_{\omega,h_1}e^{\sigma|\eta-h_1|}e^{\sigma|\operatorname{Im}\lambda|}.$$

3. Clearly, properties (i) and (ii) in Definition 11.2.5 also hold for ω_η, with h replaced with $h + |\eta|$ and h_1 replaced with $h_1 - \eta$.

In order to show property (iii) in Definition 11.2.5 we observe that, by (11.2.6) and the maximum modulus principle,

$$\begin{aligned}
M_{\omega_\eta,0,\pi}(r) + M_{\omega,0}e^{\sigma|\eta|} &\geq M_{\omega_\eta,0,\pi}(r) + \max\{|\omega(\lambda)| : |\operatorname{Re}\lambda| \leq r, \, |\operatorname{Im}\lambda| \leq |\eta|\} \\
&\geq \max\{|\omega(\lambda)| : |\lambda| = r - |\eta|, \, \operatorname{Im}\lambda \geq 0\} \\
&= M_{\omega,0,\pi}(r - |\eta|)
\end{aligned}$$

for $r > |\eta|$, which gives

$$\sigma \geq \limsup_{r\to\infty} \frac{\log M_{\omega_\eta,0,\pi}(r)}{r} \geq \limsup_{r\to\infty} \frac{\log M_{\omega,0,\pi}(r - |\eta|)}{r} = \sigma.$$

Hence the type of ω_η in the upper half-plane is σ. Similarly, we obtain that the type of ω_η in the lower half-plane is σ. $\square$

The following lemma is a special case of [172, Lemma 2].

Lemma 11.2.7. *Let ω be a sine type function. Then there is a natural number ℓ such that for each $t \in \mathbb{R}$ the number of zeros of ω inside the vertical strip $S_t = \{\lambda \in \mathbb{C} : t \leq \operatorname{Re}\lambda \leq t+1\}$, counted with multiplicity, does not exceed ℓ.*

Proof. Suppose the statement of the lemma is false. Then there is a sequence of real numbers $(t_j)_{j=1}^{\infty}$ such that for each j, the entire function $\omega_j = \omega(t_j + \cdot)$ has at least j zeros in the strip S_0. In view of Lemma 11.2.6, part 2, the sequence $(\omega_j)_{j=1}^{\infty}$ of entire functions is locally bounded. By Montel's theorem, see [55, VII.2.9], we may assume that the t_j are chosen in such a way that $(\omega_j)_{j=1}^{\infty}$ converges uniformly on compact subsets of $\mathbb{C}$ to some entire function $\tilde{\omega}$. In view of Definition 11.2.5, (ii), it follows that $|\tilde{\omega}(x+ih_1)| \geq m > 0$ for all $x \in \mathbb{R}$. Therefore $\tilde{\omega}$ is not identically zero. With h from Definition 11.2.5,(i), we choose $R > h+1$ such that $\tilde{\omega}$ has no zeros λ with $|\lambda| = R$. From Definition 11.2.5, (i), it follows that all zeros λ of ω_j in the strip S_0 satisfy $|\lambda| < R$, which implies that ω_j has at least j zeros with $|\lambda| < R$, contradicting the fact that this number must be constant for sufficiently large j by Hurwitz's theorem, see [55, VII.2.5]. $\square$

Proposition 11.2.8. *Let ω be of sine type and let $n_\omega(r)$ be the number of zeros λ of ω with $|\lambda| \leq r$, counted with multiplicity. Then:*

1. *ω has infinitely many zeros.*
2. *There is a number $d > 0$ such that $n_\omega(r) - n_\omega(r-1) \leq d$ for all $r > 1$.*
3. *There is $c \geq 0$ such that $n_\omega(r) \leq c + dr$ for all $r > 1$, where d is as in 2.*
4. *The nonzero zeros a_k of ω, $k \in \mathbb{N}$, counted with multiplicity, satisfy*

$$\sum_{k=1}^{\infty} \left| \operatorname{Im} \frac{1}{a_k} \right| < \infty. \tag{11.2.7}$$

Proof. 1. If ω has no zeros, then the standard form of ω, see (5.1.3), is $\omega(\lambda) = e^{b\lambda+c}$, $\lambda \in \mathbb{C}$, where $b, c \in \mathbb{C}$, and it is easy to see that the type of ω is $|b|$. Since ω is of sine type, there are $h_1, M \in \mathbb{R}$ such that

$$e^{\operatorname{Re} c - h_1 \operatorname{Im} b} e^{x \operatorname{Re} b} = |\omega(x + ih_1)| \leq M, \quad x \in \mathbb{R}, \tag{11.2.8}$$

whence $\operatorname{Re} b = 0$. Then $|\omega(x+iy)| = e^{-y \operatorname{Im} b} e^{\operatorname{Re} c}$, so that ω would be bounded in $\mathbb{C}^+$ or $\mathbb{C}^-$, contradicting the fact that ω must have the same positive type on $\mathbb{C}^+$ and $\mathbb{C}^-$ by Definition 11.2.5, (iii).

Now assume ω has finitely many zeros. Then the standard form of ω is $\omega(\lambda) = e^{b\lambda+c} p(\lambda)$, $\lambda \in \mathbb{C}$, where $b, c \in \mathbb{C}$, p is a nonconstant polynomial, and $b \neq 0$ since the order of ω is 1. Observing that $|e^{b(x+ih_1)+c}|$ has a positive lower bound in the set $\{x \in \mathbb{R} : x \operatorname{Re} b \geq 0\}$ and that $|p(\lambda)| \to \infty$ as $|\lambda| \to \infty$, it follows that ω is not bounded on any horizontal line $\{x + ih_1 : x \in \mathbb{R}\}$, contradicting property (ii) of Definition 11.2.5 of functions of sine type. Hence ω has infinitely many zeros.

2. Because of Definition 11.2.5, (i), and Lemma 11.2.7 it is obvious for $r > \frac{1}{2}(h^2 + 3)$ that all zeros of ω inside the annulus $\{\lambda \in \mathbb{C} : r-1 < |\lambda| \leq r\}$ lie inside the two vertical strips $-r \leq \operatorname{Re} \lambda \leq -r+2$ and $r-2 \leq \operatorname{Re} \lambda \leq r$. With ℓ from Lemma 11.2.7 it follows that $n_\omega(r) - n_\omega(r-1) \leq 4\ell$ if $r > \frac{1}{2}(h^2 + 3)$ and $n_\omega(r) - n_\omega(r-1) \leq n_\omega(r) \leq \ell(h^2 + 4)$ for $1 < r < \frac{1}{2}(h^2 + 3)$. This proves part 2 with $d = \ell(h^2 + 4)$.

3. immediately follows with $c = n_\omega(1)$ since $n_\omega(r) \le d\lfloor r \rfloor + n_\omega(1)$ by part 2.

4. If we arrange the nonzero zeros of ω in such a way that $|a_k| \le |a_{k+1}|$ for all $k \in \mathbb{N}$, then it follows from part 3 with $r = |a_k|$ that

$$k \le n(|a_k|) \le c + d|a_k|$$

and hence

$$\sum_{k=\lfloor c \rfloor + 1}^{\infty} |a_k|^{-2} \le \sum_{k=\lfloor c \rfloor + 1}^{\infty} \frac{d^2}{(k-c)^2} < \infty.$$

But

$$\left| \mathrm{Im}\, \frac{1}{a_k} \right| = \frac{|\mathrm{Im}\, \overline{a_k}|}{|a_k|^2} < \frac{h}{|a_k|^2}, \tag{11.2.9}$$

and the estimate (11.2.7) follows. □

Later in this section we will substantially improve our knowledge on the location of the zeros of ω; however, the rough estimate in Proposition 11.2.8 is needed for the proof of the refinement.

Next we will prove an auxiliary result which roughly says that a slight movement of the zeros of the above form does not change the estimates of the function too much. For this we introduce the following notation, based on Proposition 11.2.8.

Notation 11.2.9. A sequence $(b_k)_{k=1}^{\infty}$ in $\mathbb{C} \setminus \{0\}$ is called a *pseudoregular sequence* if

(i) the number $n(r)$ of $k \in \mathbb{N}$ with $|b_k| \le r$ is finite for all $r > 0$,

(ii) there is a number $d > 0$ such that $n(r) - n(r-1) \le d$ for all $r > 1$.

Remark 11.2.10. From the proof of Proposition 11.2.8 we know that a pseudoregular sequence $(b_k)_{k=1}^{\infty}$ satisfies $n(r) \le (d+1)r$ for some $d > 0$ and all $r > 1$ and that $(b_k^{-1})_{k=1}^{\infty} \in \ell^2$. Clearly, if $(b_k)_{k=1}^{\infty}$ is a pseudoregular sequence and if $(c_k)_{k=1}^{\infty}$ is a sequence in $\mathbb{C} \setminus \{0\}$ such that $(c_k - b_k)_{k=1}^{\infty}$ is bounded, then also $(c_k)_{k=1}^{\infty}$ is a pseudoregular sequence. Furthermore, since $c_k^{-1} - b_k^{-1} = (b_k - c_k)c_k^{-1}b_k^{-1}$ we have that $(c_k^{-1} - b_k^{-1})_{k=1}^{\infty} \in \ell^1$.

The following lemma is a generalization of [173, Lemma 5, p. 237].

Lemma 11.2.11. *Let $(b_k)_{k=1}^{\infty}$ and $(c_k)_{k=1}^{\infty}$ be sequences in $\mathbb{C} \setminus \{0\}$ such that $(b_k)_{k=1}^{\infty}$ is pseudoregular and such that $(c_k - b_k)_{k=1}^{\infty}$ is bounded, and consider*

$$f(\lambda) = \prod_{k=1}^{\infty} \left(1 - \frac{\lambda}{c_k}\right) \left(1 - \frac{\lambda}{b_k}\right)^{-1}, \quad \lambda \in \mathbb{C} \setminus \{b_k : k \in \mathbb{N}\}.$$

Then f converges uniformly on compact subsets of $\mathbb{C} \setminus \{b_k : k \in \mathbb{N}\}$, and for all $\delta > 0$,

$$\limsup_{r \to \infty} \frac{\log |f(re^{i\phi})|}{r} \le 0 \tag{11.2.10}$$

uniformly for all $\phi \in [-\pi, \pi]$ *on the set of* $re^{i\phi} \in \mathbb{C}$ *with* $|re^{i\phi} - b_k| \geq \delta$ *for all* $k \in \mathbb{N}$, *and*

$$\liminf_{r \to \infty} \frac{\log |f(re^{i\phi})|}{r} \geq 0 \qquad (11.2.11)$$

uniformly for all $\phi \in [-\pi, \pi]$ *on the set of* $re^{i\phi} \in \mathbb{C}$ *with* $|re^{i\phi} - c_k| \geq \delta$ *for all* $k \in \mathbb{N}$.

Proof. For $k \in \mathbb{N}$ let

$$f_k(\lambda) = \left(1 - \frac{\lambda}{c_k}\right)\left(1 - \frac{\lambda}{b_k}\right)^{-1}.$$

Then we can write

$$\log |f_k(\lambda)| = \log \left| 1 + \lambda \frac{\frac{1}{b_k} - \frac{1}{c_k}}{1 - \frac{\lambda}{b_k}} \right|.$$

From $\log |1 + z| \leq \log(1 + |z|) \leq |z|$ for all $z \in \mathbb{C} \setminus \{-1\}$ it therefore follows with $r = |\lambda|$ that

$$\frac{\log |f_k(\lambda)|}{r} \leq \frac{|c_k - b_k|}{\left| 1 - \frac{\lambda}{b_k} \right|} \frac{1}{|b_k| \, |c_k|}.$$

Since $1 - \frac{\lambda}{b_k} \to 1$ as $k \to \infty$ and since $(b_k^{-1})_{k=1}^{\infty}, (c_k^{-1})_{k=1}^{\infty} \in \ell^2$, we can deduce that the series over k on the right-hand side converges locally uniformly for all $\lambda \in \mathbb{C} \setminus \{b_k : k \in \mathbb{N}\}$, which proves the stated convergence of the infinite product f and gives the estimate

$$\frac{\log |f(\lambda)|}{r} \leq \sum_{k=1}^{\infty} \frac{|c_k - b_k|}{\left| 1 - \frac{\lambda}{b_k} \right|} \frac{1}{|b_k| \, |c_k|}. \qquad (11.2.12)$$

To prove (11.2.10), let $\delta > 0$, $m = \sup_{k=1}^{\infty} |c_k - b_k|$ and consider $\lambda \in \mathbb{C}$ such that $r = |\lambda| > m + \delta + 1$ and $|\lambda - b_k| \geq \delta$ for all $k \in \mathbb{N}$. We may assume that $|b_k| \leq |b_{k+1}|$ for all $k \in \mathbb{N}$. Then, in view of (11.2.12), we can establish the estimates

$$\frac{\log |f(\lambda)|}{r} \leq \sum_{k=1}^{\infty} \frac{m}{|c_k| \, |b_k - \lambda|}$$

$$\leq \sum_{k=1}^{n(r-\delta)} \frac{m}{|c_k| \, (r - |b_k|)} + \sum_{k=n(r-\delta)+1}^{n(r+\delta)} \frac{m}{(|b_k| - |c_k - b_k|) \, |b_k - \lambda|}$$

$$+ \sum_{k=n(r+\delta)+1}^{n(2r)} \frac{m}{|c_k| \, (|b_k| - r)} + \sum_{k=n(2r)+1}^{\infty} \frac{2m}{|c_k| \, |b_k|}. \qquad (11.2.13)$$

We are going to show that each of the four sums tends to 0 as $r \to \infty$. Let d be the number from Notation 11.2.9. Then the second sum has at most $\lceil 2\delta \rceil d$ terms, and it follows that

$$\sum_{k=n(r-\delta)+1}^{n(r+\delta)} \frac{m}{(|b_k| - |c_k - b_k|)\,|b_k - \lambda|} \leq \sum_{k=n(r-\delta)+1}^{n(r+\delta)} \frac{m}{(r - m - \delta)\delta} \xrightarrow[r\to\infty]{} 0 \quad (11.2.14)$$

since each summand tends to 0 as $r \to \infty$. Since the fourth sum is the tail of a converging series, we have

$$\sum_{k=n(2r)+1}^{\infty} \frac{2m}{|c_k|\,|b_k|} \xrightarrow[r\to\infty]{} 0. \quad (11.2.15)$$

For the first sum we observe that there are at most $\lceil m + 2 \rceil d$ indices k for which $|b_k| < m + 2$ and at most d indices k such that $\delta \leq r - |b_k| < 1$. For each of the remaining indices k we can find an integer $j \in \{1, \dots, \lfloor r - m - 2 \rfloor\}$ such that $j \leq r - |b_k| \leq j + 1$. Since $(b_k)_{k=1}^{\infty}$ is pseudoregular, for each such j there are at most d indices k with this property. For such index k, we have

$$|c_k| \geq r - (r - |b_k|) - |b_k - c_k| \geq r - j - 1 - m \geq \lfloor r - 1 - m \rfloor - j.$$

Altogether, we have the estimate

$$\sum_{k=1}^{n(r-\delta)} \frac{1}{|c_k|\,(r - |b_k|)}$$
$$\leq \frac{\lceil m + 2 \rceil d}{c_0(r - m - 2)} + \frac{d}{(r - m - 1)\delta} + d \sum_{j=1}^{\lfloor r-m-2 \rfloor} \frac{1}{(\lfloor r - 1 - m \rfloor - j)j},$$

where c_0 is the minimum of the $|c_k|$, $k \in \mathbb{N}$. We calculate

$$\sum_{j=1}^{n-1} \frac{1}{(n-j)j} = \frac{1}{n} \sum_{j=1}^{n-1} \left(\frac{1}{n-j} + \frac{1}{j} \right) = \frac{2}{n} \sum_{j=1}^{n-1} \frac{1}{j} = \frac{2}{n}(\log n + \gamma_n),$$

where $\gamma_n \to \gamma$ as $n \to \infty$ and γ is Euler's number, see [55, VII.7.5]. Therefore

$$\sum_{k=1}^{n(r-\delta)} \frac{m}{|c_k|\,(r - |b_k|)} \xrightarrow[r\to\infty]{} 0. \quad (11.2.16)$$

Finally, the third sum in (11.2.13) can be estimated in a similar way to the first sum. The corresponding c_k have modulus at least $r + \delta - m$, so that an estimate of the third sum follows indeed from that of the first sum. We do not need a better estimate, and therefore we are content to state

$$\sum_{k=n(r+\delta)+1}^{n(2r)} \frac{m}{|c_k|\,(|b_k| - r)} \xrightarrow[r\to\infty]{} 0. \quad (11.2.17)$$

The estimates (11.2.13)–(11.2.17) immediately imply (11.2.10).

Interchanging b_k with c_k will result in replacing $\log|f(\lambda)|$ with $-\log|f(\lambda)|$, and (11.2.11) will follow from (11.2.10). $\square$

As in Proposition 11.2.8, let a_k, $k \in \mathbb{N}$, be the nonzero zeros of the sine type function ω, where we assume without loss of generality that the modulus increases with the index, that is, $|a_k| \le |a_{k+1}|$ for all $k \in \mathbb{N}$. We will use the following notations:

$$U_{\omega,\delta} = \{r \in (0,\infty) : \forall\, k \in \mathbb{N}\ |r - |a_k|| \ge \delta\},\ \delta > 0, \tag{11.2.18}$$

$$E_{\omega,\delta} = (0,\infty) \setminus U_{\omega,\delta},\ \delta > 0, \tag{11.2.19}$$

$$E_{\omega,\delta,r} = E_{\omega,\delta} \cap (0,r),\ \delta, r > 0, \tag{11.2.20}$$

Corollary 11.2.12. *Let ω be a sine type function. Then there exists $\delta > 0$ such that for all $r > 1$, the set $U_{\omega,\delta} \cap (r, r+1)$ contains an interval of length at least 2δ.*

Proof. With d from Proposition 11.2.8, part 2, $(r, r+1) \cap U_{\omega,\delta}$ consists of at most $d+1$ intervals, whose total length is at least $1 - 2d\delta$. Hence every positive δ with $\delta < (4d + 2)^{-1}$ will satisfy the statement of this corollary. $\square$

Remark 11.2.13. Let ω be a sine type function and let $(\alpha_k)_{k=1}^{(\infty)}$ be its zeros, counted with multiplicity, in the upper half-plane. From Lemma 11.2.11 we know that χ_ω^+ defined by

$$\chi_\omega^+(\lambda) = \prod_{k=1}^{(\infty)} \left(1 - \frac{\lambda}{\alpha_k}\right) \left(1 - \frac{\lambda}{\overline{\alpha_k}}\right)^{-1} \tag{11.2.21}$$

is a meromorphic function which converges absolutely and uniformly on compact subset of $\mathbb{C} \setminus \{\overline{\alpha_k} : k \in \mathbb{Z}\}$, and therefore in particular on compact subsets in the closed upper half-plane. Indeed, χ_ω^+ is a function as considered in Lemma 11.2.11 if $(a_k)_{k=1}^\infty$ denotes the zeros of ω and if we put $c_k = a_k$ for all $k \in \mathbb{N}$, $b_k = a_k$ if $\operatorname{Im} a_k \le 0$, and $b_k = \overline{a_k}$ if $\operatorname{Im} a_k > 0$.

Corollary 11.2.14. *Let ω be a sine type function. Then*

$$\lim_{r \to \infty} \frac{\log|\chi_\omega^+(re^{i\phi})|}{r} = 0 \ \text{for all } \phi \in [0, \pi],$$

where for all $\delta > 0$, the convergence is uniform on all $\lambda \in \overline{\mathbb{C}^+}$ such that $|\lambda - \alpha_k| \ge \delta$ for all indices k.

Proof. For $\phi = 0$ and $\phi = \pi$, that is $\lambda \in \mathbb{R}$, we have $|\chi_\omega^+(\lambda)| = 1$, and the statement of the lemma is trivial in this case. The general result now follow from (11.2.11) and the fact that $|\chi_\omega^+(\lambda)| \le 1$ for $\lambda \in \overline{\mathbb{C}^+}$. $\square$

We need the following generalization of the Schwarz reflection principle.

Lemma 11.2.15. *Let f be analytic on $\mathbb{C}^+$ such that the absolute value of f has a continuous extension to $\overline{\mathbb{C}^+}$ which is 1 on the real axis. Then f has a meromorphic extension h to $\mathbb{C}$, and $h(\lambda) = \dfrac{1}{\overline{f(\overline{\lambda})}}$ for all $\lambda \in \mathbb{C}^-$ with $f(\overline{\lambda}) \neq 0$.*

Proof. Let $P = \{\lambda \in \mathbb{C}^- : f(\overline{\lambda}) = 0\}$ and let Ω be the set of all $\lambda \in \mathbb{C}$ such that the line segment connecting i and λ does not contain any of the points of P. It is clear that Ω is an open set containing $\overline{\mathbb{C}^+}$. For $\lambda \in \mathbb{C} \setminus (\mathbb{R} \cup P)$ let $h_0(\lambda)$ be defined by $f(\lambda)$ if $\operatorname{Im} \lambda > 0$ and by $\frac{1}{\overline{f(\overline{\lambda})}}$ otherwise. It is clear that h_0 is a meromorphic function in $\mathbb{C} \setminus \mathbb{R}$ with poles at the points in P. We are going to show that h_0 has an extension h to the real axis which is analytic at each point on the real axis. For $\lambda \in \Omega$ define

$$g(\lambda) = \int_0^1 (\lambda - i) h_0(i + t(\lambda - i)) \, dt,$$

that is, $g(\lambda)$ is the integral over h_0 along the straight line segment from i to λ. This integral is well defined since the integrand is bounded and continuous on $\mathcal{I}_\lambda$, where $\mathcal{I}_\lambda = [0, 1]$ if $\lambda \in \mathbb{C}^+$, and $\mathcal{I}_\lambda = [0, 1] \setminus \{t_0\}$ in case $\lambda \in \overline{\mathbb{C}^-}$, and where t_0 is the one value of t for which $i + t(\lambda_0 - i) \in \mathbb{R}$. For $\lambda, \lambda_0 \in \Omega$ we have

$$g(\lambda) - g(\lambda_0) = \int_0^1 \left[(\lambda - i) h_0(i + t(\lambda - i)) - (\lambda_0 - i) h_0(i + t(\lambda_0 - i)) \right] dt,$$

where the integrand is uniformly bounded for λ in any compact neighbourhood of λ_0 in Ω, and converges to 0 pointwise as $\lambda \to \lambda_0$ for all $t \in \mathcal{I}_{\lambda_0}$. In view of Lebesgue's dominated convergence theorem it follows that g is continuous on Ω.

Clearly, since f is analytic in the open upper half-plane, also g is analytic in $\mathbb{C}^+$, and a straightforward calculation shows that $g' = h_0$ in $\mathbb{C}^+$. Furthermore, if $\lambda_0, \lambda \in \Omega \cap \mathbb{C}^-$ such that the line segment $\Gamma_{\lambda_0, \lambda}$ from λ_0 to λ is contained in Ω, then

$$g(\lambda) - g(\lambda_0) = \int_{\Gamma_{\lambda_0, \lambda}} h_0(z) \, dz. \tag{11.2.22}$$

Indeed, since h_0 is analytic in $\Omega \setminus \mathbb{R}$, by a standard argument using Cauchy's integral theorem, it suffices to show that the contour integral over h_0 along the quadrilateral with vertices $a_{\pm\varepsilon} \pm i\varepsilon$ and $b_{\pm\varepsilon} \pm i\varepsilon$ with $a_{\pm\varepsilon}, b_{\pm\varepsilon} \in \mathbb{R}$ for sufficiently small $\varepsilon > 0$, $a_{\pm\varepsilon} \to a$, $b_{\pm\varepsilon} \to b$ and $a < b$, tends to 0 as $\varepsilon \to 0$. Since h_0 is locally bounded, it suffices to show that

$$I_\varepsilon = \int_a^b |h_0(t + i\varepsilon) - h_0(t - i\varepsilon)| \, dt \to 0 \text{ as } \varepsilon \to 0.$$

The integrand

$$\left| f(t + i\varepsilon) - \frac{1}{\overline{f(t + i\varepsilon)}} \right| = |f(t + i\varepsilon)| \left| 1 - \frac{1}{|f(t + i\varepsilon)|^2} \right|$$

is uniformly bounded in t and ε and converges to 0 pointwise for all $t \in [a, b]$ as $\varepsilon \to 0$. This completes the proof of (11.2.22), and the Fundamental Theorem of Calculus shows that g is analytic in $\Omega \cap \mathbb{C}^-$ with $g' = h_0$ in $\Omega \cap \mathbb{C}^-$.

Let T be a triangular path which together with its interior lies in Ω. If T is in $\mathbb{C}^+$ or $\mathbb{C}^-$, then Cauchy's integral theorem shows that

$$\int_T g(z)\,dz = 0.$$

Since g is continuous, this extends to the case when T is in $\overline{\mathbb{C}^+}$ or in in $\overline{\mathbb{C}^-}$, and therefore to arbitrary such T in Ω. But by Morera's theorem, see [55, Theorem IV.5.10], this means that g is analytic in Ω. Hence also g' is analytic in Ω, and since $g' = h_0$ on $\Omega \setminus \mathbb{R}$, it follows that h_0 has an analytic extension h to $\mathbb{C} \setminus P$, which, in particular, is analytic at each point of the real axis and which is meromorphic on $\mathbb{C}$ with $h(\lambda) = h_0(\lambda) = \frac{1}{f(\lambda)}$ for all $\lambda \in \mathbb{C}^- \setminus P$.   □

The following result is a special case of [173, Theorem 5, p. 240].

Lemma 11.2.16. *Let ω be a function of sine type σ with $h_1 = 0$. Then*

$$\log |\omega(\lambda)| = \frac{|y|}{\pi} \int_{-\infty}^{\infty} \log |\omega(t)| \frac{dt}{(t-x)^2 + y^2} + \sigma|y| + \log |\chi(\lambda)| \qquad (11.2.23)$$

for all $\lambda = x + iy \in \mathbb{C}$ with $\omega(\lambda) \neq 0$, where

$$\chi(\lambda) = \begin{cases} \chi_\omega^+(\lambda) & \text{if } \operatorname{Im}\lambda \geq 0, \\ \chi_{\overline{\omega}}^+(\lambda) & \text{if } \operatorname{Im}\lambda < 0. \end{cases}$$

Proof. It suffices to prove the results for $\operatorname{Im}\lambda \geq 0$ since also $\overline{\omega}$ is a function of sine type σ with $h_1 = 0$. The function u on $\mathbb{C}^+$ defined by

$$u(\lambda) = \frac{y}{\pi} \int_{-\infty}^{\infty} \log |\omega(t)| \frac{dt}{(t-x)^2 + y^2} + \sigma y$$

is well defined and satisfies

$$|u(\lambda) - \sigma y| \leq \max\{|\log M|, |\log m|\} =: M_0 \quad \text{for } y > 0 \qquad (11.2.24)$$

since

$$\frac{y}{\pi} \int_{-\infty}^{\infty} \frac{dt}{(t-x)^2 + y^2} = 1,$$

where m and M are the constants from Definition 11.2.5, (ii). From

$$\frac{y}{(t-x)^2 + y^2} = -\operatorname{Im} \frac{\overline{\lambda} - t}{|\lambda - t|^2} = -\operatorname{Im} \frac{1}{\lambda - t}$$

and a standard argument it follows that u is harmonic. Furthermore, for $x_0 \in \mathbb{R}$ we have

$$u(\lambda) - \sigma|y| - \log|\omega(x_0)| = \frac{y}{\pi} \int_{-\infty}^{\infty} \frac{\log|\omega(t)| - \log|\omega(x_0)|}{(t-x)^2 + y^2} \, dt,$$

so that for all $\delta > 0$ and $\lambda \in (x_0 - \delta, x_0 + \delta) \times i(0, \infty)$,

$$\begin{aligned}
\left| u(\lambda) - \sigma|y| - \log|\omega(x_0)| \right| &\leq \left| \frac{y}{\pi} \int_{|t-x_0|>2\delta} \frac{\log|\omega(t)| - \log|\omega(x_0)|}{(t-x)^2 + y^2} \, dt \right| \\
&\quad + \left| \frac{y}{\pi} \int_{|t-x_0|<2\delta} \frac{\log|\omega(t)| - \log|\omega(x_0)|}{(t-x)^2 + y^2} \, dt \right| \\
&\leq \frac{4M_0}{\pi} \left(\frac{\pi}{2} - \arctan\frac{\delta}{y} \right) + \frac{2}{\pi} \sup_{|t-x_0|<2\delta} \left| \log|\omega(t)| - \log|\omega(x_0)| \right| \arctan\frac{3\delta}{y} \\
&\leq \frac{4M_0}{\pi} \left(\frac{\pi}{2} - \arctan\frac{\delta}{y} \right) + \sup_{|t-x_0|<2\delta} \left| \log|\omega(t)| - \log|\omega(x_0)| \right|.
\end{aligned}$$

Now let $\varepsilon > 0$ and choose $\delta, \delta_1 > 0$ such that $\displaystyle\sup_{|t-x_0|<2\delta} \left| \log|\omega(t)| - \log|\omega(x_0)| \right| < \frac{\varepsilon}{2}$ and $\dfrac{4M_0}{\pi} \left(\dfrac{\pi}{2} - \arctan\dfrac{\delta}{\delta_1} \right) < \dfrac{\varepsilon}{2}$. Then it follows for $\lambda \in (x_0 - \delta, x_0 + \delta) \times i(0, \delta_1)$ that $\left| u(\lambda) - \sigma|y| - \log|\omega(x_0)| \right| < \varepsilon$. Therefore u has a continuous extension to $\overline{\mathbb{C}^+}$, given by $u(x) = \log|\omega(x)|$, $x \in \mathbb{R}$. Let $v : \mathbb{C}^+ \to \mathbb{R}$ be a harmonic conjugate of u on $\mathbb{C}^+$ and define

$$\psi(\lambda) = \omega(\lambda) e^{-u(\lambda) - iv(\lambda)}, \quad \lambda \in \mathbb{C}^+.$$

For $x \in \mathbb{R}$ we have in view of $u(x) = \log|\omega(x)|$ that $|\psi(x)| = 1$ defines a continuous extension of the absolute value of ψ to $\overline{\mathbb{C}^+}$. By the Schwarz reflection principle, see Lemma 11.2.15, ψ can be extended to a meromorphic function on $\mathbb{C}$ without poles in $\overline{\mathbb{C}^+}$. Since ω is of sine type σ, it follows from Lemma 11.2.6 that there is $M_1 > 0$ such that

$$|\psi(\lambda)| \leq M_1 e^{\sigma y - u(\lambda)} \leq M_1 e^{M_0}, \quad \lambda \in \mathbb{C}^+,$$

where we have made use of (11.2.24) in the last estimate. Hence $|\psi(\lambda)| \leq 1$ for all $\lambda \in \overline{\mathbb{C}^+}$ by the Phragmén–Lindelöf principle, see [55, Corollary VI.4.2]. With the notation from χ_ω^+, we define

$$\chi_n^+(\lambda) = \prod_{k=1}^{(n)} \left(1 - \frac{\lambda}{\alpha_k} \right) \left(1 - \frac{\lambda}{\overline{\alpha_k}} \right)^{-1},$$

where (n) means that we extend the product over all indices $k = 1, \ldots, n$ for which α_k exists. Since $|\chi_n^+(\lambda)| \to 1$ as $|\lambda| \to \infty$ and $|\chi_n^+(\lambda)| = 1$ for $\lambda \in \mathbb{R}$, it follows that

$$\psi_n = \frac{\psi}{\chi_n^+}$$

is meromorphic without poles in the closed upper half-plane and bounded in the closed upper half-plane with $|\psi_n(\lambda)| = 1$ for $\lambda \in \mathbb{R}$. Again by the Phragmén–Lindelöf principle, $|\psi_n(\lambda)| \leq 1$ for all $\lambda \in \overline{\mathbb{C}^+}$. In view of Remark 11.2.13

$$\psi_\infty := \lim_{n \to \infty} \psi_n = \frac{\psi}{\chi_\omega^+}$$

exists. Since both ψ and χ_ω^+ are analytic in the closed upper half-plane and their zeros there coincide, with multiplicity, ψ_∞ is meromorphic without zeros and poles in the closed upper half-plane. The properties of the ψ_n carry over to the limit so that $|\psi_\infty(\lambda)| \leq 1$ for $\lambda \in \overline{\mathbb{C}^+}$ and $|\psi_\infty(\lambda)| = 1$ for $\lambda \in \mathbb{R}$. By the Schwarz reflection principle, see Lemma 11.2.15, ψ_∞ has no zeros and poles in $\mathbb{C}$. Hence

$$\varphi = -i \log \psi_\infty$$

is a real entire function satisfying $\varphi(\mathbb{C}^+) \subset \overline{\mathbb{C}^+}$. If φ maps the open upper half-plane into itself, then Lemma 11.1.3 shows that φ has exactly one zero, which is real. In view of Theorem 11.1.6, there are real number α and β with $\alpha > 0$ such that $\varphi(\lambda) = \alpha\lambda + \beta$. If φ does not map the open upper half-plane into itself, then φ must be constant by the maximum principle. This case is covered if we allow $\alpha = 0$ above. It therefore follows from the definition of ψ that

$$\begin{aligned}
\log|\omega(\lambda)| &= u(\lambda) + \log|\psi(\lambda)| \\
&= u(\lambda) + \operatorname{Re}\log\psi_\infty(\lambda) + \log|\chi_\omega^+(\lambda)| \\
&= u(\lambda) - \alpha y + \log|\chi_\omega^+(\lambda)|
\end{aligned} \tag{11.2.25}$$

for all $\lambda = x + iy \in \overline{\mathbb{C}^+}$ with $\omega(\lambda) \neq 0$. To complete the proof we have to show that $\alpha = 0$.

To this end let $(\lambda_n)_{n=1}^\infty$ be a sequence in $\overline{\mathbb{C}^+}$ such that $|\lambda_n| \to \infty$ as $n \to \infty$ and

$$\lim_{n \to \infty} \frac{\log|\omega(\lambda_n)|}{|\lambda_n|} = \sigma, \tag{11.2.26}$$

which exists since ω is of exponential type σ in the upper half-plane. In view of (11.2.6) we have

$$\lim_{n \to \infty} \frac{\log|\omega(\lambda_n)|}{|\lambda_n|} \leq \sigma \liminf_{n \to \infty} \frac{\operatorname{Im}\lambda_n}{|\lambda_n|}. \tag{11.2.27}$$

From (11.2.26) and (11.2.27) it follows that

$$\lim_{n \to \infty} \frac{\operatorname{Im}\lambda_n}{|\lambda_n|} = 1.$$

Then (11.2.26), (11.2.27), (11.2.24) and Corollary 11.2.14 lead to

$$\sigma = \lim_{n \to \infty} \frac{\log|\omega(\lambda_n)|}{|\lambda_n|} = \sigma - \alpha,$$

which proves that $\alpha = 0$. $\qquad\qquad\square$

The following lemma is a particular case of a result from [172].

Lemma 11.2.17. *Let ω be a sine type function of type σ. Then there is a positive real number m_1 such that*

$$|\omega(\lambda)| \geq m_1 e^{\sigma|\operatorname{Im}\lambda|}, \quad |\operatorname{Im}\lambda| \geq h+1, \tag{11.2.28}$$

where h is chosen according to Definition 11.2.5.

Proof. Let $h_0 = h+1$. First we are going to prove that there is $m_0 > 0$ such that

$$|\omega(x + ih_0)| \geq m_0, \quad x \in \mathbb{R}. \tag{11.2.29}$$

Suppose that (11.2.29) is false. Then there exists a sequence $(x_j)_{j=1}^\infty$ of real numbers such that $\omega(x_j + ih_0) \to 0$ as $j \to \infty$. Similar to the proof of Lemma 11.2.7 we may assume that $\omega_j = \omega(x_j + \cdot)$ converges uniformly on compact subsets to an entire function $\tilde{\omega}$ which is not identically zero. By Definition 11.2.5, (i), none of the functions ω_j has a zero in the open disc with centre ih_0 and radius 1. By Hurwitz's theorem, also $\tilde{\omega}$ has no zero there. On the other hand,

$$\tilde{\omega}(ih_0) = \lim_{j\to\infty} \omega_j(ih_0) = \lim_{j\to\infty} \omega(x_j + ih_0) = 0.$$

This contradiction proves (11.2.29).

In view of Lemma 11.2.6 and (11.2.29), ω_{h_0} is a sine type function with $h_1 = 0$. Now we can apply Lemma 11.2.16 to ω_{h_0}, which does not have any zeros in the upper half-plane, so that $\chi_{\omega_{h_0}}(\lambda) = 1$ for $\operatorname{Im}\lambda \geq 0$. Then (11.2.23) and (11.2.24) show that

$$|\omega_{h_0}(\lambda)| \geq e^{-M_0} e^{\sigma \operatorname{Im}\lambda}, \quad \operatorname{Im}\lambda \geq 0.$$

This proves (11.2.28) for $\operatorname{Im}\lambda \geq h+1$. Applying this result to $\bar{\omega}$ completes the proof. $\qquad\square$

Lemma 11.2.18. *Let ω be an entire function of finite order, let θ, M and $h_1 < h_2$ be real numbers and let*

$$S = \{\lambda \in \mathbb{C} : h_1 \leq \operatorname{Im}\lambda e^{i\theta} \leq h_2\}$$

be a strip in the complex plane such that ω is bounded on the boundary ∂S of S. Then ω is bounded on S.

Proof. Let ρ be the order of ω, choose $\alpha \in (0, \frac{\pi}{2\rho})$ with $\alpha < \pi$ and define

$$S_j = \{\lambda \in S \setminus \{0\} : \arg(\lambda e^{i\theta}) \in (-\alpha + j\pi, \alpha + j\pi)\}, \quad j = 0, 1.$$

Clearly, $\partial S_j \setminus \partial S$ and $S \setminus (S_0 \cup S_1)$ are bounded sets. Therefore, ω is bounded on ∂S_j for $j = 0, 1$, and it suffices to prove that ω is bounded on S_j for $j = 0, 1$. But the boundedness of ω on S_j follows from the Phragmén–Lindelöf principle, see [55, Theorem VI.4.1], by a straightforward adaptation of the proof of [55, Corollary VI.4.2] to subsets of angular regions. $\qquad\square$

The following result is mentioned in [176, §1].

Proposition 11.2.19. *An entire function ω of finite order is of sine type if and only if there exist positive constants σ, h, m and M such that*

$$m \leq |\omega(\lambda)|e^{-\sigma|\operatorname{Im}\lambda|} \leq M \tag{11.2.30}$$

for $|\operatorname{Im}\lambda| > h$, in which case σ is the exponential type of ω.

Proof. First we prove that the condition is sufficient. Indeed, property (i) of Definition 11.2.5 is obvious since $m > 0$, whereas property (ii) holds with $h_1 = h$ and m and M replaced with $me^{\sigma h}$ and $Me^{\sigma h}$, respectively. Since ω is bounded on the lines $\operatorname{Im}\lambda = h$ and $\operatorname{Im}\lambda = -h$, Lemma 11.2.18 shows that ω is also bounded on the strip $\{\lambda \in \mathbb{C} : |\operatorname{Im}\lambda| \leq h\}$, where we may write the bound as $M'e^{\sigma h}$ with $M' \geq M$. For $|\operatorname{Im}\lambda| > h$ we have

$$|\omega(\lambda)| \leq Me^{\sigma|\operatorname{Im}\lambda|} \leq Me^{\sigma|\lambda|},$$

and therefore $M_\omega(r) \leq M'e^{\sigma r}$ for $r > h$. Hence ω is of exponential type $\sigma' \leq \sigma$. On the other hand,

$$|\omega(\pm iy)| \geq me^{\sigma y} \quad \text{for } y > h$$

shows that the exponential type of ω is at least σ both in the upper and lower half-plane, and it therefore equals σ in both half-planes. We have thus shown that property (iii) of Definition 11.2.5 holds.

The necessity of (11.2.30) immediately follows from Lemmas 11.2.6 and 11.2.17. □

The following lemma is a special case of [175, Theorem 2.2].

Lemma 11.2.20. *Let ω be a sine type function and for $\delta > 0$ let Λ_δ be the set of all complex numbers whose distance from the set of zeros of ω is less than δ, i. e.,*

$$\Lambda_\delta = \{\lambda \in \mathbb{C} : \exists \zeta \in \mathbb{C} \ \omega(\zeta) = 0 \ \text{ and } \ |\lambda - \zeta| < \delta\}. \tag{11.2.31}$$

Then for each $\delta > 0$ there exists $k_\delta > 0$ such that

$$|\omega(\lambda)| \geq k_\delta \quad \text{for } \lambda \in \mathbb{C} \setminus \Lambda_\delta. \tag{11.2.32}$$

Proof. Assume that for some $\delta > 0$ such k_δ does not exist. Then there exists a sequence of points $(\zeta_k)_{k=1}^\infty$ in $\mathbb{C} \setminus \Lambda_\delta$ such that $\lim_{k\to\infty} \omega(\zeta_k) = 0$. In view of the lower bound in (11.2.30), the sequence $(\operatorname{Im}\zeta_k)_{k=1}^\infty$ is bounded. Then $\omega_k = \omega(\cdot + \zeta_k)$ defines a sequence of entire functions $(\omega_k)_{k=1}^\infty$ which is locally bounded in $\mathbb{C}$ because of the upper bound in (11.2.30). By Montel's theorem we may assume that $(\omega_k)_{k=1}^\infty$ converges to an entire function ω_0, choosing a subsequence, if necessary. Again, the lower bound in (11.2.30) shows that ω_0 is not identically zero, whereas

$$\omega_0(0) = \lim_{k\to\infty} \omega_k(0) = \lim_{k\to\infty} \omega(\zeta_k) = 0.$$

Then by Hurwitz' theorem there exists k_0 such that for all $k > k_0$ the open disc with centre zero and radius δ contains at least one zero of ω_k, which means that ζ_k has a distance less than δ to at least one zero of ω. This contradicts $\zeta_k \in \mathbb{C} \setminus \Lambda_\delta$. □

Remark 11.2.21. Lemma 11.2.20 extends the lower bound in (11.2.30) to $\mathbb{C} \setminus \Lambda_\delta$. Indeed, it is easy to see that (11.2.30) and (11.2.32) lead to the following statement: for any $\delta > 0$ there exists $m_\delta > 0$ such that

$$e^{-\sigma |\operatorname{Im} \lambda|} |\omega(\lambda)| \geq m_\delta \quad \text{for } \lambda \in \mathbb{C} \setminus \Lambda_\delta, \tag{11.2.33}$$

where σ is the order of the sine type function ω.

In order to give a more precise statement on the location of the zeros of sine type functions, we need some preparations.

Definition 11.2.22. An analytic function f on $S = S_{\phi_1,\phi_2}$ is said to be of *exponential type* on S_{ϕ_1,ϕ_2} if

$$\sigma = \sigma_{f,\phi_1,\phi_2} = \limsup_{r \to \infty} \frac{\log M_{f,\phi_1,\phi_2}(r)}{r} < \infty, \tag{11.2.34}$$

and then the number σ is called the exponential type of f on S. The indicator function of f on S is defined by

$$h_f(\phi) = \limsup_{r \to \infty} \frac{\log |f(re^{i\phi})|}{r}, \quad \phi_1 \leq \phi \leq \phi_2.$$

We will also write $\hat{h}_f(\phi) = h_f(\phi)$ if the limit of $\dfrac{\log |f(re^{i\phi})|}{r}$ as $r \to \infty$ exists. If f is analytic on S, then $n_f(r, \phi_1, \phi_2)$ denotes the number of zeros λ of f in S with $|\lambda| \leq r$.

From (11.2.30) and (11.2.33) we immediately obtain

Proposition 11.2.23. *Let ω be a function of sine type σ. Then $\hat{h}_\omega(\phi) = \sigma |\sin \phi|$ for all angles $\phi \in \mathbb{R} \setminus \mathbb{Z}\pi$, where the convergence is uniform on each compact subset of $\mathbb{R} \setminus \mathbb{Z}\pi$. Furthermore, for each $\delta > 0$, the following limit is uniform in $\phi \in \mathbb{R}$:*

$$\lim_{\substack{r \to \infty \\ r \in U_{\omega,\delta}}} \frac{\log |\omega(re^{i\phi})|}{r} = \sigma |\sin \phi|.$$

The following result can be found in [173, pp. 142–143].

Lemma 11.2.24. *Let $\phi_1 < \phi_2 < \phi_1 + 2\pi$, let f be an analytic function on the sector $S_r = \{\lambda \in S_{\phi_1,\phi_2} : |\lambda| \leq r\}$ and assume that no zeros of f lie on the boundary of S_r. Then*

$$2\pi n_f(r, \phi_1, \phi_2) = \int_0^r \frac{1}{t} \frac{\partial \log |f(te^{i\phi})|}{\partial \phi} \bigg|_{\phi=\phi_2} dt - \int_0^r \frac{1}{t} \frac{\partial \log |f(te^{i\phi})|}{\partial \phi} \bigg|_{\phi=\phi_1} dt$$

$$+ r \int_{\phi_1}^{\phi_2} \frac{\partial \log |f(re^{i\phi})|}{\partial r} d\phi.$$

Proof. By the argument principle, an integration along the boundary of S_r gives

$$2\pi i n_f(r, \phi_1, \phi_2) = e^{i\phi_1} \int_0^r \frac{f'(te^{i\phi_1})}{f(te^{i\phi_1})}\, dt - e^{i\phi_2} \int_0^r \frac{f'(te^{i\phi_2})}{f(te^{i\phi_2})}\, dt$$

$$+ ir \int_{\phi_1}^{\phi_2} e^{i\phi} \frac{f'(re^{i\phi})}{f(re^{i\phi})}\, d\phi.$$

From

$$\frac{\partial \log f(te^{i\phi})}{\partial t} = e^{i\phi} \frac{f'(te^{i\phi})}{f(te^{i\phi})} \quad \text{and} \quad \frac{\partial \log f(te^{i\phi})}{\partial \phi} = ite^{i\phi} \frac{f'(te^{i\phi})}{f(te^{i\phi})}$$

it follows that

$$2\pi i n_f(r, \phi_1, \phi_2) = \int_0^r \frac{\partial \log f(te^{i\phi_1})}{\partial t}\, dt - \int_0^r \frac{\partial \log f(te^{i\phi_2})}{\partial t}\, dt$$

$$+ \int_{\phi_1}^{\phi_2} \frac{\partial \log f(re^{i\phi})}{\partial \phi}\, d\phi.$$

Using the polar form $\dfrac{\partial \log f}{\partial t} + \dfrac{i}{t} \dfrac{\partial \log f}{\partial \phi} = 0$ of the Cauchy–Riemann equations and taking imaginary parts completes the proof.   □

Proposition 11.2.25. *Let ω be a sine type function of sine type σ with $|\omega(0)| = 1$ and define*

$$J_\omega^r(\phi) = \frac{1}{r} \int_0^r \frac{\log |\omega(te^{i\phi})|}{t}\, dt, \quad r > 0, \ \phi \in \mathbb{R} \setminus \mathbb{Z}\pi.$$

Then

$$\lim_{r \to \infty} J_\omega^r(\phi) = \sigma |\sin \phi|, \quad \phi \in \mathbb{R} \setminus \mathbb{Z}\pi,$$

uniformly on each compact subset of $\mathbb{R} \setminus \mathbb{Z}\pi$.

Proof. Let $\varepsilon > 0$ and let Φ be a compact subset of $\mathbb{R} \setminus \mathbb{Z}\pi$. By Proposition 11.2.23, there is $r_0 > 1$ such that

$$\left| \frac{\log |\omega(te^{i\phi})|}{t} - \sigma |\sin \phi| \right| < \varepsilon \quad \text{for } t > r_0 \text{ and } \phi \in \Phi.$$

Since $\omega(0) = 1$, $\lambda \mapsto \frac{\log \omega(\lambda)}{\lambda}$ is analytic at 0, and therefore there exist $r_1 \in (0, r_0)$ and $C_1 > 0$ such that $\omega(\lambda) \neq 0$ for $|\lambda| \leq r_1$ and

$$\frac{|\log |\omega(te^{i\phi})||}{t} \leq C_1 \quad \text{for } 0 < t \leq r_1, \ \phi \in \Phi.$$

Hence

$$\int_0^{r_1} \frac{|\log |\omega(te^{i\phi})||}{t}\, dt \leq C_1 r_1, \ \phi \in \Phi.$$

Let $(a_k)_{k=1}^n$ be the tuple of all zeros λ of ω, counted with multiplicity, satisfying $|\lambda| \leq 2r_0$. Then we can write

$$\omega(\lambda) = h(\lambda) \prod_{k=1}^n (\lambda - a_k), \ \lambda \in \mathbb{C},$$

where $h(\lambda) \neq 0$ for $|\lambda| \leq 2r_0$. Hence there is $C_2 > 0$ such that $|\log|h(\lambda)|| \leq C_2$ for $|\lambda| \leq 2r_0$. Then

$$\int_{r_1}^{r_0} \frac{|\log|h(te^{i\phi})||}{t} \, dt \leq \frac{r_0 - r_1}{r_1} C_2, \ \phi \in \Phi.$$

For $k = 1, \ldots, n$ and $\phi \in \Phi$ we calculate

$$\int_{r_1}^{r_0} |\log|te^{i\phi} - a_k|| \, dt \leq \int_{|a_k|-1}^{|a_k|+1} |\log|t - |a_k||| \, dt + \int_{r_1}^{r_0} \max\{0, \log(t + |a_k|)\} \, dt$$

$$\leq 2\int_0^1 |\log t| \, dt + (r_0 - r_1)\log(3r_0)$$

$$= 2 + (r_0 - r_1)\log(3r_0).$$

Altogether, we get

$$\int_0^{r_0} \left| \frac{\log|\omega(te^{i\phi})|}{t} - \sigma|\sin\phi| \right| \leq C_1 r_1 + \frac{r_0 - r_1}{r_1} C_2 + n(2 + (r_0 - r_1)\log(3r_0)) + \sigma r_0.$$

Therefore we conclude for all $\phi \in \Phi$ that

$$\limsup_{r\to\infty} |J_\omega^r(\phi) - \sigma|\sin\phi|| \leq \limsup_{r\to\infty} \frac{1}{r}\int_0^{r_0} \left| \frac{\log|\omega(te^{i\phi})|}{t} - \sigma|\sin\phi| \right| dt$$

$$+ \limsup_{r\to\infty} \frac{1}{r}\int_{r_0}^r \varepsilon \, dt \leq \varepsilon. \qquad \square$$

The following result is a special case of [173, Theorem III.3, p. 152].

Proposition 11.2.26. *Let ω be a sine type function of sine type σ and let $\phi_1, \phi_2 \in \mathbb{R} \setminus \mathbb{Z}\pi$ such that $\phi_1 < \phi_2 < \phi_1 + 2\pi$. Then*

$$\lim_{r\to\infty} \frac{n_\omega(r, \phi_1, \phi_2)}{r} = \frac{\sigma}{\pi} n_{\phi_1,\phi_2},$$

where n_{ϕ_1,ϕ_2} is the number of elements of the set $(\phi_1, \phi_2) \cap \mathbb{Z}\pi$.

Proof. If $n_{\phi_1,\phi_2} = 0$, then the angular region S_{ϕ_1,ϕ_2} contains at most finitely many zeros of ω, and the statement is trivially true. Thus it suffices to prove the result for $n_{\phi_1,\phi_2} = 1$. The transformation $\hat{\omega}(\lambda) = \omega(-\lambda)$ gives a sine type function $\hat{\omega}$ whose zeros are obtained by rotating the zeros of ω by the angle π. Hence it suffices to consider the case $-\pi < \phi_1 < 0 < \phi_2 < \pi$. Since a shift of the variable and the multiplication of ω by a nonzero constant changes neither the assumptions nor the conclusion of this proposition, we may assume that $\omega(0) = 1$. We also may assume that ϕ_1 and ϕ_2 are chosen such that no zeros of ω lie on the rays with angle ϕ_1 and ϕ_2, respectively. Hence no zeros of ω lie on the rays with angle ϕ in $[\phi_1, \phi_1 + k]$ and $[\phi_2, \phi_2 + l]$ for sufficiently small positive k and l. Introducing the function

$$N(r, \phi, \varphi) = \int_0^r \frac{n_\omega(t, \phi, \varphi)}{t} \, dt$$

and averaging over the above intervals we obtain in view of Lemma 11.2.24 that

$$\begin{aligned}
N(r, \phi_1, \phi_2) &= \frac{1}{kl} \int_{\phi_1}^{\phi_1 + k} \int_{\phi_2}^{\phi_2 + l} N(r, \phi, \varphi) \, d\varphi \, d\phi \\
&= \frac{1}{2\pi} \int_0^r \frac{J_\omega^t(\phi_2 + l) - J_\omega^t(\phi_2)}{l} \, dt - \frac{1}{2\pi} \int_0^r \frac{J_\omega^t(\phi_1 + k) - J_\omega^t(\phi_1)}{k} \, dt \\
&\quad + \frac{1}{2\pi kl} \int_{\phi_1}^{\phi_1 + k} \int_{\phi_2}^{\phi_2 + l} \int_\phi^\varphi \log |\omega(re^{i\vartheta})| \, d\vartheta \, d\phi \, d\varphi,
\end{aligned}$$

where we have to observe that the integration over t is valid for all angles ϑ such that ω has no zero on the ray $te^{i\vartheta}$.

Let $\varepsilon > 0$. In view of Proposition 11.2.25 there is $r_0 > 0$ such that

$$|J_\omega^r(\phi) - \sigma|\sin\phi|| < \varepsilon$$

for $r > r_0$ and $\phi \in \{\phi_1, \phi_1 + k, \phi_2, \phi_2 + l\}$. In the proof of Proposition 11.2.25 we have seen that $|J_\omega^r(\phi)| \le C_1$ for $\phi \in \mathbb{R}$ and $0 < r < r_1$, with r_0 and C_1 from the proof of Proposition 11.2.25. From that proof we also conclude that there is $C_3 > 0$ such that $|J_\omega^r(\phi)| \le C_3$ for all $r \in [r_1, r_0]$ and $\phi \in \mathbb{R}$. Hence we obtain

$$\lim_{r \to \infty} \frac{1}{r} \int_0^r J_\omega^t(\phi) \, dt = \sigma|\sin\phi|$$

uniformly on each compact subset of $\mathbb{R} \setminus \mathbb{Z}\pi$. In view of Corollary 11.2.12, for sufficiently small $\delta > 0$,

$$\begin{aligned}
\lim_{\substack{r \to \infty \\ r \in U_{\omega,\delta}}} \frac{N(r, \phi_1, \phi_2)}{r} &= \frac{\sigma}{2\pi} \frac{\sin(\phi_2 + l) - \sin\phi_2}{l} + \frac{\sigma}{2\pi} \frac{\sin(\phi_1 + k) - \sin\phi_1}{k} \\
&\quad + \frac{\sigma}{2\pi kl} \int_{\phi_1}^{\phi_1 + k} \int_{\phi_2}^{\phi_2 + l} \int_\phi^\varphi |\sin\vartheta| \, d\vartheta \, d\phi \, d\varphi,
\end{aligned}$$

where we have used Proposition 11.2.23 in the last summand. Therefore

$$\lim_{\substack{r \to \infty \\ r \in U_{\omega,\delta}}} \frac{N(r, \phi_1, \phi_2)}{r} = \frac{\sigma}{\pi}. \tag{11.2.35}$$

We shall prove that

$$\lim_{r \to \infty} \frac{N(r, \phi_1, \phi_2)}{r} = \frac{\sigma}{\pi}. \tag{11.2.36}$$

To this end, let $r > 1$. Choose $\delta > 0$ according to Corollary 11.2.12. Then there are $\alpha(r) \in (r-1, r) \cap U_{\omega,\delta}$ and $\beta(r) \in (r, r+1) \cap U_{\omega,\delta}$. Since $N(\cdot, \phi_1, \phi_2)$ is an increasing nonnegative function, it follows that

$$\frac{r-1}{r} \frac{N(\alpha(r), \phi_1, \phi_2)}{\alpha(r)} \le \frac{N(r, \phi_1, \phi_2)}{r} \le \frac{r+1}{r} \frac{N(\beta(r), \phi_1, \phi_2)}{\beta(r)},$$

and therefore (11.2.36) follows from (11.2.35).

To establish the limit as $r \to \infty$ of $n_\omega(r, \phi_1, \phi_2)$ we use the monotonicity of $n_\omega(\cdot, \phi_1, \phi_2)$ to deduce for $\gamma > 1$ that

$$n_\omega(r, \phi_1, \phi_2) \log \gamma \le \int_r^{\gamma r} \frac{n_\omega(t, \phi_1, \phi_2)}{t} \, dt$$
$$= N(\gamma r, \phi_1, \phi_2) - N(r, \phi_1, \phi_2).$$

In view of (11.2.36) it follows that

$$\limsup_{r \to \infty} \frac{n_\omega(r, \phi_1, \phi_2)}{r} \le \frac{\gamma - 1}{\log \gamma} \frac{\sigma}{\pi},$$

and taking the limit as $\gamma \to 1$ on the right-hand side we obtain

$$\limsup_{r \to \infty} \frac{n_\omega(r, \phi_1, \phi_2)}{r} \le \frac{\sigma}{\pi}. \tag{11.2.37}$$

Similarly, for $0 < \gamma < 1$,

$$n_\omega(r, \phi_1, \phi_2)(-\log \gamma) \ge \int_{\gamma r}^r \frac{n_\omega(t, \phi_1, \phi_2)}{t} \, dt$$
$$= N(r, \phi_1, \phi_2) - N(\gamma r, \phi_1, \phi_2),$$

so that

$$\liminf_{r \to \infty} \frac{n_\omega(r, \phi_1, \phi_2)}{r} \ge \frac{\sigma}{\pi}. \tag{11.2.38}$$

The inequalities (11.2.37) and (11.2.38) complete the proof in case $n_{\phi_1, \phi_2} = 1$. $\square$

Lemma 11.2.27. *If ω is a sine type function of type σ, then there is a positive number M_1 such that*

$$|\omega'(\lambda)| \le M_1 e^{\sigma |\operatorname{Im} \lambda|}, \quad \lambda \in \mathbb{C},$$

and for each $\delta > 0$, the logarithmic derivative $\dfrac{\omega'}{\omega}$ of ω is bounded on $\mathbb{C} \setminus \Lambda_\delta$.

Proof. For $\lambda \in \mathbb{C}$ and $r > 0$, let $\Gamma_{\lambda,r}$ be the counter-clockwise circle with centre λ and radius r. By Cauchy's integral formula and (11.2.30),

$$|\omega'(\lambda)| = \left| \frac{1}{2\pi i} \oint_{\Gamma_{\lambda,1}} \frac{\omega(z)}{(z-\lambda)^2} \, dz \right| \leq M e^{\sigma} e^{\sigma |\operatorname{Im} \lambda|}.$$

The second statement follows from this estimate and (11.2.33). $\square$

Notation 11.2.28. In view of Proposition 11.2.26 a sine type function ω has infinitely many zeros with positive real part and infinitely many zeros with negative real part. We therefore can write the nonzero zeros of ω, counted with multiplicity, as a sequence $(\lambda_k)_{k=-\infty}^{\infty}$ which is indexed over all integers such that $\operatorname{Re}\lambda_k \leq \operatorname{Re}\lambda_{k+1}$. For convenience we may also assume that $\operatorname{Re}\lambda_k \geq 0$ if $k \geq 0$ and that $\operatorname{Re}\lambda_k \leq 0$ if $k < 0$.

Lemma 11.2.29. *Observing Notation 11.2.28, it is true that for each sine type function ω,*

$$\lim_{n \to \infty} \sum_{k=-n}^{n} \frac{1}{\lambda_k}$$

exists, and, there are $c \in \mathbb{C} \setminus \{0\}$ and $m \in \mathbb{N}_0$ such that

$$\omega(\lambda) = c\lambda^m \lim_{n \to \infty} \prod_{k=-n}^{n} \left(1 - \frac{\lambda}{\lambda_k}\right), \quad \lambda \in \mathbb{C}.$$

Proof. Let σ be the type of ω and let $b_k = |\lambda_k|$ if $k \geq 0$ and $b_k = -|\lambda_k|$ if $k < 0$. Since the imaginary parts of the λ_k are uniformly bounded, it follows that the sequence $(b_k - \lambda_k)_{k=-\infty}^{\infty}$ is bounded. Observing that $(\lambda_k)_{k=-\infty}^{\infty}$ is pseudoregular, we conclude that also $(b_k)_{k=-\infty}^{\infty}$ is pseudoregular and that $(b_k^{-1} - \lambda_k^{-1})_{k=-\infty}^{\infty} \in \ell^1$, see Remark 11.2.10. From Proposition 11.2.26 and Proposition 11.2.8, part 2, we know that

$$\lim_{n \to \infty} \frac{b_n}{n} = \lim_{n \to \infty} \frac{-b_{-n}}{n} = \frac{\pi}{\sigma}. \tag{11.2.39}$$

In order to account for repeated values in the sequence b_k, we let l_n^- be the smallest negative integer k such that $b_k \geq b_{-n}$ and l_n^+ be the largest positive integer k such that $b_k \leq b_n$. Since $-n - d + 1 \leq l_n^- \leq -n$ and $n \leq l_n^+ \leq n + d - 1$, with d from Proposition 11.2.8, part 2, we may replace the limits $-n$ and n in the sum and the product in the statement by l_n^- and l_n^+.

In view of Corollary 11.2.12 and Lemma 11.2.20, choosing $\delta > 0$ there sufficiently small, we can find for each sufficiently large positive integer n a circle γ_n whose centre lies on the real axis and which intersects the real axis inside the intervals $(b_{l_n^-} - 1, b_{l_n^-})$ and $(b_{l_n^+}, b_{l_n^+} + 1)$, respectively, such that γ_n lies in $\mathbb{C} \setminus \Lambda_\delta$. Hence, by Lemma 11.2.27, there are $n_0 > 0$ and $M_2 > 0$ such that

$$\left| \frac{\omega'(\lambda)}{\omega(\lambda)} \right| \leq M_2 \text{ for all } \lambda \in \gamma_n \text{ and all integers } n > n_0. \tag{11.2.40}$$

We may also assume that all λ_k with $l_n^- \leq k \leq l_n^+$ lie inside γ_n. Let Δ_n be the set of remaining indices k for which λ_k is inside γ_n. In view of Lemma 11.2.7, the number of elements in Δ_n has a bound which is independent of n. Therefore

$$\lim_{n \to \infty} \left| \sum_{k \in \Delta_n} \frac{1}{\lambda_k} \right| = 0.$$

In view of the residue theorem, we have

$$\eta_n := \sum_{k=l_n^-}^{l_n^+} \frac{1}{\lambda_k} = \frac{1}{2\pi i} \oint_{\gamma_n} \frac{\omega'(\lambda)}{\lambda \omega(\lambda)} \, d\lambda - \sum_{k \in \Delta_n} \frac{1}{\lambda_k} - \operatorname{res}_{\lambda=0} \frac{\omega'(\lambda)}{\lambda \omega(\lambda)},$$

where

$$\left| \frac{1}{2\pi i} \oint_{\gamma_n} \frac{\omega'(\lambda)}{\lambda \omega(\lambda)} \, d\lambda \right| \leq \frac{\max\{b_n, -b_{-n}\}}{\min\{b_n, -b_{-n}\}} M_2 \to M_2 \text{ as } n \to \infty$$

by (11.2.39) and (11.2.40). Therefore the sequence $(\eta_n)_{n=1}^\infty$ is bounded. Since ω has order 1, we know from (5.1.3) and Hadamard's factorization theorem, see [55, XI.3.4], that there are complex numbers a and b and a nonnegative integer m such that

$$\omega(\lambda) = \lambda^m e^{a\lambda+b} \prod_{k=-\infty}^\infty \left(1 - \frac{\lambda}{\lambda_k}\right) e^{\frac{\lambda}{\lambda_k}}.$$

For $n \in \mathbb{N}$ we can write

$$\omega(\lambda) = \lambda^m e^{(a+\eta_n)\lambda+b} \prod_{k=l_n^-}^{l_n^+} \left(1 - \frac{\lambda}{\lambda_k}\right) \prod_{k<l_n^-, k>l_n^+} \left(1 - \frac{\lambda}{\lambda_k}\right) e^{\frac{\lambda}{\lambda_k}}. \qquad (11.2.41)$$

In view of the boundedness of (η_n) we can choose a subsequence $(\eta_{n_j})_{j=1}^\infty$ of $(\eta_n)_{n=1}^\infty$ such that

$$\lim_{j \to \infty} \eta_{n_j} = \eta$$

exists. Since the infinite product on the right-hand side of (11.2.41) converges to 1 uniformly on each compact subset of $\mathbb{C}$, it follows that

$$\omega(\lambda) = \lambda^m e^{(a+\eta)\lambda+b} \lim_{j \to \infty} \prod_{k=l_{n_j}^-}^{l_{n_j}^+} \left(1 - \frac{\lambda}{\lambda_k}\right). \qquad (11.2.42)$$

The proof will be complete if we show that $a + \eta = 0$ because this shows that η does not depend on the chosen convergent subsequence as $n \to \infty$, so that

$$\lim_{n \to \infty} \eta_n = -a.$$

From Lemma 11.2.11 we know that

$$f(\lambda) = \prod_{k=-\infty}^{\infty} \left(1 - \frac{\lambda}{\lambda_k}\right)\left(1 - \frac{\lambda}{b_k}\right)^{-1}, \quad \lambda \in \mathbb{C} \setminus \{b_k : k \in \mathbb{Z}\},$$

defines a meromorphic function f, so that (11.2.42) leads to the representation

$$w(\lambda) = \lambda^m e^{(a+\eta)\lambda+b} f(\lambda) w_R(\lambda), \tag{11.2.43}$$

where

$$w_R(\lambda) = \lim_{j\to\infty} \prod_{k=l_{n_j}^-}^{l_{n_j}^+} \left(1 - \frac{\lambda}{b_k}\right). \tag{11.2.44}$$

We obtain for all $k \in \mathbb{Z}$ and $\lambda = re^{i\phi}$ with $0 < |\phi| < \pi$ and sufficiently large r that

$$\log\left(\left|1 - \frac{re^{i\phi}}{b_k}\right|^2\right) = \log\left(1 + \frac{r^2}{b_k^2} - \frac{2r\cos\phi}{b_k}\right)$$

$$= \log\left(1 + \frac{r^2}{b_k^2}\right) + \log\left(1 - 2\frac{\dfrac{r}{b_k}\cos\phi}{1 + \dfrac{r^2}{b_k^2}}\right). \tag{11.2.45}$$

In particular, for $\cos\phi = 0$, i.e., $\lambda = \pm ir$, we have

$$\log|w_R(\pm ir)| = \frac{1}{2}\sum_{k=-\infty}^{\infty} \log\left(1 + \frac{r^2}{b_k^2}\right). \tag{11.2.46}$$

On the other hand, it follows from (11.2.43), Proposition 11.2.23 and Lemma 11.2.11 that

$$\frac{\log|w_R(\pm ir)|}{r} = \frac{\log|w(\pm ir)|}{r} - \frac{\operatorname{Re} b + m\log r}{r} \mp \operatorname{Im}(a+\eta) - \frac{\log|f(\pm ir)|}{r}$$

$$\xrightarrow[r\to\infty]{} \sigma \mp \operatorname{Im}(a+\eta).$$

Since this limit must be the same along the positive and the negative imaginary semiaxis by (11.2.46), it follows that

$$\operatorname{Im}(a+\eta) = 0 \tag{11.2.47}$$

and that

$$\lim_{r\to\infty} \frac{1}{2r}\sum_{k=-\infty}^{\infty} \log\left(1 + \frac{r^2}{b_k^2}\right) = \sigma \tag{11.2.48}$$

exists.

We put

$$\omega_{R,n}(\lambda) = \prod_{k=l_n^-}^{l_n^+} \left(1 - \frac{\lambda}{b_k}\right), \quad \delta_n = \sum_{k=l_n^-}^{l_n^+} \frac{1}{b_k},$$

$$\sigma_n(r) = \frac{1}{2} \sum_{k=l_n^-}^{l_n^+} \log\left(1 + \frac{r^2}{b_k^2}\right), \quad \sigma_\infty(r) = \frac{1}{2} \sum_{k=-\infty}^{\infty} \log\left(1 + \frac{r^2}{b_k^2}\right),$$

we let $n(t)$ be the number of k for which $0 < b_k \le t$ for $t \ge 0$, and we let $n(t)$ be the number of k for which $0 > b_k \ge t$ for $t < 0$. Then it follows with the aid of (11.2.45) and integration by parts that

$$\log|\omega_{R,n}(re^{i\phi})| + r\delta_n\cos\phi = \int_{b_{-n}}^{b_n} \left[\frac{1}{2}\log\left(1 + \frac{r^2}{t^2} - \frac{2r\cos\phi}{t}\right) + \frac{r\cos\phi}{t}\right] dn(t)$$

$$= \sigma_n(r) + \int_{b_{-n}}^{b_n} \left[\frac{1}{2}\log\left(1 - 2\frac{\frac{r}{t}\cos\phi}{1 + \frac{r^2}{t^2}}\right) + \frac{r\cos\phi}{t}\right] dn(t)$$

$$= \sigma_n(r) + \frac{1}{2}\log\left(1 - 2\frac{\frac{r}{b_n}\cos\phi}{1 + \frac{r^2}{b_n^2}}\right)l_n^+ - \log\left(1 - 2\frac{\frac{r}{b_{-n}}\cos\phi}{1 + \frac{r^2}{b_{-n}^2}}\right)l_n^-$$

$$+ r\cos\phi\left(\frac{l_n^+}{b_n} - \frac{l_n^-}{b_{-n}}\right)$$

$$+ \int_{b_{-n}}^{b_n} \left[\frac{r(r^2 - t^2)\cos\phi}{(r^2 + t^2)(r^2 + t^2 - 2rt\cos\phi)} + \frac{r\cos\phi}{t^2}\right] n(t)\, dt.$$

At the beginning of this proof we have stated that $(b_k^{-1} - \lambda_k^{-1})_{k=-\infty}^{\infty} \in \ell^1$, and therefore

$$\delta = \lim_{j\to\infty} \delta_{n_j}$$

exists. Using (11.2.39), l'Hôpital's rule and the fact that

$$\frac{r(r^2 - t^2)\cos\phi}{(r^2 + t^2)(r^2 + t^2 - 2rt\cos\phi)} + \frac{r\cos\phi}{t^2} = O(t^{-3}) \text{ as } |t| \to \infty$$

and that $t^{-1}n(t)$ is bounded, it follows that

$$\log|\omega_R(re^{i\phi})| = -r\delta\cos\phi + \sigma_\infty(r)$$

$$+ r\cos\phi \int_{-\infty}^{\infty} \left[\frac{t(r^2 - t^2)}{(r^2 + t^2)(r^2 + t^2 - 2rt\cos\phi)} + \frac{1}{t}\right] \frac{n(t)}{t}\, dt.$$

Taking logarithms of the absolute values in (11.2.43) with $\lambda = re^{i\phi}$, substituting the above identity, dividing the resulting equation by r and taking the limit as $r \to \infty$ and observing Proposition 11.2.23, (11.2.47), Lemma 11.2.11 and (11.2.48), it follows that

$$\sigma|\sin\phi| = (a + \eta)\cos\phi - \delta\cos\phi + \sigma$$
$$+ \cos\phi \lim_{r\to\infty} \int_{-\infty}^{\infty} \left[\frac{t(r^2 - t^2)}{(r^2 + t^2)(r^2 + t^2 - 2rt\cos\phi)} + \frac{1}{t}\right] \frac{n(t)}{t}\, dt.$$

We consider integer values k for r, split off one part of the integral and substitute $t = k\tau$ for the remaining part, so that the integral becomes

$$\int_{-\infty}^{\infty} \left[\frac{\tau(1 - \tau^2)}{(1 + \tau^2)(1 + \tau^2 - 2\tau\cos\phi)} + \frac{1}{\tau}\left(1 - \chi_{[\frac{b-k}{k}, \frac{b_k}{k}]}(\tau)\right)\right] \frac{n(k\tau)}{k\tau}\, d\tau + \int_{b_{-k}}^{b_k} \frac{n(t)}{t^2}\, dt.$$

In the course of the above calculations, using integration by parts and the limiting process, we have seen that

$$\delta = \lim_{j\to\infty} \int_{b_{-n_j}}^{b_{n_j}} \frac{n(t)}{t^2}\, dt.$$

The sequence of the functions in the square brackets is absolutely integrable with integrable upper bound

$$\tau \mapsto \frac{|\tau|(1 - \tau^2)}{(1 + \tau^2)(1 + \tau^2 - 2\tau\cos\phi)} + \frac{1}{|\tau|}\left(1 - \chi_{[b_-,b_+]}(\tau)\right),$$

where

$$b_- = \sup_{k>0} \frac{b_{-k}}{k} < 0 \quad \text{and} \quad b_+ = \inf_{k>0} \frac{b_k}{k} > 0$$

in view of (11.2.39) and since $b_{-k} < 0$ and $b_k > 0$ for $k > 0$. Furthermore, the boundedness of $t^{-1}n(t)$ and Proposition 11.2.26 show that $(k\tau)^{-1}n(k\tau)$ is bounded and converges to $\pi^{-1}\sigma$ as $k \to \infty$ pointwise for all $\tau \neq 0$. Hence, by Lebesgue's dominated convergence theorem as $k = n_j \to \infty$ and by (11.2.39),

$$\sigma|\sin\phi| = (a + \eta)\cos\phi + \sigma$$
$$+ \frac{\sigma}{\pi}\cos\phi \int_{-\infty}^{\infty} \left[\frac{\tau(1 - \tau^2)}{(1 + \tau^2)(1 + \tau^2 - 2\tau\cos\phi)} + \frac{1}{\tau}\left(1 - \chi_{[-\frac{\pi}{\sigma}, \frac{\pi}{\sigma}]}(\tau)\right)\right] d\tau$$
$$= (a + \eta)\cos\phi + \sigma$$
$$+ \frac{\sigma}{\pi}\cos\phi \int_{0}^{\infty} \frac{\tau(1 - \tau^2)}{1 + \tau^2}\left[\frac{1}{1 + \tau^2 - 2\tau\cos\phi} - \frac{1}{1 + \tau^2 + 2\tau\cos\phi}\right] d\tau$$
$$= (a + \eta)\cos\phi + \sigma$$
$$+ \frac{\sigma}{\pi}\cos^2\phi \int_{0}^{\infty} \frac{4\tau^2(1 - \tau^2)}{(1 + \tau^2)[(1 + \tau^2)^2 - 4\tau^2\cos^2\phi]}\, d\tau.$$

Taking now, say, $0 < \phi < \frac{\pi}{2}$, and subtracting the corresponding expression for ϕ replaced with $\pi - \phi$, i.e., $\cos \phi$ replaced with $-\cos \phi$, we arrive at

$$0 = 2(a + \eta) \cos \phi,$$

which finally shows that $(a + \eta) = 0$. $\qquad\qquad\square$

11.3 Perturbations of sine type functions

In this section we are going to provide results on perturbations of sine type functions through perturbations of their zeros. The constant c in Lemma 11.2.29 is mostly immaterial in our considerations, so that, in general, we will consider only the case $c = 1$.

If we perturb zeros, then the conditions $\operatorname{Re} \lambda_k \leq \operatorname{Re} \lambda_{k+1}$ for all $k \in \mathbb{Z}$ will no longer be true, so that we may relax this ordering condition to the requirement that there is a positive integer k_0 such that $\operatorname{Re} \lambda_k \leq \operatorname{Re} \lambda_{k+k_0}$ for all $k \in \mathbb{Z}$. Such a condition will always hold if we replace λ_k with $\lambda_k + \mu_k$ for any bounded sequence $(\mu_k)_{k=-\infty}^\infty$ if $(\lambda_k)_{k=-\infty}^\infty$ is pseudoregular. By the same reasoning one can abandon the technical requirement that $\operatorname{Re} \lambda_k \geq 0$ if $k \geq 0$ and $\operatorname{Re} \lambda_k \leq 0$ if $k < 0$. Finally, we can omit some indices when indexing the sequence of zeros; notably, we may omit 0 so that each index k has a matching distinct index $-k$ and the product from $-n$ to n has $2n$ factors. To see that all these corresponding finite products from $-n$ to n have the same limit we just have to observe that they differ by at most a finite number n_0, independent on n, of factors of the form

$$\left(1 - \frac{\lambda}{c}\right) \quad \text{and} \quad \left(1 - \frac{\lambda}{d}\right)^{-1}$$

with $|c|, |d| \to \infty$ as $n \to \infty$ and that these factors converge to 1 uniformly on compact subsets of $\mathbb{C}$.

Furthermore, when perturbing the zeros of sine type functions, they may enter or leave the imaginary axis and in particular 0, so that it may be convenient to replace factors of the form $1 - \dfrac{\lambda}{a_k}$ with $\lambda - a_k$. If we do such a substitution for finitely many factors, then the function will differ from the original one by a nonzero factor, which is immaterial for the properties of sine type function. Hence we may do such a substitution whenever it appears to be convenient.

However, in general it will be more convenient to do the following shift of the variable. In this way, all factors look formally the same, and we will therefore assume in the following generic lemmas and propositions that no zero is located on the imaginary axis.

Lemma 11.3.1. *Let ω be an entire function of the form*

$$\omega(\lambda) = \lambda^m \lim_{n \to \infty} \prod_{k=-n}^{n} \left(1 - \frac{\lambda}{\lambda_k} \right), \quad \lambda \in \mathbb{C}.$$

Choose a number $a \in \mathbb{C}$ such that $\operatorname{Im} a \neq \operatorname{Im} \lambda_k$ for all $k \in \mathbb{Z}$. Then

$$\hat{\omega}_a(\lambda) = \left(1 + \frac{\lambda - a}{a} \right)^m \lim_{n \to \infty} \prod_{k=-n}^{n} \left(1 - \frac{\lambda - a}{\lambda_k - a} \right), \quad \lambda \in \mathbb{C}.$$

defines an entire function $\hat{\omega}_a$, and $\hat{\omega}_a(\lambda)\omega(a) = \omega(\lambda)$ for all $\lambda \in \mathbb{C}$.

Proof. The statement immediately follows from

$$\left(1 - \frac{\lambda - a}{\lambda_k - a} \right) \left(1 - \frac{a}{\lambda_k} \right) = \left(1 - \frac{\lambda}{\lambda_k} \right)$$

and

$$\left(1 + \frac{\lambda - a}{a} \right) a = \lambda. \qquad \square$$

Lemma 11.3.2. *Let $\mu \in \mathbb{C}$ be such that $\operatorname{Re} \mu \neq 0$ and define*

$$f(\lambda) = \left(1 - \frac{\lambda}{\mu} \right), \quad f_R(\lambda) = \left(1 - \frac{\lambda}{\operatorname{Re} \mu} \right).$$

Let $h > |\operatorname{Im} \mu|$. Then

$$|f_R(\lambda - ih)||f_R(-ih)|^{-1} \leq |f(\lambda)| \leq |f_R(\lambda + ih)| \tag{11.3.1}$$

for $\lambda \in \mathbb{C}$ with $\operatorname{Im} \lambda \geq h$ and

$$|f_R(\lambda + ih)||f_R(ih)|^{-1} \leq |f(\lambda)| \leq |f_R(\lambda - ih)| \tag{11.3.2}$$

for $\lambda \in \mathbb{C}$ with $\operatorname{Im} \lambda \leq -h$.

Proof. Replacing λ with $-\lambda$ and μ with $-\mu$ does not change f and f_R. Hence (11.3.2) follows from (11.3.1) via this substitution, and it suffices to prove (11.3.1). Let a, b, x be fixed real numbers with $b \geq h$ and $x \neq 0$ and consider the function

$$g(\gamma) = \frac{|(a + ib) - (x + i\gamma)|^2}{|x + i\gamma|^2} = \frac{(a - x)^2 + (b - \gamma)^2}{x^2 + \gamma^2}, \quad \gamma \in \mathbb{R}.$$

Differentiation leads to

$$\tilde{g}(\gamma) := \frac{1}{2} g'(\gamma)(x^2 + \gamma^2)^2 = (\gamma - b)(x^2 + \gamma^2) - \gamma[(a - x)^2 + (b - \gamma)^2]$$

$$= b\gamma^2 - \gamma((a - x)^2 + b^2 - x^2) - bx^2.$$

In particular,

$$\tilde{g}(0) = -bx^2 < 0 \quad \text{and} \quad \tilde{g}(b) = -b(a - x)^2 \le 0.$$

Since $\tilde{g}$ is a quadratic polynomial with positive leading term, the set where this function is not positive is a closed interval, and it follows that g is decreasing on the interval $[0, b]$. Putting $a = \operatorname{Re}\lambda$, $b = \operatorname{Im}\lambda$, $x = \operatorname{Re}\mu$ and, in turn, $\gamma = h$, $\gamma = \operatorname{Im}\mu$ and $\gamma = 0$ in case $\operatorname{Im}\mu \ge 0$ it follows that

$$\frac{|\lambda - (\operatorname{Re}\mu + ih)|}{|\operatorname{Re}\mu + ih|} \le \frac{|\lambda - \mu|}{|\mu|} \le \frac{|\lambda - \operatorname{Re}\mu|}{|\operatorname{Re}\mu|} \le \frac{|\lambda - (\operatorname{Re}\mu - ih)|}{|\operatorname{Re}\mu|},$$

where the last inequality is obvious because $\operatorname{Im}\lambda > 0$ gives $|\operatorname{Im}\lambda| < |\operatorname{Im}\lambda - (-h)|$. If now $\operatorname{Im}\mu < 0$, then the above estimates can be applied to $\overline{\mu}$, so that

$$\frac{|\lambda - (\operatorname{Re}\mu + ih)|}{|\operatorname{Re}\mu + ih|} \le \frac{|\lambda - \overline{\mu}|}{|\overline{\mu}|} \le \frac{|\lambda - \mu|}{|\mu|} \le \frac{|\lambda - \mu|}{|\operatorname{Re}\mu|} \le \frac{|\lambda - (\operatorname{Re}\mu - ih)|}{|\operatorname{Re}\mu|},$$

where the last three estimates are straightforward estimates of the numerator and denominator, respectively. This leads to common estimates for all $\mu, \lambda \in \mathbb{C}$ with $\operatorname{Re}\mu \ne 0$, $|\operatorname{Im}\mu| < h$ and $\operatorname{Im}\lambda \ge h$:

$$\frac{|\lambda - (\operatorname{Re}\mu + ih)|}{|\operatorname{Re}\mu + ih|} \le \frac{|\lambda - \mu|}{|\mu|} \le \frac{|\lambda - (\operatorname{Re}\mu - ih)|}{|\operatorname{Re}\mu|},$$

which can be written as

$$\left|1 - \frac{\lambda - ih}{\operatorname{Re}\mu}\right|\left|1 + \frac{ih}{\operatorname{Re}\mu}\right|^{-1} \le \left|1 - \frac{\lambda}{\mu}\right| \le \left|1 - \frac{\lambda + ih}{\operatorname{Re}\mu}\right|.$$

The proof of (11.3.1) is complete. $\qquad\square$

Lemma 11.3.3 ([176, Lemma 1]). *Let $(\lambda_k)_{k=-\infty}^{\infty}$ be a sequence in $\mathbb{C}$ such that $\operatorname{Re}\lambda_k \ne 0$ for all $k \in \mathbb{Z}$. Then for the infinite product*

$$\omega(\lambda) = \lim_{n \to \infty} \prod_{k=-n}^{n} \left(1 - \frac{\lambda}{\lambda_k}\right) \tag{11.3.3}$$

to converge and represent a sine type function ω it is necessary and sufficient that $\sup_{k \in \mathbb{Z}} |\operatorname{Im}\lambda_k| < \infty$ and that the infinite product

$$\omega_R(\lambda) = \lim_{n \to \infty} \prod_{k=-n}^{n} \left(1 - \frac{\lambda}{\operatorname{Re}\lambda_k}\right) \tag{11.3.4}$$

converges and represents a sine type function ω_R. In this case, the types of ω and ω_R coincide.

Proof. If either condition is satisfied, then both $(\lambda_k)_{k=-\infty}^{\infty}$ and $(\operatorname{Re}\lambda_k)_{k=-\infty}^{\infty}$ are pseudoregular and $(\lambda_k - \operatorname{Re}\lambda_k)_{k=-\infty}^{\infty}$ is bounded. By Lemma 11.2.11, the infinite product (11.3.3) converges if and only if the infinite product (11.3.4) converges. From Lemma 11.3.2 we know for $h > \sup\limits_{k\in\mathbb{Z}}|\operatorname{Im}\lambda_k| < \infty$ that

$$|\omega_R(\lambda - ih)||\omega_R(-ih)|^{-1} \le |\omega(\lambda)| \le |\omega_R(\lambda + ih)|$$

for $\lambda \in \mathbb{C}$ with $\operatorname{Im}\lambda \ge h$ and

$$|\omega_R(\lambda + ih)||\omega_R(ih)|^{-1} \le |\omega(\lambda)| \le |\omega_R(\lambda - ih)|$$

for $\lambda \in \mathbb{C}$ with $\operatorname{Im}\lambda \le -h$. This shows that ω satisfies inequalities of the form (11.2.30) if and only if ω_R does, with different constants h, m, and M, in general, but with the same σ.

Also, if one of the functions (11.3.3) or (11.3.4) represents a sine type function, then its order is 1, and (11.2.10), (11.2.11) and the maximum principle imply that then also the other function has order 1. An application of Proposition 11.2.19 completes the proof. $\qquad\square$

The following result is a special case of [176, Lemma 3].

Lemma 11.3.4. *Let ω be a sine type function with $h > 0$ as in Definition 11.2.5 (i). In the half-plane $\operatorname{Im}\lambda > h$ choose one sheet of $\arg\omega$. Then, for each $H > h$,*

$$\arg\omega_R(T + iH) - \arg\omega_R(iH) = -\pi n_\omega(T) + O(1), \quad T > 1,$$

where $n_\omega(T)$ is the number of zeros, counted with multiplicity, of ω in the rectangle $\{\lambda \in \mathbb{C} : 0 < \operatorname{Re}\lambda < T, |\operatorname{Im}\lambda| < h\}$.

Proof. According to Lemmas 11.3.3 and 11.3.1, $\omega_R(\lambda)$ is a sine type function. In view of Lemmas 11.2.12 and 11.2.20 there are $\delta > 0$ and $k_\delta > 0$ such that for each $T > 1$ there is $T' \in [T, T+1]$ such that $|\omega_R(\lambda)| \ge k_\delta$ for all $\lambda \in \mathbb{C}$ with $\operatorname{Re}\lambda = T'$. Now consider the rectangle $\{\lambda \in \mathbb{C} : 0 < \operatorname{Re}\lambda < T', |\operatorname{Im}\lambda| < H\}$ and the change of the argument of ω along its boundary. Since the change of argument of ω_R along a path equals the imaginary part of the integral over $\frac{\omega_R'}{\omega_R}$ along this path, it follows from above and from Lemma 11.2.27 that the change of argument of ω_R along the vertical sides of the rectangle has a bound which is independent of T' and thus of T. The same applies to the horizontal paths from $T \pm iH$ to $T' \pm iH$. Since ω_R is a real analytic function, it follows that

$$\arg\omega_R(T - iH) - \arg\omega_R(-iH) = -[\arg\omega_R(T + iH) - \arg\omega_R(iH)].$$

Together with the argument principle the above estimates and identities show that

$$\arg\omega_R(T + iH) - \arg\omega_R(iH) = -\pi n_{\omega_R}(T) + O(1).$$

Observing $n_{\omega_R}(T) = n_\omega(T)$ completes the proof. $\qquad\square$

Lemma 11.3.5. *Let ω be a sine type function with $h > 0$ as in Definition 11.2.5 (i). In the half-plane $\operatorname{Im}\lambda > h$ choose one sheet of $\arg\omega$. Then, for each $H > h$,*

$$\arg\omega(T + iH) - \arg\omega(iH) = -\pi n_\omega(T) + O(\log T), \quad T > 1,$$

where $n_\omega(T)$ is the number of zeros, counted with multiplicity, of ω in the rectangle $\{\lambda \in \mathbb{C} : 0 < \operatorname{Re}\lambda < T, |\operatorname{Im}\lambda| < h\}$.

Proof. It is convenient to take a particular sheet of $\arg\omega$. By Lemma 11.3.1 we may assume without loss of generality that the sine type function ω has no zeros on the imaginary axis, and we may also assume that $\omega(0) = 1$. Then letting Ω be a simply connected open neighbourhood of $\{\lambda \in \mathbb{C} : |\operatorname{Im}\lambda| \geq h\} \cup i\mathbb{R}$ which does not contain any zeros of ω, we choose the continuous branch of the argument of ω on Ω with $\arg\omega(0) = 0$. Then clearly

$$\arg\omega(\lambda) = \lim_{n\to\infty} \sum_{k=-n}^{n} \arg\left(1 - \frac{\lambda}{\lambda_k}\right), \tag{11.3.5}$$

where $\arg\left(1 - \frac{\lambda}{\lambda_k}\right)$ is the continuous branch on Ω whose argument is 0 at $\lambda = 0$.

To evaluate this latter argument, we consider $\zeta \in \mathbb{C}$ with $\operatorname{Re}\zeta \neq 0$ and $\operatorname{Im}\zeta < h$ and write

$$\arg\left(1 - \frac{\lambda}{\zeta}\right) = \arg(\zeta - \lambda) - \arg(\zeta),$$

where it is convenient to take $\arg(\zeta) \in (-\frac{3\pi}{2}, \frac{\pi}{2})$. As λ moves from 0 along the imaginary axis to iH with $H \geq h$, $\operatorname{Re}(\zeta - \lambda) = \operatorname{Re}\zeta \neq 0$, so that $\arg(\zeta - \lambda) \in (-\frac{3\pi}{2}, \frac{\pi}{2})$ for λ along the (positive) imaginary axis. Since $\operatorname{Im}(\zeta - iH) < 0$, it follows that $\arg(\zeta - iH) \in (-\pi, 0)$. Now the real part of $\zeta - (x + iH)$ is decreasing with $x \in \mathbb{R}$, while the imaginary part is a constant negative number. Hence $\arg(\zeta - (x + iH))$ is decreasing with increasing $x \in \mathbb{R}$. The following sketch

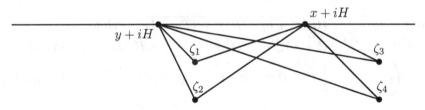

shows that for $y < x$ and $\operatorname{Im}\zeta < h$,

$$\arg(\zeta - (y + iH)) - \arg(\zeta - (x + iH)) \in (0, \pi)$$

is the angle subtended by the segment $[y + iH, x + iH]$ from the point ζ, and that for $y \leq \operatorname{Re}\zeta \leq x$, this angle decreases with decreasing $\operatorname{Im}\zeta$. However, the situation differs for $\operatorname{Re}\zeta$ outside $[y, x]$. Here the angle first increases with decreasing $\operatorname{Im}\zeta$ and then decreases.

The following sketch illustrates the generic case which we will investigate:

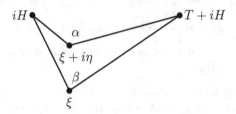

Here $T > 0$, $\eta < h < H$, and ξ is an arbitrary real number. Then

$$\arg(\xi + i\eta - (T + iH)) - \arg(\xi + i\eta - iH) = -\alpha,$$
$$\arg(\xi - (T + iH)) - \arg(\xi - iH) = -\beta.$$

Observing that the sine of the angle between two complex numbers a and b of modulus one is the imaginary part of $\bar{a}b$, it follows that

$$\sin(-\alpha - (-\beta))$$
$$= \operatorname{Im} \frac{(\xi + i\eta - (T + iH))\overline{(\xi + i\eta - iH)}\,\overline{(\xi - (T + iH))}(\xi - iH)}{|\xi + i\eta - (T + iH)|\,|\xi + i\eta - iH|\,|\xi - (T + iH)|\,|\xi - iH|}$$
$$= \operatorname{Im} \frac{[(\xi - T)\xi + (\eta - H)^2 + i(\eta - H)T]\,[(\xi - T)\xi + H^2 + iHT]}{|\xi + i\eta - (T + iH)|\,|\xi + i\eta - iH|\,|\xi - (T + iH)|\,|\xi - iH|}$$
$$= \frac{\eta T[(\xi - T)\xi + (\eta - H)H]}{|\xi + i\eta - (T + iH)|\,|\xi + i\eta - iH|\,|\xi - (T + iH)|\,|\xi - iH|}.$$

Clearly, $\beta - \alpha \in (-\pi, \pi)$ since $\alpha, \beta \in (0, \pi)$. However, we would like to have $\beta - \alpha \in (-\frac{\pi}{2}, \frac{\pi}{2})$, that is, $\cos(\beta - \alpha) > 0$. But $\cos(\beta - \alpha)$ is the real part of the above expression, so that we consider

$$\operatorname{Re}[(\xi + i\eta - (T + iH))\overline{(\xi + i\eta - iH)}\,\overline{(\xi - (T + iH))}(\xi - iH)]$$
$$= (\xi - T)^2\xi^2 + (\xi - T)\xi[(\eta - H)^2 + H^2] + (\eta - H)^2H^2 + (H - \eta)HT^2$$
$$= \left((\xi - T)\xi + \frac{1}{2}[(\eta - H)^2 + H^2]\right)^2 - \frac{1}{4}[(\eta - H)^2 - H^2]^2 + (H - \eta)HT^2,$$

which is positive if $T > 0$ is sufficiently large.

Putting now $\xi + i\eta = \lambda_k$, we obtain for

$$\gamma_k = \arg\left(1 - \frac{T + iH}{\lambda_k}\right) - \arg\left(1 - \frac{iH}{\lambda_k}\right)$$
$$- \left[\arg\left(1 - \frac{T + iH}{\operatorname{Re}\lambda_k}\right) - \arg\left(1 - \frac{iH}{\operatorname{Re}\lambda_k}\right)\right]$$

that

$$\lim_{n\to\infty} \sum_{k=-n}^{n} \gamma_k = \arg\omega(T+iH) - \arg\omega(iH) - [\arg\omega_R(T+iH) - \arg\omega_R(iH)]$$

(11.3.6)

and that

$$\sin\gamma_k =: c_k = \frac{\operatorname{Im}\lambda_k\, T[(\operatorname{Re}\lambda_k - T)\operatorname{Re}\lambda_k + (\operatorname{Im}\lambda_k - H)H]}{|\lambda_k - (T+iH)|\,|\lambda_k - iH|\,|\operatorname{Re}\lambda_k - (T+iH)|\,|\operatorname{Re}\lambda_k - iH|}.$$

With the above restrictions on H and T we can write

$$\gamma_k = \arcsin c_k = c_k + g(c_k)c_k^3,$$

where g is a bounded function on $[-1,1]$. There are positive constants C_1 and C_2 such that

$$|c_k| \le \frac{[|\operatorname{Re}\lambda_k - T| + |\operatorname{Re}\lambda_k|][C_1|\operatorname{Re}\lambda_k - T||\operatorname{Re}\lambda_k| + C_2]}{|\lambda_k - (T+iH)|\,|\lambda_k - iH|\,|\operatorname{Re}\lambda_k - (T+iH)|\,|\operatorname{Re}\lambda_k - iH|}.$$

Expanding the numerator on the right-hand side we obtain a sum of four terms where each of the factors in the numerator, apart from C_1 or C_2, is of the form $|\operatorname{Re} u|$ with a matching term $|u|$ in the denominator for some complex number u. Cancelling these pairs of numbers, the above estimate leads to

$$|c_k| \le \frac{C_1}{|\operatorname{Re}\lambda_k - iH|} + \frac{C_1}{|\operatorname{Re}\lambda_k - (T+iH)|}$$
$$+ \frac{C_2}{|\lambda_k - iH|\,|\operatorname{Re}\lambda_k - (T+iH)|\,|\operatorname{Re}\lambda_k - iH|}$$
$$+ \frac{C_2}{|\lambda_k - (T+iH)|\,|\operatorname{Re}\lambda_k - (T+iH)|\,|\operatorname{Re}\lambda_k - iH|}.$$

(11.3.7)

We observe that all factors in the denominators are bounded below by $H - h > 0$. Recall from Proposition 11.2.8, part 2, that the number of indices k with $|\lambda_k|$ inside an interval of length 1 has a bound which is independent of the location of that interval. Taking (11.2.39) into account, it follows that there is $\delta > 0$ and for each $t \in \mathbb{R}$ a number $\nu(t) \in \mathbb{Z}$ such that

$$|\lambda_k - t - iH| \ge \delta(|k - \nu(t)| + 1), \quad |\operatorname{Re}\lambda_k - t - iH| \ge \delta(|k - \nu(t)| + 1), \quad k \in \mathbb{Z}.$$

Here $\nu(t)$ is the index k for which $|\operatorname{Re}\lambda_k - t|$ assumes its minimum. We can now conclude that all four terms on the right-hand side of (11.3.7) give l_2 sequences whose l_2 norms are bounded as functions of T. Hence

$$\sum_{k=-\infty}^{\infty} |c_k|^3$$

converges with a limit which is bounded as a function of T. Also, the sums over the last two summands in (11.3.7) converge with limits which are bounded as functions of T, so that

$$\lim_{n\to\infty} \sum_{k=-n}^{n} \gamma_k = \lim_{n\to\infty} \sum_{k=-n}^{n} d_k + O(1),$$

where

$$d_k = \frac{T(\operatorname{Im}\lambda_k)(\operatorname{Re}\lambda_k - T)\operatorname{Re}\lambda_k}{|\lambda_k - (T+iH)|\,|\lambda_k - iH|\,|\operatorname{Re}\lambda_k - (T+iH)|\,|\operatorname{Re}\lambda_k - iH|}.$$

Clearly,

$$|d_k| \le \frac{C_1 T}{\delta^2(|k - \nu(T)| + 1)(|k| + 1)},$$

and

$$\sum_{k=-\infty}^{\infty} \frac{T}{(|k - \nu(T)| + 1)(|k| + 1)} = \sum_{k=0}^{\infty} \frac{T}{(k + \nu(T) + 1)(k + 1)}$$

$$+ \sum_{k=1}^{\nu(T)-1} \frac{T}{(\nu(T) - k + 1)(k + 1)} + \sum_{k=\nu(T)}^{\infty} \frac{T}{(k - \nu(T) + 1)(k + 1)}$$

$$= \frac{T}{\nu(T)} \sum_{k=0}^{\infty} \left(\frac{1}{k + 1} - \frac{1}{k + \nu(T) + 1} \right)$$

$$+ \frac{T}{\nu(T) + 2} \sum_{k=1}^{\nu(T)-1} \left(\frac{1}{\nu(T) - k + 1} + \frac{1}{k + 1} \right)$$

$$+ \frac{T}{\nu(T)} \sum_{k=\nu(T)}^{\infty} \left(\frac{1}{k - \nu(T) + 1} - \frac{1}{k + 1} \right)$$

$$= \frac{T}{\nu(T)} \sum_{k=0}^{\nu(T)-1} \frac{1}{k + 1} + \frac{2T}{\nu(T) + 2} \sum_{k=1}^{\nu(T)-1} \frac{1}{k + 1} + \frac{T}{\nu(T)} \sum_{k=\nu(T)}^{2\nu(T)-1} \frac{1}{k - \nu(T) + 1}$$

$$\le \frac{4T}{\nu(T)} \sum_{k=0}^{\nu(T)-1} \frac{1}{k + 1} \le \frac{4T}{\nu(T)} O(\log \nu(T))$$

$$= O(\log(T)),$$

where we have used that $\lim_{T\to\infty} \frac{T}{\nu(T)} = \frac{\pi}{\sigma}$ in view of (11.2.36). Therefore, (11.3.6) leads to

$$\arg\omega(T + iH) - \arg\omega(iH) = \arg\omega_R(T + iH) - \arg\omega_R(iH) + O(\log(T)),$$

which completes the proof in view of Lemma 11.3.4. $\qquad\square$

In [176, Lemma 3] the authors formulate the statement of Lemma 11.3.5 with $O(1)$ instead of $O(\log(T))$. However, the following example shows that this is incorrect and that indeed the asymptotic behaviour stated in Lemma 11.3.5 is sharp.

Example 11.3.6. We now consider the special case of a sequence $(\lambda_k)_{k=-\infty}^{\infty}$ given by $\lambda_k = 2k+1$ for $k \geq 0$ and $\lambda_k = 2k+1+i$ for $k < 0$. Choosing $H > 1$ it follows that the numbers d_k from the proof of Lemma 11.3.5 satisfy $d_k = 0$ for $k > 0$ and $d_k > 0$ for $k < 0$. Since

$$
\omega_R(\lambda) = \lim_{n\to\infty} \sum_{k=-n}^{n} \left(1 - \frac{\lambda}{2k+1}\right) = \cos\left(\frac{\pi}{2}\lambda\right),
$$

is a sine type function of type $\frac{\pi}{2}$, it follows from Lemma 11.3.3 that also

$$
\omega(\lambda) = \lim_{n\to\infty} \sum_{k=-n}^{n} \left(1 - \frac{\lambda}{\lambda_k}\right)
$$

is a sine type function of type $\frac{\pi}{2}$. From the reasoning in the proof of Lemma 11.3.5 we know that

$$
\Delta(T, H) = \arg\omega(T+iH) - \arg\omega(iH) - [\arg\omega_R(T+iH) - \arg\omega(iH)]
$$
$$
= \sum_{k=-\infty}^{-1} d_k + \delta(T, H),
$$

where δ is bounded with respect to $T > 0$ and all H in any bounded closed interval of $[2, \infty)$. Here, for $k < 0$,

$$
d_k = \frac{T(2k+1-T)(2k+1)}{|2k+1+i-(T+iH)||2k+1+i-iH||2k+1-(T+iH)||2k+1-iH|} \cdot \frac{1}{}
$$
$$
\geq \frac{T(2|k|-1+T)(2|k|-1)}{|2|k|-1+T+H||2|k|-1+H||2|k|-1+T+H||2|k|-1|}.
$$

For $k \geq H$ we conclude further that

$$
d_{-k} \geq \frac{T(2k-1+T)(2k-1)}{16|2k-1+T||2k-1||2k-1+T||2k-1|}
$$
$$
\geq \frac{T}{16(2k+T)(2k)}
$$
$$
= \frac{1}{16(2k)} - \frac{1}{16(2k+T)}.
$$

We assume now for convenience that T is a positive even integer. Then

$$\sum_{k=-\infty}^{-1} d_k \geq \frac{1}{16} \sum_{k=\lceil H \rceil}^{\infty} \left(\frac{1}{2k} - \frac{1}{2k+T} \right)$$

$$= \frac{1}{16} \sum_{k=\lceil H \rceil}^{\lceil H \rceil + \frac{T}{2} - 1} \frac{1}{2k}$$

$$= \frac{1}{32} \log T + O(1).$$

Altogether, it follows that

$$\Delta(T, H) \geq \frac{1}{40} \log T$$

for sufficiently large T, uniformly in $H \in [2, H_0]$ for any $H_0 > 2$. Next we observe that for $y > 0$ and $x \in \mathbb{R}$,

$$\cos(x + iy) = \cos x \cosh y - i \sin x \sinh y,$$

so that

$$\tan \arg \cos(x + iy) = - \tan x \tanh y.$$

Choosing a continuous branch of $\arg \cos$ in the open upper half-plane it follows that

$$\arg \cos(j\pi + iy) - \arg \cos(iy) = -j\pi$$

for all positive integers j. In particular,

$$\arg \omega_R(T + iH) - \arg \omega_R(iH) = -T \frac{\pi}{2}$$

for all even integers T. Altogether, we have

$$\arg \omega(T + iH) - \arg \omega(iH) \geq -T \frac{\pi}{2} + \frac{1}{40} \log T.$$

The claim in [176, Lemma 3] for this case is that

$$\arg \omega(T + iH) - \arg \omega(iH) = -T \frac{\pi}{2} + O(1)$$

for all $T > 0$. Hence this example shows that the claim is incorrect.

Corollary 11.3.7. *Let ω be a sine type function with $h > 0$ as in Definition 11.2.5, (i). In each of the half-planes $\operatorname{Im} \lambda > h$ and $\operatorname{Im} \lambda < -h$ choose one sheet of $\arg \omega$. Then, for each $H \in \mathbb{R}$ with $|H| > h$,*

$$\arg \omega_R(T + iH) - \arg \omega_R(iH) = - \operatorname{sgn}(TH)\pi n_\omega(T) + O(1|), \quad |T| > 1,$$
$$\arg \omega(T + iH) - \arg \omega(iH) = - \operatorname{sgn}(TH)\pi n_\omega(T) + O(\log |T|), \quad |T| > 1,$$

where $n_\omega(T)$ is the number of zeros, counted with multiplicity, of ω in the rectangle $\{\lambda \in \mathbb{C} : 0 < \operatorname{Re} \lambda < T, |\operatorname{Im} \lambda| < h\}$ if $T > 1$, $\{\lambda \in \mathbb{C} : -T < \operatorname{Re} \lambda < 0, |\operatorname{Im} \lambda| < h\}$ if $T < -1$.

Proof. The statements for $T > 1$ and $H > h$ are those of Lemmas 11.3.4 and 11.3.5. Now let $T > 1$ and $H < -h$. Observing that also $\overline{\omega}$ is a sine type function, it follows from Lemma 11.3.5 that

$$\arg \omega(T + iH) - \arg \omega(iH) = -\arg \overline{\omega}(T - iH) + \arg \overline{\omega}(-iH)$$
$$= \pi n_{\overline{\omega}}(T) + O(\log T) = \pi n_\omega(T) + O(\log T).$$

For $T < -1$ we apply the above to the function $\breve{\omega}$ defined by $\breve{\omega}(\lambda) = \omega(-\lambda)$ and obtain

$$\arg \omega(T + iH) - \arg \omega(iH) = \arg \breve{\omega}(-T - iH) - \arg \breve{\omega}(-iH)$$
$$= (\operatorname{sgn} H)\pi n_{\breve{\omega}}(-T) + O(\log(-T)) = (\operatorname{sgn} H)\pi n_\omega(T) + O(\log |T|).$$

A corresponding proof holds for ω_R. □

Taking the correction to [176, Lemma 3] in Lemma 11.3.5 into account, we obtain the following analogue of [176, Lemma 4].

Lemma 11.3.8. *Let $(\psi_k)_{k=-\infty}^\infty$ be a bounded sequence of complex numbers and let $(\lambda_k)_{k=-\infty}^\infty$ be the sequence of zeros of a sine type function ω. Assume that $\operatorname{Im} \lambda_k \neq 0$ and $\operatorname{Im}(\lambda_k + \psi_k) \neq 0$ for all $k \in \mathbb{Z}$. Then the function $\widetilde{\omega}$ defined by*

$$\widetilde{\omega}(\lambda) = \lim_{n \to \infty} \prod_{k=-n}^n \left(1 - \frac{\lambda}{\lambda_k + \psi_k}\right) \tag{11.3.8}$$

represents an entire function. The function $\widetilde{\omega}$ is a sine type function if and only if there is a line $\operatorname{Im} \lambda = H$ located in the exterior of the strip containing all the zeros of ω and $\widetilde{\omega}$ on which the real part of

$$\Phi = \log\left(\frac{\widetilde{\omega}}{\omega}\right)$$

is bounded.

If $\widetilde{\omega}$ is a sine type function, then

$$\operatorname{Im} \log\left(\frac{\widetilde{\omega}_R}{\omega_R}\right)(x+iH) = O(1) \text{ and } \operatorname{Im} \Phi(x+iH) = O(\log x) \text{ as } |x| \to \infty \tag{11.3.9}$$

on any line $\operatorname{Im} \lambda = H$ located in the exterior of the strip containing all the zeros of ω and $\widetilde{\omega}$.

Proof. It follows from Lemma 11.2.11 that $\widetilde{\omega}$ represents an entire function which has the same exponential type as ω in the upper and lower half-planes. Indeed, let f be defined as in Lemma 11.2.11 with $b_k = \lambda_k$ and $c_k = \lambda_k + \psi_k$, $k \in \mathbb{Z}$. Then $\widetilde{\omega} = f\omega$, and we may apply the notation $U_{\widetilde{\omega},\delta}$, see (11.2.18), to $\widetilde{\omega}$. Let σ be the type of ω. For all $\delta, \varepsilon > 0$ it follows from Lemmas 11.2.6 and 11.2.11 that there is $r_0 > 0$ such that

$$\log |\widetilde{\omega}(\lambda)| \leq (\sigma + \varepsilon)|\lambda|, \quad |\lambda| > r_0, \ |\lambda| \in U_{\widetilde{\omega}, \delta}.$$

Choosing $\delta > 0$ sufficiently small it follows as in the proof of Corollary 11.2.12 that for each $\lambda \in \mathbb{C}$ there is $r \in (|\lambda|, |\lambda| + 1) \cap U_{\widetilde{\omega}, \delta}$. Applying the Maximum Modulus Theorem to $\widetilde{\omega}$ on the disc with centre 0 and radius r, it follows that $\log |\widetilde{\omega}(\lambda)| \leq (\sigma + \varepsilon)r \leq (\sigma + \varepsilon)(|\lambda| + 1)$ for each $\lambda \in \mathbb{C}$ with $|\lambda| > r_0$. Hence the exponential type of $\widetilde{\omega}$ is at most σ. Finally, Lemma 11.2.11 applied to f and Lemma 11.2.17 applied to ω show that the exponential types of $\widetilde{\omega}$ in the upper and lower half-planes are at least σ.

In view of Proposition 11.2.19 and Definition 11.2.5 it is clear that $\widetilde{\omega}$ is a sine type function if and only if there are real numbers H and $0 < m < M$ such that

$$m \leq \left| \frac{\widetilde{\omega}(\lambda)}{\omega(\lambda)} \right| \leq M \quad \text{for } \lambda \in \mathbb{C} \text{ with } \operatorname{Im} \lambda = H.$$

But this condition means that the real part of Φ is bounded on the line $\operatorname{Im} \lambda = H$.

Because of Lemma 11.3.3 the same arguments apply to ω_R, $\widetilde{\omega}_R$ and Φ_R.

Now assume that $\widetilde{\omega}$ is a sine type function. Since $(\psi_k)_{k=-\infty}^{\infty}$ is bounded, it follows that $n_{\widetilde{\omega}}(x) - n_\omega(x) = O(1)$ for $|x| \to \infty$. Applying Corollary 11.3.7 to ω and $\widetilde{\omega}$ as well as ω_R and $\widetilde{\omega}_R$ we obtain (11.3.9). $\square$

Taking the above corrections to results from [176] into account, we obtain the following analogue of [176, Theorem 1].

Theorem 11.3.9. *Let $(\psi_k)_{k=-\infty}^{\infty}$ be a bounded sequence of complex numbers and let $(\lambda_k)_{k=-\infty}^{\infty}$ be the sequence of zeros of a sine type function ω of sine type σ. For the function $\widetilde{\omega}$ of the form (11.3.8) to be a sine type function it is sufficient that there exists an entire function ϕ of exponential type $\leq \sigma$ which is bounded on the real axis and which satisfies the following interpolation conditions at all points λ_k, $k \in \mathbb{Z}$, where q_k denotes the multiplicity of λ_k:*

$$\phi^{(p)}(\lambda_k) = 0 \quad \text{for } p = 0, \dots, q_k - 2, \quad \phi^{(q_k - 1)}(\lambda_k) = \frac{1}{q_k} \sum_{\substack{j=-\infty \\ \lambda_j = \lambda_k}}^{\infty} \psi_j \, \omega^{(q_k)}(\lambda_k).$$

$$(11.3.10)$$

If all ψ_k and λ_k are real, this sufficient condition is also necessary.

Proof. We start by constructing an auxiliary entire function ϕ_1 of exponential type $\leq \sigma$ which satisfies (11.3.10) for the two given sequences $(\psi_k)_{k=-\infty}^{\infty}$ and $(\lambda_k)_{k=-\infty}^{\infty}$. Replacing ω with a nonzero multiple of $\hat{\omega}$ defined by $\hat{\omega}(\lambda) = \omega(\lambda + \eta)$ for a suitable real number η, we may assume that $\lambda_k \neq 0$ and $\lambda_k + \psi_k \neq 0$ for all $k \in \mathbb{Z}$ and that ω is of the form (11.3.3). Because of $(\lambda_k^{-1})_{k=-\infty}^{\infty} \in \ell^2$,

$$\psi(\lambda) = \sum_{k=-\infty}^{\infty} \psi_k \left(\frac{1}{\lambda - \lambda_k} + \frac{1}{\lambda_k} \right) = \lambda \sum_{k=-\infty}^{\infty} \frac{\psi_k}{\lambda_k(\lambda - \lambda_k)} \qquad (11.3.11)$$

converges absolutely for all $\lambda \in \mathbb{C} \setminus \{\lambda_k : k \in \mathbb{Z}\}$. With the notation from Lemma 11.3.8 we can write

$$\Phi(\lambda) = \log \frac{\widetilde{\omega}(\lambda)}{\omega(\lambda)} = \sum_{k=-\infty}^{\infty} \left(\log \left(1 - \frac{\psi_k}{\lambda - \lambda_k} \right) - \log \left(1 + \frac{\psi_k}{\lambda_k} \right) \right)$$

for $|\operatorname{Im} \lambda| \geq H \geq h + 2A$ where $A = \sup_{k \in \mathbb{Z}} |\psi_k|$. Then $|\psi_k (\lambda - \lambda_k)^{-1}| < \frac{1}{2}$ is valid for all $k \in \mathbb{Z}$. Using that $|\log(1 - \tau) + \tau| \leq |\tau|^2$ for $|\tau| < \frac{1}{2}$ we obtain

$$|\Phi(\lambda) + \psi(\lambda)| \leq C + \sum_{k=-\infty}^{\infty} \frac{|\psi_k|^2}{|\lambda - \lambda_k|^2}, \tag{11.3.12}$$

where

$$C = \sum_{|\psi_k \lambda_k^{-1}| \geq \frac{1}{2}} \left| \log \left(1 + \frac{\psi_k}{\lambda_k} \right) - \frac{\psi_k}{\lambda_k} \right| + \sum_{k=-\infty}^{\infty} \left| \frac{\psi_k}{\lambda_k} \right|^2.$$

Denoting by $Q(\lambda)$ the sum of the series on the right-hand side of (11.3.12) and setting $\lambda = x + iy$ we obtain

$$Q(\lambda) \leq A^2 \sum_{k=-\infty}^{\infty} \frac{1}{|\lambda - \lambda_k|^2} = A^2 \sum_{s=-\infty}^{\infty} \sum_{\lambda_k \in \Pi(s,x)} \frac{1}{|\lambda - \lambda_k|^2} \tag{11.3.13}$$

where

$$\Pi(s, x) = \left\{ \zeta : x - s - \frac{1}{2} < \operatorname{Re} \zeta \leq x - s + \frac{1}{2}, |\operatorname{Im} \zeta| \leq H \right\}.$$

For all $s \in \mathbb{Z} \setminus \{0\}$ we have

$$\sum_{\lambda_k \in \Pi(s,x)} |\lambda - \lambda_k|^{-2} \leq \ell \left(\left(|s| - \frac{1}{2} \right)^2 + (|y| - h)^2 \right)^{-1},$$

where ℓ is an upper bound of the number of zeros of ω which can occur in any vertical strip of width 1, see Lemma 11.2.7. Therefore,

$$\sum_{k=-\infty}^{\infty} |\lambda - \lambda_k|^{-2} \leq \sum_{\lambda_k \in \Pi(0,x)} |\lambda - \lambda_k|^{-2} + 2\ell \sum_{s=1}^{\infty} \left(\left(s - \frac{1}{2} \right)^2 + (|y| - h)^2 \right)^{-1},$$
$$\tag{11.3.14}$$

and using, e. g., the integral test, (11.3.14) implies that there is a positive constant C_1 such that for $|y| > h$,

$$Q(x + iy) \leq \frac{C_1}{|y| - h}. \tag{11.3.15}$$

Thus it follows from (11.3.12) and (11.3.15) that the function

$$\Phi + \psi \text{ is bounded on } \{\lambda \in \mathbb{C} : |\operatorname{Im} \lambda| \geq H\}. \tag{11.3.16}$$

From (11.3.11) we find

$$|\psi(\lambda)| \le A|\lambda| \sum_{k=-\infty}^{\infty} \frac{1}{|\lambda_k||\lambda - \lambda_k|} \le A|\lambda| \left(\sum_{k=-\infty}^{\infty} \frac{1}{|\lambda - \lambda_k|^2} \right)^{\frac{1}{2}} \left(\sum_{k=-\infty}^{\infty} \frac{1}{|\lambda_k|^2} \right)^{\frac{1}{2}}.$$

For each $\delta > 0$ and $\lambda \in \mathbb{C} \setminus \Lambda_\delta$, where Λ_δ is defined in (11.2.31), we have

$$\sum_{k=-\infty}^{\infty} |\lambda - \lambda_k|^{-2} \le \sum_{\lambda_k \in \Pi(0,x)} |\lambda - \lambda_k|^{-2} + 2\ell \sum_{s=1}^{\infty} \left(s - \frac{1}{2} \right)^{-2}$$

$$\le \ell \delta^{-2} + 2\ell \sum_{s=1}^{\infty} \left(s - \frac{1}{2} \right)^{-2}, \qquad (11.3.17)$$

so that there is $C_2(\delta) > 0$ such that

$$|\psi(\lambda)| \le C_2(\delta)|\lambda| \quad \text{for } \lambda \in \mathbb{C} \setminus \Lambda_\delta.$$

From (11.3.15) we also infer $\psi(iy) = o(|y|)$ for $|y| \to \infty$. Taking Lemma 11.2.6 into account, it follows that the entire function $\phi_1 = \psi\omega$ satisfies the estimate $O(|\lambda|)$ on $\partial\Lambda_\delta$. For sufficiently small $\delta > 0$, all components of Λ_δ are bounded in view of Corollary 11.2.12, so that the Maximum Modulus Theorem gives that ϕ_1 satisfies the estimate $O(|\lambda|)$ on Λ_δ. These estimates of ϕ_1, ψ and ω lead to

$$|\phi_1(x + iy)| \le C_3|x + iy|e^{\sigma|y|}, \quad x, y \in \mathbb{R}, \qquad (11.3.18)$$

for a suitable constant C_3, and also $|\phi_1(iy)| = o(|y|e^{\sigma|y|})$ as $|y| \to \infty$. Hence ϕ_1 is a function of exponential type $\le \sigma$. Inserting the right-hand side of (11.3.11) for ψ and the Taylor expansion of ω about λ_k into $\phi_1 = \psi\omega$ proves (11.3.10) for $\phi = \phi_1$.

To prove sufficiency, let ϕ be a function of exponential type $\le \sigma$ which is bounded on the real axis and satisfies the interpolation condition (11.3.10). Then

$$\chi = \frac{\phi_1 - \phi}{\omega} = \psi - \frac{\phi}{\omega}$$

is an entire function. Since ϕ is bounded on the real axis and of exponential type $\le \sigma$, it follows from Lemma 11.2.6 that $|\phi(x + iy)| \le Ce^{\sigma|y|}$ for some $C > 0$ and all $x, y \in \mathbb{R}$. Then Lemma 11.2.20 and (11.3.18) show that $\chi(\lambda) = O(|\lambda|)$ on $\mathbb{C} \setminus \Lambda_\delta$ for each $\delta > 0$, and the Maximum Modulus Theorem gives that χ is a polynomial of degree not exceeding 1. Also, $\chi(iy) = o(|y|)$ as $|y| \to \infty$ since $\frac{\phi(iy)}{\omega(iy)} = O(1)$ and $\psi(iy) = o(|y|)$ as $|y| \to \infty$. It follows that χ is constant, say c, and therefore $\phi_1 = \phi + c\omega$ is bounded on the real axis. In view of Lemmas 11.2.6 and 11.2.20 it follows that $\psi = \frac{\phi_1}{\omega}$ is bounded on each line $\operatorname{Im} \lambda = H$ for sufficiently large $|H|$. Therefore Φ is bounded on each such line by (11.3.16), and Lemma 11.3.8 shows that $\widetilde{\omega}$ is a sine type function.

If $\widetilde{\omega}$ is of sine type and if all ψ_k and λ_k are real, then $\omega = \omega_R$ and $\widetilde{\omega} = \widetilde{\omega}_R$, and Lemma 11.3.8 gives that Φ is bounded on some line $\operatorname{Im}\lambda = H$, and therefore ψ has this property in view of (11.3.16). Since ω is also bounded on this line, see Lemma 11.2.6, it follows that $\phi = \phi_1 = \psi\omega$ is bounded on the line $\operatorname{Im}\lambda = H$. Again in view of Lemma 11.2.6 it follows that ϕ is bounded on each horizontal line and in particular on the real axis. □

Remark 11.3.10. In the proof of Theorem 11.3.9 we have seen that the function ϕ_1 is of exponential type $\leq \sigma$ and satisfies the interpolation condition (11.3.10). We also know that any entire function ϕ satisfying these properties is of the form $\phi_1 + c\omega$ for some $c \in \mathbb{C}$. Hence there exists such a ϕ which is bounded on the real axis if and only if ϕ_1 is bounded on the real axis.

Example 11.3.11. Denoting the entire functions ω and ω_R from Example 11.3.6 by $\widetilde{\omega}$ and ω, we have the assumptions of Theorem 11.3.9 satisfied with $\lambda_k = 2k - 1$ for $k \in \mathbb{Z}$, $\psi_k = i$ for $k \leq 0$ and $\psi_k = 0$ for $k > 0$. Then, again in the notation of Theorem 11.3.9, it follows that

$$\psi(\lambda) = \sum_{k=0}^{\infty} i \left(\frac{1}{\lambda + 2k + 1} - \frac{1}{2k + 1} \right)$$

and

$$\phi_1(\lambda) = i \cos\left(\frac{\pi}{2}\lambda \right) \sum_{k=0}^{\infty} \left(\frac{1}{\lambda + 2k + 1} - \frac{1}{2k + 1} \right).$$

It follows for positive integers j that

$$|\phi_1(4j)| = \left| \sum_{k=0}^{\infty} \left(\frac{1}{4j + 2k + 1} - \frac{1}{2k + 1} \right) \right|$$

$$= \sum_{k=0}^{2j-1} \frac{1}{2k + 1} \geq C \log j,$$

where C is a positive constant which is independent of j. From Remark 11.3.10 we conclude that there are sine type functions ω and $\widetilde{\omega}$ as in Theorem 11.3.9 for which there is no entire function of exponential type $\leq \sigma$ which satisfies (11.3.10) and is bounded on the real axis.

Corollary 11.3.12 ([176, Corollary, p. 85]). *If in Theorem 11.3.9, $(\psi_k)_{k=-\infty}^{\infty} \in l^p$ for some real number $p > 1$, then the function $\widetilde{\omega}$ defined by (11.3.8) is a sine type function.*

Proof. In view of Remark 11.3.10 it suffices to show that ϕ_1 is bounded on the real axis, which, by Lemma 11.2.6, is equivalent to the boundedness of ϕ_1 on any horizontal line. Since ω is a sine type function and $\phi_1 = \psi\omega$, it therefore suffices to prove that ψ defined by (11.3.11) is bounded on some line $\operatorname{Im}\lambda = H$. To this end, let $H \geq 1 + \sup_{k\in\mathbb{Z}}|\operatorname{Im}\lambda_k|$ and $x \in \mathbb{R}$. Then, by Hölder's inequality,

$$|\psi(x+iH)| \leq \left(\sum_{k=-\infty}^{\infty}|\psi_k|^p\right)^{\frac{1}{p}}\left(\sum_{k=-\infty}^{\infty}|\lambda_k|^{-p'}\right)^{\frac{1}{p'}}$$

$$+ \left(\sum_{k=-\infty}^{\infty}|\psi_k|^p\right)^{\frac{1}{p}}\left(\sum_{k=-\infty}^{\infty}|x+iH-\lambda_k|^{-p'}\right)^{\frac{1}{p'}},$$

where $\frac{1}{p} + \frac{1}{p'} = 1$. Similar to (11.3.13) we finally have that

$$\sum_{k=-\infty}^{\infty}|x+iH-\lambda_k|^{-p'} = \sum_{s=-\infty}^{\infty}\sum_{\lambda_k\in\Pi(s,x)}\frac{1}{|x+iH-\lambda_k|^{p'}}$$

$$\leq \ell + 2\ell\sum_{s=1}^{\infty}\left(\left(s-\frac{1}{2}\right)^2+1\right)^{-\frac{p'}{2}}$$

$$\leq 3\ell + 2\ell\sum_{k=1}^{\infty}k^{-p'} < \infty$$

is bounded with respect to x. $\square$

The following lemma is adapted from [174, Lecture 22.1, Theorem 1]. It should be noted that in [174], sine type functions are assumed to have simple zeros $(\lambda_k)_{k=-\infty}^{\infty}$ with $\inf\{|\lambda_k - \lambda_j| : k, j \in \mathbb{Z}, k \neq j\} > 0$.

Lemma 11.3.13. *Let* $(\lambda_k)_{k=-\infty}^{\infty}$ *be the sequence of the zeros of a sine type function* ω *of sine type* σ. *For* $d = (d_k)_{k=-\infty}^{\infty} \in l_p$, $p > 1$, *the function* ψ_d *is defined by*

$$\psi_d(\lambda) = \lim_{n\to\infty}\sum_{k=-n}^{n}\frac{d_k}{\lambda - \lambda_k},$$

where $\operatorname{Im}\lambda \geq H > A = \sup_{k=-\infty}^{\infty}\operatorname{Im}\lambda_k$. *Then* $d \mapsto \psi_d$ *is a bounded operator from* l_p
to the Hardy class H_p *in the half-plane* $\overline{\mathbb{C}}_H^+ = \{\lambda \in \mathbb{C} : \operatorname{Im}\lambda \geq H\}$, *i. e., for each* $d \in l_p$, ψ_d *is analytic in* $\overline{\mathbb{C}}_H^+$, *and there is* $C > 0$ *such that*

$$\sup_{y\geq H}\int_{-\infty}^{\infty}|\psi_d(x+iy)|^p\,dx \leq C\|d\|_p^p \tag{11.3.19}$$

for all $d \in l_p$, *where* $\|d\|_p$ *is the* l_p-*norm of the series* d. *Furthermore,* $\omega\psi_d$ *is an entire function of exponential type not exceeding* σ, *the corresponding series converges uniformly on any compact subset of* $\mathbb{C}$, *and* $\omega\psi_d|_{iy+\mathbb{R}}$ *converges in the sense of* $L_p(\mathbb{R})$ *for all* $y \in \mathbb{R}$.

Proof. In this proof we will refer to results which assume that $\operatorname{Re} \lambda_k \neq 0$ for all $k \in \mathbb{Z}$. This can be achieved by the transformation $\lambda \mapsto \lambda - a$ for a suitable real number a, which neither changes the assumptions nor the conclusions of this lemma. Similarly, in view of Lemma 11.3.3, we may apply the transformation $\lambda \mapsto \lambda + iA$, that is, we may assume that $A = 0$.

Let $d \in l_p$. The statement is trivial for $d = 0$, so that we may assume $d \neq 0$, that is, $d_k \neq 0$ for at least one $k \in \mathbb{Z}$. From Lemma 11.2.7 it follows that $|\lambda - \lambda_k|^{-p'} = O(|k|^{-p'})$ as $|k| \to \infty$ uniformly for all λ in any compact subset of $\overline{\mathbb{C}}_H^+$, and from Hölder's inequality

$$\sum_{k=-\infty}^{\infty} \left| \frac{d_k}{\lambda - \lambda_k} \right| \leq \left(\sum_{k=-\infty}^{\infty} |d_k|^p \right)^{\frac{1}{p}} \left(\sum_{k=-\infty}^{\infty} |\lambda - \lambda_k|^{-p'} \right)^{\frac{1}{p'}} \tag{11.3.20}$$

it follows that the series converges uniformly and absolutely on compact subsets of $\overline{\mathbb{C}}_H^+$. Hence ψ_d is an analytic function in $\overline{\mathbb{C}}_H^+$.

By Lemma 11.3.3, ω_R is a sine type function, and, with $\lambda = x + iy$, $x \in \mathbb{R}$, $y \geq H$,

$$\frac{\omega_R'(\lambda)}{\omega_R(\lambda)} = \sum_{k=-\infty}^{\infty} \frac{1}{\lambda - \operatorname{Re} \lambda_k}$$

$$= \sum_{k=-\infty}^{\infty} \frac{x - \operatorname{Re} \lambda_k}{(x - \operatorname{Re} \lambda_k)^2 + y^2} - i \sum_{k=-\infty}^{\infty} \frac{y}{(x - \operatorname{Re} \lambda_k)^2 + y^2},$$

is bounded in $\overline{\mathbb{C}}_H^+$ by Lemma 11.2.27. Putting

$$h(x, y) = \sum_{k=-\infty}^{\infty} \frac{1}{(x - \operatorname{Re} \lambda_k)^2 + y^2}, \quad x \in \mathbb{R}, \ y \geq H,$$

the imaginary part of the previous expression leads to

$$\eta := \sup_{x \in \mathbb{R}, \, y \geq H} y h(x, y) < \infty. \tag{11.3.21}$$

Next let

$$\psi_{d,R}(\lambda) = \lim_{n \to \infty} \sum_{k=-n}^{n} \frac{d_k}{\lambda - \operatorname{Re} \lambda_k}$$

and consider

$$\psi_{d,I}(\lambda) = \psi_d(\lambda) - \psi_{d,R}(\lambda) = \lim_{n \to \infty} \sum_{k=-n}^{n} \frac{i d_k \operatorname{Im} \lambda_k}{(\lambda - \lambda_k)(\lambda - \operatorname{Re} \lambda_k)}.$$

Clearly, $\psi_{d,R}$ is analytic in $\overline{\mathbb{C}}_H^+$ and, with $\lambda = x + iy$,

$$|\psi_{d,I}(\lambda)| \leq A_1 \sum_{k=-\infty}^{\infty} \frac{|d_k|}{(x - \operatorname{Re}\lambda_k)^2 + y^2}, \tag{11.3.22}$$

where $A_1 = \sup\{|\operatorname{Im}\lambda_k| : k \in \mathbb{Z}\}$. We are going to show that $\psi_{d,I}$ satisfies the estimate (11.3.19). By definition of h,

$$\mu_{x,y}(n) = \sum_{k=-\infty}^{n} \frac{1}{h(x,y)[(x - \operatorname{Re}\lambda_k)^2 + y^2]}$$

defines a probability measure on $\mathbb{Z}$ for all $x \in \mathbb{R}$ and $y \geq H$. Since $t \mapsto t^p$ is a convex function on $[0, \infty)$, Jensen's inequality, see [56, 19.4.13], gives

$$|\psi_{d,I}(x+iy)|^p \leq A_1^p (h(x,y))^{p-1} \sum_{k=-\infty}^{\infty} \frac{|d_k|^p}{(x - \operatorname{Re}\lambda_k)^2 + y^2},$$

and therefore, for $y \geq H$,

$$\int_{-\infty}^{\infty} |\psi_{d,I}(x+iy)|^p \, dx \leq A_1^p \left(\frac{\eta}{H}\right)^{p-1} \sum_{k=-\infty}^{\infty} |d_k|^p \int_{-\infty}^{\infty} \frac{dx}{x^2 + H^2} = C\|d\|_p^p, \tag{11.3.23}$$

where the constant C does not depend on y and d.

We still have to prove (11.3.19) for $\psi_{d,R}$, so that we may now assume that all λ_k are real. We are going to use a proof which follows along the lines of the proof for the Hilbert transform, [263, Theorem 101*], see also the proof of [189, Proposition 4.5.1]. Since each d_k can be written as $d_k = d_{k,1} - d_{k,2} + i(d_{k,3} - d_{k,4})$ with non-negative real numbers $d_{k,j}$, $j = 1, 2, 3, 4$, we may assume that all d_k are non-negative real numbers. It suffices to consider finite sums, i.e., the case that $Z = \{k \in \mathbb{Z} : d_k \neq 0\}$ is finite. The general case is easily obtained by a standard limiting process. Clearly, $\psi_d(\lambda) = O(|\lambda|^{-1})$ as $|\lambda| \to \infty$ since each of the finite number of summands has this property.

1. First we consider the case that p is not an odd integer. We set $C_1 = p2^{\frac{1}{2}(p-1)}$ and

$$C_2 = \sup \left\{ r > 0 : C_1(1 + r^{1-p}) - r \left|\cos\left(p\frac{\pi}{2}\right)\right| \geq 0 \right\} + 1,$$

which is finite since $r^{1-p} \to 0$ as $r \to \infty$ and $\cos(p\frac{\pi}{2}) \neq 0$. Then we write

$$\psi_d(\lambda) =: u(x,y) - iv(x,y),$$

where $x, y \in \mathbb{R}$, $y > H$, $\lambda = x + iy$, and u, v are real valued functions. Then

$$u(x,y) = \sum_{k \in Z} \frac{d_k(x - \lambda_k)}{(x - \lambda_k)^2 + y^2}, \quad v(x,y) = y \sum_{k \in Z} \frac{d_k}{(x - \lambda_k)^2 + y^2} > 0.$$

This shows that $\operatorname{Im} \psi_d(\lambda) < 0$ for all $\lambda \in \mathbb{C}$ with $\operatorname{Im} \lambda \geq H$. Hence

$$\lambda \mapsto (\psi_d(\lambda))^p = \exp\{p \log \psi_d(\lambda)\}$$

defines an analytic function on $\{\lambda \in \mathbb{C} : \operatorname{Im} \lambda \geq H\}$, where the argument of the logarithm is taken in $(-\pi, 0)$. For $R > 0$, we consider the contour integral

$$\oint (\psi_d(\lambda))^p \, d\lambda = 0$$

along the straight line from $-R + iy$ to $R + iy$ and along the semicircle above it. Since $\psi_d(\lambda) = O(|\lambda|^{-1})$, we obtain that

$$\int_{-R}^{R} (\psi_d(x + iy))^p \, dx = O(R^{-p}) \cdot \pi R = O(R^{1-p}) \quad \text{as } R \to \infty.$$

Hence

$$\int_{-\infty}^{\infty} (\psi_d(x + iy))^p \, dx = 0 \tag{11.3.24}$$

for all $y \geq H$. From

$$(\psi_d(x + iy))^p - (u(x, y))^p = p \int_{u(x,y)}^{\psi_d(x+iy)} z^{p-1} \, dz$$

we infer

$$\left| (\psi_d(x + iy))^p - (u(x, y))^p \right| \leq p v(x, y) \big(u(x, y)^2 + v(x, y)^2\big)^{\frac{1}{2}(p-1)}$$
$$\leq p v(x, y) \big(2 \max\{u(x, y)^2, v(x, y)^2\}\big)^{\frac{1}{2}(p-1)}$$
$$\leq C_1 \big(v(x, y) |u(x, y)|^{p-1} + v(x, y)^p\big).$$

With the aid of (11.3.24) we conclude

$$\left| \int_{-\infty}^{\infty} (u(x, y))^p \, dx \right| = \left| \int_{-\infty}^{\infty} \big((\psi_d(x + iy))^p - (u(x, y))^p\big) \, dx \right|$$
$$\leq C_1 \left(\int_{-\infty}^{\infty} v(x, y) |u(x, y)|^{p-1} \, dx + \int_{-\infty}^{\infty} v(x, y)^p \, dx \right).$$

Since

$$e^{ip\frac{\pi}{2}} u(x, y)^p = \begin{cases} e^{ip\frac{\pi}{2}} |u(x, y)|^p & \text{if } u(x, y) > 0, \\ e^{-ip\frac{\pi}{2}} |u(x, y)|^p & \text{if } u(x, y) < 0, \end{cases}$$

we obtain

$$\left| \cos\left(p\frac{\pi}{2}\right) \right| \int_{-\infty}^{\infty} |u(x, y)|^p \, dx = \left| \operatorname{Re}\left(e^{ip\frac{\pi}{2}} \int_{-\infty}^{\infty} u(x, y)^p \, dx\right) \right|$$
$$\leq \left| \int_{-\infty}^{\infty} u(x, y)^p \, dx \right|.$$

Then, by Hölder's inequality,

$$\left|\cos\left(p\frac{\pi}{2}\right)\right|\,\|u(\cdot,y)\|_p^p \leq C_1\{\|u(\cdot,y)^{\frac{p}{p'}}v(\cdot,y)\|_1 + \|v(\cdot,y)\|_p^p\}$$

$$\leq C_1\{\|u(\cdot,y)^{\frac{p}{p'}}\|_{p'}\|v(\cdot,y)\|_p + \|v(\cdot,y)\|_p^p\}$$

$$= C_1\{\|u(\cdot,y)\|_p^{p-1}\|v(\cdot,y)\|_p + \|v(\cdot,y)\|_p^p\}.$$

Dividing the above inequality by $\|u(\cdot,y)\|_p^{p-1}\|v(\cdot,y)\|_p$ if $u(\cdot,y)$ is not identically zero and setting $r = \|u(\cdot,y)\|_p\|v(\cdot,y)\|_p^{-1}$ we obtain $|\cos(p\frac{\pi}{2})|r \leq C_1(1 + r^{1-p})$. Hence $r \leq C_2 - 1$ by definition of C_2, i.e.,

$$\|u(\cdot,y)\|_p \leq (C_2 - 1)\|v(\cdot,y)\|_p, \tag{11.3.25}$$

which trivially holds if u is identically zero. Applying Hölder's inequality to

$$\left(\frac{d_k}{((x-\lambda_k)^2 + y^2)^{\frac{1}{p}}}\right)_{k\in Z} \in l^p \quad \text{and} \quad \left(\frac{1}{((x-\lambda_k)^2 + y^2)^{\frac{1}{p'}}}\right)_{k\in Z} \in l^{p'}$$

we obtain that

$$(v(x,y))^p \leq y^p \sum_{k\in Z}\frac{d_k^p}{(x-\lambda_k)^2 + y^2}\left(\sum_{k\in Z}\frac{1}{(x-\lambda_k)^2 + y^2}\right)^{p-1}$$

$$\leq \eta^{p-1}y\sum_{k\in Z}\frac{d_k^p}{(x-\lambda_k)^2 + y^2}.$$

Hence

$$\|v(\cdot,y)\|_p^p \leq \eta^{p-1}\sum_{k\in Z}d_k^p\int_{-\infty}^{\infty}\frac{y\,dx}{(x-\lambda_k)^2 + y^2} = \pi\eta^{p-1}\|d\|_p^p.$$

Together with (11.3.25) we infer

$$\|\psi_d(\cdot + iy)\|_p \leq C_2\pi^{\frac{1}{p}}\eta^{\frac{1}{p'}}\|d\|_p.$$

2. Now let p be an odd integer. Then the result follows from part 1 due to the Riesz convexity theorem, see [70, 6.10.11].

To prove the last statement, we can apply the transformation $\lambda \mapsto \lambda + iy$, so that we may assume $y = 0$. Putting $\chi = \omega\psi_d$, it follows as in (11.3.20) from

$$|\chi(\lambda)| \leq \left(\sum_{k=-\infty}^{\infty}|d_k|^p\right)^{\frac{1}{p}}\left(\sum_{k=-\infty}^{\infty}|\omega(\lambda)(\lambda-\lambda_k)^{-1}|^{p'}\right)^{\frac{1}{p'}} \tag{11.3.26}$$

that χ is an entire function. Here we have used that

$$\left|\frac{\omega(\lambda)}{\lambda - \lambda_k}\right| \leq \sup_{|\mu - \lambda_k| < |\lambda - \lambda_k|}|\omega'(\mu)|, \tag{11.3.27}$$

which shows in view of the boundedness of ω' on horizontal strips, see Lemma 11.2.27, that on each set $\{\lambda \in \mathbb{C} : |\lambda - \lambda_k| \le \delta\}$ the function $\lambda \mapsto |\omega(\lambda)(\lambda - \lambda_k)^{-1}|$ is bounded by a constant which is independent of $k \in \mathbb{Z}$ and $\delta \in (0, 1)$.

From

$$\frac{\omega(\lambda)}{\lambda - \lambda_k} - \frac{\omega(\lambda + iH)}{\lambda + iH - \lambda_k} = \frac{\omega(\lambda) - \omega(\lambda + iH)}{\lambda + iH - \lambda_k} + iH \frac{\omega(\lambda)}{(\lambda - \lambda_k)(\lambda + iH - \lambda_k)}$$

we see that $\chi - \chi(\cdot + iH) = \chi_1 + \chi_2$, where

$$\chi_1(\lambda) = (\omega(\lambda) - \omega(\lambda + iH)) \sum_{k=-\infty}^{\infty} \frac{d_k}{\lambda - (\lambda_k - iH)},$$

$$\chi_2(\lambda) = iH \sum_{k=-\infty}^{\infty} \frac{d_k \omega(\lambda)}{(\lambda - \lambda_k)(\lambda - (\lambda_k - iH))}.$$

The first part of this lemma shows that $\chi(\cdot + iH)$ is an L_p-function on the real axis. Since ω and $\omega(\cdot + iH)$ are sine type functions and therefore bounded on the real axis, we obtain that also $\chi_1 = (\omega - \omega(\cdot + iH))\chi(\cdot + iH)$ is an L_p-function on the real axis. In view of (11.3.27) and the boundedness of ω on horizontal strips we can find $C_1 > 0$ such that, for all $k \in \mathbb{Z}$ and $x \in \mathbb{R}$, $|\omega(x)(x - \lambda_k)^{-1}| \le C_1$ if $|x - \operatorname{Re} \lambda_k| < H - A$ and $|\omega(x)(x - \lambda_k)^{-1}| \le C_1 |x - \lambda_k|^{-1}$ if $|x - \operatorname{Re} \lambda_k| \ge H - A$. It is now easy to see that

$$|\omega(x)(x - \lambda_k)^{-1}| \le C_3[(x - \operatorname{Re} \lambda_k)^2 + (H - A)^2]^{-\frac{1}{2}}$$

for some $C_3 > 0$ and all $k \in \mathbb{Z}$ and $x \in \mathbb{R}$. Therefore,

$$|\chi_2(x)| \le C_3 \sum_{k=-\infty}^{\infty} \frac{|d_k|}{(x - \operatorname{Re} \lambda_k)^2 + (H - A)^2}, \quad x \in \mathbb{R}.$$

But the right-hand side is as in (11.3.22), so that (11.3.23) shows that also χ_2 is an L_p-function on the real axis.

Altogether, we have proved that $\chi = \chi(\cdot + iH) + \chi_1 + \chi_2$ is an L_p-function on the real axis, and the norm estimates in Lemma 11.3.13 which we have used in this proof show that the series $\omega \psi_d$ converges in the L_p-norm on the real axis.

Finally we are going to prove that the entire function $\chi = \omega \psi_d$ is of exponential type $\le \sigma$. Estimating the second factor on the right-hand side of (11.3.26), without $\omega(\lambda)$, as in (11.3.17), we obtain constants $C_2, C_3 > 0$ such that

$$\sum_{k=-\infty}^{\infty} \frac{1}{|\lambda_k - \lambda|^{p'}} \le C_2 \sum_{j=1}^{\infty} j^{-p'} = C_3 \tag{11.3.28}$$

for $\lambda \in \mathbb{C} \setminus \Lambda_1$. The same estimate is clearly true for $\lambda \in \Lambda_1$ if one omits the at most ℓ indices k from the sum for which $|\lambda - \lambda_k| < 1$. In view of (11.3.27), each of the remaining terms $\omega(\lambda)(\lambda - \lambda_k)^{-1}$ is bounded by the bound of ω' on Λ_1. Hence (11.3.26) shows that χ is of exponential type $\le \sigma$. $\qquad \square$

Theorem 11.3.14 ([175, Theorem A]). *Let ω be a sine type function of type σ with only simple zeros. Let $(\lambda_k)_{k=-\infty}^{\infty}$ be the sequence of the zeros of ω and assume that $(a_k)_{k=-\infty}^{\infty} \in l_p$, where $p > 1$. If $\inf\{|\lambda_k - \lambda_j| : k, j \in \mathbb{Z},\ k \neq j\} > 0$, then the series*

$$\omega(\lambda) \sum_{k=-\infty}^{\infty} \frac{a_k}{\omega'(\lambda_k)(\lambda - \lambda_k)}, \quad \lambda \in \mathbb{C}, \tag{11.3.29}$$

converges uniformly on any compact subset of $\mathbb{C}$ to an entire function of exponential type not exceeding σ and converges in the sense of $L_p(\mathbb{R})$ for real λ.

Proof. By assumption, there is $\delta > 0$ such that $|\lambda_k - \lambda_j| > 2\delta$ whenever $k \neq j$. In particular, all zeros of ω are simple. Hence, for all $k \in \mathbb{Z}$,

$$\frac{1}{\omega'(\lambda_k)} = \lim_{\lambda \to \lambda_k} \frac{\lambda - \lambda_k}{\omega(\lambda)} = \operatorname{res}\left(\frac{1}{\omega}, \lambda_k\right) = \frac{1}{2\pi i} \oint_{|\lambda - \lambda_k| = \delta} \frac{d\lambda}{\omega(\lambda)},$$

and hence it follows from Lemma 11.2.20 that there is a positive number k_δ, which is independent of k, such that

$$\left|\frac{1}{\omega'(\lambda_k)}\right| \leq \frac{\delta}{k_\delta}.$$

Therefore, setting

$$d_k = \frac{a_k}{\omega'(\lambda_k)}, \quad k \in \mathbb{Z},$$

we find that $(d_k)_{k=-\infty}^{\infty} \in l_p$. Applying Lemma 11.3.13 completes the proof. $\square$

Lemma 11.3.15 ([176, Lemma 5]). *Let $(\lambda_k)_{k=-\infty}^{\infty}$ be the sequence of the zeros of a sine type function ω of sine type σ. Assume that $\lambda_k \neq 0$ for all $k \in \mathbb{Z}$. Let $(\psi_k)_{k=-\infty}^{\infty}$ be a sequence of the form*

$$\psi_k = a\lambda_k^{-n} + b_k\lambda_k^{-n}, \tag{11.3.30}$$

where $n \in \mathbb{N}$, $a \in \mathbb{C}$, and $(b_k)_{k=-\infty}^{\infty} \in l_p$, $p > 1$. If $\lambda_k + \psi_k \neq 0$ for all $k \in \mathbb{Z}$, then the function $\widetilde{\omega}$ defined by

$$\widetilde{\omega}(\lambda) = \lim_{m \to \infty} \prod_{k=-m}^{m} \left(1 - \frac{\lambda}{\lambda_k + \psi_k}\right)$$

is a sine type function of sine type σ, which has a representation of the form

$$\widetilde{\omega}(\lambda) = C\omega(\lambda)(1 + B_1\lambda^{-1} + \cdots + B_n\lambda^{-n}) - Ca\omega'(\lambda)\lambda^{-n} + f_n(\lambda)\lambda^{-n}, \tag{11.3.31}$$

where

$$C = \prod_{k=-\infty}^{\infty} \left(1 + \psi_k\lambda_k^{-1}\right)^{-1},$$

$B_j \in \mathbb{C}$ for $j = 1, \ldots, n$, and f_n is an entire function of exponential type $\leq \sigma$, whose restriction to the real axis belongs to $L_p(\mathbb{R})$.

Proof. The infinite product C converges absolutely since $(\psi_k \lambda_k^{-1})_{k=-\infty}^{\infty} \in l_1$ in view of Lemma 11.2.7 and (11.3.30). The function $\widetilde{\omega}$ is a sine type function of type σ by Theorem 11.3.9 and Corollary 11.3.12 since $(\psi_k)_{k=-\infty}^{\infty} \in l_q$ for any $q > 1$. Let $A = \sup\limits_{k=-\infty} (|\operatorname{Im} \lambda_k| + |\psi_k|)$ and fix some $H > A$. Choosing suitable sheets of the logarithms considered below we have for $|\operatorname{Im} \lambda| \geq H$ that

$$\log \frac{\widetilde{\omega}(\lambda)}{C\omega(\lambda)} = \sum_{k=-\infty}^{\infty} \log \frac{\lambda_k + \psi_k - \lambda}{\lambda_k - \lambda} = \sum_{k=-\infty}^{\infty} \int_{\lambda_k}^{\lambda_k + \psi_k} \frac{d\tau}{\tau - \lambda},$$

where the integration is taken along the line segment joining the points λ_k and $\lambda_k + \psi_k$. From

$$\frac{1}{\tau - \lambda} = -\sum_{j=0}^{n-1} \frac{\tau^j}{\lambda^{j+1}} + \frac{\tau^n}{\lambda^n(\tau - \lambda)}$$

it follows that

$$\int_{\lambda_k}^{\lambda_k + \psi_k} \frac{d\tau}{\tau - \lambda} = -\sum_{j=0}^{n-1} \int_{\lambda_k}^{\lambda_k + \psi_k} \frac{\tau^j}{\lambda^{j+1}} d\tau + \frac{1}{\lambda^n} \frac{a}{\lambda_k - \lambda}$$
$$+ \frac{1}{\lambda^n} \left(\int_{\lambda_k}^{\lambda_k + \psi_k} \frac{\tau^n \, d\tau}{\tau - \lambda} - \frac{a}{\lambda_k - \lambda} \right).$$

In view of (11.3.30) we have

$$\int_{\lambda_k}^{\lambda_k + \psi_k} \tau^j \, d\tau = \frac{1}{j+1} [(\lambda_k + \psi_k)^{j+1} - \lambda_k^{j+1}] = \lambda_k^j \psi_k \left(1 + O\left(\frac{1}{|\lambda_k|} \right) \right)$$
$$= O(|\lambda_k|^{-2}), \quad j = 0, \dots, n-2;$$

$$\int_{\lambda_k}^{\lambda_k + \psi_k} \tau^{n-1} \, d\tau = \lambda_k^{n-1} \psi_k \left(1 + O\left(\frac{1}{|\lambda_k|} \right) \right) = \frac{a}{\lambda_k} + \frac{b_k}{\lambda_k} + O(|\lambda_k|^{-2}),$$

and therefore the numbers

$$A_{j+1} = \lim_{m \to \infty} \sum_{k=-m}^{m} \int_{\lambda_k}^{\lambda_k + \psi_k} \tau^j \, d\tau, \quad j = 0, \dots, n-1,$$

exist, where we have taken Proposition 11.2.26 and Lemma 11.2.29 into account. Observing that

$$\lim_{m \to \infty} \sum_{k=-m}^{m} \frac{a}{\lambda_k - \lambda} = -a \frac{\omega'(\lambda)}{\omega(\lambda)}$$

and setting

$$q(\lambda) = \lim_{m \to \infty} \sum_{k=-m}^{m} \left(\int_{\lambda_k}^{\lambda_k + \psi_k} \frac{\tau^n \, d\tau}{\tau - \lambda} - \frac{a}{\lambda_k - \lambda} \right),$$

it follows that

$$\log \frac{\widetilde{\omega}(\lambda)}{C\omega(\lambda)} = -\sum_{j=1}^{n} A_j \lambda^{-j} - a\lambda^{-n} \frac{\omega'(\lambda)}{\omega(\lambda)} + q(\lambda)\lambda^{-n}. \tag{11.3.32}$$

Next we are going to show that q belongs to the Hardy class H_p in the half-plane $\overline{\mathbb{C}}_H^+$. To this end we write $q = q_1 + q_2$, where

$$q_1(\lambda) = \lim_{m \to \infty} \sum_{k=-m}^{m} \int_{\lambda_k}^{\lambda_k + \psi_k} \frac{\tau^n (\lambda_k - \tau) d\tau}{(\tau - \lambda)(\lambda_k - \lambda)}, \quad q_2(\lambda) = \lim_{m \to \infty} \sum_{k=-m}^{m} \frac{c_k}{\lambda - \lambda_k},$$

with

$$c_k = a - \int_{\lambda_k}^{\lambda_k + \psi_k} \tau^n \, d\tau.$$

To estimate q_1, we note that

$$\left| \int_{\lambda_k}^{\lambda_k + \psi_k} \frac{\tau^n (\lambda_k - \tau) d\tau}{(\tau - \lambda)(\lambda_k - \lambda)} \right| \leq \frac{(|\lambda_k| + |\psi_k|)^n |\psi_k|}{|\lambda_k - \lambda|^2} \int_{\lambda_k}^{\lambda_k + \psi_k} \left| \frac{\lambda_k - \lambda}{\tau - \lambda} \right| |d\tau|. \tag{11.3.33}$$

For $\operatorname{Im} \lambda \geq H$, $k \in \mathbb{Z}$ and τ on the line segment from λ_k to $\lambda_k + \psi_k$, which can be written as $\tau = \lambda_k + t\psi_k$ with $0 \leq t \leq 1$, we conclude

$$|\lambda_k - \lambda| \geq H - |\operatorname{Im} \lambda_k| \geq H - A + |\psi_k|$$

and thus

$$\left| \frac{\lambda_k - \lambda}{\tau - \lambda} \right| = \left| 1 + t \frac{\psi_k}{\lambda_k - \lambda} \right|^{-1} \leq \left(1 - \frac{|\psi_k|}{|\lambda_k - \lambda|} \right)^{-1}$$

$$\leq \left(1 - \frac{|\psi_k|}{H - A + |\psi_k|} \right)^{-1} = \frac{H - A + |\psi_k|}{H - A}$$

$$\leq \frac{H}{H - A}.$$

It follows that the integral on the right-hand side of (11.3.33) is bounded by $H(H - A)^{-1} |\psi_k|$. Taking (11.3.30) into account, there is $C_1 > 0$ such that

$$|q_1(\lambda)| \leq \frac{H}{H - A} \sum_{k=-\infty}^{\infty} \frac{(1 + |\psi_k| \, |\lambda_k|^{-1})^n (|a| + |b_k|)^2}{|\lambda_k|^n |\lambda_k - \lambda|^2} \leq C_1 \sum_{k=-\infty}^{\infty} \frac{1}{|\lambda_k|^n |\lambda_k - \lambda|^2},$$

and applying Hölder's inequality with $\frac{1}{p} + \frac{1}{p'} = 1$ leads to

$$|q_1(\lambda)| \leq C_1 \left(\sum_{k=-\infty}^{\infty} \frac{1}{|\lambda_k|^{np} |\lambda_k - \lambda|^p} \right)^{\frac{1}{p}} \left(\sum_{k=-\infty}^{\infty} \frac{1}{|\lambda_k - \lambda|^{p'}} \right)^{\frac{1}{p'}}. \tag{11.3.34}$$

Clearly, each factor on the right-hand side is decreasing with increasing $\operatorname{Im}\lambda \geq H$, so that it suffices to prove that the right-hand side represents an L_p-function on the line $\operatorname{Im}\lambda = H$. The second factor on the right-hand side has been estimated in (11.3.28).

Hence we obtain

$$|q_1(\lambda)|^p \leq C_1^p C_3^{\frac{p}{p'}} \sum_{k=-\infty}^{\infty} \frac{1}{|\lambda_k|^{np}|\lambda_k - \lambda|^p}. \tag{11.3.35}$$

Integrating (11.3.35) along the line $\operatorname{Im}\lambda = H$ and using Fubini's theorem, see [108, (21.12) or (21.13)], applied to the Lebesgue measure on $\mathbb{R}$ and the counting measure on $\mathbb{Z}$, we can interchange integration and summation to arrive at

$$\int_{-\infty}^{\infty} |q_1(x+iH)|^p\, dx \leq C_1^p C_3^{\frac{p}{p'}} \sum_{k=-\infty}^{\infty} \frac{1}{|\lambda_k|^{np}} \int_{-\infty}^{\infty} \frac{dx}{|\lambda_k - x - iH|^p}$$

$$= C_1^p C_3^{\frac{p}{p'}} \sum_{k=-\infty}^{\infty} \frac{1}{|\lambda_k|^{np}} \int_{-\infty}^{\infty} \frac{dx}{|x + i(H - \operatorname{Im}\lambda_k)|^p}$$

$$\leq C_1^p C_3^{\frac{p}{p'}} \sum_{k=-\infty}^{\infty} \frac{1}{|\lambda_k|^{np}} \int_{-\infty}^{\infty} \frac{dx}{|x + i(H - A)|^p}.$$

Since $p > 1$ and $n \geq 1$, both the sum and the integral converge. We have thus shown that $q_1 \in H_p$ in $\overline{\mathbb{C}}_H^+$.

Turning our attention now to q_2, we obtain

$$c_k = a - \int_{\lambda_k}^{\lambda_k + \psi_k} \tau^n\, d\tau = a - \frac{1}{n+1}\left[(\lambda_k + \psi_k)^{n+1} - \lambda_k^{n+1}\right]$$

$$= a - \lambda_k^n \psi_k - \frac{1}{n+1} \sum_{j=0}^{n-1} \binom{n+1}{j} \lambda_k^j \psi_k^{n+1-j} = -b_k + O(|\lambda_k|^{-n-1}).$$

Hence $(c_k)_{k=-\infty}^{\infty} \in l^p$. Therefore q_2 belongs to the H_p class in $\overline{\mathbb{C}}_H^+$ by Lemma 11.3.13.

Since q belongs to the H_p class, q is bounded in $\overline{\mathbb{C}}_H^+$. This boundedness also follows immediately for q_1 from (11.3.35) since

$$\sum_{k=-\infty}^{\infty} \frac{1}{|\lambda_k|^{np}|\lambda_k - \lambda|^p} \leq \frac{1}{(H-A)^p} \sum_{k=-\infty}^{\infty} \frac{1}{|\lambda_k|^{np}} < \infty$$

and for q_2 from the estimate (11.3.20). Since $\frac{\omega'}{\omega}$ is bounded in $\overline{\mathbb{C}}_H^+$ by Lemma 11.2.27, we therefore conclude from (11.3.32) that

$$\log \frac{\widetilde{\omega}(\lambda)}{C\omega(\lambda)} = O(|\lambda|^{-1}).$$

Applying the exponential function, its Taylor series expansion leads to

$$\frac{\widetilde{\omega}(\lambda)}{C\omega(\lambda)} = 1 + \sum_{k=1}^{n} \frac{1}{k!} \left[\log \frac{\widetilde{\omega}(\lambda)}{C\omega(\lambda)} \right]^k + \frac{\vartheta(\lambda)}{\lambda^n}, \tag{11.3.36}$$

where ϑ is analytic in $\overline{\mathbb{C}}_H^+$ and satisfies $O(|\lambda|^{-1})$ there, which implies that ϑ is of H_p class and is bounded in $\overline{\mathbb{C}}_H^+$. Substituting (11.3.32) into the right-hand side of the above identity, a reasoning as above shows that

$$\frac{\widetilde{\omega}(\lambda)}{C\omega(\lambda)} = 1 + B_1\lambda^{-1} + \cdots + B_n\lambda^{-n} - a\lambda^{-n}\frac{\omega'(\lambda)}{\omega(\lambda)} + \frac{\kappa(\lambda)}{\lambda^n},$$

where κ belongs to the H_p class and is bounded in $\overline{\mathbb{C}}_H^+$ and $B_1, \ldots, B_n$ are complex numbers. This gives the equation (11.3.31) with $f_n = C\omega\kappa$. On multiplying (11.3.31) by λ^n we see that f_n is an entire function of exponential type not exceeding σ.

Replacing λ with $-\lambda$, we obtain a representation of $\widetilde{\omega}$ in the half-plane defined by $\operatorname{Im}\lambda \leq -H$. We observe that, with a suitable choice of the logarithm, we obtain the same values for A_j and thus B_j, $j = 1, \ldots, n$, as above. Therefore, the corresponding function f_n in this half-plane $\operatorname{Im}\lambda \leq -H$ is represented by the same entire function as in the half-plane $\overline{\mathbb{C}}_H^+$. In particular, f_n belongs to H_p in the half-plane $\operatorname{Im}\lambda \leq -H$.

Since κ is bounded in $\overline{\mathbb{C}}_H^+$, f_n is bounded on the line $\operatorname{Im}\lambda = H$. Then it follows in view of Lemma 11.2.6, part 2, that the entire function g_n defined by $g_n(\lambda) = f_n(\lambda)e^{i\sigma\lambda}$ is bounded in the half-plane $\overline{\mathbb{C}}_{-H}^+$. Therefore, for $\lambda = x + iy$ with $y > -H$,

$$g_n(\lambda) = \frac{1}{\pi} \int_{-\infty}^{\infty} \frac{y + H}{(x - t)^2 + (y + H)^2} g_n(t - iH)\, dt,$$

see [144, p. 107]. Since f_n is an L_p-function on the line $\operatorname{Im}\lambda = -H$, so is g_n. But then the above integral representation of g_n shows that $g_n(\cdot + iy) \in L_p(\mathbb{R})$ for all $y > -H$, see [144, pp. 111-112]. In particular, $g_n|_{\mathbb{R}}$, and then also $f_n|_{\mathbb{R}}$, are L_p-functions. $\square$

Remark 11.3.16. If in Lemma 11.3.15 we have $\lambda_k = 0$ or $\lambda_k + \psi_k = 0$ for some k and if ω and $\widetilde{\omega}$ are sine type functions with their sequences of zeros being $(\lambda_k)_{k=-\infty}^{\infty}$ and $(\lambda_k + \psi_k)_{k=-\infty}^{\infty}$, respectively, then (11.3.31) still holds, but with a differently defined nonzero complex number C. To see this, we observe the general representation of sine type functions given in Lemma 11.2.29. Then we apply Lemma 11.3.15 to the sequences of zeros $(\lambda_k + b)_{k=-\infty}^{\infty}$ and $(\lambda_k + \psi_k + b)_{k=-\infty}^{\infty}$ of $\omega(\cdot - b)$ and $\widetilde{\omega}(\cdot - b)$, respectively, for a suitable real number b. This leads to (11.3.31) with λ^{-j} replaced by $(\lambda + b)^{-j}$. Finally, we expand $(\lambda + b)^j$ in terms of λ^{-j} and observe that a bounded function of the form $O(|\lambda|^{-1})$ belongs to L_p.

Chapter 12

Inverse Sturm–Liouville Problems

We will need representations of solutions of the Sturm–Liouville equation and algorithms for recovering its potential q from two of its spectra, corresponding to two distinct sets of separated boundary conditions. These results are due to [178], see also [177], [180]. For the convenience of the reader and easy reference we recall these results from V.A. Marchenko [180], thereby adapting them to our notation and considering Sturm–Liouville problems on intervals $[0, a]$ with arbitrary $a > 0$. Other presentations of the inverse Sturm–Liouville problem can be found, e. g., in [177], [235], [282], [80].

12.1 Riemann's formula

This section is a rewrite of [180, Section 1.1]. The main improvement is that we allow for integrable potentials. This generalization is an exercise in [180] and there is no proof in [180].

Lemma 12.1.1. *Let $\alpha < \beta$ be real numbers and put*

$$D_0 = \{(\xi, \eta, \xi_0, \eta_0) : \alpha \le \eta_0 \le \eta \le \xi \le \xi_0 \le \beta\}.$$

Let q_0 be a locally integrable function on $\mathbb{R}^2$. For $f \in L_\infty(D_0)$ define

$$(Tf)(\xi, \eta, \xi_0, \eta_0) = \int_\xi^{\xi_0} \int_{\eta_0}^{\eta} q_0(\sigma, \tau) f(\sigma, \tau, \xi_0, \eta_0) \, d\tau \, d\sigma, \quad (\xi, \eta, \xi_0, \eta_0) \in D_0.$$

$$(12.1.1)$$

Then T is a bounded linear operator on $L_\infty(D_0)$ and $I + T$ is invertible. Denoting by 1 the function which is identically 1 on D_0, it follows that $g = (I + T)^{-1}1$ is the unique solution of $f = 1 - Tf$ on D_0. The function g is continuous.

Proof. Clearly, T is a linear operator, and Fubini's theorem shows that T maps $L_\infty(D_0)$ into itself. For each $M \in \mathbb{R}$, the standard norm on $L_\infty(D_0)$ is equivalent to the weighted norm given by

$$\|f\|_M = \operatorname{ess\,sup}\{|f(\xi,\eta,\xi_0,\eta_0)|e^{-M((\xi_0-\xi)+(\eta-\eta_0))} : (\xi,\eta,\xi_0,\eta_0) \in D_0\}.$$

For $f \in L_\infty(D_0)$ and $(\xi,\eta,\xi_0,\eta_0) \in D_0$ we estimate

$$|(Tf)(\xi,\eta,\xi_0,\eta_0)|e^{-M((\xi_0-\xi)+(\eta-\eta_0))}$$

$$\leq \int_\xi^{\xi_0} \int_{\eta_0}^\eta |q_0(\sigma,\tau)|\,|f(\sigma,\tau,\xi_0,\eta_0)|e^{-M((\xi_0-\xi)+(\eta-\eta_0))}\,d\tau\,d\sigma$$

$$\leq \int_\xi^{\xi_0} \int_{\eta_0}^\eta |q_0(\sigma,\tau)|e^{-M((\sigma-\xi)+(\eta-\tau))}\,d\tau\,d\sigma\,\|f\|_M.$$

Letting $\chi_{\xi,\eta,\xi_0,\eta_0}$ be the characteristic function of the set $\{(\sigma,\tau) : \eta_0 \leq \tau \leq \eta \leq \xi \leq \sigma \leq \xi_0\}$, and observing that $\chi_{\xi',\eta',\xi_0',\eta_0'}(\sigma,\tau) \to \chi_{\xi,\eta,\xi_0,\eta_0}(\sigma,\tau)$ as $(\xi',\eta',\xi_0',\eta_0') \to (\xi,\eta,\xi_0,\eta_0)$ for almost all (σ,τ), it follows from Lebesgue's dominated convergence theorem that the function f_M defined by

$$f_M(\xi,\eta,\xi_0,\eta_0) := \int_\xi^{\xi_0} \int_{\eta_0}^\eta |q_0(\sigma,\tau)|e^{-M((\sigma-\xi)+(\eta-\tau))}\,d\tau\,d\sigma$$

$$= e^{M(\xi-\eta)} \int_\xi^{\xi_0} \int_{\eta_0}^\eta |q_0(\sigma,\tau)|e^{-M(\sigma-\tau)}\,d\tau\,d\sigma$$

is continuous on D_0. For $(\xi,\eta),(\xi',\eta')$ with $\eta_0 \leq \eta' \leq \eta \leq \xi \leq \xi' \leq \xi_0$ we have

$$f_0(\xi,\eta,\xi_0,\eta_0) - f_0(\xi',\eta',\xi_0,\eta_0) \geq \int_\xi^{\xi'} \int_{\eta'}^\eta |q_0(\sigma,\tau)|\,d\tau\,d\sigma,$$

as can be easily seen from the following sketch:

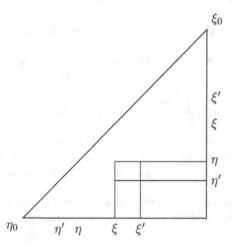

Since f_0 is a continuous function on the compact set D_0, f_0 is uniformly continuous, and it follows that there is $\delta > 0$ such that for all $(\xi, \eta, \xi_0, \eta_0), (\xi', \eta', \xi_0, \eta_0) \in D_0$ with $0 \leq \eta - \eta' \leq \delta$ and $0 \leq \xi' - \xi \leq \delta$ we have

$$\int_\xi^{\xi'} \int_{\eta'}^\eta |q_0(\sigma, \tau)| \, d\tau \, d\sigma \leq \frac{1}{4}.$$

Now let $(\xi, \eta, \xi_0, \eta_0) \in D_0$ and put $\eta' = \max\{\eta - \delta, \eta_0\}$ and $\xi' = \min\{\xi + \delta, \xi_0\}$. Denoting the rectangle with opposite vertices (ξ, η_0) and (ξ_0, η) by D_1 and the rectangle with opposite vertices (ξ, η') and (ξ', η) by D_2, it follows that $\sigma - \xi \geq \delta$ or $\eta - \tau \geq \delta$ for $(\sigma, \tau) \in D_1 \setminus D_2$. Hence we obtain

$$f_M(\xi, \eta, \xi_0, \eta_0) \leq \frac{1}{4} + \int_\xi^{\xi_0} \int_{\eta_0}^\eta |q_0(\sigma, \tau)| e^{-M\delta} \, d\tau \, d\sigma$$

$$\leq \frac{1}{4} + e^{-M\delta} \int_\alpha^\beta \int_\alpha^\sigma |q_0(\sigma, \tau)| \, d\tau \, d\sigma$$

$$\to \frac{1}{4} \text{ as } M \to \infty.$$

Hence we may choose M such that

$$f_M(\xi, \eta, \xi_0, \eta_0) \leq \frac{1}{2}, \quad (\xi, \eta, \xi_0, \eta_0) \in D_0.$$

Combining the above estimates we have shown that T is a contractive operator. Hence the operator $(I + T)$ is invertible, and the unique solution g of $f = 1 - Tf$ has the representation

$$g = (I + T)^{-1} 1 = \sum_{j=0}^\infty (-T)^j 1.$$

Since clearly T maps continuous function into continuous function, since 1 is continuous and since the set of continuous functions $C(D_0)$ is closed in $L_\infty(D_0)$, it follows that g is continuous. $\qquad \square$

Lemma 12.1.2. *Let q_0 be a locally integrable function on $\mathbb{R}^2$, let η_0 and ξ_0 be real numbers with $\eta_0 < \xi_0$ and let $D(\xi_0, \eta_0) = \{(\xi, \eta) : \eta_0 \leq \eta \leq \xi \leq \xi_0\}$. Then the problem*

$$r_{\xi\eta} - q_0 r = 0 \quad \text{on } D(\xi_0, \eta_0), \tag{12.1.2}$$

$$r(\xi_0, \eta) = r(\xi, \eta_0) = 1 \quad \text{for } \xi, \eta \in [\xi_0, \eta_0], \tag{12.1.3}$$

has a unique continuous solution r on $D(\xi_0, \eta_0)$. Furthermore, r_ξ, r_η, $r_{\xi\eta}$ and $r_{\eta\xi}$ exist and belong to $L_1(D(\xi_0, \eta_0))$, and $r_{\eta\xi} = r_{\xi\eta}$. If q_0 is continuously differentiable, then r has continuous second derivatives.

Proof. From Lemma 12.1.1 we know that the integral equation

$$r(\xi, \eta) = 1 - \int_{\xi}^{\xi_0} \int_{\eta_0}^{\eta} q_0(\sigma, \tau) r(\sigma, \tau) \, d\tau \, d\sigma. \tag{12.1.4}$$

has a solution g, and we write

$$r(\xi, \eta) = r(\xi, \eta; \xi_0, \eta_0) = g(\xi, \eta, \xi_0, \eta_0).$$

The existence of r_ξ and $r_{\xi\eta}$, their properties, and (12.1.2) and (12.1.3) easily follow from (12.1.4), so that this r is indeed a solution of (12.1.2) and (12.1.3).

Conversely, if r is a continuous solution r of problem (12.1.2), (12.1.3) where the partial derivates in (12.1.2) exist, integration of (12.1.2) with respect to η and taking into account that (12.1.3) implies $r_\xi(\xi, \eta_0) = 0$ for all $\xi \in [\eta_0, \xi_0]$ gives

$$r_\xi(\xi, \eta) = \int_{\eta_0}^{\eta} q_0(\xi, \tau) r(\xi, \tau) \, d\tau. \tag{12.1.5}$$

Integration with respect to ξ and (12.1.3) lead to (12.1.4). With fixed ξ_0 and η_0, the operator T from the proof of Lemma 12.1.1 becomes a contraction T_{ξ_0, η_0} on $L_\infty(D(\xi_0, \eta_0))$, and the uniqueness of the solution r of (12.1.2), (12.1.3) follows. $\qquad\square$

For real x_0 and y_0 with $y_0 \geq 0$ let D be the triangular region whose vertices are (x_0, y_0), $(x_0 - y_0, 0)$, $(x_0 + y_0, 0)$. We put $\xi_0 = x_0 + y_0$, $\eta_0 = x_0 - y_0$ and

$$q_0(\xi, \eta) = \frac{1}{4} \left[q_1 \left(\frac{\xi + \eta}{2} \right) - q_2 \left(\frac{\xi - \eta}{2} \right) \right], \quad \eta_0 \leq \eta \leq \xi \leq x_0. \tag{12.1.6}$$

The linear transformation $\xi = x + y$, $\eta = x - y$, maps the triangle with vertices (x_0, y_0), $(x_0 - y_0, 0)$, $(x_0 + y_0, 0)$ into the triangle with vertices (ξ_0, η_0), (η_0, η_0), (ξ_0, ξ_0), that is, it maps the triangle D to the triangle $D(\xi_0, \eta_0)$ defined in Lemma 12.1.2. Then let $(\xi, \eta) \mapsto r(\xi, \eta; \xi_0, \eta_0)$ be the solution according to Lemma 12.1.2 and define

$$R(x, y; x_0, y_0) = r(x + y, x - y; x_0 + y_0, x_0 - y_0), \quad (x, y) \in D. \tag{12.1.7}$$

The following theorem is a generalization of Riemann's theorem as stated in [180, Theorem 1.1.1].

Theorem 12.1.3. *Let q_1 and q_2 be locally integrable on $\mathbb{R}$ and let φ and ψ be continuous functions on $\mathbb{R}$. Let $u \in W_1^2(D)$ be a solution of*

$$u_{xx} - q_1(x)u = u_{yy} - q_2(y)u \tag{12.1.8}$$

such that u_x and u_y are continuous on D. Assume that u satisfies the initial conditions

$$u(x, 0) = \varphi(x), \quad u_y(x, 0) = \psi(x), \quad x_0 - y_0 \leq x \leq x_0 + y_0. \tag{12.1.9}$$

Then

$$u(x_0, y_0) = \frac{\varphi(x_0 + y_0) + \varphi(x_0 - y_0)}{2}$$

$$+ \frac{1}{2} \int_{x_0 - y_0}^{x_0 + y_0} \left(\psi(x) R(x, 0; x_0, y_0) - \varphi(x) R_y(x, 0; x_0, y_0) \right) dx. \quad (12.1.10)$$

Proof. We are going to use the transformation $\xi = x + y$, $\xi_0 = x_0 + y_0$, $\eta = x - y$, $\eta_0 = x_0 - y_0$. Expressing u as function $\tilde{u}$ in these new variables, i.e., $\tilde{u}(\xi, \eta) = u(x, y)$, we get

$$u_{xx} = \tilde{u}_{\xi\xi} + 2\tilde{u}_{\xi\eta} + \tilde{u}_{\eta\eta},$$
$$u_{yy} = \tilde{u}_{\xi\xi} - 2\tilde{u}_{\xi\eta} + \tilde{u}_{\eta\eta}.$$

Observing that the continuity of the partial derivatives of u and hence of $\tilde{u}$ gives $\tilde{u}_{\xi\eta} = \tilde{u}_{\eta\xi}$, we obtain

$$\tilde{u}_{\xi\eta} = \frac{1}{4}(u_{xx} - u_{yy}) = \frac{1}{4}(q_1(x) - q_2(y))u = q_0\tilde{u}. \quad (12.1.11)$$

We recall from Lemma 12.1.2 that (12.1.2), (12.1.3) with q_0 given by (12.1.6) has a unique solution r. Multiplying equations (12.1.11) and (12.1.2) by r and $\tilde{u}$, respectively, and then subtracting the second equation from the first equation we obtain

$$\tilde{u}_{\xi\eta}r - \tilde{u}r_{\xi\eta} = 0. \quad (12.1.12)$$

Observing

$$\frac{\partial}{\partial\eta}(\tilde{u}_\xi r) = \tilde{u}_{\xi\eta}r + \tilde{u}_\xi r_\eta, \quad \frac{\partial}{\partial\xi}(\tilde{u}_\eta r) = \tilde{u}_{\eta\xi}r + \tilde{u}_\eta r_\xi,$$

$$\frac{\partial}{\partial\eta}(\tilde{u}r_\xi) = \tilde{u}_\eta r_\xi + \tilde{u}r_{\xi\eta}, \quad \frac{\partial}{\partial\xi}(\tilde{u}r_\eta) = \tilde{u}_\xi r_\eta + \tilde{u}r_{\eta\xi},$$

we conclude that

$$\tilde{u}_{\xi\eta}r - \tilde{u}r_{\xi\eta} = \frac{1}{2}\left(\frac{\partial}{\partial\eta}(\tilde{u}_\xi r - \tilde{u}r_\xi) + \frac{\partial}{\partial\xi}(\tilde{u}_\eta r - \tilde{u}r_\eta)\right) \quad (12.1.13)$$

Integrating both sides of (12.1.12) over $D(\xi_0, \eta_0)$ and taking (12.1.13) into account we get

$$\iint_{D(\xi_0,\eta_0)} \left[\frac{\partial}{\partial\eta}(\tilde{u}_\xi r - \tilde{u}r_\xi) + \frac{\partial}{\partial\xi}(\tilde{u}_\eta r - \tilde{u}r_\eta)\right] d\xi \, d\eta = 0. \quad (12.1.14)$$

By Fubini's theorem we can integrate componentwise, and therefore the left-hand side is the sum of the two integrals

$$I_1 := \int_{\eta_0}^{\xi_0} [(\tilde{u}_\xi(\xi,\xi)r(\xi,\xi) - \tilde{u}(\xi,\xi)r_\xi(\xi,\xi))$$
$$- (\tilde{u}_\xi(\xi,\eta_0)r(\xi,\eta_0) - \tilde{u}(\xi,\eta_0)r_\xi(\xi,\eta_0))]d\xi$$
$$=: I_{11} - I_{12},$$

$$I_2 := \int_{\eta_0}^{\xi_0} [(\tilde{u}_\eta(\xi_0,\eta)r(\xi_0,\eta) - \tilde{u}(\xi_0,\eta)r_\eta(\xi_0,\eta))$$
$$- (\tilde{u}_\eta(\eta,\eta)r(\eta,\eta) - \tilde{u}(\eta,\eta)r_\eta(\eta,\eta))]d\eta$$
$$=: I_{21} - I_{22}.$$

Here we have used that, e. g., $\frac{\partial}{\partial \eta}(\tilde{u}_\xi r - \tilde{u}r_\xi)$ is integrable with respect to η for almost all ξ and that $\tilde{u}_\xi r - \tilde{u}r_\xi$ is continuous with respect to η for these ξ in view of (12.1.5) and the continuity assumption on u and its partial derivatives.

Integrating by parts and observing that $r(\xi,\eta) = 1$ if $\xi = \xi_0$ or $\eta = \eta_0$ we get

$$I_{12} = 2 \int_{\eta_0}^{\xi_0} \tilde{u}_\xi(\xi,\eta_0)r(\xi,\eta_0)\, d\xi - \tilde{u}(\xi_0,\eta_0)r(\xi_0,\eta_0) + \tilde{u}(\eta_0,\eta_0)r(\eta_0,\eta_0)$$
$$= 2 \int_{\eta_0}^{\xi_0} \tilde{u}_\xi(\xi,\eta_0)\, d\xi - \tilde{u}(\xi_0,\eta_0) + \tilde{u}(\eta_0,\eta_0)$$
$$= \tilde{u}(\xi_0,\eta_0) - \tilde{u}(\eta_0,\eta_0)$$
$$I_{21} = 2 \int_{\eta_0}^{\xi_0} \tilde{u}_\eta(\xi_0,\eta)r(\xi_0,\eta)\, d\eta - \tilde{u}(\xi_0,\xi_0)r(\xi_0,\xi_0) + \tilde{u}(\xi_0,\eta_0)r(\xi_0,\eta_0)$$
$$= 2 \int_{\eta_0}^{\xi_0} \tilde{u}_\eta(\xi_0,\eta)\, d\eta - \tilde{u}(\xi_0,\xi_0) + \tilde{u}(\xi_0,\eta_0)$$
$$= \tilde{u}(\xi_0,\xi_0) - \tilde{u}(\xi_0,\eta_0).$$

For $\xi = \eta$ we have $x + y = x - y$, so that $y = 0$ and $x = \xi$. Hence

$$I_{21} - I_{12} = -2\tilde{u}(\xi_0,\eta_0) + \tilde{u}(\xi_0,\xi_0) + \tilde{u}(\eta_0,\eta_0)$$
$$= -2u(x_0,y_0) + u(x_0 + y_0, 0) + u(x_0 - y_0, 0)$$
$$= -2u(x_0,y_0) + \varphi(x_0 + y_0) + \varphi(x_0 - y_0). \tag{12.1.15}$$

From

$$u_y(x,y) = \tilde{u}_\xi(x+y, x-y)\frac{\partial \xi}{\partial y} + \tilde{u}_\eta(x+y, x-y)\frac{\partial \eta}{\partial y}$$
$$= \tilde{u}_\xi(x+y, x-y) - \tilde{u}_\eta(x+y, x-y) \tag{12.1.16}$$

and the same equation for R defined by (12.1.7) and r we find

$$I_{11} - I_{22}$$

$$= \int_{\eta_0}^{\xi_0} [\tilde{u}_\xi(\xi,\xi)r(\xi,\xi) - \tilde{u}(\xi,\xi)r_\xi(\xi,\xi) - \tilde{u}_\eta(\xi,\xi)r(\xi,\xi) + \tilde{u}(\xi,\xi)r_\eta(\xi,\xi)]\,d\xi$$

$$= \int_{\eta_0}^{\xi_0} [\tilde{u}_\xi(\xi,\xi) - \tilde{u}_\eta(\xi,\xi)]r(\xi,\xi)\,d\xi - \int_{\eta_0}^{\xi_0} \tilde{u}(\xi,\xi)[r_\xi(\xi,\xi) - r_\eta(\xi,\xi)]\,d\xi$$

$$= \int_{\eta_0}^{\xi_0} u_y(x,0)R(x,0)\,dx - \int_{\eta_0}^{\xi_0} u(x,0)R_y(x,0)\,dx$$

$$= \int_{x_0-y_0}^{x_0+y_0} \psi(x)R(x,0)\,dx - \int_{x_0-y_0}^{x_0+y_0} \varphi(x)R_y(x,0)\,dx. \qquad (12.1.17)$$

Recall that $I_1 + I_2 = 0$ by (12.1.14). Hence the sum of (12.1.15) and (12.1.17) is zero, and solving this equation for $u(x_0, y_0)$ completes the proof. $\qquad\square$

For the solution r of (12.1.2), (12.1.3) we have already used the four variable notation $r(\xi, \eta; \xi_0, \eta_0)$, and r_ξ will denote the derivative with respect to the first variable, even if the first variable is denoted by a different symbol.

Corollary 12.1.4. *Let q_1 and q_2 be locally integrable on $\mathbb{R}$ and let φ be a continuously differentiable function on $\mathbb{R}$. Let $u \in W_1^2(D)$ be a solution of*

$$u_{xx} - q_1(x)u = u_{yy} - q_2(y)u \qquad (12.1.18)$$

such that u_x and u_y are continuous on D. Assume that u satisfies the initial conditions

$$u(x,0) = \varphi(x), \quad u_y(x,0) = \varphi'(x), \quad x_0 - y_0 \le x \le x_0 + y_0. \qquad (12.1.19)$$

Let r be the unique solution of (12.1.2), (12.1.3) with q_0 given by (12.1.6). Then

$$u(x_0, y_0) = \varphi(x_0 + y_0) - \int_{x_0-y_0}^{x_0+y_0} \varphi(x)r_\xi(x, x; x_0 + y_0, x_0 - y_0)\,dx. \qquad (12.1.20)$$

Proof. By Theorem 12.1.3, u has the representation (12.1.10) with $\psi = \varphi'$. Recall that R has been defined in (12.1.7), where r is the unique solution of (12.1.2), (12.1.3) with q_0 given by (12.1.6). Since $x_0 + y_0 = \xi_0$ and $x_0 - y_0 = \eta_0$, it follows that

$$R(x_0 \pm y_0, 0; x_0, y_0) = r(x_0 \pm y_0, x_0 \pm y_0; x_0 + y_0, x_0 - y_0) = 1.$$

As we have argued in the proof of Theorem 12.1.3, we may used integration by parts to arrive at

$$\int_{x_0-y_0}^{x_0+y_0} \varphi'(x)R(x,0;x_0,y_0)\,dx = \varphi(x_0 + y_0) - \varphi(x_0 - y_0)$$

$$- \int_{x_0-y_0}^{x_0+y_0} \varphi(x)R_x(x,0;x_0,y_0)\,dx.$$

From (12.1.16) for R_y and the corresponding formula for R_x we see that

$$R_x(x, 0; x_0, y_0) + R_x(x, 0; x_0, y_0) = 2r_\xi(x, x; x_0 + y_0, x_0 - y_0).$$

Substitution of these identities into (12.1.10) gives (12.1.20).   □

12.2   Solutions of Sturm–Liouville problems

Lemma 12.2.1 ([180, Lemma 1.4.3]). *Let $(a_k)_{k=-\infty}^{\infty}$ be a sequence of complex numbers of the form $a_k = \frac{2\pi}{a}k + b + h_k$, where $b \in \mathbb{C}$ and $h_k = O(k^{-1})$ for $k \to \pm\infty$, let $f \in L_2(0, a)$ and let*

$$\tilde{f}(\lambda) := \int_0^a f(x)e^{-i\lambda x}\,dx$$

be its Fourier transform. Then

$$\tilde{f}(a_k) = \tilde{f}\left(\frac{2\pi}{a}k + b\right) + k^{-1}g(k)$$

with $\left(\tilde{f}\left(\frac{2\pi}{a}k + b\right)\right)_{k=-\infty}^{\infty} \in l_2$ and $(g(k))_{k=-\infty}^{\infty} \in l_2$.

Proof. From the equality

$$\tilde{f}(a_k) = \int_0^a f(x)e^{-i(\frac{2\pi}{a}k+b)x}e^{-ih_k x}\,dx = \int_0^a f(x)e^{-i(\frac{2\pi}{a}k+b)x}[1 - ih_k x + O(h_k^2)]\,dx$$

it follows that

$$\tilde{f}(a_k) = \tilde{f}\left(\frac{2\pi}{a}k + b\right) + h_k\tilde{f}'\left(\frac{2\pi}{a}k + b\right) + O(h_k^2) = \tilde{f}\left(\frac{2\pi}{a}k + b\right) + k^{-1}g(k),$$

where

$$g(k) = kh_k\tilde{f}'\left(\frac{2\pi}{a}k + b\right) + k^{-1}O(k^2h_k^2).$$

Since

$$\tilde{f}\left(\frac{2\pi}{a}k + b\right) = \int_0^a f(x)e^{-ibx}e^{-2i\frac{\pi}{a}x}\,dx$$

and

$$\tilde{f}'\left(\frac{2\pi}{a}k + b\right) = -i\int_0^a f(x)xe^{-ibax}e^{-2i\frac{\pi}{a}kx}\,dx$$

are the Fourier coefficients of the functions $x \mapsto f(x)e^{-ibx}$ and $x \mapsto -if(x)xe^{-ibx}$, which belong to $L_2(0, a)$, Bessel's inequality implies that

$$\sum_{k=-\infty}^{\infty}\left|\tilde{f}\left(\frac{2\pi}{a}k + b\right)\right|^2 < \infty, \quad \sum_{k=-\infty}^{\infty}\left|\tilde{f}'\left(\frac{2\pi}{a}k + b\right)\right|^2 < \infty,$$

and hence also that $\sum_{k=-\infty}^{\infty}|g(k)|^2 < \infty$, because by assumption, $\sup_{k\in\mathbb{Z}}|kh_k| < \infty$.   □

Definition 12.2.2 ([281, Section 2.5]). An entire function ω of exponential type $\leq \sigma$ is said to belong to the Paley–Wiener class $\mathcal{L}^\sigma$ if its restriction to the real axis belongs to $L_2(-\infty, \infty)$.

Remark 12.2.3. For an entire function ω, let $\omega_e = \frac{1}{2}(\omega + \breve{\omega})$ and $\omega_o = \frac{1}{2}(\omega - \breve{\omega})$ be the even and odd parts of ω, where $\breve{\omega}(\lambda) = \omega(-\lambda)$. Clearly, ω belongs to $\mathcal{L}^\sigma$ if and only if ω_e and ω_o belong to $\mathcal{L}^\sigma$. We denote the sets of even and odd functions in $\mathcal{L}^\sigma$ by $\mathcal{L}^\sigma_e$ and $\mathcal{L}^\sigma_o$, respectively.

Lemma 12.2.4 (Plancherel's theorem). *The function ω belongs to $\mathcal{L}^\sigma$ if and only if it is of the form*

$$\omega(\lambda) = \int_0^\sigma \xi(t) \cos \lambda t\, dt + i \int_0^\sigma \zeta(t) \sin \lambda t\, dt, \quad \lambda \in \mathbb{C},$$

where $\xi, \zeta \in L_2(0, \sigma)$. Furthermore, $\lim\limits_{|\lambda| \to \infty} e^{-|\operatorname{Im} \lambda| a}|\omega(\lambda)| = 0$ *if $\omega \in \mathcal{L}^a$.*

Proof. The first statement can be found in [263, Theorems 48 and 50] or [281, Theorem 2.18], while the second statement easily follows from [180, Lemma 1.3.1]. □

Remark 12.2.5. In the notation of Lemma 12.2.4,

$$\omega_e(\lambda) = \int_0^\sigma \xi(t) \cos \lambda t\, dt, \quad \omega_o(\lambda) = i \int_0^\sigma \zeta(t) \sin \lambda t\, dt, \quad \lambda \in \mathbb{C}.$$

Consider the Sturm–Liouville equation

$$y'' - q(x)y + \lambda^2 y = 0 \tag{12.2.1}$$

on the interval $(0, a)$, where $0 < a < \infty$, $q \in L_2(0, a)$ is a real-valued function and λ is a complex parameter. Let $e_0(\lambda, x)$ denote the solution of equation (12.2.1) with initial data

$$e_0(\lambda, 0) = 1, \quad e_0'(\lambda, 0) = -i\lambda. \tag{12.2.2}$$

Theorem 12.2.6 ([180, Theorem 1.2.1]). *The solution $e_0(\lambda, \cdot)$ of the initial value problem (12.2.1), (12.2.2) admits the representation*

$$e_0(\lambda, x) = e^{-i\lambda x} + \int_{-x}^x \tilde{K}(x, t) e^{-i\lambda t} dt, \ 0 \leq x \leq a, \tag{12.2.3}$$

where

$$\tilde{K}(x, t) = -r_\xi(t, t; x, -x), \ 0 \leq x \leq a, \ |t| \leq x, \tag{12.2.4}$$

and r is the function defined in Lemma 12.1.2 with

$$q_0(\xi, \eta) = -\frac{1}{4} q\left(\frac{\xi - \eta}{2}\right). \tag{12.2.5}$$

Proof. The function

$$u(x, y) = e^{-i\lambda x} e_0(\lambda, y)$$

belongs locally to W_2^2, is continuously differentiable for $-\infty < x < \infty$, $0 \le y \le a$, and solves the Cauchy problem

$$u_{xx} = u_{yy} - q(y)u \tag{12.2.6}$$

with initial conditions

$$u(x, 0) = e^{-i\lambda x}, \quad u_y(x, 0) = -i\lambda e^{-i\lambda x}.$$

Corollary 12.1.4 gives that the value of the function u at (x_0, y_0) is given by

$$e^{-i\lambda x_0} e_0(\lambda, y_0) = e^{-i\lambda(x_0 + y_0)} - \int_{x_0 - y_0}^{x_0 + y_0} r_\xi(x, x; x_0 + y_0, x_0 - y_0) e^{-i\lambda x} dx.$$

Letting $x_0 = 0$ we get

$$e_0(\lambda, y_0) = e^{-i\lambda y_0} - \int_{-y_0}^{y_0} r_\xi(x, x; y_0, -y_0) e^{-i\lambda x} dx.$$

An obvious change in notation now proves (12.2.3). Since (12.2.6) is equation (12.1.18) with $q_1 = 0$ and $q_2 = q$, (12.2.5) follows from (12.1.6). $\qquad\square$

Proposition 12.2.7.

$$e_0(\lambda, x) = e^{-i\lambda x} + \int_0^x \frac{\sin \lambda(x - t)}{\lambda} q(t) e_0(\lambda, t) \, dt. \tag{12.2.7}$$

Proof. It is easy to see that

$$y(x) = e^{-i\lambda x} + \int_0^x \frac{\sin \lambda(x - t)}{\lambda} g(t) \, dt \tag{12.2.8}$$

is the solution of the differential equation $y'' + \lambda^2 y = g$ with $g \in L_2(0, a)$ subject to the initial condition $y(0) = 1$, $y'(0) = -i\lambda$. For $g = q e_0(\lambda, \cdot)$ it follows that y is the unique solution of (12.2.1), (12.2.2). Since this solution is $e_0(\lambda, \cdot)$, the equation (12.2.7) follows. $\qquad\square$

Lemma 12.2.8. *Let $p \in \mathbb{N}_0$ and $q \in W_2^p(0, a)$. Then $\tilde{K}$ defined in Theorem 12.2.6 is continuous, and for all $0 \le x \le a$ and $|t| \le x$,*

$$\tilde{K}(x, t) = \frac{1}{2} \int_0^{\frac{x+t}{2}} q(u) \, du + \int_0^{\frac{x+t}{2}} \int_0^{\frac{x-t}{2}} q(\alpha + \beta) \tilde{K}(\alpha + \beta, \alpha - \beta) \, d\beta \, d\alpha. \tag{12.2.9}$$

Furthermore, $\frac{\partial^{p_1} \partial^{p_2}}{\partial x^{p_1} \partial t^{p_2}} \tilde{K}(a, \cdot) \in W_2^{p+1-p_1-p_2}(-a, a)$ whenever $p_1 + p_2 \le p + 1$. If q is real valued, then also $\tilde{K}$ is real valued.

Proof. From (12.2.4), (12.1.4) and Lemma 12.1.1 we conclude that $\tilde{K}$ is continuous. If q is real valued, then also q_0 given by (12.2.5) is real valued, and hence the real part of r also satisfies (12.1.4). From the uniqueness of the solution of (12.1.4) we conclude that r is real valued. Substituting (12.2.3) into (12.2.7) we arrive at

$$\int_{-x}^{x} \tilde{K}(x,t) e^{-i\lambda t} dt = \int_0^x \frac{\sin \lambda(x-t)}{\lambda} q(t) e^{-i\lambda t} dt$$

$$+ \int_0^x \frac{\sin \lambda(x-t)}{\lambda} q(t) \int_{-t}^t \tilde{K}(t,\xi) e^{-i\lambda \xi} d\xi \, dt. \qquad (12.2.10)$$

Next we express the right-hand side of the above equation as a Fourier transform. Since

$$\frac{\sin \lambda(x-t)}{\lambda} e^{-i\lambda \xi} = \frac{1}{2} \int_{\xi-(x-t)}^{\xi+(x-t)} e^{-i\lambda u} \, du, \qquad (12.2.11)$$

it follows that

$$\int_0^x \frac{\sin \lambda(x-t)}{\lambda} q(t) e^{-i\lambda t} dt = \frac{1}{2} \int_0^x q(t) \int_{2t-x}^x e^{-i\lambda u} \, du \, dt$$

$$= \frac{1}{2} \int_{-x}^x e^{-i\lambda u} \int_0^{\frac{x+u}{2}} q(t) \, dt \, du. \qquad (12.2.12)$$

Using equation (12.2.11) once more, we obtain the equality

$$\int_0^x \frac{\sin \lambda(x-t)}{\lambda} q(t) \int_{-t}^t \tilde{K}(t,\xi) e^{-i\lambda \xi} \, d\xi \, dt$$

$$= \frac{1}{2} \int_0^x q(t) \left(\int_{-t}^t \tilde{K}(t,\xi) \int_{\xi-(x-t)}^{\xi+(x-t)} e^{-i\lambda u} \, du \, d\xi \right) dt.$$

Interchanging variables, the inner double integral becomes

$$\int_{-t}^t \tilde{K}(t,\xi) \int_{\xi-(x-t)}^{\xi+(x-t)} e^{i\lambda u} \, du \, d\xi = \int_{-x}^x e^{i\lambda u} \int_{\max\{-t,u-(x-t)\}}^{\min\{t,u+(x-t)\}} \tilde{K}(t,\xi) \, d\xi \, du.$$

Consequently,

$$\int_0^x q(t) \int_{-t}^t \tilde{K}(t,\xi) \int_{\xi-(x-t)}^{\xi+(x-t)} e^{-i\lambda u} \, du \, d\xi \, dt$$

$$= \int_{-x}^x e^{-i\lambda u} \int_0^x q(t) \int_{\max\{-t,u-(x-t)\}}^{\min\{t,u+(x-t)\}} \tilde{K}(t,\xi) \, d\xi \, dt \, du.$$

It follows that

$$\int_0^x \frac{\sin \lambda(x-t)}{\lambda} q(t) \int_{-t}^t \tilde{K}(t,\xi) e^{-i\lambda \xi} \, d\xi \, dt \qquad (12.2.13)$$

$$= \frac{1}{2} \int_{-x}^x e^{-i\lambda t} \int_0^x q(u) \int_{\max\{-u,t-(x-u)\}}^{\min\{-u,t+(x-u)\}} \tilde{K}(u,\xi) \, d\xi \, du \, dt.$$

Equations (12.2.12) and (12.2.13) show that (12.2.10) leads to

$$\int_{-x}^{x} \tilde{K}(x,t)e^{-i\lambda t}\,dt$$

$$= \frac{1}{2}\int_{-x}^{x}\left(\int_{0}^{\frac{x+t}{2}} q(u)du + \int_{0}^{x} q(u)\int_{\max\{-u,t-(x-u)\}}^{\min\{u,t+(x-u)\}} \tilde{K}(u,\xi)\,d\xi\,du\right)e^{-i\lambda t}\,dt.$$

Taking the inverse Fourier transform, we arrive at

$$\tilde{K}(x,t) = \frac{1}{2}\int_{0}^{\frac{x+t}{2}} q(u)du + \frac{1}{2}\int_{0}^{x} q(u)\int_{\max\{-u,t-(x-u)\}}^{\min\{u,t+(x-u)\}} \tilde{K}(u,\xi)\,d\xi\,du. \quad (12.2.14)$$

Performing the change of variables

$$u + \xi = 2\alpha \text{ and } u - \xi = 2\beta$$

in this integral, the region of the double integral,

$$-u \le \xi \le u, \quad t-(x-u) \le \xi \le t+(x-u), \quad 0 \le u \le x,$$

becomes

$$0 \le \alpha, \quad 0 \le \beta, \quad 2\beta \le x-t, \quad 2\alpha \le x+t, \quad 0 \le \alpha+\beta \le x,$$

where the last condition is redundant since it follows from the first four. Hence (12.2.9) follows. □

From [180, Corollary after Theorem 1.2.1, Theorem 1.2.2 and (1.2.18)] we obtain

Theorem 12.2.9.

1. *If the potential q of the Sturm–Liouville equation*

$$-y''(x) + q(x)y(x) = \lambda^2 y(x), \quad x \in (0,a), \quad (12.2.15)$$

belongs to $L_2(0,a)$, then the solutions of the initial value problems $s(\lambda,0)=0$, $s'(\lambda,0)=1$ and $c(\lambda,0)=1$, $c'(\lambda,0)=0$ can be expressed as

$$s(\lambda,x) = \frac{\sin\lambda x}{\lambda} + \int_{0}^{x} K(x,t)\frac{\sin\lambda t}{\lambda}\,dt, \quad (12.2.16)$$

$$c(\lambda,x) = \cos\lambda x + \int_{0}^{x} B(x,t)\cos\lambda t\,dt, \quad (12.2.17)$$

where $K(x,t) = \tilde{K}(x,t) - \tilde{K}(x,-t)$, $B(x,t) = \tilde{K}(x,t) + \tilde{K}(x,-t)$, and $\tilde{K}(x,t)$ is the unique solution of the integral equation

$$\tilde{K}(x,t) = \frac{1}{2}\int_{0}^{\frac{x+t}{2}} q(s)\,ds + \int_{0}^{\frac{x+t}{2}}\int_{0}^{\frac{x-t}{2}} q(s+p)\tilde{K}(s+p,s-p)\,dp\,ds$$

$$(12.2.18)$$

on the triangular region $\{(x,t) \in [0,a] \times [-a,a] : |t| \leq x\}$. In particular, it is true that $K(x,0) = 0$ and

$$B(x,x) = K(x,x) = \tilde{K}(x,x) = \frac{1}{2}\int_0^x q(s)ds. \tag{12.2.19}$$

2. If $q \in W_2^n(0,a)$, then K and B have partial derivatives up to $(n+1)$th order which belong to $L_2(0,a)$.

Proof. We observe that

$$s(\lambda,x) = \frac{e_0(-\lambda,x) - e_0(\lambda,x)}{2i\lambda} \tag{12.2.20}$$

for all $x \in [0,a]$ and all nonzero complex numbers λ since both functions solve the initial value problem (12.2.15) subject to the boundary condition $y(0) = 0$, $y'(0) = 1$, see (12.2.1) and (12.2.2) for e_0. Theorem 12.2.6 shows that

$$s(\lambda,x) = \frac{\sin \lambda x}{\lambda} + \frac{1}{2i\lambda}\int_{-x}^x \left(\tilde{K}(x,t)e^{i\lambda t} - \tilde{K}(x,t)e^{-i\lambda t}\right) dt$$

$$= \frac{\sin \lambda x}{\lambda} + \frac{1}{2i\lambda}\int_0^x \left([\tilde{K}(x,t) - \tilde{K}(x,-t)]e^{i\lambda t} - (\tilde{K}(x,t) - \tilde{K}(x,-t)e^{-i\lambda t}\right) dt$$

$$= \frac{\sin \lambda x}{\lambda} + \int_0^x [\tilde{K}(x,t) - \tilde{K}(x,-t)]\frac{\sin \lambda t}{\lambda} dt,$$

which proves (12.2.17). Similarly, (12.2.18) follows from

$$c(\lambda,x) = \frac{e_0(\lambda,x) + e_0(-\lambda,x)}{2}. \tag{12.2.21}$$

The properties of K and B now follow immediately from their definition and Lemma 12.2.8. □

Observe that $-\sin^{(j+1)}$ is an antiderivative of $\sin^{(j)}$ for all $j \in \mathbb{N}_0$. We also note that K is an odd function with respect to t, whence $\frac{\partial^j}{\partial t^j}K(a,0) = 0$ and $\frac{\partial^j}{\partial t^j}K_x(a,0) = 0$ for all even nonnegative integers. Similarly, since B is an even function with respect to t, it follows that $\frac{\partial^j}{\partial t^j}B(a,0) = 0$ and $\frac{\partial^j}{\partial t^j}B_x(a,0) = 0$ for all odd nonnegative integers. Then integration by parts and differentiation with respect to x, respectively, followed by n integrations by parts in (12.2.16) and (12.2.17) leads to

Corollary 12.2.10. *If $n \in \mathbb{N}_0$ and $q \in W_2^n(0, a)$, then*

$$s(\lambda, a) = \frac{\sin \lambda a}{\lambda} - \sum_{j=0}^{n} \frac{\partial^j}{\partial t^j} K(a, a) \frac{\sin^{(j+1)} \lambda a}{\lambda^{j+2}} + \int_0^a \frac{\partial^{n+1}}{\partial t^{n+1}} K(a, t) \frac{\sin^{(n+1)} \lambda t}{\lambda^{n+2}} \, dt,$$

$$(12.2.22)$$

$$s'(\lambda, a) = \cos \lambda a + K(a, a) \frac{\sin \lambda a}{\lambda} - \sum_{j=1}^{n} \frac{\partial^{j-1}}{\partial t^{j-1}} K_x(a, a) \frac{\sin^{(j)} \lambda a}{\lambda^{j+1}}$$

$$+ \int_0^a \frac{\partial^n}{\partial t^n} K_x(a, t) \frac{\sin^{(n)} \lambda t}{\lambda^{n+1}} \, dt, \qquad\qquad (12.2.23)$$

$$c(\lambda, a) = \cos \lambda a + \sum_{j=0}^{n} \frac{\partial^j}{\partial t^j} B(a, a) \frac{\sin^{(j)} \lambda a}{\lambda^{j+1}} - \int_0^a \frac{\partial^{n+1}}{\partial t^{n+1}} B(a, t) \frac{\sin^{(n)} \lambda t}{\lambda^{n+1}} \, dt,$$

$$(12.2.24)$$

$$c'(\lambda, a) = -\lambda \sin \lambda a + B(a, a) \cos \lambda a + \sum_{j=1}^{n} \frac{\partial^{j-1}}{\partial t^{j-1}} B_x(a, a) \frac{\sin^{(j-1)} \lambda a}{\lambda^j}$$

$$+ \int_0^a \frac{\partial^n}{\partial t^n} B_x(a, t) \frac{\sin^{(n+1)} \lambda t}{\lambda^n} \, dt, \qquad\qquad (12.2.25)$$

where $\frac{\partial^{n+1}}{\partial t^{n+1}} K(a, \cdot)$, $\frac{\partial^n}{\partial t^n} K_x(a, \cdot)$, $\frac{\partial^{n+1}}{\partial t^{n+1}} B(a, \cdot)$, $\frac{\partial^n}{\partial t^n} B_x(a, \cdot)$ belong to $L_2(0, a)$.

Corollary 12.2.11. *The entire functions $\lambda \mapsto \lambda s(\lambda, a)$, $s'(\cdot, a)$, $c(\cdot, a)$, and $\lambda \mapsto \lambda^{-1} c'(\lambda, a)$ are sine type functions of type a.*

Proof. The first term of the representation in Corollary 12.2.10 of each of these functions can be estimated as

$$\frac{1}{4} e^{|\operatorname{Im} \lambda| a} \le |\sin \lambda a| \le e^{|\operatorname{Im} \lambda| a} \quad \text{or} \quad \frac{1}{4} e^{|\operatorname{Im} \lambda| a} \le |\cos \lambda a| \le e^{|\operatorname{Im} \lambda| a}$$

for sufficiently large $|\operatorname{Im} \lambda|$, whereas the remaining terms satisfy

$$O(\lambda^{-1}) e^{|\operatorname{Im} \lambda| a}.$$

Hence each of these functions is a sine type function by Proposition 11.2.19. □

12.3 Representations of some sine type functions

Let b and c be real numbers. Then the solution ω of the initial value problem

$$y'' + \left(\lambda^2 - 2ib\lambda - c \right) y = 0, \quad x \in (0, a), \quad y(\lambda, 0) = 0, \quad y'(\lambda, 0) = 1, \quad (12.3.1)$$

has the representation $\omega(\lambda, x) = \tau(\lambda)^{-1} \sin \tau(\lambda) x$, and $\omega'(\lambda, x) = \cos \tau(\lambda) x$, where $\tau(\lambda) = \sqrt{\lambda^2 - 2ib\lambda - c}$. Since both ω and ω' are even functions with respect to τ, the representation is unambiguous. The next lemma gives an asymptotic representation of these two functions in terms of $\sin \lambda a$ and $\cos \lambda a$ for $x = a > 0$.

Lemma 12.3.1. *Let $b, c \in \mathbb{R}$ and $a > 0$. Then there are $R > 0$ and analytic functions $f_{j,k}$ on $\{z \in \mathbb{C} : |z| < R^{-1}\}$ for $j, k = 1, 2$, satisfying $f_{j,k}(-\bar{z}) = \overline{f_{j,k}(z)}$, $f_{1,1}(0) = f_{2,1}(0) = \cosh ba$ and $-f_{1,2}(0) = f_{2,2}(0) = \sinh ba$ such that for the solution $\omega(\cdot, a)$ of the initial value problem (12.3.1) at $x = a$ and its derivative $\omega'(\cdot, a)$ at $x = a$ we have the representations*

$$\omega(\lambda, a) = \frac{\sin \tau(\lambda) a}{\tau(\lambda)} = f_{1,1}(\lambda^{-1}) \frac{\sin \lambda a}{\lambda} + i f_{1,2}(\lambda^{-1}) \frac{\cos \lambda a}{\lambda}, \qquad (12.3.2)$$

$$\omega'(\lambda, a) = \cos \tau(\lambda) a = f_{2,1}(\lambda^{-1}) \cos \lambda a + i f_{2,2}(\lambda^{-1}) \sin \lambda a, \qquad (12.3.3)$$

for $|\lambda| > R$.

Proof. Let $r > 0$ such that $2|b|r + cr^2 < 1$. Let h_1 be the unique analytic branch of $z \mapsto \sqrt{1 - 2ibz - cz^2}$ on $\{z \in \mathbb{C} : |z| < r\}$ with $h_1(0) = 1$, i.e.,

$$h_1(z) = \sum_{j=0}^{\infty} \binom{\frac{1}{2}}{j} (-2ibz - cz^2)^j, \ |z| < r. \qquad (12.3.4)$$

Note that $h_1(z) \neq 0$ for all $|z| < r$. For $|\lambda| > R = \frac{1}{r}$ we will now choose the branch of τ such that

$$\frac{\tau(\lambda)}{\lambda} = h_1(\lambda^{-1}).$$

For $|z| < r$ we define

$$h_2(z) = \frac{h_1(z) - 1}{z}. \qquad (12.3.5)$$

Clearly, h_2 is analytic with $h_2(0) = -ib$. For $|\lambda| > R$ we conclude that

$$\tau(\lambda) - \lambda = \lambda \left(\frac{\tau(\lambda)}{\lambda} - 1 \right) = h_2(\lambda^{-1}).$$

It follows that

$$\frac{\sin \tau(\lambda) a}{\tau(\lambda)} = \frac{\cos h_2(\lambda^{-1}) a}{h_1(\lambda^{-1})} \frac{\sin \lambda a}{\lambda} + \frac{\sin h_2(\lambda^{-1}) a}{h_1(\lambda^{-1})} \frac{\cos \lambda a}{\lambda},$$

$$\cos \tau(\lambda) a = \cos h_2(\lambda^{-1}) a \cos \lambda a - \sin h_2(\lambda^{-1}) a \sin \lambda a.$$

For $|z| < r$ we define

$$f_{1,1}(z) = \frac{\cos h_2(z) a}{h_1(z)}, \ f_{1,2}(z) = -i \frac{\sin h_2(z) a}{h_1(z)}, \qquad (12.3.6)$$

$$f_{2,1}(z) = \cos h_2(z) a, \ f_{2,2}(z) = i \sin h_2(z) a, \qquad (12.3.7)$$

which proves the representations (12.3.2) and (12.3.3). We also obtain $f_{1,1}(0) = f_{2,1}(0) = \cos h_2(0) a = \cosh ba$ and $-f_{1,2}(0) = f_{2,2}(0) = i \sinh h_2(0) a = \sinh ba$. The symmetry of these functions follows from

$$h_1(-\bar{z}) = \sqrt{1 + 2ib\bar{z} - c\bar{z}^2} = \overline{h_1(z)}$$

and $h_2(-\bar{z}) = -\overline{h_2(z)}$ for $|z| < r$. $\qquad \square$

Corollary 12.3.2. *Under the assumptions of Lemma 12.3.1, for each $n \in \mathbb{N}$ there are polynomials $f_{j,k,n}$ of degree $\leq n$ and entire functions $\psi_{j,n} \in \mathcal{L}^a$, $j, k = 1, 2$, such that $f_{j,k,n}(-\bar{z}) = \overline{f_{j,k,n}(z)}$, $f_{j,1,n}(0) = \cosh ba$, $(-1)^j f_{j,2,n}(0) = \sinh ba$, $\psi_{j,n}(-\bar{z}) = (-1)^{n+j}\overline{\psi_{j,n}(z)}$ and such that*

$$\frac{\sin \tau(\lambda)a}{\tau(\lambda)} = f_{1,1,n}(\lambda^{-1})\frac{\sin \lambda a}{\lambda} + if_{1,2,n}(\lambda^{-1})\frac{\cos \lambda a}{\lambda} + \frac{\psi_{1,n}(\lambda)}{\lambda^{n+1}}, \qquad (12.3.8)$$

$$\cos \tau(\lambda)a = f_{2,1,n}(\lambda^{-1})\cos \lambda a + if_{2,2,n}(\lambda^{-1})\sin \lambda a + \frac{\psi_{2,n}(\lambda)}{\lambda^n}. \qquad (12.3.9)$$

Proof. Let $f_{j,k,n}$, $j, k = 1, 2$, be the Taylor polynomial about 0 of order n of the function $f_{j,k}$ from Lemma 12.3.1. Defining the function $\psi_{1,n}$ by (12.3.8), we have

$$\psi_{1,n}(\lambda) = \lambda^{n+1}\frac{\sin \tau(\lambda)a}{\tau(\lambda)} - \lambda^n f_{1,1,n}(\lambda^{-1})\sin \lambda a - i\lambda^n f_{1,2,n}(\lambda^{-1})\cos \lambda a$$

$$= \lambda^n \left(f_{1,1}(\lambda^{-1}) - f_{1,1,n}(\lambda^{-1})\right)\sin \lambda a + i\lambda^n \left(f_{1,2}(\lambda^{-1}) - f_{1,2,n}(\lambda^{-1})\right)\cos \lambda a,$$

where the second identity follows from (12.3.2). The first of these representations shows that $\psi_{1,n}$ is an entire function with the stated symmetry. Finally, the second representation shows that $\psi_{1,n}$ is of the form $O(|\lambda|^{-1})\sin \lambda a + O(|\lambda|^{-1})\cos \lambda a$, and therefore $\psi_{1,n}$ is of exponential type $\leq a$ and $O(|\lambda|^{-1})$ on the real axis. Hence $\psi_{1,n} \in \mathcal{L}^a$. The proof for $\psi_{2,n}$ is similar. $\square$

Lemma 12.3.3 ([180, Lemma 3.4.2]). *For the functions u and v to have the representations*

$$u(\lambda) = \frac{\sin \lambda a}{\lambda} - \frac{4\pi^2 Aa \cos \lambda a}{4\lambda^2 a^2 - \pi^2} + \frac{f(\lambda)}{\lambda^2}, \qquad (12.3.10)$$

$$v(\lambda) = \cos \lambda a + B\pi^2 \frac{\sin \lambda a}{\lambda a} + \frac{g(\lambda)}{\lambda}, \qquad (12.3.11)$$

where $A, B \in \mathbb{C}$, $f \in \mathcal{L}_e^a$, $f(0) = 0$, $g \in \mathcal{L}_o^a$, it is necessary and sufficient that

$$u(\lambda) = a \prod_{k=1}^{\infty} \left(\frac{\pi k}{a}\right)^{-2}(u_k^2 - \lambda^2), \quad u_k = \frac{\pi k}{a} + \frac{\pi A}{ak} + \frac{\alpha_k}{k}, \qquad (12.3.12)$$

$$v(\lambda) = \prod_{k=1}^{\infty} \left(\frac{\pi}{a}\left(k - \frac{1}{2}\right)\right)^{-2}(v_k^2 - \lambda^2), \quad v_k = \frac{\pi}{a}\left(k - \frac{1}{2}\right) + \frac{\pi B}{ak} + \frac{\beta_k}{k}, \qquad (12.3.13)$$

where $(\alpha_k)_{k=1}^{\infty} \in l_2$ and $(\beta_k)_{k=1}^{\infty} \in l_2$.

Proof. We define

$$\varphi_s(\lambda) = i\lambda^2 u(\lambda).$$

In view of

$$\frac{4\pi^2 a\lambda^2}{4\lambda^2 a^2 - \pi^2} - \frac{\pi^2}{a} = \frac{\pi^4}{(4\lambda^2 a^2 - \pi^2)a},$$

φ_s is an even entire function of the form φ as in (7.1.4) with $\sigma = 0$, $\alpha = 1$, $M = 0$ and $N = \frac{\pi^2 A}{a}$. Then the representation of u_k in (12.3.13) follows from part 1 in Lemma 7.1.3 if we observe that u is even and that we have to omit a double zeros at 0 from the sequence $(\tilde{\lambda}_k)_{k=-\infty, k \neq 0}^{\infty}$. Since $\lambda \mapsto \lambda \sin \lambda a$ is a sine type function, it follows from Lemma 11.2.29 that

$$u(\lambda) = a' \prod_{k=1}^{\infty} \left(\frac{\pi k}{a} \right)^{-2} (u_k^2 - \lambda^2)$$

with some $a' \neq 0$. From the product representation of the sine function we infer that

$$\frac{\lambda u(\lambda)}{\sin \lambda a} = \frac{a'}{a} \prod_{k=1}^{\infty} \left(\frac{\pi k}{a} \right)^{-2} (u_k^2 - \lambda^2) \left(1 - \frac{\lambda^2 a^2}{\pi^2 k^2} \right)^{-1}$$

$$= \frac{a'}{a} \prod_{k=1}^{\infty} \frac{u_k^2 - \lambda^2}{\frac{\pi^2 k^2}{a^2} - \lambda^2}$$

$$= \frac{a'}{a} \prod_{k=1}^{\infty} \left(1 + \frac{u_k^2 - \frac{\pi^2 k^2}{a^2}}{\frac{\pi^2 k^2}{a^2} - \lambda^2} \right).$$

Since $u_k^2 - \frac{\pi^2 k^2}{a^2}$ is bounded with respect to k and since $|\frac{\pi^2 k^2}{a^2} - \lambda^2| \geq \frac{\pi^2 k^2}{a^2}$ for λ on the imaginary axis, it follows that the right-hand side converges to $\frac{a'}{a}$ as $\lambda \to \infty$ along the imaginary axis, whereas the corresponding limit on the left-hand side is 1. Hence $a' = a$ and we have shown that the function given by (12.3.10) has the representation (12.3.12).

Conversely, assume that u is given by (12.3.12). Putting $\lambda_k = \frac{\pi k}{a}$, we have

$$u_k = \lambda_k + \frac{\pi^2 A}{a^2} \lambda_k^{-1} + \frac{\pi \alpha_k}{a} \lambda_k^{-1}.$$

Since $(\lambda_k)_{k \in \mathbb{Z}}$ is the sequence of the zeros of $\lambda \mapsto \sin \lambda a$, it follows from Remark 11.3.16 that

$$\lambda u(\lambda) = C_0 \sin \lambda a \left(1 + \frac{B_1}{\lambda} \right) - C_0 \frac{\pi^2 A}{a \lambda} \cos \lambda a + \frac{f_2(\lambda)}{\lambda}, \qquad (12.3.14)$$

where $C_0 \neq 0$, B_1 is a constant and $f_2 \in \mathcal{L}^a$. Taking into account that u is an even entire function, we obtain $B_1 = 0$, $f_2 \in \mathcal{L}_e^a$ and $f_2(0) = 0$. From the first part of the proof we conclude that $C_0 = 1$. Hence u is of the form (12.3.10).

Similarly, the function defined by

$$\varphi_c(\lambda) = \lambda v(\lambda)$$

is an odd entire function of the form φ as in (7.1.4) with $\sigma = 1$, $\alpha = 0$, $M = \frac{\pi^2 B}{a}$ and $N = 0$. Arguing as for φ_s it follows that (12.3.11) and (12.3.13) are equivalent. $\square$

Lemma 12.3.4. *For the entire functions u and v to admit the representations*

$$u(\lambda) = \frac{\sin \lambda a}{\lambda} - \frac{4\pi^2 Aa \cos \lambda a}{4\lambda^2 a^2 - \pi^2} + C \frac{\sin \lambda a}{\lambda^3} + \frac{f(\lambda)}{\lambda^3}, \tag{12.3.15}$$

$$v(\lambda) = \cos \lambda a + B\pi^2 \frac{\sin \lambda a}{\lambda a} + D \frac{\cos \lambda a}{4\lambda^2 a^2 - \pi^2} + \frac{g(\lambda)}{\lambda^2}, \tag{12.3.16}$$

where $A, B, C, D \in \mathbb{C}$, $f \in \mathcal{L}_o^a$, $g \in \mathcal{L}_e^a$, it is necessary and sufficient that

$$u(\lambda) = a \prod_{k=1}^{\infty} \left(\frac{\pi k}{a} \right)^{-2} (u_k^2 - \lambda^2), \quad u_k = \frac{\pi k}{a} + \frac{\pi A}{ak} + \frac{\alpha_k}{k^2}, \tag{12.3.17}$$

$$v(\lambda) = \prod_{k=1}^{\infty} \left(\frac{\pi}{a} \left(k - \frac{1}{2} \right) \right)^{-2} (v_k^2 - \lambda^2), \quad v_k = \frac{\pi}{a} \left(k - \frac{1}{2} \right) + \frac{\pi B}{ak} + \frac{\beta_k}{k^2}, \tag{12.3.18}$$

where $(\alpha_k)_{k=1}^{\infty} \in l_2$ and $(\beta_k)_1^{\infty} \in l_2$.

Proof. First assume that (12.3.15) or (12.3.16) hold. We are going to prove that the representations of the zeros u_k and v_k given in (12.3.12) and (12.3.13) can be written in the form (12.3.17) and (12.3.18). In case (12.3.15) we define

$$\tilde{\chi}(\mu) = \mu^2 u(\mu), \tag{12.3.19}$$

while in case (12.3.16) we define

$$\tilde{\chi}(\mu) = -\mu v \left(\mu + \frac{\pi}{2a} \right). \tag{12.3.20}$$

It is easy to see that in either case,

$$\tilde{\chi}(\mu) = (\mu + B_2 \mu^{-1}) \sin \mu a + A_1 \cos \mu a + \Psi_1(\mu) \mu^{-1}$$

with $\Psi_1 \in \mathcal{L}^a$. Indeed, in case (12.3.15) we have $B_2 = C$ and $A_1 = -\pi^2 Aa^{-1}$, while in case (12.3.16) we have $B_2 = \frac{D}{4a^2}$ and $A_1 = -B\pi^2 a^{-1}$. Hence $\tilde{\chi}$ is of the form as considered in Lemma 7.1.5 with $n = 1$, $B_0 = 1$, $B_1 = 0$, $A_2 = 0$, except that we do not require that A_1 and B_1 are real and that Ψ_n is symmetric. But it is easy to see that these requirements are only used to guarantee that the zeros of $\tilde{\chi}$ can be indexed properly; the asymptotic representation (7.1.13) of the zeros holds without these requirements. We therefore conclude from Lemma 7.1.5 that the zeros of $\tilde{\chi}$ have the asymptotic representation

$$\mu_k = \frac{\pi k}{a} - \frac{A_1}{k\pi} + \frac{b_k^{(1)}}{k^2},$$

where $(b_k^{(1)})_{k=1}^{\infty} \in l_2$. Observing that $u_k = \mu_k$ and $v_k = \mu_k - \frac{\pi}{2a}$, the representations (12.3.17) and (12.3.18) follow.

Conversely, assume that (12.3.17) holds. By the first part of this proof, the zeros $(\lambda_k^{(0)})_{k\in\mathbb{Z}\setminus\{0\}}$ of the entire function u_0 defined by

$$u_0(\lambda) = \frac{\sin\lambda a}{\lambda} - \frac{4\pi^2 Aa\cos\lambda a}{4\lambda^2 a^2 - \pi^2} \tag{12.3.21}$$

have the asymptotic behaviour

$$\lambda_k^{(0)} = \frac{\pi k}{a} + \frac{\pi A}{ak} + \frac{\alpha_{0,k}}{k^2}, \tag{12.3.22}$$

where $(\alpha_{0,k})_{k=1}^\infty \in l_2$ and $\alpha_{0,-k} = -\alpha_{0,k}$ for all $k \in \mathbb{N}$. Comparing (12.3.22) with (12.3.17) we obtain

$$u_k = \lambda_k^{(0)} + \frac{\gamma_k}{(\lambda_k^{(0)})^2},$$

where $(\gamma_k)_{k\in\mathbb{Z}\setminus\{0\}} \in l_2$. It is easy to see that $\lambda \mapsto \lambda u_0(\lambda)$ is a sine-type function of type a. In view of Remark 11.3.16 we obtain

$$\lambda u(\lambda) = C_0 \lambda u_0(\lambda)\left(1 + \frac{B_1}{\lambda} + \frac{C}{\lambda^2}\right) + \frac{f_2(\lambda)}{\lambda^2}, \tag{12.3.23}$$

where $C_0 \neq 0$, B_1, C are constants and $f_2 \in \mathcal{L}^a$. Taking into account that u and u_0 are even functions, we obtain $B_1 = 0$ and $f_2 \in \mathcal{L}_o^a$. Since u satisfies the representation (12.3.10) in Lemma 12.3.3, it follows that $C_0 = 1$. Substituting (12.3.21) into (12.3.23) and observing $C_0 = 1$ and $B_1 = 0$, we obtain (12.3.15).

Finally assume that (12.3.18) holds. By the first part of this proof, the zeros $(\lambda_k^{(0)})_{k\in\mathbb{Z}\setminus\{0\}}$ of the entire function v_0 defined by

$$v_0(\lambda) = \cos\lambda a + B\pi^2 \frac{\sin\lambda a}{\lambda a} \tag{12.3.24}$$

have the asymptotic behaviour

$$\lambda_k^{(0)} = \frac{\pi}{a}\left(|k| - \frac{1}{2}\right)\operatorname{sgn} k + \frac{\pi B}{ak} + \frac{\beta_{0,k}}{k^2}, \tag{12.3.25}$$

where $(\beta_{0,k})_{k=1}^\infty \in l_2$ and $\beta_{0,-k} = -\beta_{0,k}$ for all $k \in \mathbb{N}$. Comparing (12.3.25) with (12.3.18) we obtain

$$v_k = \lambda_k^{(0)} + \frac{\gamma_k}{(\lambda_k^{(0)})^2},$$

where $(\gamma_k)_{k\in\mathbb{Z}\setminus\{0\}} \in l_2$. It is easy to see that v_0 is a sine-type function of type a. In view of Remark 11.3.16 we obtain

$$v(\lambda) = C_0 v_0(\lambda)\left(1 + \frac{B_1}{\lambda} + \frac{D}{4a^2\lambda^2}\right) + \frac{f_2(\lambda)}{\lambda^2}, \tag{12.3.26}$$

where $C_0 \neq 0$, B_1, D are constants and $f_2 \in \mathcal{L}^a$. Taking into account that v and v_0 are even functions, we obtain $B_1 = 0$ and $f_2 \in \mathcal{L}_e^a$. Since v satisfies the representation (12.3.11) in Lemma 12.3.3, it follows that $C_0 = 1$. Substituting (12.3.24) into (12.3.26) and observing $C_0 = 1$ and $B_1 = 0$, we obtain (12.3.16). $\quad\square$

12.4   The fundamental equation

Throughout this section let u and v be as in (12.3.10) and (12.3.11) with $A = B$ and such that the numbers u_k and v_k in (12.3.12) and (12.3.13) are real or pure imaginary for all $k \in \mathbb{N}$ and satisfy

$$v_1^2 < u_1^2 < v_2^2 < u_2^2 \cdots .$$

We consider the entire function χ defined by

$$\chi(\lambda) = v(\lambda) + i\lambda u(\lambda), \quad \lambda \in \mathbb{C}. \tag{12.4.1}$$

Lemma 12.4.1. *The function χ is of* SSHB *class.*

Proof. By definition, u and v are even functions, and we can write $u(\lambda) = Q(\lambda^2)$ and $v(\lambda) = P(\lambda^2)$ with entire functions Q and P, where the sets $\{u_k^2 : k \in \mathbb{N}\}$ and $\{v_k^2 : k \in \mathbb{N}\}$ are the sets of the zeros of Q and P, respectively, and all zeros of Q and P are simple and interlace. Since Q and P have the representations (12.3.12) and (12.3.13), it is easy to see that $Q(v_1^2)$ is an infinite product of positive numbers, whereas $P'(v_1^2)$ is the negative of an infinite product of positive numbers. It follows that $Q'(x)P(x) - Q(x)P'(x) > 0$ for all $x \in \mathbb{R}$ sufficiently close to v_1^2. Hence there are $x \in \mathbb{R}$ such that $P(x) \neq 0$ and such that $\theta = \frac{Q}{P}$ satisfies $\theta'(x) > 0$. Therefore θ is a Nevanlinna function by Theorem 11.1.6 and Remark 11.1.7. Then $\theta \in \mathcal{N}_+^{\mathrm{ep}}$ by Corollary 5.2.3, and thus χ is of SSHB class by Definition 5.2.6. $\qquad\square$

We further define the function ψ by

$$\psi(\lambda) = e^{-i\lambda a}\chi(\lambda) \tag{12.4.2}$$

and the function S by

$$S(\lambda) = \frac{\psi(\lambda)}{\psi(-\lambda)}. \tag{12.4.3}$$

Proposition 12.4.2. *Let $\kappa := \#\{k \in \mathbb{N} : v_k^2 < 0\}$. Then S is meromorphic on $\mathbb{C}$ and analytic on $\mathbb{R}$, and S has exactly κ poles in the open upper half-plane. All poles in the open upper half-plane are simple and lie on the imaginary axis. Furthermore, for $\lambda \in \mathbb{C}$ such that $\psi(\lambda) \neq 0$ and $\psi(-\lambda) \neq 0$ we have*

$$S(-\bar{\lambda}) = \frac{\psi(-\bar{\lambda})}{\psi(\bar{\lambda})} = \overline{S(\lambda)}, \quad \text{and} \quad \frac{1}{S(\lambda)} = \frac{\psi(-\lambda)}{\psi(\lambda)} = S(-\lambda). \tag{12.4.4}$$

In particular,

$$S(-\lambda) = \overline{S(\lambda)}, \quad |S(\lambda)| = 1, \quad \lambda \in \mathbb{R}. \tag{12.4.5}$$

Proof. The function S is the quotient of two nonzeros entire functions and hence meromorphic on $\mathbb{C}$. Let $\check{\chi}(\lambda) = \chi(-\lambda)$, $\lambda \in \mathbb{C}$. Since χ and $\check{\chi}$ do not have common nonzero zeros, the statement on the poles and the analyticity of S on $\mathbb{R} \setminus \{0\}$ immediately follows from Theorem 5.2.9. Furthermore, the possible singularity of S at 0 is removable since $\lim_{\lambda \to 0} S(\lambda) = -1$ if 0 is a (simple) zero of χ. The identities (12.4.4) and (12.4.5) are immediate consequences of the fact that χ is real on the imaginary axis. $\qquad\square$

For $b \geq 0$ we define S_b by

$$S_b(\lambda) = 1 - S(ib + \lambda), \quad \lambda \in \mathbb{C}, \ \psi(-ib - i\lambda) \neq 0. \tag{12.4.6}$$

Lemma 12.4.3.

1. *Let $b \geq 0$ such that ib is not a pole of S. Then $S_b \in L_2(\mathbb{R})$.*
2. *Let $\gamma \geq 0$ such that S is analytic on $\{\lambda \in \mathbb{C} : \operatorname{Im} \lambda \geq \gamma\}$ and define*

$$F(x) = \frac{e^{-\gamma x}}{2\pi} \int_{-\infty}^{\infty} S_\gamma(\lambda) e^{i\lambda x} \, d\lambda, \quad x \in \mathbb{R}. \tag{12.4.7}$$

Then the function F is real valued and independent of γ. Furthermore, the function $x \mapsto e^{\gamma x} F(x)$ is the inverse Fourier transform of S_γ and can be represented as $F_1 + F_2$, where $F_1 \in L_2(\mathbb{R})$, $F_2 \in W_2^1(\mathbb{R})$ are real-valued functions, $F_1(x) = 0$ for $x > 0$, and $F_2(x) = 0$ for $x > 2a$.

Proof. 1. We conclude from the representations (12.3.10) and (12.3.11) of u and v that

$$\chi(\lambda) = e^{i\lambda a} + \frac{A\pi^2}{a} \frac{\sin \lambda a}{\lambda} - \frac{4\pi^2 i\lambda Aa \cos \lambda a}{4\lambda^2 a^2 - \pi^2} + \frac{g(\lambda) + if(\lambda)}{\lambda},$$

which gives

$$\psi(\lambda) = 1 + \frac{A\pi^2}{a} \frac{\sin \lambda a}{\lambda} e^{-i\lambda a} - \frac{4\pi^2 i\lambda Aa \cos \lambda a}{4\lambda^2 a^2 - \pi^2} e^{-i\lambda a} + \frac{g(\lambda) + if(\lambda)}{\lambda} e^{-i\lambda a} \tag{12.4.8}$$

and

$$\frac{1}{\psi(-\lambda)} = 1 - \frac{A\pi^2}{a} \frac{\sin \lambda a}{\lambda} e^{i\lambda a} - \frac{4\pi^2 i\lambda Aa \cos \lambda a}{4\lambda^2 a^2 - \pi^2} e^{i\lambda a} + \frac{g(\lambda) - if(\lambda)}{\lambda} e^{i\lambda a} + O(\lambda^{-2}) \tag{12.4.9}$$

for λ in the closed upper half-plane. Therefore, the function S satisfies

$$S(\lambda) = 1 - 2\frac{A\pi^2 i}{a} \frac{\sin^2 \lambda a}{\lambda} - \frac{8\pi^2 i\lambda Aa \cos^2 \lambda a}{4\lambda^2 a^2 - \pi^2} + \frac{2}{\lambda}(g(\lambda) \cos \lambda a - f(\lambda) \sin \lambda a)$$
$$+ O(\lambda^{-2}) e^{2 \operatorname{Im} \lambda a}.$$

Observing that

$$\frac{4a^2}{4\lambda^2 a - \pi^2} = \frac{1}{\lambda^2} + O(\lambda^{-4}),$$

this representation can be written in the form

$$1 - S(\lambda) = 2\frac{A\pi^2 i}{\lambda a} - \frac{2}{\lambda}(g(\lambda)\cos\lambda a - f(\lambda)\sin\lambda a) + O(\lambda^{-2})e^{2\operatorname{Im}\lambda a}. \quad (12.4.10)$$

By Proposition 12.4.2 we know that poles of S in the closed upper half-plane lie on the positive imaginary axis, and from (12.4.10) we thus infer that $1 - S$ is analytic and square integrable on each line $\operatorname{Im}\lambda = b$ with $b \geq 0$ for which ib is not a pole of S.

2. Using (12.4.4) we have

$$S_\gamma(-\lambda) = 1 - S(i\gamma - \lambda) = 1 - \overline{S(i\gamma + \lambda)} = \overline{S_\gamma(\lambda)}, \quad \lambda \in \mathbb{R},$$

so that

$$F(x) = \frac{e^{-\gamma x}}{\pi}\int_0^\infty \operatorname{Re}\left(S_\gamma(\lambda)e^{i\lambda x}\right)d\lambda, \quad x \in \mathbb{R},$$

which shows that F is real valued. We can write

$$F(x) = \frac{1}{2\pi}\int_{-\infty}^\infty (1 - S(i\gamma + \lambda))e^{i(i\gamma+\lambda)x}\,d\lambda$$

$$= \frac{1}{2\pi}\int_{\operatorname{Im}\lambda=\gamma}(1 - S(\lambda))e^{i\lambda x}\,d\lambda =: F_\gamma(\lambda).$$

If we now take $\gamma' > \gamma$, then it follows from Cauchy's theorem that

$$|F_{\gamma'}(x) - F_\gamma(x)| \leq \frac{1}{2\pi}\limsup_{R\to\infty}\left(\int_\gamma^{\gamma'}|(1 - S(R + it))e^{i(R+it)x}|dt\right.$$

$$\left.+ \int_\gamma^{\gamma'}|(1 - S(-R + it))e^{i(-R+it)x}|dt\right)$$

$$= \limsup_{R\to\infty}O(R^{-1}) = 0,$$

and therefore the function F is independent of the special choice of γ.

The first term of the representation (12.4.10) of $1 - S$ has a pole at $\lambda = 0$. But it is easy to see that (12.4.10) can be written as

$$1 - S(\lambda) = \frac{2A\pi^2 i}{a(\lambda + i)} + \frac{\omega(\lambda)}{\lambda + i}, \quad \lambda \in \mathbb{C},$$

where ω is analytic in the closed half-plane $\{\lambda \in \mathbb{C} : \operatorname{Im}\lambda \geq \gamma\}$, $\omega(i\gamma+\cdot) \in L_2(\mathbb{R})$, and $\lambda \mapsto \omega(\lambda)e^{-2\operatorname{Im}\lambda a}$ is bounded in that half-plane. We can therefore write $e^{\gamma x}F(x) = F_1(x) + F_2(x)$ with

$$F_1(x) = \frac{1}{2\pi}\int_{-\infty}^\infty \frac{2A\pi^2 i}{a(\lambda + i\gamma + i)}e^{i\lambda x}\,d\lambda, \quad F_2(x) = \frac{1}{2\pi}\int_{-\infty}^\infty \frac{\omega(\lambda + i\gamma)}{\lambda + i\gamma + i}e^{i\lambda x}\,d\lambda.$$

For $x > 0$ and $R > 0$ it follows from Cauchy's integral theorem and Lebesgue's dominated convergence theorem that

$$\int_{-R}^{R} \frac{e^{i\lambda x}}{\lambda + i\gamma + i}\, d\lambda = -\int_0^{\pi} \frac{e^{Re^{i(\theta + \frac{\pi}{2})}x}}{Re^{i\theta} + i\gamma + i} Re^{i\theta}\, d\theta \to 0 \text{ as } R \to \infty.$$

Similarly, for $x < 0$ and $R > \gamma + 1$ we use the residue theorem and Lebesgue's dominated convergence theorem to conclude that

$$\int_{-R}^{R} \frac{e^{i\lambda x}}{\lambda + i\gamma + i}\, d\lambda = \int_{\pi}^{2\pi} \frac{e^{Re^{i(\theta + \frac{\pi}{2})}x}}{Re^{i\theta} + i\gamma + i} Re^{i\theta}\, d\theta - 2\pi i \operatorname{res}_{-i\gamma - i} \frac{e^{i\lambda x}}{\lambda + i\gamma + i}$$

$$\to -2\pi i e^{(\gamma + 1)x} \text{ as } R \to \infty.$$

Therefore

$$F_1(x) = \begin{cases} 0 & \text{if } x > 0, \\ \frac{2A\pi^2}{a} e^{(\gamma+1)x} & \text{if } x < 0. \end{cases}$$

The function F_2 is differentiable with

$$F_2'(x) = \frac{i}{2\pi} \int_{-\infty}^{\infty} \frac{\lambda}{\lambda + i\gamma + i} \omega(\lambda + i\gamma) e^{i\lambda x}\, d\lambda, \quad x \in \mathbb{R}.$$

Hence F_2 and F_2' are inverse Fourier transforms of functions in $L_2(\mathbb{R})$, which shows that $F_2 \in W_2^1(\mathbb{R})$. For $x > 2a$ and $R > 0$ it follows from Cauchy's integral theorem, the boundedness of $\lambda \mapsto \omega(\lambda)e^{2i\lambda a}$ in the closed half-plane $\{\lambda \in \mathbb{C} : \operatorname{Im}\lambda \geq \gamma\}$, and Lebesgue's dominated convergence theorem that

$$\int_{-R}^{R} \frac{\omega(\lambda + i\gamma)}{\lambda + i\gamma + i} e^{i\lambda x}\, d\lambda = -\int_0^{\pi} \frac{\omega(Re^{i\theta} + i\gamma)}{Re^{i\theta} + i\gamma + i} e^{Re^{i(\theta + \frac{\pi}{2})}(2a - x)} Re^{i\theta}\, d\theta \to 0$$

as $R \to \infty$. Therefore, $F_2(x) = 0$ for $x > 2a$. $\qquad \square$

The fundamental equation, see [180, (3,2,10), (3.3.7)] will formally be defined as

$$F(x + y) + H(x, y) + \int_x^{\infty} H(x, t) F(y + t)\, dt = 0, \quad 0 \leq x \leq y. \tag{12.4.11}$$

In order to have the limits of integration independent of x, we substitute $y + x$ for y and $t + x$ for t, see [180, (3.3.7')], and the fundamental equation becomes

$$F(2x + y) + H(x, x + y) + \int_0^{\infty} H(x, x + t) F(2x + y + t)\, dt = 0, \quad x, y \geq 0. \tag{12.4.12}$$

We observe that in view of $F(x) = 0$ for $x > 2a$, the equation (12.4.12) can be written as

$$F(2x + y) + H(x, x + y) + \int_0^{2a - 2x - y} H(x, x + t) F(2x + y + t)\, dt = 0, \quad (x, y) \in D, \tag{12.4.13}$$

where $D = \{(x, y) \in \mathbb{R}^2 : 0 \leq x \leq a, 0 \leq y \leq 2(a - x)\}$ since for any other values of $x, y \geq 0$, $H(x, x + y) = 0$ is necessary and sufficient for (12.4.12) to hold. However, it will be more convenient to use a rectangular region, and therefore we will consider the region $[0, a] \times [0, 2a]$ instead of D.

For $0 \leq x \leq a$ define

$$(\mathbb{F}_x f)(y) = \int_0^{2a} F(2x + y + t)f(t)\, dt, \quad f \in L_2(0, 2a), \ y \in [0, 2a]. \quad (12.4.14)$$

The following result is a special case of [180, Lemmas 3.3.1, 3.3.2 and 3.3.3].

Lemma 12.4.4. For $0 \leq x \leq a$, the operator $\mathbb{F}_x$ is a self-adjoint compact operator in the space $L_2(0, 2a)$. If $v_1^2 > 0$, then $I + \mathbb{F}_x \gg 0$.

Proof. Since $(y, t) \mapsto F(2x + y + t)$ is a real-valued and symmetric continuous function on $[0, 2a] \times [0, 2a]$, the operator $\mathbb{F}_x$ is self-adjoint and compact, see, e. g., [109, p. 240].

Now let $v_1^2 > 0$. Then S is analytic in the closed upper half-plane, and we can take $\gamma = 0$ in (12.4.7). In particular, $F \in L_2(\mathbb{R})$. Let $f \in L_2(0, 2a)$. Putting $f(t) = (\mathbb{F}_x f)(t) = 0$ for $t \in \mathbb{R} \setminus [0, 2a]$ and using the notations $\check{f}(t) = f(-t)$, $\tau_b(t) = y + b$ for $x, b \in \mathbb{R}$, the right-hand side of (12.4.14) can be written as a convolution, so that we arrive at

$$\mathbb{F}_x f = (F \circ \tau_{2x}) * \check{f} \ \text{on} \ [0, \infty).$$

Hence there is a function g with support in $(-\infty, 0]$ such that

$$\mathbb{F}_x f + g = (F \circ \tau_{2x}) * \check{f} \ \text{on} \ \mathbb{R}. \quad (12.4.15)$$

For $y < 0$ we have

$$|g(y)|^2 = \left| \int_0^{2a} F(2x + y + t)f(t)\, dt \right|^2 \leq \int_0^{2a} |F(2x + y + t)|^2\, dt\, \|f\|^2,$$

where $\| \cdot \|$ denotes the norm in $L_2(\mathbb{R})$. Hence

$$\|g\| = \int_{-\infty}^0 |g(y)|^2\, dy \leq \int_{-\infty}^0 \int_0^{2a} |F(2x + y + t)|^2\, dt\, dy\, \|f\|^2$$

$$= \int_0^{2a} \int_{-\infty}^0 |F(2x + y + t)|^2\, dy\, dt\, \|f\|^2 \leq 2a\|F\|^2\|f\|^2,$$

which shows that $g \in L_2(-\infty, 0)$.

Taking the inner product with f in (12.4.15) and observing that $g\overline{f} = 0$, we arrive at

$$(\mathbb{F}_x f, f) = ((F \circ \tau_{2x}) * \check{f}, f).$$

Let $\hat{f}$ denote the Fourier transformation of f. Taking the Fourier transforms of the functions in the above inner product and observing Parseval's formula gives

$$((\mathbb{F}_x f)\check{}, \hat{f}) = ((F \circ \tau_{2x})\check{}\hat{f}, \hat{f}),$$

see, e.g., [108, (21.41)]. It is well known and easy to check that $\hat{f}$ is an entire function. We observe that for any function $h \in L_2(\mathbb{R})$ and $b \in \mathbb{R}$,

$$(h \circ \tau_b)\check{}(\lambda) = \int_{-\infty}^{\infty} h(b+t)e^{-i\lambda t}\, dt = \int_{-\infty}^{\infty} h(t)e^{-i\lambda(t-b)}\, dt = e^{i\lambda b}\hat{h}(\lambda). \quad (12.4.16)$$

Since $\hat{F} = 1 - S$ by definition of F in (12.4.7), we conclude that

$$((\mathbb{F}_x f)\check{}, \hat{f}) = \int_{-\infty}^{\infty} e^{2i\lambda x}(1 - S(\lambda))\hat{f}(-\lambda)\overline{\hat{f}(\lambda)}\, d\lambda.$$

Again from (12.4.16) and Parseval's identity we conclude that

$$\int_{-\infty}^{\infty} e^{2i\lambda x}\hat{f}(-\lambda)\overline{\hat{f}(\lambda)}\, d\lambda = \int_{-\infty}^{\infty} e^{i\lambda x}\hat{f}(-\lambda)\overline{e^{-i\lambda x}\hat{f}(\lambda)}\, d\lambda$$

$$= \int_{-\infty}^{\infty} (f \circ \tau_{-x})(-t)\overline{(f \circ \tau_{-x})(t)}\, dt.$$

But $(f \circ \tau_{-x})(t) = f(t-x) = 0$ if $t < 0$ since f is zero outside $[0, 2a]$, and therefore the integrand on the right-hand side is zero. We conclude that

$$(\mathbb{F}_x f, f) = \frac{1}{2\pi}((\mathbb{F}_x f)\check{}, \hat{f}) = -\frac{1}{2\pi}\int_{-\infty}^{\infty} e^{2i\lambda x} S(\lambda)\hat{f}(-\lambda)\overline{\hat{f}(\lambda)}\, d\lambda. \quad (12.4.17)$$

Observing (12.4.5) it follows from the Cauchy–Schwarz–Bunyakovskii inequality that

$$(\mathbb{F}_x f, f) \geq -\frac{1}{2\pi}\int_{-\infty}^{\infty} |\hat{f}(-\lambda)|\, |\hat{f}(\lambda)|\, d\lambda \geq -\frac{1}{2\pi}\|\hat{f}\|^2 = -\|f\|^2. \quad (12.4.18)$$

Altogether, we conclude that

$$((I + \mathbb{F}_x)f, f) \geq 0 \quad (12.4.19)$$

for all $f \in L_2[0, 2a]$. Hence $I + \mathbb{F}_x \geq 0$.

Since $\mathbb{F}_x$ is compact, $I + \mathbb{F}_x$ is a Fredholm operator of index 0, and for $I + \mathbb{F}_x \gg 0$ it remains to show that $I + \mathbb{F}_x$ is injective. Hence let $f \in L_2(0, 2a)$ such that $(I + \mathbb{F}_x)f = 0$. For this f, (12.4.19) becomes and inequality, and hence also the Cauchy–Schwarz–Bunyakovskii inequality (12.4.18) is an equality. But this happens if and only if one function in the inner product is a nonnegative multiple of the other function, i.e., when there is $\alpha \geq 0$ such that

$$-e^{2i\lambda x} S(\lambda)\hat{f}(-\lambda) = \alpha\hat{f}(\lambda), \quad \lambda \in \mathbb{R}.$$

Observing that $S(0) = 1$ and that in case $f \neq 0$ the Taylor expansion of the entire function $\hat{f}$ about 0 leads to

$$\lim_{\lambda \to 0} \left| \frac{\hat{f}(\lambda)}{\hat{f}(-\lambda)} \right| = 1,$$

it follows that $\alpha = 1$, and the identity theorem gives

$$\hat{f}(\lambda) + e^{2i\lambda x} S(\lambda) \hat{f}(-\lambda) = 0, \quad \lambda \in \mathbb{C}, \ S(\lambda) \neq 0, \tag{12.4.20}$$

which is also trivially true in case $f = 0$.

First let $x = 0$. Then (12.4.20) holds with $x = 0$. Hence all poles of S must be cancelled by zeros of $\hat{f}$, and since the poles of S are the zeros of $\check{\psi}$, there is an entire function ω such that

$$\hat{f} = \omega \psi.$$

Multiplying (12.4.20) by $\check{\psi}$ we arrive at

$$\omega \psi \check{\psi} = \hat{f} \check{\psi} = -\check{\hat{f}} \psi = -\check{\omega} \check{\psi} \psi,$$

which shows that ω is an odd entire function. We have seen at the beginning of this section that ψ has no zeros in the closed lower half-plane and satisfies the estimate (12.4.9). Hence $\frac{1}{\psi}$ is a bounded analytic function in the closed lower half-plane. Since $\hat{f}$ is the Fourier transform of a function in $L_2[0, 2a]$, also $\hat{f}$ is bounded in the closed lower half-plane. Indeed, it is easy to see that the integral over $|f|$ is such a bound. Hence $\omega = \frac{\hat{f}}{\psi}$ is bounded in the closed lower half-plane and then also bounded in the closed upper half-plane due to the symmetry of ω. By Liouville's theorem, ω is constant. But $\frac{1}{\psi}$ is bounded on $\mathbb{R}$, see (12.4.9), and $\hat{f}$ is an L_2 function on $\mathbb{R}$, so that this constant must be zero. We have shown that $\omega = 0$, and $\hat{f} = 0$ follows. Therefore $I + \mathbb{F}_0$ is injective, and $I + \mathbb{F}_0 \gg 0$ is proved.

Now let $x > 0$. Recall that we consider $(I + \mathbb{F}_x)f = 0$. Since $(\mathbb{F}_x f)(y) = 0$ for $y > 2(a - x)$, this implies $f(y) = 0$ for $y > 2(a - x)$. It follows for $0 < h < x$ that the function

$$f_h = \frac{1}{2}(f \circ \tau_{-x+h} + f \circ \tau_{-x-h})$$

has support in $[x - h, 2a - x + h]$, and therefore $f_h \in L_2[0, 2a]$. In view of (12.4.16) we conclude that

$$\hat{f}_h(\lambda) = \hat{f}(\lambda) e^{-i\lambda x} \cos \lambda h. \tag{12.4.21}$$

A substitution of (12.4.21) into (12.4.20) leads to

$$\hat{f}_h(\lambda) - S(\lambda) \hat{f}_h(-\lambda) = 0, \quad \lambda \in \mathbb{R},$$

which means that $((I + \mathbb{F}_0)f_h, f_h) = 0$ for $0 < h < a$. In view of

$$((I + \mathbb{F}_0)f_h, f_h) = ((I + \mathbb{F}_0)^{\frac{1}{2}} f_h, (I + \mathbb{F}_0)^{\frac{1}{2}} f_h)$$

we conclude that $(I + \mathbb{F}_0)f_h = 0$. But from the case $x = 0$ we already know that this implies $f_h = 0$. Then (12.4.21) gives $\hat{f} = 0$ and thus $f = 0$.   □

Lemma 12.4.5. *Let $v_1^2 > 0$. For $0 \leq x \leq a$, define the operator $\mathbb{F}_x^0$ as the restriction of $\mathbb{F}_x$ to $C[0, 2a]$. Then $\mathbb{F}_x^0$ is a compact operator in the space $C[0, 2a]$, and $I + \mathbb{F}_x^0$ is invertible.*

Proof. We begin by defining an auxiliary operator T by

$$((Tg)f)(y) = \int_0^{2a} g(y+t)f(t)\,dt, \quad g \in L_2(0, 2a), \ f \in C[0, 2a], \ y \in [0, 2a].$$

Here we set $g(x) = 0$ for $x > 2a$. Clearly, $(Tg)f$ is a measurable function on $[0, 2a]$ for all $g \in L_2(0, 2a)$ and $f \in C[0, 2a]$, and

$$|((Tg)f)(y)| \leq \|g\| \, \|f\| \leq \sqrt{2a} \|g\| \, \|f\|_0,$$

where $\|\cdot\|_0$ is the maximum norm in the Banach space $C[0, 2a]$. This shows that $T \in L(L_2(0, 2a), L(C[0, 2a], L_\infty(0, 2a)))$. Clearly, for continuous g with $g(2a) = 0$, also $(Tg)f$ is continuous. Observing that the set of such functions g is dense in $L_2(0, 2a)$, that $C[0, 2a]$ is closed in $L_\infty(0, 2a)$ and that therefore $L(C[0, 2a], C[0, 2a])$ is closed in $L(C[0, 2a], L_\infty(0, 2a))$, it follows that $T \in L(L_2(0, 2a), L(C[0, 2a]))$. Furthermore, if $g \in W_2^1(0, 2a)$ with $g(2a) = 0$, then we can write

$$g(x) = -\int_x^{2a} g'(\tau)\,d\tau, \quad x \geq 0,$$

see, e. g., [189, Proposition 2.1.5], and therefore

$$((Tg)f)(y) = -\int_0^{2a} \int_{y+t}^{2a} g'(\tau)\,d\tau\, f(t)\,dt \quad = -\int_0^{2a} \int_y^{2a} g'(\tau + t)\,d\tau\, f(t)\,dt$$

$$= -\int_y^{2a} \int_0^{2a} g'(\tau + t)\,dt\, f(t)\,d\tau = -\int_y^{2a} ((Tg')f)(\tau)\,d\tau.$$

Since $(Tg')f$ is continuous by what we have already shown, it follows that $(Tg)f$ is differentiable with continuous derivative $((Tg)f)' = (Tg')f$. Then the norm of $(Tg)f$ in $C^1[0, 2a]$ is

$$\|(Tg)f\|_0 + \|((Tg)f)'\|_0 = \|(Tg)f\|_0 + \|(Tg')f\|_0 \leq \sqrt{2a}(\|g\| + \|g'\|)\|f\|_0,$$

which shows that $Tg \in L(C[0, 2a], C^1[0, 2a])$ if $g \in W_2^1[0, 2a]$ with $g(2a) = 0$. But since the embedding from $C^1[0, 2a]$ into $C[0, 2a]$ is compact, see, e. g., [189, Proposition 2.1.7 and Lemma 2.4.1], it follows that Tg is a compact operator on $C[0, 2a]$.

As we have seen in the proof of Lemma 12.4.4, we can take $\gamma = 0$ in (12.4.7). For $0 \leq x \leq a$ we now apply the above auxiliary result to the operators

$$\mathbb{F}_x^0 = T(F \circ \tau_{2x}),$$

which proves that $\mathbb{F}_x^0$ is a compact operator in $C[0, 2a]$. Therefore $I + \mathbb{F}_x^0$ is a Fredholm operator with index 0. But $N(I + \mathbb{F}_x^0) \subseteq N(I + \mathbb{F}_x)$ and $I + \mathbb{F}_x$ is injective by Lemma 12.4.4. It follows that $I + \mathbb{F}_x$ is invertible. $\qquad \square$

Lemma 12.4.6. *The operator functions* $x \mapsto \mathbb{F}_x$ *and* $x \mapsto \mathbb{F}_x^0$ *are differentiable on* $[0, a]$. *The derivative* $\mathbb{F}_x'$ *of* $x \mapsto \mathbb{F}_x$ *at* x *is the operator*

$$(\mathbb{F}_x' f)(y) = 2 \int_0^{2a} F'(2x + y + t) f(t)\, dt, \quad f \in L_2(0, 2a), \ y \in [0, 2a], \quad (12.4.22)$$

and $(\mathbb{F}_x^0)'$ *is the restriction of* $\mathbb{F}_x'$ *to* $C[0, 2a]$. *For each* $x \in [0, a]$ *and for all* $f \in L_2(a, b)$, *the function* $\mathbb{F}_x f$ *is differentiable, and* $(\mathbb{F}_x f)' = \frac{1}{2} \mathbb{F}_x' f$.

Proof. Arguing as at the beginning of the proof of Lemma 12.4.4 we see that $\mathbb{F}_x'$ is a bounded operator in $L_2(0, 2a)$. For $x, x' \in [0, a]$ we define the auxiliary operator

$$\mathbb{F}_{x, x'} = \mathbb{F}_{x'} - \mathbb{F}_x - (x' - x)\mathbb{F}_x'.$$

Then it follows for $f \in L_2(0, 2a)$ and $y \in [0, 2a]$ that

$$
\begin{aligned}
&(\mathbb{F}_{x, x'} f)(y) \\
&\quad = \int_0^{2a} [F(2x' + y + t) - F(2x + y + t) - 2(x' - x)F'(2x + y + t)] f(t)\, dt.
\end{aligned}
$$

Since $F \in W_2^1(0, a)$ by Lemma 12.4.3, we can write

$$F(2x' + y + t) - F(2x + y + t) = \int_{2x}^{2x'} F'(\tau + y + t)\, d\tau,$$

see, e. g., [189, Proposition 2.1.5]. Therefore

$$(\mathbb{F}_{x, x'} f)(y) = \int_0^{2a} \int_{2x}^{2x'} [F'(\tau + y + t) - F'(2x + y + t)] f(t)\, d\tau\, dt.$$

Let $\varepsilon > 0$. Since the set of continuous functions on $[0, 2a]$ is dense in $L_2(0, a)$, there is a continuous function g on $[0, 2a]$ such that $\|F' - g\| < \varepsilon$. Then

$$
\begin{aligned}
&\left| \int_0^{2a} \int_{2x}^{2x'} [(F' - g)(\tau + y + t) - (F' - g)(2x + y + t)] f(t)\, d\tau\, dt \right| \\
&\qquad \leq \left| \int_{2x}^{2x'} \int_0^{2a} |(F' - g)(\tau + y + t)|\, |f(t)|\, dt\, d\tau \right| \\
&\qquad\quad + \left| \int_{2x}^{2x'} \int_0^{2a} |(F' - g)(2x + y + t)|\, |f(t)|\, dt\, d\tau \right| \\
&\qquad \leq 4|x' - x|\, \|F' - g\|\, \|f\| < 4\varepsilon|x' - x|\, \|f\|.
\end{aligned}
$$

Since g is continuous and therefore uniformly continuous, there exists $\delta > 0$ such that $|g(t) - g(t')| < \varepsilon$ for all $t, t' \in [0, 2a]$ with $|t - t'| < 2\delta$. Hence it follows for

$|x' - x| < \delta$ that

$$\left| \int_0^{2a} \int_{2x}^{2x'} [g(\tau + y + t) - g(2x + y + t)] f(t)\, dt \right| \le \varepsilon \left| \int_{2x}^{2x'} \int_0^{2a} |f(t)|\, dt\, d\tau \right|$$

$$\le \varepsilon \sqrt{2a} |x' - x| \|f\|.$$

Altogether, we conclude that

$$\|\mathbb{F}_{x,x'} f\|_\infty \le \varepsilon (4 + \sqrt{2a}) |x' - x| \|f\|.$$

This shows that $\mathbb{F}_x$ is differentiable as an operator function from $L_2(0, 2a)$ to $L_\infty(0, 2a)$ with derivative $\mathbb{F}'_x$, see, e.g., [66, Section 8.1]. By the product rule, see, e.g., [66, 8.3.1], the same is clearly true if these operators are considered as operators into $L_2(0, 2a)$, that is, multiplied by the constant embedding from $L_\infty(0, 2a)$ to $L_2(0, 2a)$.

The same reasoning as above applies to the operator function $x \mapsto \mathbb{F}_x^0$. We only have to restrict f to functions in $C[0, 2a]$ and replace $\|f\|$ with $\sqrt{2a} \|f\|_0$.

In (12.4.14), we can interchange integration and differentiation with respect to y, and $(\mathbb{F}_x f)' = \frac{1}{2} \mathbb{F}'_x f$ is therefore an immediate consequence of (12.4.22). □

Proposition 12.4.7. *Let $v_1^2 > 0$ and let the function F be as defined in (12.4.7). Then for every $x \in [0, a]$, $(I + \mathbb{F}_x) g = -F \circ \tau_{2x}$ has a unique solution $g = G(x, \cdot)$ in $L_2(0, 2a)$. The function G is continuous on $[0, a] \times [0, 2a]$ and $G(\cdot, y) \in W_2^1(0, a)$ for $y \in [0, 2a]$.*

Proof. As we have seen in the proof of Lemma 12.4.4, we can take $\gamma = 0$ in (12.4.7). The existence and uniqueness of G follows immediately from the invertibility of $I + \mathbb{F}_x$ for all $x \in [0, a]$, which was shown in Lemma 12.4.4, and we have

$$G(x, y) = -((I + \mathbb{F}_x)^{-1}(F \circ \tau_{2x}))(y)$$

for all $(x, y) \in [0, a] \times [0, 2a]$. Since $F \circ \tau_{2x}$ is continuous, Lemma 12.4.5 shows that we can also write

$$G(x, y) = -((I + \mathbb{F}_x^0)^{-1}(F \circ \tau_{2x}))(y),$$

and therefore $G(x, \cdot)$ is continuous. By Lemma 12.4.6, $x \mapsto \mathbb{F}_x^0$ is differentiable and therefore continuous on $[0, a]$, so that also $x \mapsto (I + \mathbb{F}_x^0)^{-1}$ is continuous. For $x, x' \in [0, a]$ we therefore conclude

$$|G(x', y) - G(x, y)| \le |((I + \mathbb{F}_{x'}^0)^{-1}(F \circ \tau_{2x'}))(y) - ((I + \mathbb{F}_x^0)^{-1}(F \circ \tau_{2x'}))(y)|$$
$$+ |((I + \mathbb{F}_x^0)^{-1}(F \circ \tau_{2x'}))(y) - ((I + \mathbb{F}_x^0)^{-1}(F \circ \tau_{2x}))(y)|$$
$$\le \|(I + \mathbb{F}_{x'}^0)^{-1} - (I + \mathbb{F}_x^0)^{-1}\| \|(F \circ \tau_{2x'})\|$$
$$+ \|(I + \mathbb{F}_x^0)^{-1}\| \|F \circ \tau_{2x'} - F \circ \tau_{2x}\|$$
$$\to 0 \text{ as } x' \to x.$$

In the last step we also have used that F is uniformly continuous. We have thus shown that $G(x, \cdot)$ and $G(\cdot, y)$ are continuous for all $x \in [0, a]$ and $y \in [0, 2a]$, and a standard argument shows that G is continuous on $[0, a] \times [0, 2a]$.

Since $x \mapsto \mathbb{F}_x$ is differentiable by Lemma 12.4.6 and since $I + \mathbb{F}_x$ is invertible for all $x \in [0, a]$ by Lemma 12.4.4, $x \mapsto (I + \mathbb{F}_x)^{-1}$ is differentiable on $[0, a]$ by the quotient rule, see [66, 8.3.2], and

$$\frac{d}{dx}(I + \mathbb{F}_x)^{-1} = -(I + \mathbb{F}_x)^{-1}\mathbb{F}'_x(I + \mathbb{F}_x)^{-1}.$$

Furthermore, also $x \mapsto F \circ \tau_{2x}$ is differentiable with derivative $2F' \circ \tau_{2x}$, and the product rule, see [66, 8.3.1], gives

$$\frac{\partial}{\partial x}G(x, \cdot) = (I + \mathbb{F}_x)^{-1}\mathbb{F}'_x(I + \mathbb{F}_x)^{-1}(F \circ \tau_{2x}) - 2(I + \mathbb{F}_x)^{-1}(F' \circ \tau_{2x}). \quad (12.4.23)$$

Since $F \circ \tau_{2x}$ is continuous, we know from Lemma 12.4.5 that the first summand can be written as

$$G_1(x, \cdot) := (I + \mathbb{F}^0_x)^{-1}(\mathbb{F}^0_x)'(I + \mathbb{F}^0_x)^{-1}(F \circ \tau_{2x}),$$

and since

$$x \mapsto (I + \mathbb{F}^0_x)^{-1}(\mathbb{F}^0_x)'(I + \mathbb{F}^0_x)^{-1}$$

is a continuous operator function in $L(C[0, 2a])$, it follows like we have shown for G above that also G_1 is continuous on $[0, a] \times [0, 2a]$. Next we consider the auxiliary operator G_0 defined by

$$(G_0 g)(x, \cdot) := (I + \mathbb{F}_x)^{-1}(g \circ \tau_{2x}), \quad g \in L_2(0, 2a), x \in [0, a].$$

Again for continuous g, we can replace $\mathbb{F}_x$ with $\mathbb{F}^0_x$, and the above considerations show that $G_0 g$ is continuous on $[0, 2a]$ and therefore square integrable. We calculate

$$\left| \int_0^a \int_0^{2a} \left((I + \mathbb{F}_x)^{-1}(g \circ \tau_{2x}) \right)(y)\, dy\, dx \right| \leq \int_0^a \|(I + \mathbb{F}_x)^{-1}\|\, \|g\|\, dx$$

$$\leq a \max_{x \in [0, a]} \|(I + \mathbb{F}_x)^{-1}\|\, \|g\|.$$

By continuity, this extends to all $g \in L_2(0, 2a)$, and we obtain that G_0 is a bounded operator from $L_2(0, a)$ to $L_2((0, a) \times (0, 2a))$. Therefore, $G_2 := G_0 F'$ belongs to $L_2((0, a) \times (0, 2a))$. Finally, let $y \in [0, 2a]$. Then

$$((I + \mathbb{F}_x)G_2)(x, \cdot) = F' \circ \tau_{2x}$$

gives

$$G_2(x, y) = -\int_0^a F(2x + y + t)G_2(x, t)\, dt + F'(2x + y).$$

Since $(x, t) \mapsto F(2x + y + t)G_2(x, t)$ is a square integrable kernel, it follows that the function $x \mapsto \int_0^a F(2x + y + t)G_2(x, t)\, dt$ is square integrable on $(0, a)$, see, e. g., [109, p. 240]. Altogether, $\frac{\partial}{\partial x}G(\cdot, y) = G_1(\cdot, y) - 2G_2(\cdot, y) \in L_2(0, a)$ follows. $\square$

Proposition 12.4.8 ([180, Theorem 3.3.1]). *Let $v_1^2 > 0$ and consider the function F defined in (12.4.7). Then the fundamental equation (12.4.11) has a unique continuous solution H on $D_0 = \{(x, y) \in \mathbb{R}^2 : 0 \le x \le y\}$. Furthermore, $q \in L_2(0, a)$, where*

$$q(x) := -2\frac{d}{dx}H(x, x), \quad x \in (0, a), \tag{12.4.24}$$

H and q are real valued, and H satisfies the integral equation

$$H(x, y) = \frac{1}{2}\int_{\frac{x+y}{2}}^{a+\frac{y-x}{2}}\int_{x+|y-\tau|}^{a-|a-\tau|} q(\sigma)H(\sigma, \tau)\, d\sigma\, d\tau + \frac{1}{2}\int_{\frac{x+y}{2}}^{2a} q(\sigma)\, d\sigma. \tag{12.4.25}$$

Proof. As we have seen in the proof of Lemma 12.4.4, we can take $\gamma = 0$ in (12.4.7). Recall that H satisfies the fundamental equation if and only if $(x, y) \mapsto H(x, x+y)$ satisfies (12.4.13). But from Proposition 12.4.7 we know that (12.4.13) has a unique solution G. Hence the fundamental equation has the unique continuous solution H given by $H(x, x + y) = G(x, y)$ for $(x, y) \in D_0$. Since F is real valued, Also H is real valued. In view of $H(x, x) = G(x, 0)$, $q \in L_2(0, a)$ is a consequence of $G(\cdot, 0) \in W_2^1(0, a)$, which was shown in Proposition 12.4.7.

To prove (12.4.25), we first consider the case that F is a twice continuously differentiable function on $(0, \infty)$ with support in $[0, 2a]$ such that $I + \mathbb{F}_x^0$ has an inverse for all $x \in [0, a]$. We observe that $\mathbb{F}_x^0$ depends continuously on x and hence a compactness argument shows that the norm of $(I + \mathbb{F}_x^0)^{-1}$ is uniformly bounded for $x \in [0, a]$. In that case, the operator function $x \mapsto \mathbb{F}_x^0$ is twice differentiable and we can differentiate once more on both sides of (12.4.22). Adapting the proof of Proposition 12.4.7 to this case we see that G_x and G_{xx} exist and are continuous. It is also immediately clear from (12.4.13) that G_y, G_{xy} and G_{yy} exist and are continuous. Furthermore, for $x \in [0, a]$, $y \in [0, 2a]$ and $h \in \mathbb{R}$ such that $x+h \in [0, a]$ and $y + 2h \in [0, 2a]$ we have

$$
\begin{aligned}
&G(x, y + 2h) - G(x, y) \\
&= ((I + \mathbb{F}_x^0)^{-1}(F \circ \tau_{2x}))(y+2h) - ((I + \mathbb{F}_x^0)^{-1}(F \circ \tau_{2x}))(y) \\
&= ((I + \mathbb{F}_x^0)^{-1}(F \circ \tau_{2(x+h)}))(y) - ((I + \mathbb{F}_x^0)^{-1}(F \circ \tau_{2x}))(y) \\
&= ((I + \mathbb{F}_x^0)^{-1}[F \circ \tau_{2(x+h)} - F \circ \tau_{2x}])(y).
\end{aligned}
$$

A reasoning as in the proof of Lemma 12.4.6 shows that

$$G_y(x, y) = \frac{1}{2}((I + \mathbb{F}_x^0)^{-1}(F' \circ \tau_{2x}))(y).$$

Arguing as in the proof of Proposition 12.4.7 and as above in this proof we see that also G_{yx} exists and is continuous. We have shown that G is twice continuously differentiable, and therefore also H is twice continuously differentiable.

Differentiating the fundamental equation (12.4.11) twice with respect to x and twice with respect to y we obtain

$$F''(x+y) + H_{xx}(x,y) - \frac{d}{dx}[H(x,x)F(x+y)]$$
$$- H_x(x,x)F(x+y) + \int_x^\infty H_{xx}(x,t)F(y+t)\,dt = 0$$

and

$$F''(x+y) + H_{yy}(x,y) + \int_x^\infty H(x,t)F''(y+t)\,dt = 0.$$

Integrating by parts twice we have

$$\int_x^\infty H(x,t)F''(y+t)\,dt = -H(x,x)F'(x+y) + H_y(x,x)F(x+y)$$
$$+ \int_x^\infty H_{yy}(x,t)F(y+t)\,dt.$$

Taking now the difference of the two second-order partial derivatives we arrive at

$$H_{xx}(x,y) - H_{yy}(x,y) + q(x)F(x+y)$$
$$+ \int_x^\infty [H_{xx}(x,t) - H_{yy}(x,t)]F(y+t)\,dt = 0.$$

With $y = x + \tilde{y}$ this equation becomes

$$H_{xx}(x, x+\tilde{y}) - H_{yy}(x, x+\tilde{y}) + q(x)F(2x+\tilde{y})$$
$$+ \int_0^\infty [H_{xx}(x, x+t) - H_{yy}(x, x+t)]F(2x+\tilde{y}+t)\,dt = 0. \quad (12.4.26)$$

Since the fundamental equation in the form (12.4.12) gives

$$q(x)(F \circ \tau_{2x}) = -q(x)(I + \mathbb{F}_x)H(x, x + \cdot) = -(I + \mathbb{F}_x)(q(x)H(x, x + \cdot)),$$

equation (12.4.26) can be written as

$$(I + \mathbb{F}_x)[H_{xx}(x, x + \cdot) - H_{yy}(x, x + \cdot)) - q(x)H(x, x + \cdot)] = 0.$$

Defining

$$\varphi(x,y) = H_{xx}(x,y) - H_{yy}(x,y) - q(x)H(x,y),$$

$\varphi(x, x + \cdot)$ is continuous for all $x \in [0, a]$ and satisfies

$$(I + \mathbb{F}_x)\varphi(x, x + \cdot) = 0.$$

In view of Lemma 12.4.4 we conclude that $\varphi = 0$.

Now put $\xi = 2a + x - y$ and $\eta = 2a - x - y$ and define

$$\tilde{u}(\xi, \eta) := H(x, y) = H\left(\frac{\xi - \eta}{2}, 2a - \frac{\xi + \eta}{2}\right). \qquad (12.4.27)$$

We observe that $0 \le x \le y \le 2a - x$ gives the domain $0 \le \eta \le \xi \le 2a$ for ξ and η. As in the proof of Theorem 12.1.3 we obtain

$$\tilde{u}_{\xi\eta} = -\frac{1}{4}\left(H_{xx}\left(\frac{\xi - \eta}{2}, 2a - \frac{\xi + \eta}{2}\right) - H_{yy}\left(\frac{\xi - \eta}{2}, 2a - \frac{\xi + \eta}{2}\right)\right),$$

and $\varphi = 0$ leads to

$$\tilde{u}_{\xi\eta}(\xi, \eta) = -\frac{1}{4}q\left(\frac{\xi - \eta}{2}\right)\tilde{u}(\xi, \eta).$$

With

$$U(\xi, \eta) := \frac{1}{4}\int_{\xi}^{2a}\int_{0}^{\eta} q\left(\frac{\sigma - \tau}{2}\right)\tilde{u}(\sigma, \tau)\, d\tau\, d\sigma$$

we conclude that

$$\tilde{u}_{\xi\eta} - U_{\xi\eta} = 0.$$

Hence there are continuous functions f and g on $[0, 2a]$ such that

$$\tilde{u}(\xi, \eta) = U(\xi, \eta) + f(\xi) + g(\eta), \quad 0 \le \eta \le \xi \le 2a.$$

For definiteness, we may assume that $g(0) = 0$. For $\xi \in [0, 2a]$ we have

$$\tilde{u}(\xi, 0) = H\left(\frac{\xi}{2}, 2a - \frac{\xi}{2}\right) = 0 \quad \text{and} \quad U(\xi, 0) = 0,$$

which shows that $f = 0$. Similarly,

$$\tilde{u}(2a, \eta) = H\left(a - \frac{\eta}{2}, a - \frac{\eta}{2}\right) = \frac{1}{2}\int_{a-\frac{\eta}{2}}^{a} q(\sigma)\, d\sigma \quad \text{and} \quad U(2a, \eta) = 0,$$

which shows that

$$g(\eta) = \frac{1}{2}\int_{a-\frac{\eta}{2}}^{a} q(\sigma)\, d\sigma.$$

Altogether, we have shown that

$$\tilde{u}(\xi, \eta) = \frac{1}{4}\int_{\xi}^{2a}\int_{0}^{\eta} q\left(\frac{\sigma - \tau}{2}\right)\tilde{u}(\sigma, \tau)\, d\tau\, d\sigma + \frac{1}{2}\int_{a-\frac{\eta}{2}}^{a} q(\sigma)\, d\sigma. \qquad (12.4.28)$$

Now let F be defined by (12.4.7). In $W_2^1(0, \infty)$ we can approximate the restriction of F to $[0, \infty)$ by a sequence $(F_n)_{n\in\mathbb{N}}$ of twice differentiable functions on $[0, \infty)$ with support in $[0, 2a]$. Consider the corresponding operators $\mathbb{F}^0_{n,x}$, which converge as $n \to \infty$ and depend continuously on x. Then a compactness argument shows that we may assume without loss of generality that $I + \mathbb{F}^0_{n,x}$ is invertible for all $n \in \mathbb{N}$ and $x \in [0, a]$. Furthermore, $(I + \mathbb{F}^0_{n,x})^{-1}$ converges uniformly in x to $(I + \mathbb{F}^0_x)^{-1}$. Since also $F_n \to F$ uniformly as $n \to \infty$, a standard argument shows that $G_n \to G$ uniformly as $n \to \infty$, where the functions G_n are defined as

$$G_n(x, y) = -((I + \mathbb{F}^0_{n,x})^{-1}(F_n \circ \tau_x))(y).$$

Hence also $H_n \to H$ uniformly as $n \to \infty$ and the corresponding functions $\tilde{u}_n$ defined by (12.4.27) with H_n converge uniformly to $\tilde{u}$ as $n \to \infty$. Since the functions $\tilde{u}_n$ satisfy the integral equation (12.4.28), it follows that also $\tilde{u}$ satisfies (12.4.28). Substituting H for $\tilde{u}$ in (12.4.28) shows that H satisfies the integral equation (12.4.25). Here we have to observe that the region of integration is determined by

$$2a + x - y \le 2a + \sigma - \tau \le 2a \quad \text{and} \quad 0 \le 2a - \sigma - \tau \le 2a - x - y,$$

which can be rewritten as

$$\frac{x + y}{2} \le \tau \le a + \frac{y - x}{2}, \quad \tau + x - y \le \sigma \le \tau, \quad \text{and} \quad x + y - \tau \le \sigma \le 2a - \tau. \qquad \square$$

Lemma 12.4.9. *Let $q \in L_2(0, a)$. Then the integral equation*

$$H(x, y) = \frac{1}{2} \int_{\frac{x+y}{2}}^{a + \frac{y-x}{2}} \int_{x+|y-\tau|}^{a - |a-\tau|} q(\sigma) H(\sigma, \tau) \, d\sigma \, d\tau + \frac{1}{2} \int_{\frac{x+y}{2}}^{2a} q(\sigma) \, d\sigma \qquad (12.4.29)$$

has a unique solution H on $\{(x, y) \in \mathbb{R}^2 : 0 \le x \le y \le 2a - x\}$.

Proof. Putting $\xi = 2a + x - y$ and $\eta = 2a - x - y$ and

$$\tilde{u}(\xi, \eta) := H(x, y) = H\left(\frac{\xi - \eta}{2}, 2a - \frac{\xi + \eta}{2}\right),$$

we have seen in the proof of Proposition 12.4.8 that the integral equation (12.4.29) becomes equivalent to the integral equation

$$\tilde{u}(\xi, \eta) = \frac{1}{4} \int_\xi^{2a} \int_0^\eta q\left(\frac{\sigma - \tau}{2}\right) \tilde{u}(\sigma, \tau) \, d\tau \, d\sigma + \frac{1}{2} \int_{a - \frac{\eta}{2}}^a q(\sigma) \, d\sigma$$

for $0 \le \eta \le \xi \le 2a$. From the proof of Lemma 12.1.1 we see that its statement remains true for fixed ξ_0 and η_0. Thus the above integral equation can be written as $\tilde{u} = -\tilde{T}\tilde{u} + \tilde{g}$ with invertible operator $I + \tilde{T}$ on $C(\{(\xi, \eta) \in \mathbb{R}^2 : 0 \le \eta \le \xi \le 2a\})$. Hence the integral equation for $\tilde{u}$ has a unique continuous solution, and it follows that (12.4.29) has a unique continuous solution H. $\qquad \square$

12.5 Two spectra and the fundamental equation

Lemma 12.5.1. *The spectrum of the Sturm–Liouville problem*

$$-y'' + q(x)y = \lambda y, \quad 0 \le x \le a, \tag{12.5.1}$$

$$y(0) = 0, \quad \cos \beta y(a) - \sin \beta y'(a) = 0, \tag{12.5.2}$$

with a real potential $q \in L_2(0, a)$ and $\beta \in [0, \pi]$ consists of an increasing sequence of simple real eigenvalues $(\lambda_k(\beta))_{k=1}^\infty$ which tend to ∞. For $0 < \beta' < \beta \le \pi$, the eigenvalues interlace as follows:

$$\lambda_1(\beta') < \lambda_1(\beta) < \lambda_2(\beta') < \lambda_2(\beta) < \cdots.$$

Proof. The result is well known. Indeed, it is easy to see from Theorems 10.3.5, 10.3.8 and the spectral theorem for compact operators that the spectrum consists of real eigenvalues which are bounded below and tend to ∞. Since the initial value problem (12.5.1), $y(0) = 0$ has a one-dimensional solution space, it follows that all eigenvalues are simple and that $\lambda_k(\beta) \ne \lambda_j(\beta')$ for all $k, j \in \mathbb{N}$ and $0 < \beta' < \beta \le \pi$. From [143, Theorem 4.2, (4.5)] we know that $\lambda_k(\beta)$ depends continuously on β and is strictly increasing as a function of β for all k, and the stated interlacing property of the eigenvalues follows. □

Corollary 12.5.2. *Let $q \in L_2(0, a)$ be real valued. Then the spectra of the Sturm–Liouville problem (12.5.1) subject to the boundary conditions $y(0) = y(a) = 0$ and $y(0) = y'(a) = 0$, respectively, consist of two sequences of real eigenvalues which interlace as follows:*

$$\zeta_1 < \xi_1 < \zeta_2 < \xi_2 < \cdots$$

and obey the asymptotic formulae

$$\xi_k = \frac{\pi^2 k^2}{a^2} - 2\frac{\pi^2 A}{a^2} + \alpha_k, \quad \zeta_k = \frac{\pi^2}{a^2}\left(k - \frac{1}{2}\right)^2 - 2\frac{\pi^2 A}{a^2} + \beta_k, \tag{12.5.3}$$

where $A \in \mathbb{R}$, $(\alpha_k)_{k=1}^\infty \in l_2$ and $(\beta_k)_{k=1}^\infty \in l_2$ are real-valued sequences.

Proof. The first part is an immediate consequence of Lemma 12.5.1 with $\lambda_k(\pi) = \xi_k$ and $\lambda_k(\frac{\pi}{2}) = \zeta_k$. For the asymptotic expansion of the eigenvalues we recall from Corollary 12.2.10 that

$$s(\lambda, a) = \frac{\sin \lambda a}{\lambda} - K(a, a)\frac{\cos \lambda a}{\lambda^2} + \frac{f(\lambda)}{\lambda^2},$$

$$s'(\lambda, a) = \cos \lambda a + K(a, a)\frac{\sin \lambda a}{\lambda} + \frac{g(\lambda)}{\lambda},$$

with $f, g \in \mathcal{L}^a$. An application of Lemma 12.3.3 and (12.2.19) proves (12.5.3) with

$$A = -\frac{a}{\pi^2}K(a, a) = -\frac{a}{2\pi^2}\int_0^a q(x)\, dx. \tag{12.5.4}$$

□

Let $q \in L_2(0, a)$ and define $q_a(x) = q(a - x)$. Clearly, $q_a \in L_2(0, a)$.

Proposition 12.5.3. *Let e_a be the solution of (12.2.1), (12.2.2) with respect to q_a. For $\lambda \in \mathbb{C}$ define*

$$e(\lambda, x) = \begin{cases} e^{-i\lambda a} e_a(-\lambda, a - x) & \text{if } 0 \le x \le a, \\ e^{-i\lambda x} & \text{if } a < x. \end{cases} \tag{12.5.5}$$

Then the function e is the Jost solution of (12.2.1) as defined in Section 2.1.

Proof. Clearly, $e(\lambda, \cdot)$ satisfies the differential equation (12.2.1) on $(0, a)$ and

$$e(\lambda, a) = e^{-i\lambda a} e_a(-\lambda, 0) = e^{-i\lambda a},$$
$$e'(\lambda, a) = -e^{-i\lambda a} e_a'(-\lambda, 0) = -i\lambda e^{-i\lambda a}$$

shows that e is indeed the Jost solution. $\square$

Proposition 12.5.4. *Let K_a be the function $\tilde{K}$ from Theorem 12.2.6 with respect to the potential q_a and define*

$$K_\infty(x, t) = \begin{cases} K_a(a - x, a - t) & \text{if } 0 \le x \le t \le 2a - x, \\ 0 & \text{for all other } x, t \in \mathbb{R}. \end{cases} \tag{12.5.6}$$

Then K_∞ satisfies the integral equation (12.4.29),

$$e(\lambda, x) = e^{-i\lambda x} + \int_x^\infty K_\infty(x, t) e^{-i\lambda t}\, dt, \quad \lambda \in \mathbb{C}, \ x \ge 0, \tag{12.5.7}$$

and

$$K_\infty(x, x) = \frac{1}{2} \int_x^a q(t)\, dt, \quad x \in [0, a]. \tag{12.5.8}$$

Proof. With the aid of (12.2.18) we calculate

$$K_\infty(x, y) = K_a(a - x, a - y) = \frac{1}{2} \int_0^{a - \frac{x+y}{2}} q(a - s)\, ds$$

$$+ \int_0^{a - \frac{x+t}{2}} \int_0^{\frac{y-x}{2}} q(a - s - p) K_\infty(a - s - p, a - s + p)\, dp\, ds$$

$$= \frac{1}{2} \int_{\frac{x+y}{2}}^{a + \frac{y-x}{2}} \int_{x + |y - \tau|}^{a - |a - \tau|} q(\sigma) K_\infty(\sigma, \tau)\, d\sigma\, d\tau + \frac{1}{2} \int_{\frac{x+y}{2}}^{2a} q(\sigma)\, d\sigma.$$

From Proposition 12.5.3, Theorem 12.2.6 and with the aid of the transformation $\tau = a - t$ we infer for $0 \le x \le a$ that

$$e(\lambda, x) = e^{-i\lambda a} e_a(-\lambda, a - x)$$

$$= e^{-i\lambda a} e^{i\lambda(a-x)} + \int_{x-a}^{a-x} K_a(a - x, t) e^{i\lambda t} e^{-i\lambda a} \, dt$$

$$= e^{-i\lambda x} + \int_{x}^{2a-x} K_a(a - x, a - t) e^{-i\lambda t} \, dt,$$

and (12.5.7) follows in view of (12.5.6). For $x > a$, (12.5.7) is obvious. Finally, we conclude from (12.2.19) that

$$K_\infty(x, x) = K_a(a - x, a - x) = \frac{1}{2} \int_0^{a-x} q_a(\sigma) \, d\sigma$$

$$= \frac{1}{2} \int_x^{a} q_a(a - \tau) \, d\tau = \frac{1}{2} \int_x^{a} q(t) \, dt. \qquad \square$$

The function

$$v_a(\lambda, \cdot) := \frac{e(\lambda, \cdot) e^{i\lambda a} + e(-\lambda, \cdot) e^{-i\lambda a}}{2}$$

is the solution of (12.2.1) satisfying $v_a(\lambda, a) = 1$ and $v'_a(\lambda, a) = 0$, whereas the function

$$u_a(\lambda, \cdot) := \frac{e(-\lambda, \cdot) e^{-i\lambda a} - e(\lambda, \cdot) e^{i\lambda a}}{2i\lambda}$$

is the solution of (12.2.1) satisfying $u_a(\lambda, a) = 0$ and $u'_a(\lambda, a) = 1$. Observe that $e(\lambda, \cdot) = e^{-i\lambda a}(v_a(\lambda, \cdot) - i\lambda u_a(\lambda, \cdot))$. Since $x \mapsto q(a - x)$ belongs to $L_2(0, a)$, v_a and u_a have a representation like c and $-s$ in Theorem 12.2.9, respectively, with x replaced by $a - x$.

Clearly, the zeros of $u_a(\cdot, 0)$ are the eigenvalues of (12.5.1) with the boundary conditions $y(0) = y(a) = 0$, whereas the zeros of $v_a(\cdot, 0)$ are the eigenvalues of (12.2.1) with the boundary conditions $y(0) = y'(a) = 0$. With ζ_k and ξ_k according to Corollary 12.5.2 we put $v_k = \zeta_k^{\frac{1}{2}}$, $u_k = \xi_k^{\frac{1}{2}}$ for $k \in \mathbb{Z}$. Then it follows from Corollary 12.5.2 that these numbers satisfy the assumptions posed at the beginning of Section 12.4. Hence the entire function u defined there has exactly the same zeros as $u_a(\cdot, 0)$, and the entire function v defined there has exactly the same zeros as $v_a(\cdot, 0)$. The function v has the representation (12.3.11), and by the above discussion, also $v_a(\cdot, 0)$ has such a representation, with the same leading term as v. By Corollary 12.2.11 and its proof, both v and $v_a(\cdot, 0)$ are sine type functions. Hence they are multiples of each other by Lemma 11.2.29, and since they have the same leading terms, $v_a(\cdot, 0) = v$ follows. A corresponding argument for the function $\lambda \mapsto \lambda u(\lambda)$ and $\lambda \mapsto -\lambda u_a(\lambda, 0)$ gives $u_a(\cdot, 0) = -u$.

Hence the function ψ defined in (12.4.2) satisfies

$$\psi(\lambda) = e^{-i\lambda a}(v_a(\lambda, 0) - i\lambda u_a(\lambda, 0)) = e(\lambda, 0), \qquad (12.5.9)$$

and it follows that the function S defined in (12.4.3) has the representation

$$S(\lambda) = \frac{e(\lambda, 0)}{e(-\lambda, 0)}, \qquad \lambda \in \mathbb{R}. \qquad (12.5.10)$$

Lemma 12.5.5 ([180, Lemma 3.1.5]). *For $\lambda \neq 0$ and $e(-\lambda, 0) \neq 0$,*

$$-\frac{2i\lambda s(\lambda, \cdot)}{e(-\lambda, 0)} = e(\lambda, \cdot) - S(\lambda)e(-\lambda, \cdot).$$

Proof. The function $s(\lambda, \cdot)$ is the solution of (12.2.1) which satisfies the initial conditions $s(\lambda, 0) = 0$, $s'(\lambda, 0) = 1$. On the other hand,

$$w(\lambda, \cdot) := e(-\lambda, 0)e(\lambda, \cdot) - e(\lambda, 0)e(-\lambda, \cdot)$$

is a solution of (12.2.1) with $w(\lambda, 0) = 0$. It is well known that the Wronskian

$$W(e(-\lambda, \cdot), e(\lambda, \cdot)) = e(-\lambda, \cdot)e'(\lambda, \cdot) - e'(-\lambda, \cdot)e(\lambda, \cdot)$$

is constant, and its value at a is $-2i\lambda$, so that $w'(\lambda, 0) = -2i\lambda$. Hence we have shown that $w(\lambda, \cdot) = -2i\lambda s(\lambda, \cdot)$. $\square$

The proof of the following lemma is extracted from [180, pp. 204–206].

Lemma 12.5.6. *Let $\gamma \geq 0$ such that S is analytic on $\{\lambda \in \mathbb{C} : \operatorname{Im} \lambda \geq \gamma\}$ and define F by*

$$F(x) = \frac{e^{-\gamma x}}{2\pi} \int_{-\infty}^{\infty} S_\gamma(\lambda)e^{i\lambda x}\, d\lambda, \qquad x \in \mathbb{R}, \qquad (12.5.11)$$

where S_γ is defined by (12.4.6). Then the fundamental equation

$$F(x + y) + K_\infty(x, y) + \int_x^\infty K_\infty(x, t)F(y + t)\, dt = 0, \qquad 0 \leq x \leq y, \qquad (12.5.12)$$

is satisfied.

Proof. In view of Lemma 12.5.5 and (12.5.7) we have

$$-\frac{2i\lambda s(\lambda, x)}{e(-\lambda, 0)} = e^{-i\lambda x} - e^{i\lambda x} + \int_x^\infty K_\infty(x, t)e^{-i\lambda t}\, dt - \int_x^\infty K_\infty(x, t)e^{i\lambda t}\, dt$$

$$+ (1 - S(\lambda))\left(e^{i\lambda x} + \int_x^\infty K_\infty(x, t)e^{i\lambda t}\, dt\right),$$

which can be rewritten as

$$-2i\lambda s(\lambda, x) \left(\frac{1}{e(-\lambda, 0)} - 1 \right) + 2i(\sin \lambda x - \lambda s(\lambda, x))$$
$$= \int_x^\infty K_\infty(x, t)e^{-i\lambda t}\, dt - \int_{-\infty}^{-x} K_\infty(x, -t)e^{-i\lambda t}\, dt$$
$$+ (1 - S(\lambda)) \left(e^{i\lambda x} + \int_x^\infty K_\infty(x, t)e^{i\lambda t}\, dt \right). \qquad (12.5.13)$$

We multiply both sides of (12.5.13) by $\frac{1}{2\pi}e^{i\lambda y}$, $y \in \mathbb{R}$, and integrate along Im $\lambda = \gamma$, resulting in an identity which we formally write as $I_l(x, y) = I_r(x, y)$. Then $I_r(x, y)$ is the sum of the 4 integrals

$$I_1(x, y) = \frac{1}{2\pi} \int_{-\infty}^\infty \int_x^\infty K_\infty(x, t)e^{-i(i\gamma+\lambda)t}\, dt\, e^{i(i\gamma+\lambda)y}\, d\lambda,$$

$$I_2(x, y) = -\frac{1}{2\pi} \int_{-\infty}^\infty \int_{-\infty}^{-x} K_\infty(x, -t)e^{-i(i\gamma+\lambda)t}\, dt\, e^{i(i\gamma+\lambda)y}\, d\lambda,$$

$$I_3(x, y) = \frac{1}{2\pi} \int_{-\infty}^\infty S_\gamma(\lambda)e^{i(i\gamma+\lambda)x}e^{i(i\gamma+\lambda)y}\, d\lambda,$$

$$I_4(x, y) = \frac{1}{2\pi} \int_{-\infty}^\infty S_\gamma(\lambda) \int_x^\infty K_\infty(x, t)e^{i(i\gamma+\lambda)t}e^{i(i\gamma+\lambda)y}\, dt\, d\lambda.$$

The Fourier inversion formula gives

$$I_1 = \frac{1}{2\pi} \int_{-\infty}^\infty \int_x^\infty e^{\gamma(t-y)}K_\infty(x, t)e^{-i\lambda t}\, dt\, e^{i\lambda y}\, d\lambda = K_\infty(x, y),$$

and similarly

$$I_2 = -K_\infty(x, -y) = 0 \quad \text{for } y > x.$$

We further calculate

$$I_3 = \frac{e^{-\gamma(x+y)}}{2\pi} \int_{-\infty}^\infty S_\gamma(\lambda)e^{i\lambda(x+y)}\, d\lambda = F(x + y)$$

and

$$I_4(x, y) = \frac{1}{2\pi} \int_x^\infty K_\infty(x, t)e^{-\gamma(y+t)} \int_{-\infty}^\infty S_\gamma(\lambda)e^{i\lambda(y+t)}\, d\lambda\, dt$$
$$= \int_x^\infty K_\infty(x, t)F(y + t)\, dt.$$

Hence we obtain

$$I_r(x,y) = K_\infty(x,y) + F(x+y) + \int_x^\infty K_\infty(x,t)F(y+t)\,dt, \quad 0 \le x < y. \quad (12.5.14)$$

To prove the fundamental equation, it remains to prove that $I_l(x,y) = 0$ for $0 \le x \le a$ and $y > x$.

Therefore, let $x \in [0,a]$ and $y > x$. We define the functions g_1 and g_2 by

$$g_1(\lambda) = \lambda s(\lambda,x)\left(\frac{1}{e(-\lambda,0)} - 1\right)e^{i\lambda y}, \quad g_2(\lambda) = (\sin\lambda x - \lambda s(\lambda,x))e^{i\lambda y}. \quad (12.5.15)$$

The function g_2 is an entire function and the function g_1 is analytic on the set $\{\lambda \in \mathbb{C} : \operatorname{Im}\lambda \ge \gamma\}$ since the poles of S are the zeros of $\lambda \mapsto e(-\lambda,0)$. By Lemma 12.2.8 applied to K_a we conclude that $\frac{\partial}{\partial t}K_\infty(0,\cdot) \in L_2(-a,a)$. Therefore we have in view of (12.5.7) that

$$e(-\lambda,0) - 1 = \int_0^{2a} K_\infty(0,t)e^{i\lambda t}\,dt$$
$$= \frac{1}{i\lambda}\left[K_\infty(0,2a)e^{2i\lambda a} - K_a(0,0)\right] - \frac{1}{i\lambda}\int_0^{2a}\frac{\partial}{\partial t}K_\infty(0,t)e^{i\lambda t}\,dt$$
$$= O\left(\lambda^{-1}\right) \text{ for } \operatorname{Im}\lambda \ge 0.$$

Hence it follows that

$$\frac{1}{e(-\lambda,0)} - 1 = \frac{1}{1+O(\lambda^{-1})} - 1 = O(\lambda^{-1}) \text{ for } \operatorname{Im}\lambda \ge 0. \quad (12.5.16)$$

From (12.2.16) we conclude that $\lambda \mapsto \lambda s(\lambda,x)$ is bounded on each horizontal line and that $g_2(\lambda) = O(\lambda^{-1})$ for λ on any horizontal line. Together with (12.5.16) we conclude that g_1 and g_2 are square integrable on the line $\operatorname{Im}\lambda = \gamma$.

Up to a constant factor, $I_l(x,y)$ is the difference of the integrals $I_5(x,y)$ and $I_6(x,y)$ given by

$$I_{4+j}(x,y) = \int_{-\infty}^\infty g_j(\lambda)\,d\lambda, \quad j = 1,2.$$

Clearly, Corollary 12.2.11 holds with a there replaced by any $x > 0$, and we can conclude in view of Lemma 11.2.6 and (12.5.16) that there are constants $M_1 > 0$ and $M_2 > 0$ such that

$$|g_j(i\gamma + Re^{i\theta})| \le M_j R^{-1}e^{-R(y-x)\sin\theta}, \quad R > 0, \ 0 \le \theta \le \pi, \ j = 1,2.$$

Hence it follows from Cauchy's theorem and Lebesgue's dominated convergence theorem that

$$I_j = -i\lim_{R\to\infty}\int_0^\pi g_j(i\gamma + Re^{i\theta})Re^{i\theta}\,d\theta = 0, \quad j = 5,6. \qquad \square$$

12.6   The potential and two spectra

The following result is well known, but for convenience we will present its proof.

Lemma 12.6.1. *Let h be a bounded measurable function on $[0, 2a] \times [0, 2a]$ and define*

$$(\mathbb{H}f)(x) = \int_x^{2a} h(x,t)f(t)\,dt, \quad f \in L_2(0, 2a), \ 0 \le x \le 2a. \tag{12.6.1}$$

Then the operator $\mathbb{H}$ is a Volterra operator on $L_2(0, 2a)$, i. e., $\mathbb{H}$ is compact and its spectral radius is 0.

Proof. With

$$M = \sup\{|h(x,t)| : 0 \le x, t \le 2a\},$$

the estimate

$$\int_0^{2a} \int_0^{2a} |h(x,t)|^2 \, dt \, dx \le 4aM < \infty$$

shows that $\mathbb{H}$ is an integral operator with L_2 kernel and therefore compact, see, e. g., [109, p. 240].

Since the spectral radius of the adjoint $\mathbb{H}^*$ equals the spectral radius of $\mathbb{H}$, it suffices to show that the spectral radius of $\mathbb{H}^*$ is 0. For $f, g \in L_2(0, 2a)$ we calculate

$$(\mathbb{H}^* f, g) = (f, \mathbb{H}g) = \int_0^{2a} f(x) \int_x^{2a} \overline{h(x,t)\,g(t)} \, dt \, dx$$

$$= \int_0^{2a} \int_0^t \overline{h(x,t)} f(x) \, dx \, \overline{g(t)} \, dt,$$

which shows that the adjoint $\mathbb{H}^*$ of $\mathbb{H}$ has the representation

$$(\mathbb{H}^* f)(t) = \int_0^t \overline{h(x,t)} f(x) \, dx, \quad f \in L_2(0, 2a), \ 0 \le t \le 2a.$$

Let $m > 0$ and define the norm $\| \cdot \|_m$ on $L_2(0, 2a)$ by

$$\|f\|_m^2 = \int_0^{2a} |f(x)|^2 e^{-2mx} \, dx, \quad f \in L_2(0, 2a),$$

which is clearly equivalent to the standard L_2-norm. Then we obtain for each $f \in L_2(0, 2a)$ and $0 \le t \le 2a$ that

$$|(\mathbb{H}^* f)(t)|^2 e^{-2mt} = \left| \int_0^t \overline{h(x,t)} e^{-m(t-x)} f(x) e^{-mx} \, dx \right|^2$$

$$\le \int_0^t |h(x,t)|^2 e^{-2m(t-x)} \, dx \int_0^t |f(x)|^2 e^{-2mx} \, dx$$

$$\le \frac{M^2}{2m} \|f\|_m^2,$$

which gives

$$\|\mathbb{H}^* f\|_m^2 \leq \frac{aM^2}{m} \|f\|_m^2.$$

Since the spectral radius of a bounded operator is bounded by the norm of the operator, it follows that the spectral radius of $\mathbb{H}^*$ is less or equal to $M\sqrt{am^{-1}}$. But $m > 0$ was arbitrary, and it follows that the spectral radius of $\mathbb{H}^*$ is 0.   □

Theorem 12.6.2 ([180, Theorem 3.4.1, p. 248]). *For two sequences $(\xi_k)_{k=1}^\infty$ and $(\zeta_k)_{k=1}^\infty$ of real numbers to be the spectra of the boundary value problems generated by the Sturm–Liouville equation*

$$-y'' + q(x)y = \lambda y \text{ on } [0, a], \tag{12.6.2}$$

with a real potential $q \in L_2(0, a)$ and the boundary conditions $y(0) = y(a) = 0$ and $y(0) = y'(a) = 0$, respectively, it is necessary and sufficient that the sequences interlace:

$$\zeta_1 < \xi_1 < \zeta_2 < \xi_2 < \cdots$$

and obey the asymptotic formulae

$$\xi_k = \frac{\pi^2 k^2}{a^2} - 2\frac{\pi^2 A}{a^2} + \alpha_k, \quad \zeta_k = \frac{\pi^2}{a^2}\left(k - \frac{1}{2}\right)^2 - 2\frac{\pi^2 A}{a^2} + \beta_k,$$

where $A \in \mathbb{R}$, $(\alpha_k)_{k=1}^\infty \in l_2$ and $(\beta_k)_{k=1}^\infty \in l_2$. The potential q is uniquely determined by the sequences $(\xi_k)_{k=1}^\infty$ and $(\zeta_k)_{k=1}^\infty$.

Proof. The necessity of the interlacing property and the asymptotic distribution of the eigenvalues was shown in Corollary 12.5.2.

Next we are going to show that the potential is uniquely determined by the two spectra. Let $q \in L_2(a, b)$ be real valued and let $(\xi_k)_{k=1}^\infty$ and $(\zeta_k)_{k=1}^\infty$ be the corresponding spectra. Without loss of generality we may assume that $\zeta_1 > 0$, which can be achieved by a shift of the eigenvalue parameter λ in (12.6.2). Putting $u_k = (\xi_k)^{\frac{1}{2}}$ and $v_k = (\zeta_k)^{\frac{1}{2}}$ for $k \in \mathbb{N}$, we consider the two functions S and F defined by (12.4.3) and (12.4.7). It remains to show that q is uniquely determined by F. Indeed, in view of (12.5.8), q is uniquely determined by K_∞. From Lemma 12.5.6 we know that the function K_∞ associated with the potential q satisfies the fundamental equation (12.5.12), which is the same as (12.4.11). Then it follows from Proposition 12.4.8 that K_∞ is uniquely determined by F. Altogether, we have shown that the potential is uniquely determined by the two sequences.

Now let two sequences $(\xi_k)_{k=1}^\infty$ and $(\zeta_k)_{k=1}^\infty$ with the required properties be given. Shifting all elements in these two sequences by the same real number $b_0 > -\zeta_1$ if $\zeta_1 \leq 0$, we obtain that $\zeta_1 > 0$. With the two sequences $(\xi_k)_{k=1}^\infty$ and $(\zeta_k)_{k=1}^\infty$ we associate the functions S and F defined by (12.4.3) and (12.4.7). By Proposition 12.4.8, there is a unique solution H of the fundamental equation and a potential q defined by (12.4.24). With this q we associate the two sequences $(\tilde{\xi}_k)_{k=1}^\infty$ and

$(\tilde{\zeta}_k)_{k=1}^{\infty}$ representing the Dirichlet spectrum and the Dirichlet–Neumann spectrum, respectively, of (12.6.2). With these two sequences we can now associate functions $\tilde{S}$ and $\tilde{F}$ defined by (12.4.3) and (12.4.7). Let K_{∞} be the function defined in (12.5.6) with respect to q. By Propositions 12.4.8 and 12.5.4, both H and K_{∞} satisfy (12.4.29) with the same q. But since the solution of the integral equation (12.4.29) is unique by Proposition 12.4.8, $K_{\infty} = H$ follows. Let

$$(\mathbb{H}f)(x) = \int_x^{2a} H(x,t)f(t)\,dt, \quad f \in L_2(0,2a), \; x \in [0,2a].$$

Since $I + \mathbb{H}$ is invertible by Lemma 12.6.1, we obtain from the fundamental equations (12.4.11) and (12.5.12) for $y \in [0,2a]$ that

$$F \circ \tau_y = -(I+\mathbb{H})^{-1}H(\cdot,y) = -(I+\mathbb{H})^{-1}K_{\infty}(\cdot,y) = \tilde{F} \circ \tau_y.$$

Hence $F = \tilde{F}$, and since the definition of F in (12.4.7) is independent of $\gamma \geq 0$, we can take the same γ in F and $\tilde{F}$. But S_{γ} is the Fourier transform of $x \mapsto e^{\gamma x}F(x)$, which show that S_{γ} and therefore S is uniquely determined by F. Thus we have that $S = \tilde{S}$.

Next we will show that the sequences $(\xi_k)_{k=1}^{\infty}$ and $(\zeta)_{k=1}^{\infty}$ are uniquely determined by S. Indeed, it follows from (12.4.1), (12.4.2), (12.4.3) and the proof of Lemma 12.4.1 that

$$\frac{P(\lambda^2) + i\lambda Q(\lambda^2)}{P(\lambda^2) - i\lambda Q(\lambda^2)} = S(\lambda)e^{-2i\lambda a}.$$

The sequences $(\xi_k)_{k=1}^{\infty}$ and $(\zeta_k)_{k=1}^{\infty}$ interlace and are the zeros of the entire functions P and Q, respectively, with a corresponding result for the sequences $(\tilde{\xi}_k)_{k=1}^{\infty}$ and $(\tilde{\zeta}_k)_{k=1}^{\infty}$ and entire functions $\tilde{Q}$ and $\tilde{P}$. Hence it follows for $\lambda \neq 0$ that $Q(\lambda^2) = 0$ if and only if $S(\lambda)e^{-2i\lambda a} = 1$, that $P(\lambda^2) = 0$ if and only if $S(\lambda)e^{-2i\lambda a} = -1$, that $\tilde{Q}(\lambda^2) = 0$ if and only if $S(\lambda)e^{-2i\lambda a} = 1$, and that $\tilde{P}(\lambda^2) = 0$ if and only if $S(\lambda)e^{-2i\lambda a} = -1$. Since $\zeta_1 > 0$, it follows that the two sequences $(\xi_k)_{k=1}^{\infty}$ and $(\zeta_k)_{k=1}^{\infty}$ are indeed uniquely determined by S. Furthermore, the nonzero zeros of $\tilde{P}$ and $\tilde{Q}$ coincide with the nonzero zeros of P and Q, respectively. Hence $0 < \zeta_1 < \xi_1$ and $0 \leq \tilde{\zeta}_1 < \tilde{\xi}_1$, so that $\tilde{\xi}_1 = \xi_1$ since they are positive and the smallest zeros of $\tilde{Q}$ and Q, respectively. Now ζ_1 is a positive zero of P, and therefore also a positive zero of $\tilde{P}$, and $\tilde{\zeta}_1 = \zeta_1$ follows. Hence also the two sequences $(\tilde{\xi}_k)_{k=1}^{\infty}$ and $(\tilde{\zeta}_k)_{k=1}^{\infty}$ are uniquely determined by S. We have shown that the two spectra of the differential equation (12.6.2) with the q given by Proposition 12.4.8, after a possible backshift by b_0, are indeed the two sequences $(\xi_k)_{k=1}^{\infty}$ and $(\zeta_k)_{k=1}^{\infty}$. $\quad\square$

Bibliography

[1] Yu.Sh. Abramov, *Variatsionnye metody v teorii operatornykh puchkov*, Leningrad. Univ., Leningrad, 1983 (Russian).

[2] R.A. Adams, *Sobolev spaces*, Academic Press, New York-London, 1975.

[3] V. Adamyan, H. Langer, and M. Möller, *Compact perturbation of definite type spectra of self-adjoint quadratic operator pencils*, Integral Equations Operator Theory **39** (2001), no. 2, 127–152.

[4] V. Adamyan, R. Mennicken, and V. Pivovarchik, *On the spectral theory of degenerate quadratic operator pencils*, Recent advances in operator theory (Groningen, 1998), Oper. Theory Adv. Appl., vol. 124, Birkhäuser, Basel, 2001, pp. 1–19.

[5] V. Adamyan and V. Pivovarchik, *On the spectra of some classes of quadratic operator pencils*, (Vienna, 1995), Oper. Theory Adv. Appl., vol. 106, Birkhäuser, Basel, 1998, pp. 23–36.

[6] V. Adamjan, V. Pivovarchik, and C. Tretter, *On a class of non-self-adjoint quadratic matrix operator pencils arising in elasticity theory*, J. Operator Theory **47** (2002), no. 2, 325–341.

[7] N.I. Ahiezer and M. Krein, *Some questions in the theory of moments*, Translations of Mathematical Monographs, vol. 2, American Mathematical Society, Providence, R.I., 1962.

[8] T. Aktosun, M. Klaus, and C. van der Mee, *Wave scattering in one dimension with absorption*, J. Math. Phys. **39** (1998), no. 4, 1957–1992.

[9] _____, *Inverse scattering in one-dimensional nonconservative media*, Integral Equations Operator Theory **30** (1998), no. 3, 279–316.

[10] _____, *Wave scattering in 1-D nonconservative media*, Spectral and scattering theory (Newark, DE, 1997), Plenum, New York, 1998, pp. 1–18.

[11] S. Alexander, *Superconductivity of networks. A percolation approach to the effects of disorder*, Phys. Rev. B (3) **27** (1983), no. 3, 1541–1557.

[12] V. Ambarzumian, *Über eine Frage der Eigenwerttheorie*, Z. f. Physik **53** (1929), 690–695 (German).

[13] N. Aronszajn and W.F. Donoghue, *On exponential representations of analytic functions in the upper half-plane with positive imaginary part*, J. Analyse Math. **5** (1956/57), 321–388.

[14] _____, *A supplement to the paper on exponential representations of analytic functions in the upper half-plane with positive imaginary part*, J. Analyse Math. **12** (1964), 113–127.

[15] D.Z. Arov, *Realization of a canonical system with a dissipative boundary condition at one end of the segment in terms of the coefficient of dynamical compliance*, Sibirsk. Mat. Ž. **16** (1975), no. 3, 440–463, 643 (Russian); English transl., Siberian Math. J. **16** (1975), no. 3, 335–352.

[16] F.V. Atkinson, *Discrete and continuous boundary problems*, Mathematics in Science and Engineering, vol. 8, Academic Press, New York, 1964.

[17] F.V. Atkinson and A.B. Mingarelli, *Asymptotics of the number of zeros and of the eigenvalues of general weighted Sturm–Liouville problems*, J. Reine Angew. Math. **375/376** (1987), 380–393.

[18] T.Ya. Azizov and I.S. Iokhvidov, *Linear operators in spaces with an indefinite metric*, Pure and Applied Mathematics (New York), John Wiley & Sons Ltd., Chichester, 1989.

[19] A. Bamberger, J. Rauch, and M. Taylor, *A model for harmonics on stringed instruments*, Arch. Rational Mech. Anal. **79** (1982), no. 4, 267–290.

[20] R. Band, G. Berkolaiko, H. Raz, and U. Smilansky, *The number of nodal domains on quantum graphs as a stability index of graph partitions*, Comm. Math. Phys. **311** (2012), no. 3, 815–838.

[21] R. Band, T. Shapira, and U. Smilansky, *Nodal domains on isospectral quantum graphs: the resolution of isospectrality?*, J. Phys. A **39** (2006), no. 45, 13999–14014.

[22] V. Barcilon, *Explicit solution of the inverse problem for a vibrating string*, J. Math. Anal. Appl. **93** (1983), no. 1, 222–234.

[23] L. Barkwell and P. Lancaster, *Overdamped and gyroscopic vibrating systems*, Trans. ASME J. Appl. Mech. **59** (1992), no. 1, 176–181.

[24] L. Barkwell, P. Lancaster, and A.S. Markus, *Gyroscopically stabilized systems: a class of quadratic eigenvalue problems with real spectrum*, Canad. J. Math. **44** (1992), no. 1, 42–53.

[25] H. Baumgärtel, *Analytic perturbation theory for matrices and operators*, Operator Theory: Advances and Applications, vol. 15, Birkhäuser Verlag, Basel, 1985.

[26] J. von Below, *Can one hear the shape of a network?*, Partial differential equations on multistructures (Luminy, 1999), Lecture Notes in Pure and Appl. Math., vol. 219, Dekker, New York, 2001, pp. 19–36.

[27] C.M. Bender and S. Kuzhel, *Unbounded c-symmetries and their nonuniqueness*, J. Phys. A **45** (2012), no. 44, 444005, 14 pp.

[28] G. Berkolaiko and P. Kuchment, *Introduction to quantum graphs*, Mathematical Surveys and Monographs, vol. 186, American Mathematical Society, Providence, RI, 2013.

[29] T. Betcke, N.J. Higham, V. Mehrmann, C. Schröder, and F. Tisseur, *NLEVP: a collection of nonlinear eigenvalue problems*, ACM Trans. Math. Software **39** (2013), no. 2, Art. 7, 28 pp.

[30] M. Biehler, *Sur une classe d'équations algébriques dont toutes les racines sont réelles*, J. Reine Angew. Math. **87** (1879), 350–352.

[31] V.V. Bolotin, *Nonconservative problems of the theory of elastic stability*, A Pergamon Press Book. The Macmillan Co., New York, 1963.

[32] G. Borg, *Uniqueness theorems in the spectral theory of $y'' + (\lambda - q(x))y = 0$*, Den 11te Skandinaviske Matematikerkongress, Trondheim, 1949, Johan Grundt Tanums Forlag, Oslo, 1952, pp. 276–287.

[33] O. Boyko and V. Pivovarchik, *The inverse three-spectral problem for a Stieltjes string and the inverse problem with one-dimensional damping*, Inverse Problems **24** (2008), no. 1, 015019, 13 pp.

[34] _____, *Inverse spectral problem for a star graph of Stieltjes strings*, Methods Funct. Anal. Topology **14** (2008), no. 2, 159–167.

[35] L. de Branges, *Local operators on Fourier transforms*, Duke Math. J. **25** (1958), 143–153.

[36] _____, *Some mean squares of entire functions*, Proc. Amer. Math. Soc. **10** (1959), 833–839.

[37] _____, *Some Hilbert spaces of entire functions*, Proc. Amer. Math. Soc. **10** (1959), 840–846.

[38] _____, *Some Hilbert spaces of entire functions*, Trans. Amer. Math. Soc. **96** (1960), 259–295.

[39] _____, *Some Hilbert spaces of entire functions. II*, Trans. Amer. Math. Soc. **99** (1961), 118–152.

[40] _____, *Some Hilbert spaces of entire functions. III*, Trans. Amer. Math. Soc. **100** (1961), 73–115.

[41] _____, *Some Hilbert spaces of entire functions. IV*, Trans. Amer. Math. Soc. **105** (1962), 43–83.

[42] _____, *Hilbert spaces of entire functions*, Prentice-Hall Inc., Englewood Cliffs, N.J., 1968.

[43] J. Bronski, M.A. Johnson, and T. Kapitula, *An instability index theory for quadratic pencils and applications*, Comm. Math. Phys. **327** (2014), no. 2, 521–550.

[44] B.M. Brown and R. Weikard, *A Borg–Levinson theorem for trees*, Proc. R. Soc. Lond. Ser. A Math. Phys. Eng. Sci. **461** (2005), no. 2062, 3231–3243.

[45] R.A. Brualdi, *From the Editor-in-Chief. Addendum to: "Generalizations of the Hermite–Biehler theorem" by M.-T. Ho, A. Datta and S.P. Bhattacharyya*, Linear Algebra Appl. **320** (2000), no. 1-3, 214–215.

[46] R. Carlson and V. Pivovarchik, *Ambarzumian's theorem for trees*, Electron. J. Differential Equations (2007), no. 142, 9 pp. (electronic).

[47] _____, *Spectral asymptotics for quantum graphs with equal edge lengths*, J. Phys. A **41** (2008), no. 14, 145202, 16 pp.

[48] C. Cattaneo, *The spectrum of the continuous Laplacian on a graph*, Monatsh. Math. **124** (1997), no. 3, 215–235.

[49] W. Cauer, *Die Verwirklichung von Wechselstromwiderständen vorgeschriebener Frequenzabhängigkeit*, Arch. Elektrotech. **17** (1926), no. 4, 355–388.

[50] N.G. Čebotarëv, *On Hurwitz's problem for transcendent functions*, C. R. (Doklady) Acad. Sci. URSS (N.S.) **33** (1941), 479–481.

[51] _____, *On entire functions with real interlacing roots*, C. R. (Doklady) Acad. Sci. URSS (N. S.) **35** (1942), 195–197.

[52] N.G. Četaev, *Ustoichivost dvizheniya. Raboty po analiticheskoimekhanike*, Izdat. Akad. Nauk SSSR, Moscow, 1962 (Russian).

[53] F.R.K. Chung, *Spectral graph theory*, CBMS Regional Conference Series in Mathematics, vol. 92, Published for the Conference Board of the Mathematical Sciences, Washington, DC; by the American Mathematical Society, Providence, RI, 1997.

[54] E.A. Coddington and N. Levinson, *Theory of ordinary differential equations*, McGraw-Hill Book Company, Inc., New York-Toronto-London, 1955.

[55] J.B. Conway, *Functions of one complex variable. I*, 2nd ed., Graduate Texts in Mathematics, vol. 11, Springer-Verlag, New York, 1978.

[56] _____, *Functions of one complex variable. II*, Graduate Texts in Mathematics, vol. 159, Springer-Verlag, New York, 1995.

[57] R. Courant and D. Hilbert, *Methods of mathematical physics. Vol. I*, Interscience Publishers, Inc., New York, N.Y., 1953.

[58] S.J. Cox, M. Embree, and J.M. Hokanson, *One can hear the composition of a string: experiments with an inverse eigenvalue problem*, SIAM Rev. **54** (2012), no. 1, 157–178.

[59] S.J. Cox and A. Henrot, *Eliciting harmonics on strings*, ESAIM Control Optim. Calc. Var. **14** (2008), no. 4, 657–677.

[60] S. Cox and E. Zuazua, *The rate at which energy decays in a damped string*, Comm. Partial Differential Equations **19** (1994), no. 1-2, 213–243.

[61] G. Csordas and R.S. Varga, *Fourier transforms and the Hermite–Biehler theorem*, Proc. Amer. Math. Soc. **107** (1989), no. 3, 645–652.

[62] S. Currie and B.A. Watson, *Eigenvalue asymptotics for differential operators on graphs*, J. Comput. Appl. Math. **182** (2005), no. 1, 13–31.

[63] D.M. Cvetković, M. Doob, and H. Sachs, *Spectra of graphs*, 3rd ed., Johann Ambrosius Barth, Heidelberg, 1995.

[64] E.B. Davies, *An inverse spectral theorem*, J. Operator Theory **69** (2013), no. 1, 195–208.

[65] D.B. DeBra and R.H. Delp, *Rigid body attitude stability and natural frequencies in a circular orbit*, J. Astronaut. Sci. **8** (1961), 14–17.

[66] J. Dieudonné, *Foundations of modern analysis*, Academic Press, New York-London, 1969.

[67] P.D. Dotsenko, *Intrinsic oscillations of rectilinear pipelines with liquid*, Prikl. Mekh. **15** (1979), no. 1, 69–75 (Russian); English transl., Sov. Appl. Mech. **15** (1979), 52–57.

[68] P.D. Dotsenko and V.N. Zefirov, *Fundamental vibrations of pipes with an intermediate elastic damping*, Samoletostroenie. Tekhnika Vozdushnogo Flota (Aircraft Design for the Air Forces): Collected papers, Vischa Shkola, Kharkov, 1978, pp. 66–71 (Russian).

[69] R.J. Duffin, *A minimax theory for overdamped networks*, J. Rational Mech. Anal. **4** (1955), 221–233.

[70] N. Dunford and J.T. Schwartz, *Linear operators. Part I*, Wiley Classics Library, John Wiley & Sons Inc., New York, 1988.

[71] H. Dym and H.P. McKean, *Gaussian processes, function theory, and the inverse spectral problem*, Academic Press, New York, 1976.

[72] V.M. Eni, *Analytic perturbations of the characteristic numbers and eigenvectors of a polynomial pencil*, Mat. Issled. **1** (1966), no. vyp. 1, 189–195 (Russian).

[73] _____, *Stability of the root-number of an analytic operator-function and on perturbations of its characteristic numbers and eigenvectors*, Dokl. Akad. Nauk SSSR **173** (1967), 1251–1254 (Russian); English transl., Soviet Math. Dokl. **8** (1967), 542–545.

[74] D. Eschwé and H. Langer, *Triple variational principles for eigenvalues of self-adjoint operators and operator functions*, SIAM J. Math. Anal. **34** (2002), no. 1, 228–238 (electronic).

[75] D. Eschwé and M. Langer, *Variational principles for eigenvalues of self-adjoint operator functions*, Integral Equations Operator Theory **49** (2004), no. 3, 287–321.

[76] P. Exner, *Contact interactions on graph superlattices*, J. Phys. A **29** (1996), no. 1, 87–102.

[77] _____, *Magnetoresonances on a lasso graph*, Found. Phys. **27** (1997), no. 2, 171–190.

[78] M. Faierman, A. Markus, V. Matsaev, and M. Möller, *On n-fold expansions for ordinary differential operators*, Math. Nachr. **238** (2002), 62–77.

[79] V.I. Feodosiev, *Vibrations and stability of a pipe containing a flowing liquid*, Inzh. sbornik **10** (1951), 169–170 (Russian).

[80] G. Freiling and V. Yurko, *Inverse Sturm–Liouville problems and their applications*, Nova Science Publishers, Inc., Huntington, NY, 2001.

[81] P. Freitas, *Spectral sequences for quadratic pencils and the inverse spectral problem for the damped wave equation*, J. Math. Pures Appl. (9) **78** (1999), no. 9, 965–980.

[82] L. Friedlander, *Genericity of simple eigenvalues for a metric graph*, Israel J. Math. **146** (2005), 149–156.

[83] J. Friedman and J.-P. Tillich, *Wave equations for graphs and the edge-based Laplacian*, Pacific J. Math. **216** (2004), no. 2, 229–266.

[84] C.T. Fulton and S.A. Pruess, *Eigenvalue and eigenfunction asymptotics for regular Sturm–Liouville problems*, J. Math. Anal. Appl. **188** (1994), no. 1, 297–340.

[85] F.P. Gantmacher and M.G. Krein, *Oscillation matrices and kernels and small vibrations of mechanical systems*, Revised edition, AMS Chelsea Publishing, Providence, RI, 2002.

[86] M.G. Gasymov and G.Š. Guseĭnov, *Determination of a diffusion operator from spectral data*, Akad. Nauk Azerbaĭdzhan. SSR Dokl. **37** (1981), no. 2, 19–23 (Russian, with English and Azerbaijani summaries).

[87] J. Genin and J.S. Maybee, *Mechanical vibration trees*, J. Math. Anal. Appl. **45** (1974), 746–763.

[88] N.I. Gerasimenko, *The inverse scattering problem on a noncompact graph*, Teoret. Mat. Fiz. **75** (1988), no. 2, 187–200 (Russian, with English summary); English transl., Theoret. and Math. Phys. **75** (1988), no. 2, 460–470.

[89] N.I. Gerasimenko and B.S. Pavlov, *A scattering problem on noncompact graphs*, Teoret. Mat. Fiz. **74** (1988), no. 3, 345–359 (Russian, with English summary); English transl., Theoret. and Math. Phys. **74** (1988), no. 3, 230–240.

[90] F. Gesztesy and H. Holden, *The damped string problem revisited*, J. Differential Equations **251** (2011), no. 4-5, 1086–1127.

[91] F. Gesztesy and B. Simon, *On the determination of a potential from three spectra*, Differential operators and spectral theory, Amer. Math. Soc. Transl. Ser. 2, vol. 189, Amer. Math. Soc., Providence, RI, 1999, pp. 85–92.

[92] G. M.L. Gladwell, *Inverse problems in vibration*, 2nd ed., Solid Mechanics and its Applications, vol. 119, Kluwer Academic Publishers, Dordrecht, 2004.

[93] ――――, *Matrix inverse eigenvalue problems*, Dynamical inverse problems: theory and application, CISM Courses and Lectures, vol. 529, Springer, New York, Vienna, 2011, pp. 1–28.

[94] I. Gohberg, S. Goldberg, and M.A. Kaashoek, *Classes of linear operators. Vol. I*, Operator Theory: Advances and Applications, vol. 49, Birkhäuser Verlag, Basel, 1990.

[95] I.C. Gohberg and M.G. Kreĭn, *Introduction to the theory of linear nonselfadjoint operators*, Translated from the Russian by A. Feinstein. Translations of Mathematical Monographs, vol. 18, American Mathematical Society, Providence, R.I., 1969.

[96] ――――, *Theory and applications of Volterra operators in Hilbert space*, Translations of Mathematical Monographs, vol. 24, American Mathematical Society, Providence, R.I., 1970.

[97] I. Gohberg and J. Leiterer, *Holomorphic operator functions of one variable and applications*, Operator Theory: Advances and Applications, vol. 192, Birkhäuser Verlag, Basel, 2009.

[98] I.C. Gohberg and E.I. Sigal, *An operator generalization of the logarithmic residue theorem and Rouché's theorem*, Mat. Sb. (N.S.) **84(126)** (1971), 607–629 (Russian); English transl., Math. USSR-Sb. **13** (1971), 603–625.

[99] S. Goldberg, *Unbounded linear operators: Theory and applications*, McGraw-Hill Book Co., New York, 1966; reprinted in Dover Publications Inc., New York, 1985.

[100] I.V. Gorokhova, *Small transversal vibrations of elastic rod with point mass at one end subject to viscous friction*, Zh. Mat. Fiz. Anal. Geom. **5** (2009), no. 4, 375–385, 439 (English, with English and Ukrainian summaries).

[101] G.M. Gubreev and V. Pivovarchik, *Spectral analysis of the Regge problem with parameters*, Funktsional. Anal. i Prilozhen. **31** (1997), no. 1, 70–74 (Russian); English transl., Funct. Anal. Appl. **31** (1997), no. 1, 54–57.

[102] E.A. Guillemin, *Synthesis of passive networks. Theory and methods appropriate to the realization and approximation problems*, John Wiley and Sons, Inc., New York, 1958.

[103] B. Gutkin and U. Smilansky, *Can one hear the shape of a graph?*, J. Phys. A **34** (2001), no. 31, 6061–6068.

[104] S. Hansen and E. Zuazua, *Exact controllability and stabilization of a vibrating string with an interior point mass*, SIAM J. Control Optim. **33** (1995), no. 5, 1357–1391.

[105] G.H. Hardy and E.M. Wright, *An introduction to the theory of numbers*, Oxford, at the Clarendon Press, 1954.

[106] M. Harmer, *Inverse scattering on matrices with boundary conditions*, J. Phys. A **38** (2005), no. 22, 4875–4885.

[107] Ch. Hermite, *Extrait d'une lettre de Mr. Ch. Hermite de Paris à Mr. Borchardt de Berlin sur le nombre des racines d'une équation algébrique comprises entre des limites données*, reprinted in his Œuvres, Vol. I, Gauthier-Villars, Paris, 1905, pp. 397–414, 1856, pp. 39-51 pp.

[108] E. Hewitt and K. Stromberg, *Real and abstract analysis*, Springer-Verlag, New York, 1975.

[109] F. Hirsch and G. Lacombe, *Elements of functional analysis*, Graduate Texts in Mathematics, vol. 192, Springer-Verlag, New York, 1999.

[110] M.-T. Ho, A. Datta, and S.P. Bhattacharyya, *Generalizations of the Hermite–Biehler theorem*, Linear Algebra Appl. **302/303** (1999), 135–153.

[111] ———, *Generalizations of the Hermite–Biehler theorem: the complex case*, Linear Algebra Appl. **320** (2000), no. 1-3, 23–36.

[112] O. Holtz, *Hermite–Biehler, Routh–Hurwitz, and total positivity*, Linear Algebra Appl. **372** (2003), 105–110.

[113] O. Holtz and M. Tyaglov, *Structured matrices, continued fractions, and root localization of polynomials*, SIAM Rev. **54** (2012), no. 3, 421–509.

[114] L. Hörmander, *The analysis of linear partial differential operators. II*, Grundlehren der Mathematischen Wissenschaften, vol. 257, Springer-Verlag, Berlin, 1983.

[115] S.V. Hruščev, *The Regge problem for strings, unconditionally convergent eigenfunction expansions, and unconditional bases of exponentials in $L^2(-T, T)$*, J. Operator Theory **14** (1985), no. 1, 67–85.

[116] R.O. Hryniv, W. Kliem, P. Lancaster, and C. Pommer, *A precise bound for gyroscopic stabilization*, ZAMM Z. Angew. Math. Mech. **80** (2000), no. 8, 507–516 (English, with English and German summaries).

[117] R. Hryniv and P. Lancaster, *Stabilization of gyroscopic systems*, ZAMM Z. Angew. Math. Mech. **81** (2001), no. 10, 675–681 (English, with English and German summaries).

[118] R.O. Hryniv, P. Lancaster, and A.A. Renshaw, *A stability criterion for parameter-dependent gyroscopic systems*, Trans. ASME J. Appl. Mech. **66** (1999), no. 3, 660–664.

[119] R. Hryniv and N. Pronska, *Inverse spectral problems for energy-dependent Sturm–Liouville equations*, Inverse Problems **28** (2012), no. 8, 085008, 21 pp.

[120] A. Hurwitz, *Über die Bedingungen, unter welchen eine Gleichung nur Wurzeln mit negativen reellen Teilen besitzt*, Math. Ann. **46** (1895), 273–284.

[121] I.S. Iohvidov and M.G. Kreĭn, *Spectral theory of operators in space with indefinite metric. I*, Trudy Moskov. Mat. Obšč. **5** (1956), 367–432 (Russian).

[122] S.A. Ivanov, *The Regge problem for vector-functions*, Operator theory and function theory, No. 1, Leningrad. Univ., Leningrad, 1983, pp. 68–86 (Russian).

[123] M. Jaulent, *Inverse scattering problems in absorbing media*, J. Mathematical Phys. **17** (1976), no. 7, 1351–1360.

[124] C.R. Johnson and A. Leal Duarte, *On the possible multiplicities of the eigenvalues of a Hermitian matrix whose graph is a tree*, Linear Algebra Appl. **348** (2002), 7–21.

[125] I.S. Kats, *The spectral theory of a string*, Ukraïn. Mat. Zh. **46** (1994), no. 3, 155–176 (Russian, with English and Ukrainian summaries); English transl., Ukrainian Math. J. **46** (1994), no. 3, 159–182 (1995).

[126] I.S. Kac and M.G. Kreĭn, *R-functions–analytic functions mapping the upper half-plane into itself.* Supplement I to the Russian edition of F.V. Atkinson: *Discrete and continuous boundary value problems*, Mir, Moscow, 1968 (Russian); English transl., Amer. Math. Transl. (2) **103** (1974), 1–18.

[127] _____, *On the spectral functions of the string.* Supplement II to the Russian edition of F.V. Atkinson: *Discrete and continuous boundary value problems*, Mir, Moscow, 1968 (Russian); English transl., Amer. Math. Transl. (2) **103** (1974), 19–102.

[128] I. Kac and V. Pivovarchik, *On multiplicity of a quantum graph spectrum*, J. Phys. A **44** (2011), no. 10, 105301, 14 pp.

[129] _____, *On the density of the mass distribution of a string at its origin*, Integral Equations Operator Theory **81** (2015), no. 4, 581–519.

[130] D. Kalman, *A Matrix Proof of Newton's Identities*, Math. Mag. **73** (2000), no. 4, 313–315.

[131] M. Kaltenbäck, H. Winkler, and H. Woracek, *Generalized Nevanlinna functions with essentially positive spectrum*, J. Operator Theory **55** (2006), no. 1, 17–48.

[132] _____, *Strings, dual strings, and related canonical systems*, Math. Nachr. **280** (2007), no. 13-14, 1518–1536.

[133] T.R. Kane, E.L. Marsh, and W.G. Wilson, *Letter to Editor*, J. Austronaut. Sci. **9** (1962), no. 4, 108–109.

[134] I.M. Karabash, *Nonlinear eigenvalue problem for optimal resonances in optical cavities*, Math. Model. Nat. Phenom. **8** (2013), no. 1, 143–155.

[135] _____, *Optimization of quasi-normal eigenvalues for 1-D wave equations in inhomogeneous media; description of optimal structures*, Asymptot. Anal. **81** (2013), no. 3-4, 273–295.

[136] T. Kato, *On the perturbation theory of closed linear operators*, J. Math. Soc. Japan **4** (1952), 323–337.

[137] _____, *Perturbation theory for linear operators*, Die Grundlehren der mathematischen Wissenschaften, Band 132, Springer-Verlag New York, Inc., New York, 1966.

[138] M.V. Keldyš, *On the characteristic values and characteristic functions of certain classes of non-self-adjoint equations*, Doklady Akad. Nauk SSSR (N.S.) **77** (1951), 11–14 (Russian).

[139] _____, *The completeness of eigenfunctions of certain classes of nonselfadjoint linear operators*, Uspehi Mat. Nauk **26** (1971), no. 4(160), 15–41 (Russian).

[140] W.T. Kelvin and P.G. Tait, *Treatise on natural philosophy.* 1, Cambridge University Press, Cambridge, 1921.

[141] L. Kobyakova, *Spectral problem generated by the equation of smooth string with piece-wise constant friction*, Zh. Mat. Fiz. Anal. Geom. **8** (2012), no. 3, 280–295, 298, 301 (English, with English, Russian and Ukrainian summaries).

[142] B.L. Kogan, *The double completeness of the system of eigen- and associated functions of the Regge problem*, Funkcional. Anal. i Priložen. **5** (1971), no. 3, 70–74 (Russian); English transl., Functional Anal. Appl. **5** (1971), 229–232.

[143] Q. Kong and A. Zettl, *Eigenvalues of regular Sturm–Liouville problems*, J. Differential Equations **131** (1996), no. 1, 1–19.

[144] P. Koosis, *Introduction to H_p spaces*, 2nd ed., Cambridge Tracts in Mathematics, vol. 115, Cambridge University Press, Cambridge, 1998.

[145] N.D. Kopachevsky and S.G. Krein, *Operator approach to linear problems of hydrodynamics. Vol.* 1, Operator Theory: Advances and Applications, vol. 128, Birkhäuser Verlag, Basel, 2001.

[146] _____, *Operator approach to linear problems of hydrodynamics. Vol.* 2, Operator Theory: Advances and Applications, vol. 146, Birkhäuser Verlag, Basel, 2003.

[147] N.D. Kopachevskiĭ and V. Pivovarchik, *On a sufficient condition for the instability of convective motions of a fluid in an open vessel*, Zh. Vychisl. Mat. i Mat. Fiz. **33** (1993), no. 1, 101–118 (Russian, with Russian summary); English transl., Comput. Math. Math. Phys. **33** (1993), no. 1, 89–102.

[148] E. Korotyaev, *Inverse resonance scattering on the half line*, Asymptot. Anal. **37** (2004), no. 3-4, 215–226.

[149] A.G. Kostyuchenko and M.B. Orazov, *The problem of oscillations of an elastic half cylinder and related selfadjoint quadratic pencils*, Trudy Sem. Petrovsk. **6** (1981), 97–146 (Russian, with English summary).

[150] A.O. Kravickiĭ, *The two-fold expansion of a certain non-selfadjoint boundary value problem in series of eigenfunctions*, Differencial'nye Uravnenija **4** (1968), 165–177 (Russian).

[151] M.G. Kreĭn, *On some new problems of the theory of oscillations of Sturmian systems*, Akad. Nauk SSSR. Prikl. Mat. Meh. **16** (1952), 555–568 (Russian).

[152] _____, *On a generalization of investigations of Stieltjes*, Doklady Akad. Nauk SSSR (N.S.) **87** (1952), 881–884 (Russian).

[153] M.G. Kreĭn and H. Langer, *On the theory of quadratic pencils of self-adjoint operators*, Dokl. Akad. Nauk SSSR **154** (1964), 1258–1261 (Russian); English transl., Soviet Math. Dokl. **5** (1964), 266–269.

[154] _____, *Certain mathematical principles of the linear theory of damped vibrations of continua*, Appl. Theory of Functions in Continuum Mechanics (Proc. Internat. Sympos., Tbilisi, 1963), Vol. II, Fluid and Gas Mechanics, Math. Methods (Russian), Izdat. "Nauka", Moscow, 1965, pp. 283–322 (Russian); English transl., 1978, pp. 364–399, 539–566.

[155] _____, *Über einige Fortsetzungsprobleme, die eng mit der Theorie hermitescher Operatoren im Raume Π_κ zusammenhängen. I. Einige Funktionenklassen und ihre Darstellungen*, Math. Nachr. **77** (1977), 187–236.

[156] M.G. Kreĭn and A.A. Nudel′man, *Direct and inverse problems for the frequencies of boundary dissipation of a nonuniform string*, Dokl. Akad. Nauk SSSR **247** (1979), no. 5, 1046–1049 (Russian).

[157] _____, *Some spectral properties of a nonhomogeneous string with a dissipative boundary condition*, J. Operator Theory **22** (1989), no. 2, 369–395 (Russian).

[158] S.G. Kreĭn, *Oscillations of a viscous fluid in a container*, Dokl. Akad. Nauk SSSR **159** (1964), 262–265 (Russian); English transl., Soviet Math. Dokl. **5** (1964), 1467–1471.

[159] P. Kurasov, *Inverse scattering for lasso graph*, J. Math. Phys. **54** (2013), no. 4, 042103, 14 pp.

[160] P. Lancaster and W. Kliem, *Comments on stability properties of conservative gyroscopic systems*, Trans. ASME J. Appl. Mech. **66** (1999), no. 1, 272–273.

[161] P. Lancaster, A.S. Markus, and F. Zhou, *A wider class of stable gyroscopic systems*, Linear Algebra Appl. **370** (2003), 257–267.

[162] P. Lancaster and J. Maroulas, *Inverse eigenvalue problems for damped vibrating systems*, J. Math. Anal. Appl. **123** (1987), no. 1, 238–261.

[163] P. Lancaster and U. Prells, *Inverse problems for damped vibrating systems*, J. Sound Vibration **283** (2005), no. 3-5, 891–914.

[164] P. Lancaster and A. Shkalikov, *Damped vibrations of beams and related spectral problems*, Canad. Appl. Math. Quart. **2** (1994), no. 1, 45–90.

[165] P. Lancaster and P. Zizler, *On the stability of gyroscopic systems*, Trans. ASME J. Appl. Mech. **65** (1998), no. 2, 519–522.

[166] L.D. Landau and E.M. Lifshitz, *Quantum mechanics: non-relativistic theory. Course of Theoretical Physics, vol. 3*, Addison-Wesley Series in Advanced Physics, Pergamon Press Ltd., London-Paris, 1958.

[167] H. Langer, M. Langer, and C. Tretter, *Variational principles for eigenvalues of block operator matrices*, Indiana Univ. Math. J. **51** (2002), no. 6, 1427–1459.

[168] H. Langer and H. Winkler, *Direct and inverse spectral problems for generalized strings*, Integral Equations Operator Theory **30** (1998), no. 4, 409–431.

[169] Y. Latushkin and V. Pivovarchik, *Scattering in a forked-shaped waveguide*, Integral Equations Operator Theory **61** (2008), no. 3, 365–399.

[170] C.-K. Law and E. Yanagida, *A solution to an Ambarzumyan problem on trees*, Kodai Math. J. **35** (2012), no. 2, 358–373.

[171] A. Leal Duarte, *Construction of acyclic matrices from spectral data*, Linear Algebra Appl. **113** (1989), 173–182.

[172] B.Ja. Levin, *On bases of exponential functions in $L^2(-\pi, \pi)$*, Zap. Mekh.-Mat. Fak. Khar'kov. Gos. Univ. Khar'kov. Mat. Obšč. **27** (1961), 39–48 (Russian).

[173] _____, *Distribution of zeros of entire functions*, Revised edition, Translations of Mathematical Monographs, vol. 5, American Mathematical Society, Providence, R.I., 1980.

[174] _____, *Lectures on entire functions*, Translations of Mathematical Monographs, vol. 150, American Mathematical Society, Providence, RI, 1996.

[175] B.Ja. Levin and Ju.I. Ljubarskiĭ, *Interpolation by entire functions belonging to special classes and related expansions in series of exponentials*, Izv. Akad. Nauk SSSR Ser. Mat. **39** (1975), no. 3, 657–702, 704 (Russian); English transl., Math. USSR-Izv. **9** (1975), no. 3, 621–662.

[176] B.Ja. Levin and Ĭ.V. Ostrovs'kiĭ, *Small perturbations of the set of roots of sine-type functions*, Izv. Akad. Nauk SSSR Ser. Mat. **43** (1979), no. 1, 87–110, 238 (Russian); English transl., Math. USSR-Izv. **14** (1979), no. 1, 79–101 (1980).

[177] B.M. Levitan, *Inverse Sturm–Liouville problems*, VSP, Zeist, 1987.

[178] B.M. Levitan and M.G. Gasymov, *Determination of a differential equation by two spectra*, Uspehi Mat. Nauk **19** (1964), no. 2 (116), 3–63 (Russian).

[179] J. Liouville, *Troisième mémoire sur le développement des fonctions ou parties de fonctions en séries dont le divers termes sont assujétis à satisfaire à une même équation différentielle du second ordre, contenant un paramètre variable*, J. de math. pures et appl., Ser. 1, **2** (1837), 418–436.

[180] V.A. Marchenko, *Sturm–Liouville operators and applications*, revised edition, AMS Chelsea Publishing, Providence, RI, 2011.

[181] _____, *Introduction to inverse problems of spectral analysis*, Akta, Kharkov, 2005 (Russian).

[182] A.S. Markus, *Introduction to the spectral theory of polynomial operator pencils*, Translations of Mathematical Monographs, vol. 71, American Mathematical Society, Providence, RI, 1988.

[183] A.S. Markus and V.I. Matsaev, *The property of being a basis of some part of the eigen- and associated vectors of a selfadjoint operator pencil*, Mat. Sb. (N.S.) **133(175)** (1987), no. 3, 293–313, 415 (Russian); English transl., Math. USSR-Sb. **61** (1988), no. 2, 289–307.

[184] A.S. Markus, V.I. Macaev, and G.I. Russu, *Certain generalizations of the theory of strongly damped pencils to the case of pencils of arbitrary order*, Acta Sci. Math. (Szeged) **34** (1973), 245–271 (Russian).

[185] A.I. Markushevich, *Theory of functions of a complex variable. Vol. III*, revised English edition, Prentice-Hall Inc., Englewood Cliffs, N.J., 1967.

[186] C. van der Mee and V. Pivovarchik, *Inverse scattering for a Schrödinger equation with energy dependent potential*, J. Math. Phys. **42** (2001), no. 1, 158–181.

[187] _____, *Some properties of the eigenvalues of a Schrödinger equation with energy-dependent potential*, Mathematical results in quantum mechanics (Taxco, 2001), Contemp. Math., vol. 307, Amer. Math. Soc., Providence, RI, 2002, pp. 305–310.

[188] N.N. Meĭman, *On the distribution of the zeros of an integer function*, C. R. (Dokl.) Acad. Sci. URSS **40** (1943), 179–181 (Russian).

[189] R. Mennicken and M. Möller, *Non-self-adjoint boundary eigenvalue problems*, North-Holland Mathematics Studies, vol. 192, North-Holland Publishing Co., Amsterdam, 2003.

[190] R. Mennicken and V. Pivovarchik, *An inverse problem for an inhomogeneous string with an interval of zero density*, Math. Nachr. **259** (2003), 51–65.

[191] A.I. Miloslavskii, *On stability of linear pipes*, Dinamika Sistem, Nesuschih Podvizh-nuyu Raspredelennuyu Nagruzku, Collected papers, vol. 192, Kharkov Aviation Institute, Kharkov, 1980, pp. 34–47 (Russian).

[192] A.I. Miloslavskiĭ, *On the instability spectrum of an operator pencil*, Mat. Zametki **49** (1991), no. 4, 88–94, 159 (Russian); English transl., Math. Notes **49** (1991), no. 3-4, 391–395.

[193] S.S. Mirzoev, *Multiple completeness of a part of the root vectors of fourth-order polynomial operator pencils with a normal smooth part*, Spectral theory of operators, No. 4, "Èlm", Baku, 1982, pp. 148–161 (Russian).

[194] K. Mochizuki and I. Trooshin, *Spectral problems and scattering on noncompact star-shaped graphs containing finite rays*, J. Inverse Ill-Posed Probl. **23** (2015), no. 1, 23–40.

[195] M. Möller and V. Pivovarchik, *Spectral properties of a fourth-order differential equation*, Z. Anal. Anwend. **25** (2006), no. 3, 341–366.

[196] M. Möller and A. Zettl, *Semi-boundedness of ordinary differential operators*, J. Differential Equations **115** (1995), no. 1, 24–49.

[197] M. Möller and B. Zinsou, *Spectral asymptotics of self-adjoint fourth-order differential operators with eigenvalue parameter dependent boundary conditions*, Complex Anal. Oper. Theory **6** (2012), no. 3, 799–818.

[198] _____, *Spectral asymptotics of self-adjoint fourth-order boundary value problems with eigenvalue parameter dependent boundary conditions*, Bound. Value Probl. (2012), 2012: 106, 18 pp.

[199] _____, *Asymptotics of the eigenvalues of a self-adjoint fourth-order boundary value problem with four eigenvalue parameter dependent boundary conditions*, J. Funct. Spaces Appl. (2013), Art. ID 280970, 8 pp.

[200] A.A. Movchan, *On a problem of stability of a pipe with a fluid flowing through it*, Prikl. Mat. Mekh. **29** (1965), 760–762 (Russian); English transl., PMM, J. Appl. Math. Mech. **29** (1965), 902–904.

[201] L.P. Nizhnik, *Inverse eigenvalue problems for nonlocal Sturm–Liouville operators on a star graph*, Methods Funct. Anal. Topology **18** (2012), no. 1, 68–78.

[202] P. Nylen and F. Uhlig, *Realizations of interlacing by tree-patterned matrices*, Linear and Multilinear Algebra **38** (1994), no. 1-2, 13–37.

[203] V.A. Oliveira, M.C.M. Teixeira, and L. Cossi, *Stabilizing a class of time delay systems using the Hermite–Biehler theorem*, Linear Algebra Appl. **369** (2003), 203–216.

[204] M.B. Orazov, *Completeness of elementary solutions in a problem on steady vibrations of a finite cylinder*, Uspekhi Mat. Nauk **35** (1980), no. 5(215), 237–238 (Russian).

[205] I. Oren and R. Band, *Isospectral graphs with identical nodal counts*, J. Phys. A **45** (2012), no. 13, 135203, 12 pp.

[206] M.P. Païdoussis and N.T. Issid, *Dynamic stability of pipes conveying fluid*, J. Sound Vibrat. **33** (1974), no. 3, 267–294.

[207] V. Pivovarchik, *The spectrum of quadratic operator pencils in the right half-plane*, Mat. Zametki **45** (1989), no. 6, 101–103 (Russian).

[208] ———, *On the total algebraic multiplicity of the spectrum in the right half-plane for a class of quadratic operator pencils*, Algebra i Analiz **3** (1991), no. 2, 223–230 (Russian); English transl., St. Petersburg Math. J. **3** (1992), no. 2, 447–454.

[209] ———, *Necessary conditions for gyroscopic stabilization in a problem of mechanics*, Mat. Zametki **53** (1993), no. 6, 89–96, 159 (Russian, with Russian summary); English transl., Math. Notes **53** (1993), no. 5-6, 622–627.

[210] ———, *A lower bound of the instability index in the vibration problem for an elastic fluid-conveying pipe*, Russian J. Math. Phys. **2** (1994), no. 2, 267–272.

[211] ———, *On positive spectra of one class of polynomial operator pencils*, Integral Equations Operator Theory **19** (1994), no. 3, 314–326.

[212] ———, *Inverse problem for a smooth string with damping at one end*, J. Operator Theory **38** (1997), no. 2, 243–263.

[213] ———, *Direct and inverse problems for a damped string*, J. Operator Theory **42** (1999), no. 1, 189–220.

[214] ———, *An inverse Sturm–Liouville problem by three spectra*, Integral Equations Operator Theory **34** (1999), no. 2, 234–243.

[215] ———, *Inverse problem for the Sturm–Liouville equation on a simple graph*, SIAM J. Math. Anal. **32** (2000), no. 4, 801–819 (electronic).

[216] ———, *Scattering in a loop-shaped waveguide*, Recent advances in operator theory (Groningen, 1998), Oper. Theory Adv. Appl., vol. 124, Birkhäuser, Basel, 2001, pp. 527–543.

[217] ———, *Necessary conditions for stability of elastic pipe conveying liquid*, Methods Funct. Anal. Topology **11** (2005), no. 3, 270–274.

[218] ———, *Ambartsumyan's theorem for the Sturm–Liouville boundary value problem on a star-shaped graph*, Funktsional. Anal. i Prilozhen. **39** (2005), no. 2, 78–81 (Russian); English transl., Funct. Anal. Appl. **39** (2005), no. 2, 148–151.

[219] ———, *Symmetric Hermite–Biehler polynomials with defect*, Operator theory in inner product spaces, Oper. Theory Adv. Appl., vol. 175, Birkhäuser, Basel, 2007, pp. 211–224.

[220] _____, *On spectra of a certain class of quadratic operator pencils with one-dimensional linear part*, Ukraïn. Mat. Zh. **59** (2007), no. 5, 702–716 (English, with English and Ukrainian summaries); English transl., Ukrainian Math. J. **59** (2007), no. 5, 766–781.

[221] _____, *Inverse problem for the Sturm–Liouville equation on a star-shaped graph*, Math. Nachr. **280** (2007), no. 13-14, 1595–1619.

[222] _____, *Existence of a tree of Stieltjes strings corresponding to two given spectra*, J. Phys. A **42** (2009), no. 37, 375213, 16 pp.

[223] V. Pivovarchik and C. van der Mee, *The inverse generalized Regge problem*, Inverse Problems **17** (2001), no. 6, 1831–1845.

[224] V. Pivovarchik, N. Rozhenko, and C. Tretter, *Dirichlet–Neumann inverse spectral problem for a star graph of Stieltjes strings*, Linear Algebra Appl. **439** (2013), no. 8, 2263–2292.

[225] V. Pivovarchik and O. Taystruk, *On characteristic functions of operators on equilateral graphs*, Methods Funct. Anal. Topology **18** (2012), no. 2, 189–197.

[226] V. Pivovarchik and C. Tretter, *Location and multiplicities of eigenvalues for a star graph of Stieltjes strings*, J. Difference Equ. Appl., posted on 2015, DOI 10.1080/10236198.2014.992425.

[227] V. Pivovarchik and H. Woracek, *Shifted Hermite–Biehler functions and their applications*, Integral Equations Operator Theory **57** (2007), no. 1, 101–126.

[228] _____, *The square-transform of Hermite–Biehler functions. A geometric approach*, Methods Funct. Anal. Topology **13** (2007), no. 2, 187–200.

[229] _____, *Sums of Nevanlinna functions and differential equations on star-shaped graphs*, Oper. Matrices **3** (2009), no. 4, 451–501.

[230] _____, *Eigenvalue asymptotics for a star-graph damped vibrations problem*, Asymptot. Anal. **73** (2011), no. 3, 169–185.

[231] Yu.V. Pokornyĭ, O.M. Penkin, V.L. Pryadiev, A.V. Borovskikh, K.P. Lazarev, and S.A. Shabrov, *Differentsialnye uravneniya na geometricheskikh grafakh*, Fiziko-Matematicheskaya Literatura, Moscow, 2005 (Russian, with Russian summary).

[232] L. Pontrjagin, *On zeros of some transcendental functions*, Bull. Acad. Sci. URSS. Sér. Math. **6** (1942), 115–134 (Russian, with English summary); English transl., Amer. Math. Soc. Transl. (2) **1** (1955), 95–110.

[233] _____, *Hermitian operators in spaces with indefinite metric*, Bull. Acad. Sci. URSS. Sér. Math. **8** (1944), 243–280 (Russian, with English summary).

[234] I.Yu. Popov and A.V. Strepetov, *Completeness of the system of eigenfunctions of the two-sided Regge problem*, Vestnik Leningrad. Univ. Mat. Mekh. Astronom. 1983, vyp. 3, 25–31 (Russian, with English summary).

[235] J. Pöschel and E. Trubowitz, *Inverse spectral theory*, Pure and Applied Mathematics, vol. 130, Academic Press, Inc., Boston, MA, 1987.

[236] N. Pronska, *Reconstruction of energy-dependent Sturm–Liouville equations from two spectra*, Integral Equations Operator Theory **76** (2013), no. 3, 403–419.

[237] G.V. Radzīevs′kiĭ, *The problem of completeness of root vectors in the spectral theory of operator-valued functions*, Uspekhi Mat. Nauk **37** (1982), no. 2(224), 81–145, 280 (Russian).

[238] T. Regge, *Construction of potentials from resonance parameters*, Nuovo Cimento, X. Ser. **9** (1958), 491–503.

[239] F. Riesz and B. Sz.-Nagy, *Functional analysis*, Frederick Ungar Publishing Co., New York, 1955.

[240] W. Rundell and P. Sacks, *Numerical technique for the inverse resonance problem*, J. Comput. Appl. Math. **170** (2004), no. 2, 337–347.

[241] _____, *Inverse eigenvalue problem for a simple star graph*, J. Spectr. Theory, to appear.

[242] B.E. Sagan, *The symmetric group*, 2nd ed., Graduate Texts in Mathematics, vol. 203, Springer-Verlag, New York, 2001.

[243] A.G. Sergeev, *The asymptotic behavior of the Jost function and of the eigenvalues of the Regge problem*, Differencial′nye Uravnenija **8** (1972), 925–927 (Russian).

[244] C.-L. Shen, *On the Barcilon formula for the string equation with a piecewise continuous density function*, Inverse Problems **21** (2005), no. 2, 635–655.

[245] A.A. Shkalikov, *On the basis property of the eigenvectors of quadratic operator pencils*, Mat. Zametki **30** (1981), no. 3, 371–385, 462 (Russian); English transl., Math. Notes **30** (1981), no. 3–4, 676–684.

[246] _____, *Boundary value problems for ordinary differential equations with a parameter in the boundary conditions*, Trudy Sem. Petrovsk. **9** (1983), 190–229 (Russian, with English summary).

[247] _____, *Operator pencils arising in elasticity and hydrodynamics: the instability index formula*, Recent developments in operator theory and its applications (Winnipeg, MB, 1994), Oper. Theory Adv. Appl., vol. 87, Birkhäuser, Basel, 1996, pp. 358–385.

[248] _____, *Spectral analysis of the Redge problem*, Sovrem. Mat. Prilozh. **35** (2005), 90–97 (Russian); English transl., J. Math. Sci. (N.Y.) **144** (2007), no. 4, 4292–4300.

[249] M.A. Shubov and M. Rojas-Arenaza, *Four-branch vibrational spectrum of double-walled carbon nanotube model*, Proc. R. Soc. Lond. Ser. A Math. Phys. Eng. Sci. **467** (2011), no. 2125, 99–126.

[250] A.S. Silbergleit and Yu.I. Kopilevich, *Spectral theory of guided waves*, Institute of Physics Publishing, Bristol, 1996.

[251] B. Simon, *Resonances in one dimension and Fredholm determinants*, J. Funct. Anal. **178** (2000), no. 2, 396–420.

[252] V.Ya. Skorobogat′ko, *Branched continued fractions and convergence acceleration problems*, Rational approximation and applications in mathematics and physics (Łańcut, 1985), Lecture Notes in Math., vol. 1237, Springer, Berlin, 1987, pp. 46–50.

[253] C.B. Soh, *Generalization of the Hermite–Biehler theorem and applications*, IEEE Trans. Automat. Control **35** (1990), no. 2, 222–225.

[254] T.-J. Stieltjes, *Sur la réduction en fraction continue d'une série procédant suivant les puissances descendantes d'une variable*, Ann. Fac. Sc. Toulouse **3** (1889), H1–H17 (French).

[255] A.V. Strepetov, *Completeness and basis property of the system of eigenfunctions of the two-sided Regge problem for a polar operator*, Vestnik Leningrad. Univ. Mat. Mekh. Astronom. 1983, vyp. 4, 44–50 (Russian, with English summary).

[256] _____, *The Regge problem with scattering by spirals*, Izv. Vyssh. Uchebn. Zaved. Mat. **4** (1989), 76–83 (Russian); English transl., Soviet Math. (Iz. VUZ) **33** (1989), no. 4, 94–102.

[257] J.D. Tamarkin, *Some general problems of the theory of ordinary linear differential equations and expansion of an arbitrary function in series of fundamental functions*, Editor Frolova, Petrograd, 1917 (Russian).

[258] _____, *Some general problems of the theory of ordinary linear differential equations and expansion of an arbitrary function in series of fundamental functions*, Math. Z. **27** (1928), 1–54.

[259] A.E. Taylor and D.C. Lay, *Introduction to functional analysis*, 2nd ed., John Wiley & Sons, New York-Chichester-Brisbane, 1980.

[260] I.D. Tazehkand and A.J. Akbarfam, *Determination of Sturm–Liouville operator on a three-star graph from four spectra*, Acta Univ. Apulensis Math. Inform. **32** (2012), 147–172.

[261] W.T. Thompson, *Spin stabilization of attitude against gravity torque*, J. Austronaut. Sci. **9** (1962), no. 1, 31–33.

[262] F. Tisseur and K. Meerbergen, *The quadratic eigenvalue problem*, SIAM Rev. **43** (2001), no. 2, 235–286.

[263] E.C. Titchmarsh, *Introduction to the theory of Fourier integrals*, Clarendon Press, Oxford, 1937.

[264] M. Tyaglov, *Generalized Hurwitz polynomials*, arxiv:1005.3032v1.

[265] K. Veselić, *On linear vibrational systems with one-dimensional damping*, Appl. Anal. **29** (1988), no. 1-2, 1–18.

[266] _____, *On linear vibrational systems with one-dimensional damping. II*, Integral Equations Operator Theory **13** (1990), no. 6, 883–897.

[267] _____, *On the stability of rotating systems*, Z. Angew. Math. Mech. **75** (1995), no. 4, 325–328.

[268] _____, *Damped oscillations of linear systems*, Lecture Notes in Mathematics, vol. 2023, Springer, Heidelberg, 2011.

[269] F. Visco-Comandini, M. Mirrahimi, and M. Sorine, *Some inverse scattering problems on star-shaped graphs*, J. Math. Anal. Appl. **378** (2011), no. 1, 343–358.

[270] H.S. Wall, *Analytic Theory of Continued Fractions*, D. Van Nostrand Company, Inc., New York, N.Y., 1948.

[271] J. Weidmann, *Spectral theory of ordinary differential operators*, Lecture Notes in Mathematics, vol. 1258, Springer-Verlag, Berlin, 1987.

[272] S. Yakubov, *A uniformly well-posed Cauchy problem for abstract hyperbolic equations*, Izv. Vyssh. Uchebn. Zaved., Mat. **103** (1970), no. 12, 108–113 (Russian).

[273] S. Yakubov and Y. Yakubov, *Differential-operator equations*, Chapman & Hall / CRC Monographs and Surveys in Pure and Applied Mathematics, vol. 103, Chapman & Hall / CRC, Boca Raton, FL, 2000.

[274] V.Ya. Yakubov, *The Cauchy problem for the Sturm–Liouville equation and estimates for its solution with respect to the parameter λ*, Dokl. Akad. Nauk **360** (1998), no. 3, 320–323 (Russian).

[275] M. Yamamoto, *Inverse eigenvalue problem for a vibration of a string with viscous drag*, J. Math. Anal. Appl. **152** (1990), no. 1, 20–34.

[276] C.-F. Yang and Z.-Y. Huang, *Spectral asymptotics and regularized traces for Dirac operators on a star-shaped graph*, Appl. Anal. **91** (2012), no. 9, 1717–1730.

[277] C.-F. Yang, Z.-Y. Huang, and X.-P. Yang, *Ambarzumyan-type theorems for the Sturm–Liouville equation on a graph*, Rocky Mountain J. Math. **39** (2009), no. 4, 1353–1372.

[278] C.-F. Yang, V. Pivovarchik, and Z.-Y. Huang, *Ambarzumyan-type theorems on star graphs*, Oper. Matrices **5** (2011), no. 1, 119–131.

[279] C.-F. Yang and J.-X. Yang, *Large eigenvalues and traces of Sturm–Liouville equations on star-shaped graphs*, Methods Appl. Anal. **14** (2007), no. 2, 179–196.

[280] C.-F. Yang and X.-P. Yang, *Some Ambarzumyan-type theorems for Dirac operators*, Inverse Problems **25** (2009), no. 9, 095012, 13 pp.

[281] R.M. Young, *An introduction to nonharmonic Fourier series*, Pure and Applied Mathematics, vol. 93, Academic Press Inc., New York, 1980.

[282] V.A. Yurko, *Inverse spectral problems for differential operators and their applications*, Analytical Methods and Special Functions, vol. 2, Gordon and Breach Science Publishers, Amsterdam, 2000.

[283] E.E. Zajac, *The Kelvin–Tait–Chetaev Theorem and Extensions*, J. Astronaut. Sci. **11** (1964), 46–49.

[284] V.N. Zefirov, V.V. Kolesov, and A.I. Miloslavskii, *A study of the natural frequencies of a rectilinear pipe*, Izv. Akad. Nauk SSSR, MTT **1** (1985), 179–188 (Russian).

[285] A. Zettl, *Sturm–Liouville theory*, Mathematical Surveys and Monographs, vol. 121, American Mathematical Society, Providence, RI, 2005.

[286] H. Ziegler, *Linear elastic stability. A critical analysis of methods*, Z. Angew. Math. Physik **4** (1953), 89–121, 167–185.

[287] M. Zworski, *Distribution of poles for scattering on the real line*, J. Funct. Anal. **73** (1987), no. 2, 277–296.

Index

Index of Notation

$\mathbb{C}$	the set of complex numbers
$\overline{\mathbb{C}}$	$\mathbb{C} \cup \{\infty\}$
$\mathbb{C}^+$	the open upper half-plane
$\overline{\mathbb{C}^+}$	the closed upper half-plane
$\mathbb{C}^-$	the open lower half-plane
$\overline{\mathbb{C}^-}$	the closed lower half-plane
$C^j[a,b]$	the set of j times continuously differentiable functions on the interval $[a,b]$
$D(A)$	the domain of an operator A
$\det M$	the determinant of a square matrix M
∂S	the boundary of a set S
e_j	the jth unit vector in $\mathbb{C}^n$
I	the identity operator or the identity matrix
$(\ ,\ )$	the inner product in Hilbert spaces and particularly in L_2
$l_p = l_p(Z)$	the set of sequences $(a_k)_{k \in Z}$ with $\sum_{k \in Z} |a_k|^p < \infty$, where $Z \subset \mathbb{Z}$ and $a_k \in \mathbb{C}$
$L(X,Y)$	the Banach space of bounded linear operators from a Banach space X to a Banach space Y
$L(X)$	the Banach space $L(X,X)$
$\mathbb{N}$	the set of positive integers
$\mathbb{N}_0$	$\mathbb{N} \cup \{0\}$, the set of nonnegative integers
$N(A)$	the nullspace (kernel) of an operator, or a matrix, A
$\mathbb{R}$	the set of real numbers
$R(A)$	the range of an operator, or a matrix, A
X'	the dual space $L(X, \mathbb{C})$ of a Banach space X
$\mathbb{Z}$	the set of integers
$\mathbb{Z}a$	the set of integer multiplies of the number a

Classes of functions

$C_0^\infty(\mathcal{I})$, 270
$\mathcal{D}'(a,b)$, 270
$\mathcal{D}'(\mathbb{R})$, 270
HB, 120, 123
$\overline{HB}$, 122
H_p, 334
$L_p(\mathcal{I})$, 269
$\mathcal{M}$, 131
$\mathcal{N}$, 130

$\hat{\mathcal{N}}$, 130
$\mathcal{N}^{\mathrm{ep}}$, 131
$\mathcal{N}_+^{\mathrm{ep}}$, 131
$\mathcal{N}_-^{\mathrm{ep}}$, 131
SHB, 128
$\overline{\mathrm{SHB}}$, 128
SSHB, 135
$\overline{\mathrm{SSHB}}$, 136
SSHB_κ, 135

$\overline{\mathrm{SSHB}}_\kappa$, 136
$\mathcal{S}$, 131
$\mathcal{S}^{-1}$, 131
$\hat{\mathcal{S}}$, 131
$\mathcal{S}_0$, 131
$W_2^k(a,b)$, 270
$W_2^{-k}[a,b]$, 271
$W_2^{-k}(a,b)$, 271

Special Symbols

$(\ ,\ )_0$, 270
$(\ ,\ )_{2,k}$, 271
$\|\ \|_{2,k}$, 270
$\gg$, 9
$[\ ,\ ]$, 273
$[\ ,\ ]_1$, 273
$\mathcal{B}$, 215
$\mathcal{B}^-$, 215
$\mathcal{B}^+$, 215
$B(x,t)$, 356
χ_ω^+, 302
$\overline{\mathbb{C}}_H^+$, 334
$c(\lambda,x)$, 356
$E_{\omega,\delta}$, 302
$E_{\omega,\delta,r}$, 302
f_e, 271
I_a, 121
$\mathrm{I}\omega$, 119
$K(x,t)$, 356
$\tilde{K}(x,t)$, 356
ℓ, 272
$(\lambda - a_0)_+$, 121
$(\lambda - b_0)_-$, 121
ℓ_j, 274

$\mathcal{L}^\sigma$, 353
$\mathcal{L}_e^\sigma$, 353
$\mathcal{L}_o^\sigma$, 353
$m_0(\lambda)$, 23
M_{f,ϕ_1,ϕ_2}, 296
$m_I(\lambda)$, 23
$m(\lambda)$, 9
$m(\lambda,\eta)$, 9
$\mathcal{M}^+$, 230
$\mathcal{M}^-$, 230
M_ω, 295
$m(\Omega)$, 6
$m(\Omega,\eta)$, 9
n_{ϕ_1,ϕ_2}, 311
$n_\omega(r)$, 298
$n_S(\Omega)$, 92
$\overline{\omega}$, 119
ω_e, 353
ω_η, 296
ω_I, 128
ω_o, 353
Ω^*, 119
Φ, 329
Π_κ, 91

$\mathcal{Q}$, 238
$\mathrm{R}\omega$, 119
ρ_ω, 295
$\rho(T)$, 4
σ_{f,ϕ_1,ϕ_2}, 309
$\sigma(T)$, 4
$\sigma_0(T)$, 4
$\sigma_{\mathrm{app}}(T)$, 5
$\sigma_{\mathrm{ess}}(T)$, 4
$\mathcal{S}_\infty$, 3
$s(\lambda,x)$, 356
S_{ϕ_1,ϕ_2}, 296
$v \otimes u$, 258
T', 4
$T^{(j)}$, 4
U_1, 275
U_2, 275
U_3, 278
$U_{\omega,\delta}$, 302
$((w - w_0)^{\frac{1}{p_k}})_j$, 251
W_n, 75
W_n^0, 75
$y^{[j]}$, 272

Printed in the United States
By Bookmasters